Modules and rings

Modules and rings

JOHN DAUNS

Tulane University

CAMBRIDGE
UNIVERSITY PRESS

Published by the Press Syndicate of the University of Cambridge
The Pitt Building, Trumpington Street, Cambridge CB2 1RP
40 West 20th Street, New York, NY 10011-4211, USA
10 Stamford Road, Oakleigh, Melbourne 3166, Australia

First published 1994

Printed in the United States of America

Library of Congress Cataloging-in-Publication Data

Dauns, John.
    Modules and rings / John Dauns.
        p.   cm.
    Includes bibliographical references and index.
    ISBN 0-521-46258-4
    1. Modules (Algebra)   2. Rings (Algebra)   I. Title.
QA247.D28   1994
512'.4--dc20   93-49759
                                                                                    CIP

A catalog record for this book is available from the British Library

ISBN 0-521-46258-4 Hardback

Contents

Preface

In his or her journey through the book, how is the novice reader to identify the major results, which are like milestones marking one's progress along the way? First of all, some of them carry special names. Thus items named after a person are important, e.g. Wedderburn Theorems, Baer criterion, Jacobson's radical, Schanuel's Lemma, Fitting's Lemma, Krull-Remak-Schmidt Theorem, etc. To further help the reader, the paragraph headings "theorem", "proposition", and "construction" are used infrequently as opposed to the more numerous usage of "lemma", "corollary", "consequence", "observation", an unlabeled paragraph, and "remark". Here these terms are in descending order of importance, "theorem" being the most and "remark" the least important. In order to especially facilitate the actual identification of major results, the word "theorem" in this book is used more sparingly than in any comparable text known to the author. For example, with the exception of the preliminary review of basics in Chapter 0, the first time this word occurs is in Chapter 3 (in 3-3.3). In this particular case, it deserves the name "theorem" because it is a result that is difficult to prove, not at all obvious, but a result which once grasped is like an open passageway allowing free and easy access to the whole theory of injectivity. Also, it (Theorem 3-3.3) is an important and useful tool in more advanced module theory. Here in this text, also the word "proposition" marks significant mathematical statements and is used grudgingly. To summarize – in other texts the reader has to decide which theorems are important and which are not, but not in this book. There just are no unimportant theorems.

This text has more material than can be covered in a one year course. Although there are no real prerequisites aside from linear algebra, an introductory abstract algebra course, which usually includes groups, fields, and some commutative rings, might be helpful. Thus the book is intended for upper level undergraduates and beginning graduate students, as well as anyone else wishing to learn ring and module theory from the ground up.

Students taking a course prefer to follow a textbook without much skipping, omitting and jumping from chapter to chapter. For this reason the order in which the topics are presented here is the same in which they are usually taught in many introductory courses in many universities.

The first eight chapters with some possible omissions are designed for a one semester introductory course for advanced undergraduates or first year graduate students. This begins with free, projective, and injective modules, and tensor products of modules as well as algebras. Then free, tensor, and finite dimensional exterior algebras are discussed. The more abstract construction of arbitrary exterior algebras as quotients of tensor algebras could be omitted. At this point the course assumes a definite sense of direction: simple modules, primitive rings, the Jacobson radical, and subdirect products. In the second semester, from then on logically (with primitive replaced by prime and Jacobson radical replaced by prime radical), the next step would be to go on to primes, semiprimes, and the prime radical. Another possibility is to go directly to the Wedderburn Theorems instead. The above two semester course is not merely a theoretical possibility, but has been taught several times by the author, however, with the following optional material omitted: Appendix 5; 3-5 Noetherian Rings; 10-2 Projective Dimension; and possibly even omitting also 10-1.14 through 10-1.21. 6-6 More on Density and Simples; 7-3 Local Rings; 9-4 Nil Radicals; 9-4 Primes and Semiprimes in Derived Rings; alternate approaches to Wedderburn Theorems, 10-3 Direct Proofs; 10-4 Uniqueness; and 10-5 Rings with D.C.C. and Idempotents.

This book does not dwell too long on any one topic and thus is suitable for courses where a wide range of topics have to be covered quickly. This is also the reason why the chapters on category theory, functors, module categories, and more complicated facts about tensor products are at the end of the book.

One often hears from one's colleagues and students that in all too many books it is not possible to read and understand one topic only, without being forced to read the whole book and to decipher unnecessarily intertwined chapters and notation. Here in this book a deliberate attempt is made to make each chapter as self contained as is possible. In each chapter the main results are obtained as quickly, and as simply as possible. At the end of each chapter some more specialized and peripheral material which can be omitted is found. For example, sometimes several self contained and distinct proofs of the same theorem are given; thus providing the reader some real behind the scenes insight. In this way the book may also be of service to the student and working mathematician alike, who may wish to learn or review some less elementary topic quickly, e.g. category theory, simple algebras, or hereditary rings.

Although Abelian categories would suffice for applications to modules, nevertheless category theory is covered in greater generality so as to make it also applicable to other fields, such as topology or partially ordered systems, where greater generality is required.

In Chapters 6-12, the ring need not necessarily have an identity with the exception of the first two sections of Chapter 10 (i.e. 10-1 and 10-2). An identity is assumed in the sections on free, projective, and particularly injective modules. In these isolated cases, where an identity is not assumed once an adequate notation and terminology is established, it is just as easy to prove everything for a general ring. If all our theorems were proved for rings with an identity, they could not be applied to subrings of a ring when they arise naturally during the course of some proof. Some theorems should be tools capable of proving other theorems. And occasionally we point out that in some arguments an identity element is not needed. However, throughout almost the whole book, the ring is assumed to have an identity. Whenever a ring has an identity element, we always assume that all modules are unital. Finally, if the reader wishes, he or she can simply assume throughout that all rings have an identity.

This book also tries to teach how to do ring and module theory, rather than strive for encyclopedic coverage. There is a whole chapter which develops useful module construction techniques, like pullbacks, pushouts, and applies these to give an alternate quite different proof of the existence of the injective hull of a module. Simiarly, the chapter on module categories and the chapter on flat modules develop quickly some of the more important facts, and do not cover these subjects exhaustively.

The last chapter on systems of equations in modules, pure projectivity, pure injectivity, and pure injective hulls is somewhat more advanced. At this point in time, the author does not know of any textbook in print which develops this subject logically from the beginning as is done here.

As already stated some of the great theorems are mentioned in the introductions to the chapters and the chapters themselves. For the benefit of the newcomer to module and ring theory we will mention a few more here next. Just what is the definition of a "major result or theorem"? The author does not have an absolute universally valid definition. It may even vary from one person to the next, depending upon one's needs and goals in module and ring theory, and it may vary with the passage of time for the same person. The reader may find it an interesting and profitable exercise to list a half a dozen or so general criteria which characterize and identify major mathematical results. For these reasons the reader should not take any list, and in particular the one here, as the final absolute truth, but rather as a road map to guide him or her through the text. The author found one such criterion on a poster on the undergraduate mathematics bulletin board at M.I.T . "Beware of a theorem that counts something." It could possibly be more important than you think because in all likelyhood it can be used in novel ways and in circumstances that are beyond your imagination when you first read such a theorem. The first Wedderburn Theorem (7-1.39) counts something. It says that a simple ring R with the descending chain condition on right ideals is isomorphic $R \cong M_n(D)$ to an $n \times n$ matrix ring $M_n(D)$ over a division ring D. If in addition R happens to be a finite dimensional algebra over a commutative field F, then there are two intergers, not only n but also the dimension of D over F. To the future development of ring and module theory, the Wedderburn or Wedderburn-Artin Theorems were what Columbus was to the development of the Americas. A few of the original papers in which it appeared are the following: [Wedderburn 08], [Artin 27], [Noether 29], and [Hopkins 39]. For a possibly more complete list, see [Goodearl 76; p. 184].

Unlike other sciences, where the objects to be studied may be un-equivocally given, in mathematics they have to be invented or discovered by means of lengthy constructions. Such constructions are particularly important, because they expand the reader's repertoire of mathematical objects thus giving a new framework within which to understand mathe-matics and physics. Such constructions enlarge one's mental categories or concepts. Thus the acquisition of a construction is like obtaining a new previously missing sense modality, such as touch, hearing, or sight. Many such constructions involve universal mapping properties, such as free modules (Chapter 1), tensor products of modules, tensor products of algebras, exterior algebras (Theorem 5-2.11, Theorem 5-3.19), pullbacks and pushouts. If one stops and examines the idea of a universal property it is really a surprising device frequently used in 20th century algebra. An object or set is determined or defined in terms of entities of a totally different nature—functions. Coming from this perspective for the begin-ning novice reader at least, the Universal Mapping Theorem (4-3.5) is noteworthy. It gives necessary and sufficient conditions on an algebra C to be the algebra tensor product of given algebras A and B. Here the algebra tensor product C of A and B is defined in terms of linear disjointness of A and B as subsets of C. Many other texts either omit tensor products of algebras altogether, or merely regard them as tensor products of vector spaces with additional properties.

The beginning student may find the construction of tensor and exterior algebras illuminating and useful in Chapter 5. Aside from finite matrix rings, these two are among the most frequently used algebras in areas other than algebra, and even in physics. In applications many students find that these two algebras are used without an adequate explanation of what they really are thus making the whole application to appear logically unsound. Here these two algebras are developed from two different perspectives. Also in various applications these algebras are used in different ways which are suited to either one or the other of our perspectives.

This is a book about modules just as much as about rings, and the construction and properties of the injective hull of a module (Theorem 3-4.13, 3-3.3) is one of the major results. The latter theorem immediately implies that any module can be embedded in an injective module, and that any given module is injective if and only if it is a direct summand of any module containing it as a submodule. It is this last property which explains the importance of injective modules in module and ring theory as a whole. Frequently, the presence of direct summands either simplifies or more accurately makes certain module and ring theoretic proofs even possible to begin with.

The dual of an injective module is a projective module, and there is a dual development of the theory also for projective modules. The proofs tend to be a little shorter and easier, and for this reason none of the results in the sections on projectives (10-1, 10-2) is labelled as a theorem. However, the main results here are the equivalent characterization of projective modules (Proposition 10-1.5) and the proof that any module has a well defined projective dimension (10-2.6).

The Jacobson Density Theorem (6-5.7) says that if any ring R has a faithful simple module V, that then V is a vector space over a certain division ring D and that R is almost equal to the full ring of D-linear transformations of this vector space. Even in the early 1990's new and more complicated versions of it are the subjects of research papers and talks. What makes this theorem (and others in general) interesting is that it takes something abstract and on the surface not very informative, like R being a primitive ring, and shows that this abstract condition is completely equivalent to R being something that is quite familiar, and well understood. Historically, it possibly may be one of the more influential single theorems, such as the Hilbert Nullstellensatz (9-4.10) and Wedderburn-Artin Theorem (7-1.39). In fact, the density theorem may be viewed as a successor of the Wedderburn Theorem.

Unquestionably, the whole theory of the Jacobson radical was one of the most influential developments in the evolution of ring theory (chapter 7). Here it is hard to single out a few theorems. It is the logical coherent development of the whole theory that counts. Here to single out the most important theorem would be like asking for the most important part of a painting by Monét.

Theorem 9-2.8 characterizes the prime radical of an arbitrary ring, while Theorem 9-2.16 characterizes semiprime rings. These two theorems show that the rich and elaborate theory of primeness for commutative rings, surprisingly also has an analogous theory for noncommutative rings. During the early part of the twentieth century ring and module theorists tended to devote most of their efforts to rather specialized rings like finite dimensional algebras, commutative rings, or division algebras. It was the success of results such as the Jacobson radical and the theory of primeness developed in the context of an arbitrary associative ring which established noncommutative ring and module theory as a subject in its own right.

For a right Noetherian ring R, the prime radical is the sum of all the nilpotent right (or left) ideals of R (11-3.7) and by a theorem of Levitzki (11-3.6) is nilpotent. A right Artinian ring R is automatically right Noetherian (11-3.5). Let us ask for some kind of converse of this. Suppose that R is right Noetherian to start with. What additional hypotheses on R will force R to be also right Artinian? The answer is given by the Hopkins-Levitzki Theorem (11-3.8) which in some sense rounds out and complements the Wedderburn-Artin Theory by also looking at Noetherian rings. Noetherian rings are diverse and complicated, and in general not neatly describable. And the Hopkins-Levitzki Theorem gives us a clear picture of a natural class of right Noetherian rings.

The Krull-Schmidt Theorem (11-4.9) and a generalization (Theorem 11-4.8) gives conditions under which a module is a direct sum indecomposable modules which is unique up to isomorphism of the indecomposable summands. These theorems also describe certain interesting substitution properties, whereby some direct summands may be replaced by others. This whole theory culminates in the difficult to prove and significant Krull-Remak-Schmidt Theorem 11-4.11 which gives conditions when a module can be decomposed as a direct sum of indecomposable modules in an essentially unique way. The reader will find that the main results throughout are indicated by a syntactic (sparse) use of the word "theorem", and they are identified in the introductions to the chapters, and in the chapters themselves. This last part of the preface was for the benefit of novice student in order to guide that student well into the text.

Note to the reader

First a brief guide is given to the reader who may not wish to read the whole book but only certain select chapters. Roughly speaking, the dependance or independance of various chapters is summarized in the following diagram.

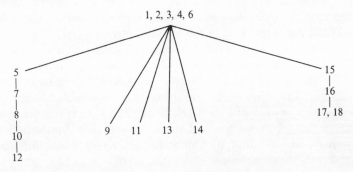

Moreover, in Chapter 1, section 1-5. Direct and Inverse Limits may be omitted, except for Chapters 16–18, where these are used. In later applications of Chapter 7, only 7-1.1 through 7-1.36 is used. Also Chapter 9 is almost independent from the other chapters, although some defintions and a few easy facts from Chapter 9 are occasionally used in Chapter 10.

At the expense of being overly repetitious, throughout, if the ring has an identity element, then always all modules are unital. If one wishes to take a middle ground between assuming that all rings have an identity, or that as a rule rings need not have an identity, then one should assume an identity element for all chapters – except 6 and 7.

There is some benefit in knowing some mathematical facts, without having the ability to prove these facts. For this reason, the reader should read the exercises even if he or she does not wish to do them.

Throughout an attempt has been made to display results and theorems by means of formulas and logical symbols, and to even incorporate logical symbols in English sentences where this seems to give more clarity and emphasis. The reader should expect to find symbols such as $\exists$, $\forall \Leftrightarrow$, and $\Rightarrow$ in some sentences, particularly in statements of lemmas and proposition, because the meaning of these symbols is always clear. Words can be hard to grasp and sometimes even ambiguous. Also, for the sake of brevity, some of the standard abbreviations of terminology are used, such as "map" for right R-module homomorphism, and "$G \to G/N$" as an abbreviation for the natural quotient homomorphism.

Many important topics are not covered in the book, particularly those which are a bit more advanced and well presented in other books. Thus if one wishes to pursue some topic further beyond this book, or to see it treated from another perspective, the bibliography is a good place to start. No attempt has been made to give credit for most theorems and proofs. The inclusion of an author in the bibliography should not be interpreted to mean that this author is the main contributor necessarily to this area. And certainly many major ring and module theorists and authors are not mentioned in the bibliography.

Two appendices are provided. Appendix 0 gives a review of some basic facts such as equivalence relations, Zorn's lemma, ordinals, cardinals, and the four isomorphism theorems. Appendix 5 to Chapter 5 gives additional information about tensor algebras and their quotients.

The author would like to thank Susan P. Q. Lam for her excellent typing. I also wish to thank Lauren Cowles, the Mathematics and Computer Science Editor of Cambridge University Press, for her help and many useful suggestions.

CHAPTER 1

Modules

Introduction

The basic nomenclature for modules and module homomorphisms is defined. Direct sums and products of modules are introduced. Split short exact sequences are discussed. Existence and universal properties of direct and inverse limits are established.

Direct limits generalize direct sums, inverse limits – direct products. This topic is covered in Chapter 26-6, but could very well be covered at the end of the present Chapter 1. The construction of direct and inverse limits of modules and rings is a good exercise in using all the concepts introduced in this Chapter 1. Furthermore, they are a rich source of non-trivial examples of modules and rings.

1-1   Definitions

Throughout, R is an arbitrary ring with or without an identity element.

1-1.1   Definition. An additive abelian group M with addition denoted by $+$ is a right *R-module* if there is a function $M \times R \to M$, $(m, r) \to mr$, for $m \in M$, $r \in R$, such that for any $x, y \in M$ and any $a, b \in R$ the following hold:

(i) $(x + y)a = xa + yb$, $x(a + b) = xa + yb$;
(ii) $x(ab) = (xa)b$.

Notation. The notation $M = M_R$ will mean that R is a ring and M is a

1

right R-module (and similarly $V = {}_R V$ for left modules). The zero module will be denoted by either one of the three $\{0\} = (0) = 0$.

For the remainder of this section, $M = M_R$.

1-1.2   Definition. $M = M_R$ is *unital* if (i) $1 \in R$, i.e. the ring has an identity element 1, and (ii) $x1 = x$ for all $x \in M$.

1-1.3   Definition. For any $M = M_R$, an additive subgroup $N \subseteq M$ is a right R-*submodule* if $nr \in N$ for all $n \in N$ and all $r \in R$. Submodules will be denoted by $N \leqslant M$ or $N < M$. The symbols $\leqslant$ and $<$ will be used exactly the same way as $\subseteq$ and $\subset$. I.e., if the case $N = M$ is trivial, we sometimes will write $N < M$; $\leqslant$ and $<$ are used for special emphasis, and whenever important, the fact that $N \neq M$ will be explicitly stated.

A submodule $N < M$ is *proper* if $N \neq M$. If $M \neq (0)$, then $(0) < M$ is proper.

Terminology. By a module will be meant a right module, by a submodule — a right submodule.

Remark. If R is commutative, the distinction between right and left disappears.

1-1.4.   Definition. For rings E and R suppose that $M = M_R$ and also that $M = {}_E M$. Then M is left E, right R-*bimodule* if for any $\psi \in E$, any $r \in R$, and any $m \in M$, $(\psi m)r = \psi(mr)$.

Example. If $M = R$ and $E = R$, then R is a left R, right R-bimodule.

1-1.5   Definition. For any ring R, for any $M = M_R$ and $W = W_R$, an additive abelian group homomorphism $f: M \to W$ is a right R-module *homomorphism* (also called module map, or R-map, or R-homomorphism) if $f(mr) = (fm)r$ for all $m \in M$ and $r \in R$.

If f is one-to-one it is called a *monomorphism*, or monic; if f is onto — *epimorphism*, or epic. If f is both one-to-one and onto, then it is an *isomorphism* and denoted by $M \cong W$. The symbols "$\rightarrowtail$", "$\twoheadrightarrow$", and "$\rightarrowtail\!\!\!\rightarrow$" will occasionally be used to indicate monics, epics and isomorphisms. A homomorphism of a module M into itself is called an *endomorphism*.

The symbol $\mathrm{Hom}_R(M, W)$ denotes the additive abelian group of all right R-module homomorphisms $f, g: M \to W$, $(f + g)m = fm + gm$,

$m \in M$; its zero element is the zero homomorphism $0: M \to (0)$. If $W = M$, define $\operatorname{End}_R M \equiv \operatorname{Hom}_R(M, M)$. Then $\operatorname{End}_R M$ is a ring under the above addition and composition; $f, g: M \to M$, $(fg)(m) \equiv f(gm)$. The identity endomorphism $1: M \to M$, $1m = m$, is the identity element of $\operatorname{End}_R M$, and M is a unitial left $\operatorname{End}_R M$ module. Furthermore, M is a left $\operatorname{End}_R M$ and right R-bimodule.

1-1.6   Definition. For a right R-submodule $N \subseteq M$, let $\bar{M} = M/N$ be the quotient abelian group. Define $\bar{M} \times R \to \bar{M}$ by $(m + N)r = mr + N$, $m \in M$, $r \in R$. Then $\bar{M}$ is a right R-module called the quotient module. The R-module map $\pi: M \to \bar{M}$, $\pi m = m + N$, $m \in M$, is called the natural projection.

1-1.7   Let $f: M \to W$ be an R-module homomorphism, and let $N \leqslant M$ and $V \leqslant W$ be submodules. Then $fN = \{fn \mid n \in N\} \leqslant W$ and $f^{-1}V = \{m \in M \mid fm \in V\} \leqslant M$ are submodules. In particular, if $N = M$ then the *image* of f is $fM = \text{image } f = \operatorname{im} f$; if $V = (0)$, then the *kernel* of f is the submodule $f^{-1}(0) = \text{kernel } f = \ker f = f^{-1}0$. The *cokernel* of f is coker $f = W/fM$.

1-1.8   Lemma. *Let $f: M \to W$ be an epic homomorphism of R-modules. Define $N = \text{kernel} f$. Let $\pi: M \to M/N$ be the natural R-module homomorphism, and define $\bar{f}: M/N \to W$ by setting $\bar{f}\bar{x} = fy$, where $\bar{x} \in M/N$ and $y \in \bar{x}$ is arbitrary. Then $\bar{f}$ is an R-module isomorphism, $f = \bar{f}\pi$, and $M/\ker f \cong fM = W$.*

1-1.9   Notation. For subsets $A \subset M = M_R$ and $B \subset R$, the set $AB \subset M$ is defined as $AB = \{\Sigma_{i=1}^{n} a_i b_i \mid n = 0, 1, \ldots; a_i \in A, b_i \in B\}$. By convention, when $n = 0$, the empty sum is $0 \in M$, and usually $AB \neq \{ab \mid a \in A, b \in B\} \neq AB \backslash \{0\}$. If $A = \{a\}$, write $\{a\}B \equiv aB$.

The next definition gives a large source of modules which will play an important role in later development.

1-1.10 Definition. Regard $R = R_R$ as a right R-module. A *right ideal* is an abelian subgroup $B \subseteq R$ which is a right R-submodule of R. (For $R = {}_R R$, left ideals are defined analogously.)

An abelian subgroup $A \subset R$ is a right ideal if and only if $AR \subseteq A$.

Thus the notation for a right ideal is $B < R$ or $B \leqslant R$.

A set $I \subseteq R$ is an *ideal* if I is both a right and a left R-submodule of R, denoted by $I \lhd R$. The symbol "$\lhd$" denotes ideals of any ring whatever, where equality is allowed.

1-1.11   Annihilators. For any subset $Y \subseteq M = M_R$, define $Y^\perp \subseteq R$ by $Y^\perp = \{r \in R \mid \forall\, y \in Y,\ yr = 0\}$, i.e. $Y^\perp = \{r \mid Yr = \{0\}\}$. Then $Y^\perp$ is a right ideal of R called the *annihilator* of Y.

Three special cases will occur frequently:

(i) $m \in M$, $Y = \{m\}$; define $m^\perp = \{m\}^\perp = \{r \in R \mid mr = 0\}$. Left multiplication by m defines a right R-module homomorphism $R_R \twoheadrightarrow mR \subseteq M$, $r \to mr$, $r \in R$. Since the kernel is $m^\perp$, $mR \cong R/m^\perp$.

(ii)   if $Y \subseteq M$ is a submodule, then $Y^\perp \lhd R$ is an ideal.

(iii)   In particular, if $Y = M$, then $M^\perp \lhd R$ is an ideal. Define $\bar{R} = R/M^\perp$. Then M is also a right $\bar{R}$-module, where for $m \in M$ and $\bar{a} \in \bar{R}$, $m\bar{a} \equiv mb$ for any $b \in \bar{a}$. According to the next definition, $M_{\bar{R}}$ will be a faithful $\bar{R}$-module.

1-1.12   Definition. A module $V = V_R$ is *faithful* if for any $r \in R$, $Vr = (0)$ forces $r = 0$. Alternatively, V is faithful if and only if for any $0 \neq r \in R$, also $Vr \neq (0)$.

Remark. Let $\mathbb{Z} = 0, \pm 1, \pm 2, \ldots$. The endomorphism ring of the additive abelian group V is $\operatorname{End}_{\mathbb{Z}} V = \operatorname{Hom}_{\mathbb{Z}}(V, V) \supseteq \operatorname{Hom}_R(V, V)$. There is a ring homomorphism $R \to \operatorname{End}_{\mathbb{Z}} V$, where $r \in R$ gives the map $r\colon V \to V$, $v \to vr$ for $v \in V$. Thus V_R is faithful as an R-module if and only if this ring homomorphism is monic, in which case R can be viewed as a subring in $R \subseteq \operatorname{End}_{\mathbb{Z}} V$.

1-1.13   Suppose that M and W are given R-modules which are annihilated by an ideal $B \lhd R$, i.e. $MB = 0$ and $WB = 0$, and suppose that $f\colon M \to W$ is an R-module homomorphism. Then both M and W are right $\bar{R} = R/B$-modules in a natural way (as in 1-1.11(iii)), and furthermore, f is also an $\bar{R}$-homomorphism.

In particular, for any R-map $f\colon M \to W$ whatever, since $(fM)M^\perp = 0$, the corestriction $f: M \to fM$ of f to $fM \leqslant W$ is a homomorphism of $R/M^\perp$-modules M and fM.

1-1.14 Examples. Let $\mathbb{Z} = 0, \pm 1, \pm 2, \ldots$.

1. Any abelian group A is a $\mathbb{Z}$-module, where for $0 < n \in \mathbb{Z}$, $a \in A$, $na = an = a + \cdots + a$, and $(-n)a = a(-n) = (-a) + \cdots + (-a)$, n-times.

2. Every additive abelian group is a left $\text{End}_{\mathbb{Z}} A$ module. For example, if elements of $A = \mathbb{Z} \times \mathbb{Z}$ are viewed as column vectors, then $\text{End}_{\mathbb{Z}} A$, which consists of 2×2 matrices over $\mathbb{Z}$, is noncommutative.

3. Let V be a finite dimensional vector space over a field F and $T: V \to V$, $v \to vT$; $v \in V$, a linear transformation. Then V is a right module over the ring $R = F[x]$ of polynomials over F, where for $v \in V$ and $p(x) \in F[x]$, $vp(x) = vp(T)$. The R-submodules of $W \leqslant V$ are precisely the T-invariant subspaces, i.e. ordinary vector subspaces W such that $WT \subseteq W$.

Let $m(W; x)$ be the minimal polynomial of the restriction $T \mid W$ of T to W, and $m(v; x)$ the minimal polynomial of v. then

(i) $v^\perp = Rm(v; x) \lhd R$;
(ii) $W^\perp = Rm(W; x) \lhd R$.

The ring $F[x]$ is a Euclidean domain, hence a principal ideal domain, which in turn implies that it is a unique factorization domain.

1-2   Direct products and sums

1-2.1   For an arbitrary family of modules M_i, $i \in I$, indexed by an arbitrary index set, the product $\Pi\{M_i \mid i \in I\} \equiv \Pi M_i$ is defined as the set of all functions $\alpha, \beta: I \to \cup\{M_i \mid i \in I\}$ such that $\alpha(i) \in M_i$ for all i, which becomes an R-module under pointwise operations, $(\alpha - \beta)(i) = \alpha(i) - \beta(i)$, and $(\alpha r)(i) = \alpha(i)r$ for $r \in R$.

(ii) Alternatively, the product can be viewed as consisting of all strings or sets

$$x = \{x_i \mid i \in I\} \equiv (x_i)_{i \in I} \equiv (x_i) \equiv (\underline{\quad}, x_i, \underline{\quad}), \ x_i \in M_i; \ i\text{-th.}$$

If $x = (x_i)$, $y = (y_i)$, and $r \in R$, then $x - y = (x_i - y_i)$, and $xr = (x_i r)$.

The direct sum $\oplus\{M_i \mid i \in I\} \equiv \oplus M_i$ is defined as the submodule $\oplus M_i \subseteq \Pi M_i$ consisting of all those elements $x = (x_i) \in \Pi M_i$ having at most a finite numbers of nonzero *coordinates* or *components* x_i. Sometimes $\oplus M_i \subseteq \Pi M_i$ are called the external direct sum and external direct product.

In general, whenever the index set is suppressed in symbols like ΠM_i, $\oplus M_i$, $\cup M_i$, (or $\cap M_i$), it will be understood that i ranges over the largest possible index set, namely all of I, and never over some subset of I.

The R-module maps $j_k: M_k \to \Pi M_i$, $j_k(m) = (0 \underline{\quad\quad} 0, m, 0 \underline{\quad\quad} 0)$, where the k-th coordinate is $m \in M_k$, and $\pi_k: \Pi M_i \to M_k$, $\pi_k[(x_i)_{i\in I}] = x_k$ are called the inclusion and projection maps. Throughout, $\delta_{ik} = 1$ if $i = k$ and 0 if $i \neq k$. Thus $\pi_k j_k = 1: M_k \to M_k$ is the identity map, and $\pi_i j_k = \delta_{ik}$.

If $\Pi\{N_i \mid i \in I\}$ is another direct product of modules N_i over the same index set I, and $f_i: M_i \to N_i$ are any R-maps, then $\Pi\{f_i \mid i \in I\} = \Pi f_i: \Pi M_i \to \Pi N_i$ is the R-map defined by $(\Pi f_i)(x_i)_{i\in I} = (f_i x_i)_{i\in I}$. Since $\Pi f_i (\oplus M_i) \subseteq \oplus N_i$, define $\oplus\{f_i \mid i \in I\} = \oplus f_i$ to be the restriction and corestriction of Πf_i to the direct sums, i.e. $\oplus f_i: \oplus M_i \to \oplus N_i$. In case all $N_i = N$ are the same module the following notations are frequently used: $\Pi N_i = \Pi\{N \mid I\} = N^I$, and $\oplus N_i = \oplus\{N \mid I\} = N^{(I)}$. Define an R-map $s: \oplus\{N \mid I\} \to N$ by $s[(n_i)_{i\in I}] = \Sigma\{n_i \mid i \in I\} \in N$, where all $n_i \in N$. In this special case, the composite of $\oplus f_i$ followed by s is the map $s \circ (\oplus f_i): \oplus M_i \to N$, $s \circ (\oplus f_i)(x_i)_{i\in I} = \Sigma\{f_i x_i \mid i \in I\}$, where $x_i \in M_i \subset \oplus M_i$.

Note that $\oplus M_i = \Pi M_i$ if and only if $|I| < \infty$, the index set is finite.

1-2.2   Universal properties of products and sums. Let $\oplus M_i \subseteq \Pi M_i$, j_k, and π_k be as before. (i) For any module A and any set of R-homomorphisms $g_k: A \to M_k$, for all $k \in I$, there exists a unique R-homomorphism $g: A \to \Pi M_i$ such that $\pi_k g = g_k$ for all $k \in I$.

(ii) For any module B and R-maps $h_k: M_k \to B$, $k \in I$, there exists a unique R-map $h: \oplus M_i \to B$ such that $hj_k = h_k$ for every $k \in I$.

Note that (i) and (ii) say that there are commutative diagrams and that there is a "duality" – either diagram can be obtained from the other by reversing all arrows and interchanging the sum with the product below.

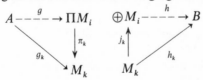

Note also that for $a \in A$, $ga = \{g_i a \mid i \in I\} = (g_i a)$, and $h[(x_i)] = \Sigma\{h_i x_i \mid i \in I\}$ is a well defined element of B because the sum has only a finite number of nonzero terms if $(x_i) \in \oplus M_i$.

1-2.3   Remark. The previous universal properties uniquely determine up to isomorphism, not only the direct product and sums $\oplus M_i \subseteq \Pi M_i$ as modules, but also the maps j_k and π_k.

Suppose that P and S are modules, and $p_k : P \to M_k$ and $s_k : M_k \to S$ maps, and that P and S have the exact same universal properties as the product ΠM_i and sum $\oplus M_i$ respectively. Then there are commutative diagrams where the horizontal maps are isomorphisms.

The two frequently used symbols "$\langle\ \rangle$", "Σ", and internal direct sum are next defined.

1-2.4   Definition. For a module M, and for any subset $T \subseteq M$, the R-submodule generated by T is denoted by $\langle T \rangle$, and is defined as $\langle T \rangle = \cap \{ N \mid T \subseteq N, \ N \leqslant M \}$. If $T = \{x\}$ is a singleton, abbreviate $\langle \{x\} \rangle = \langle x \rangle$.

1-2.5   If I is any index set and $M_i \subset M$, $i \in I$, an indexed set or family of submodules, then define $\Sigma \{ M_i \mid i \in I \} = \langle \cup \{ M_i \mid i \in I \} \rangle$. This will frequently be abbreviated as $\Sigma M_i = \langle \cup M_i \rangle$, where whenever the index set is suppressed, it will always be assumed that i ranges over the largest possible index set, that is all of I, and never over some subset of I. Thus

$$\sum_{i \in I} M_i = \left\{ x_1 + \cdots + x_n \left| \begin{array}{l} \forall\, 0 \leqslant n \in \mathbb{Z}; \quad i(1), \ldots, i(n) \in I; \\ \forall\, x_k \in M_{i(k)}, \quad k = 1, \ldots, n \end{array} \right. \right\}$$

1-2.6   The sum ΣM_i is an *internal direct sum* if for any $j \in I$, $M_j \cap \langle \cup \{ M_i \mid i \in I, \ i \neq j \} \rangle = (0)$, written as $\Sigma M_i = \oplus \{ M_i \mid i \in I \} = \oplus M_i$. Since it will be amply clear from the context whether an internal or external direct sum is meant, the two will be denoted by the same symbol.

More explicitly, the above sum by definition is direct if for any $1 \leqslant k$, $n \in \mathbb{Z}$, and for any n-distinct unequal indices $i(1), \ldots, i(n) \in I$, $M_{i(k)} \cap [M_{i(1)} + \cdots + M_{i(k-1)} + M_{i(k+1)} + \cdots + M_{i(n)}] = (0)$ for all $k = 1, \ldots, n$.

1-2.7   **Modular law.** Let A, B, and C be submodules of a module. Then

(i) $A \subseteq C \subseteq A + B \Rightarrow C \cap (A + B) = A + (C \cap B)$;
(ii) $A \subseteq C \subseteq A \oplus B \Rightarrow C \cap (A \oplus B) = A \oplus (C \cap B)$.

1-2.8   **Observations.** For $T \subseteq M$ and $\mathbb{Z} = 0, \pm 1, \pm 2, \ldots$ the following hold.

1. $\langle T \rangle$ is the unique smallest submodule of M containing T.
2. $\langle M \rangle = M$, $\langle M \backslash \{0\} \rangle = M$.
3. For $x \in M$, $\langle x \rangle = \{nx + xr \mid n \in \mathbb{Z}, r \in R\} = \mathbb{Z}_x + xR$. If $M = R_R$, then $\langle x \rangle$ is the principal right ideal generated by x.
4. $\langle T \rangle = \Sigma \{\mathbb{Z}T + tR \mid t \in T\}$; if $1 \in R$, $\langle T \rangle = \Sigma \{tR \mid t \in T\}$.
5. If $T = \varnothing$, then $\langle \varnothing \rangle = (0)$. In this case when $T = \varnothing$, the empty sum by definition is equal to $0 \in M$. (Later the empty intersection will have to be interpreted as the whole space.)

1-2.9   **Direct sum conditions.** Let $M = M_R$, let I be an arbitrary index set, and $M_i \subset M$, $i \in I$, right R-submodules. The following are all equivalent.

(i) ΣM_i is an internal direct sum.
(ii) For any $1 \leqslant n$ and any choice of $n + 1$ distinct indices $i(0)$, $i(1), \ldots, i(n) \in I$, $M_{i(0)} \cap [M_{i(1)} + \cdots + M_{i(n)}] = (0)$.
(iii) Each $0 \neq m \in M$ has a unique representation $m = m_{i(1)} + \cdots + m_{i(n)}$, all $0 \neq m_{i(k)} \in M_{i(k)}$.
(iv) For any $1 \leqslant n$, and any distinct $i(1), \ldots, i(n) \in I$, if $m_{i(k)} \in M_{i(k)}$ and $m_{i(1)} + \cdots + m_{i(n)} = 0$, then all $m_{i(k)} = 0$.

1-2.10   **Corollary 1.** *Assume in addition that I is well ordered by* "$\leqslant$". *Then the above conditions (i)–(iv) are equivalent to (v).*

(v)   For any $2 \leqslant n$, and any choice of $i(1) < i(2) < \cdots < i(n) \in I$, $M_{i(n)} \cap [M_{i(1)} + \cdots + M_{i(n-1)}] = (0)$.

1-2.11   **Corollary 2.** *If $I = \{1, \ldots, n\}$ is finite, then (i)–(iv) are equivalent to (v)*

(v)   for any $2 \leqslant j \leqslant n$, $M_j \cap [M_1 + \cdots + M_{j-1}] = (0)$.

1-3   Adjunction of 1 to R

Throughout this section $\mathbb{Z}$ is the ring $\mathbb{Z} = 0, \pm 1, \pm 2, \ldots$ of integers.

1-3.1   For any ring R whatever $\mathbb{Z} \times R$ becomes a ring under

$$(n, a) + (m, b) = (n + m, a + b).$$
$$(n, a)(m, b) = (nm, nb + ma + ab) \quad n, m \in \mathbb{Z}; a, b \in R.$$

The identity element of $\mathbb{Z} \times R$ is $e \equiv (1, 0)$ irrespective of whether R has one or not. If one identifies $R \equiv \{0\} \times R \lhd \mathbb{Z} \times R$, $r \equiv (0, r)$, $r \in R$, then the above become

$$(ne + a) + (me + b) = (n + m)e + (a + b),$$
$$(ne + a)(me + b) = nme + nb + ma + ab.$$

1-3.2   Let R be any ring M and N any R-modules, and $f: M \to N$ an R-homomorphism. Then M (and N) becomes a unital $\mathbb{Z} \times R$-module under the definition $x(n, a) = x(ne + a) = nx + xa$, $x \in M$, $(n, a) \in \mathbb{Z} \times R$. Furthermore, f is also a $\mathbb{Z} \times R$-homomorphism. The set of R-submodules of M is exactly the set of $\mathbb{Z} \times R$-submodules of M.

1-3.3   **Definition.** For any ring R, define R^1 to be the ring

$$R^1 = \begin{cases} R & \text{if } 1 \in R; \\ \mathbb{Z} \times R & \text{if } 1 \notin R. \end{cases}$$

1-3.4   **Remarks.** (1) For any $M = M_R$, and any $x \in M$, $\langle x \rangle = xR^1$. (2) For any $x \in R$, the smallest ideal containing x is $\mathbb{Z}x + xR + Rx + RxR = R^1 x R^1$.

1-3.5   **Observation.** Let $1 \in R$ and $M = M_R$. Define $M^0 = \{m \in M \mid m1 = 0\}$ and $M^1 = \{m \in M \mid m1 = m\}$. Then

  (i) $M^0 = \{m - m1 \mid m \in M\}$ and $M^1 = \{m1 \mid m \in M\}$ are submodules;
  (ii) $M = M^0 \oplus M^1$;
  (iii) $M^0 R = (0)$ while M^1 is unital;
  (iv)   UNIQUE: If $m = P \oplus Q$ with $PR = 0$ and Q unital, then $P = M^0$ and $Q = M^1$.

Proof. Conclusions (i) and (iii) are easy as well as that $M^0 \cap M^1 = (0)$. (ii) For $m \in M$, $m = (m - m1) + m1$ and (i) show that $M = M^0 \oplus M^1$. (iv) Since $P1 \subseteq PR = (0)$, $P \subseteq M^0$. Hence $M^1 = M1 = (P \oplus Q)1 = P1 + Q1 = Q1$ and $Q = Q1 = M^1$. By the modular law, $M^0 = M^0 \cap (P \oplus Q) = P + M^0 \cap Q = P$ because $M^0 \cap Q = M^0 \cap M^1 = 0$.

1-3.6   Important convention. From now on it is always assumed that if the ring R has an identity, then all modules are unital.

1-4   Sequences of modules

1-4.1   Definition. Let $\{M_i | i \in I\}$ be a set of modules together with module maps $f_i: M_i \to M_{i-1}$ indexed by a finite or infinite convex subset $I \subseteq \mathbb{Z} = \{0, \pm 1, \pm 2, \dots\}$. Then the sequence

$$\dots M_{i+1} \xrightarrow{f_{i+1}} M_i \xrightarrow{f_i} M_{i-1} \longrightarrow \dots$$

is *exact* if kernel f_i = image f_{i+1} for all i except possibly the smallest and largest one. In sequences, abbreviate $(0) = 0$.

Remarks. Let $f: M \to W$ be an R-map.
1.   By omitting the 0 and pushing "$0 \to M$" and "$W \to 0$" through the M and W and shortening the arrows the following notations for monics and epics have arisen:

   f is monic $\Leftrightarrow 0 \to M \to W$ is exact $(M \rightarrowtail M)$,
   f is epic $\Leftrightarrow M \to W \to 0$ is exact $(M \twoheadrightarrow W)$.

2.   For any f, the following is exact

   $0 \to \ker f \to M \to \operatorname{im} f \to \operatorname{coker} f$.

1-4.2   Definition. An exact sequence with five modules of the form

$$0 \longrightarrow A \xrightarrow{\alpha} B \xrightarrow{\beta} C \longrightarrow 0$$

is called a *short exact* sequence.

Consequences. 1. $A \cong \operatorname{im} \alpha$.   2. $C \cong B/\ker \beta$.   3. by use of 1 and 2, frequently a general short exact sequence of the above form may be replaced by $A \subset B \twoheadrightarrow C = B/A$.

1-4.3   Definition. An exact sequence $0 \to A \xrightarrow{\alpha} B$ *splits* if there exists an R-map $p: B \to A$ such that $p\alpha = 1_A$, the identity on A.
   Similarly, an exact sequence $B \xrightarrow{\beta} C \to 0$ *splits* if there exists $q: C \to B$ such that $\beta q = 1_C$, the identity on C.

1-4.4   Splitting lemma. *For a short exact sequence* $0 \to A \xrightarrow{\alpha} B \xrightarrow{\beta} C \to 0$ *the following are all equivalent*

(i) $0 \to A \to B$ *splits*;

(ii) $\exists D \leqslant B$ *such that* $B = \alpha A \oplus D$ (*in which case automatically* $\beta \,|\, D$ *is an isomorphism, i.e.* $D \cong C$ *under* $\beta|D$).

(iii) $B \to C \to 0$ *splits*.

1-4.5   Corollary. *The above* (i), (ii) *and* (iii) *are equivalent to the following condition* (iv).

(iv) *There exists a commutative diagram, where* φ *is an isomorphism, and* i *and* j *the natural inclusion and projection*

$$
\begin{array}{ccccc}
A & \xrightarrow{\;\alpha\;} & B & \xrightarrow{\;\beta\;} & C \\[2pt]
\| & & \downarrow{\scriptstyle\varphi} & & \| \\[2pt]
A & \xrightarrow{\;i\;} & A \oplus C & \xrightarrow{\;j\;} & C
\end{array}
$$

Proof of 4.4. (ii)($\Rightarrow$)(i). Let $p: \alpha A \oplus D \to \alpha A$ be the natural projection followed by the isomorphism $\alpha A \to A$, $\alpha a \to a$; i.e. $p(\alpha a + d) = a$ for $a \in A$, $d \in D$. Since $p \alpha a = p(\alpha a + 0) = a$, $p\alpha = 1_A$.

(i) $\Rightarrow$ (ii). Given $p: B \to A$ with $p\alpha = 1_A = 1$, set $D = (1 - \alpha p)B$. Any arbitrary $b \in B$ is of the form $b = \alpha a + (1 - \alpha p)b$ where $a = pb$. Thus $B = \alpha A + D$. If $z \in \alpha A \cap D$, then $z = \alpha x = (1 - \alpha p)y$ for some $x \in A$ and $y \in B$. Since $p\alpha = 1_A$, $pz = p\alpha x = 1_A x = x = py - p\alpha py = py - 1_A py = 0$. But then also $z = \alpha x = \alpha 0 = 0$. Thus $B = \alpha A \oplus D$.

(ii) $\Rightarrow$ (iii). Since $\ker \beta = \alpha A$, the restriction $(\beta | D): D \to C$ is an isomorphism. Let q be the isomorphism $(\beta | D)^{-1}: C \to D$ followed by the natural inclusion map $D \to B$. I.e. for $c \in C$, $qc = d \in D \subset B$, where d is the unique element of D such that $\beta d = c$. Hence $\beta q c = \beta d = c$, and $\beta q = 1_c$.

(iii) $\Rightarrow$ (ii). Given is $q: C \to B$ such that $\beta q = 1_C = 1$. First note that for any $b \in B$, the element $b - q\beta b \in \ker \beta$, because $\beta(b - q\beta b) = \beta b - 1_C \beta b = 0$. But by hypothesis, $\ker \beta = \alpha A$. Thus $b - q\beta b = \alpha a$ for some (unique) $a \in A$. Set $D = qC$. Then $\beta q = 1$ guarantees that $\beta | D: D \to C$ is an isomorphism. Every $b \in B$ is of the form $b = \alpha a + d$ for $a \in A$ as above and with $d = q\beta b \in qC = D$. If $z \in \alpha A \cap D$, then $z = qc$ for some $c \in D$. In view of $z \in \alpha A$ and $\beta \alpha = 0$, it follows that $0 = \beta z = \beta q c = 1_C c = c$. But then $z = qc = 0$, and $B = \alpha A \oplus D$.

Proof of 4.5. (ii) $\Rightarrow$ (iv). Define an R-isomorphism $\varphi: B = \alpha A \oplus D \to A \oplus C$

by $\varphi(\alpha a + d) = [a, (\beta \mid D)d]$. Thus $\varphi\alpha a = [a, 0] = ia$, or $\varphi\alpha = i$; and $\beta(\alpha a + d) = \beta d = j[a, (\beta \mid D)d] = j\varphi(\alpha a + d)$, or $\beta = j\varphi$.

(iv) $\Rightarrow$ (i). Let $h: A \oplus C \to A$ be the natural projection. Since $hi = 1_A$, and $\varphi\alpha = i$, the map $p = h\varphi: B \to A$ satisfies $p\alpha = h\varphi\alpha = hi = 1_A$.

1-5   Exercises

Notation. Let $\mathbb{Z} = 0, \pm 1, \pm 2, \ldots$; $Z_n = \mathbb{Z}/n\mathbb{Z}$ for $0 < n \in \mathbb{Z}$; $\mathbb{Q}$ denotes the rationals. For additive groups A, B, C, D and $a \in A, b \in B, c \in C, d \in D$ write

$$[a, b; c, d] = \begin{vmatrix} a & b \\ c & d \end{vmatrix}, \quad [A, B; C, D] = \begin{vmatrix} A & B \\ C & D \end{vmatrix}$$

where the latter denotes the additive group of all 2×2 matrices with entries in the indicated groups.

In problems 1–6 the following matrix rings R act on right R-modules M whose elements are row vectors. Find (i) all submodules of M; (ii) all right ideals of R; and (iii) the annihilators $a^\perp$ and $b^\perp$ of the elements a and b. (iv) Indicate which of the right ideals in (ii) and (iii) are ideals.

1. $R = [Z_2, Z_2; 0, Z_2]$, $M = Z_2 \oplus Z_2$;
   $a = [0, \bar{1}; 0, 0]$, $b = [0, \bar{1}; 0, \bar{1}]$.
2. $R = [Z_4, Z_4; Z_4, Z_4]$, $M = \bar{2}Z_4 \oplus \bar{2}Z_4$;
   $a = [0, \bar{1}; 0, 0]$, $b = [0, 0; 0, \bar{1}]$
3. $R = [\mathbb{Z}, \mathbb{Z}; \mathbb{Z}, \mathbb{Z}]$, $M = Z_2 \oplus Z_2$;
   $a = [0, 1; 0, 1]$, $b = [1, -1; 0, 0]$
4. $R = [Z_4, 2Z_4; 0, Z_4]$, $M = Z_4 \oplus Z_4$;
   $a = [0, \bar{2}; 0, 0]$, $b = [0, \bar{2}; 0, \bar{2}]$
5. $R = [\mathbb{Z}, \mathbb{Q}; 0, \mathbb{Z}]$, $M = \mathbb{Z} \oplus \mathbb{Q}$;
   $a = [1, 0; 0, 0]$, $b = [0, 0; 0, 1]$
6. Let $\mathbb{Q}[x]$ be the ring of polynomials over $\mathbb{Q}$.
   $R = [\mathbb{Q}[x], \mathbb{Q}[x]; 0, \mathbb{Q}[x]]$, $M = \mathbb{Q}[x] \oplus \mathbb{Q}[x]$;
   $a = [1, -x; 0, 0]$, $b = [x, -1; 0, 0]$
7. Let $R = F$ be a field and $V = F \oplus F$ an F vector space. Find submodules ($=$ vector subspaces) $A, B, C \subset V$ such that

   $A \cap B + C \subsetneq (A + C) \cap (B + C)$ and
   $A \cap C + B \cap C \subsetneq (A + B) \cap C$

8. Suppose that a ring R has a positive *characteristic* n ($nR = 0$ and n is the unique smallest integer $n > 0$ with this property). Prove that $Z_n \times R$ is a ring under

   $(\bar{i}, a) + (\bar{j}, b) = (\bar{i} + \bar{j}, a + b)$

$$(\bar{i}, a)(\bar{j}, b) = (\overline{ij}, \overline{ib} + \overline{ja}) \quad (\bar{i}, \bar{j} \in Z_n; a, b \in R$$

Show that $(\bar{1}, 0)$ is the identity element of $Z_n \times R$ and that $R \cong \{0\} \times R \lhd Z_n \times R$.

9. If $R = \mathbb{Z}$, find $\mathrm{Hom}_{\mathbb{Z}}(\mathbb{Q}, \mathbb{Q})$, $\mathrm{Hom}_{\mathbb{Z}}(\mathbb{Z}, \mathbb{Q})$, and $\mathrm{Hom}_{\mathbb{Z}}(\mathbb{Q}, \mathbb{Z})$.

10. If $R = [\mathbb{Z}, \mathbb{Z}; \mathbb{Z}, \mathbb{Z}]$, $A = [\mathbb{Z}, \mathbb{Z}; 0, 0]$, and $B = [0, 0; \mathbb{Z}, \mathbb{Z}]$, Find $\mathrm{Hom}_R(A, B)$.

11. If $R = [\mathbb{Z}, \mathbb{Q}; 0, \mathbb{Q}]$, $a = [0, 1; 0, 0]$ and $b = [0, 0; 0, 1]$, show that $a^{\perp} = b^{\perp}$. Find explicitly an isomorphism of right R-modules $R/a^{\perp} \to bR$.

For exercises 12–14, see Exercise 0.9–0.14.

12. Let X be an infinite set and $\{F_x \mid x \in X\}$ any indexed family of nonzero rings. Set $R = \Pi\{F_x \mid x \in X\}$. For $r = (r_x) \in R$, define the *support* of r to be the set $\mathrm{supp}\, r = \{x \in X \mid r_x \neq 0\}$, and $r^{-1}(0) = \{x \in X \mid r_x = 0\}$. (a) Let F be a filter on X. Show that $I[F] = \{r \in R \mid r^{-1}(0) \in F\} \lhd R$ is a proper ideal, $I[F] \neq R$. (b) If $F \subset G$ are filters on X, show that $I[F] \subseteq I[G]$. (c) For each $x \in X$, show that $F_x = \{Y \subseteq X \mid x \in Y\}$ is an ultrafilter and that $R/I[F_x] \cong F_x$.

13. With the same notation as in the previous problem, in addition let $F_x = \mathbb{C}$ be the complex numbers for all x and $R = \Pi\{\mathbb{C} \mid X\}$. (a) If $I \lhd R$ is an ideal, $I \neq R$, show that $F[I] = \{r^{-1}(0) \mid r \in I\}$ is a filter. (b) Prove that I is a prime ideal ($rs \in I \Rightarrow r \in I$, or $s \in I$) if and only if $F[I]$ is a prime filter. (c) Show that every filter F and proper ideal I of R satisfy

$$I[F[I]]] = I, \quad F[I[F]] = F.$$

(d) For a filter F, Suppose that $\cap\{E \mid E \in F\} \in F$. Show that then $I[F] = \{r \in R \mid r^{-1}(0) \supseteq E\}$. Give an example of a filter not of this type.

14. A ring R with $1 \in R$ is called a *Boolean ring* if $a^2 = a$ for all $a \in R$. Prove that any Boolean ring is commutative and of characteristic two. If $0 \neq a \neq 1$ in R, show that a is a divisor of zero. If $M \lhd R$ is a maximal ideal, then $R/M \cong Z_2$, the field of two elements.

15. For a nonempty set X, let $P(X)$ be the ring of all subsets of X $(A + B = A \cup B \backslash A \cap B$, $A \cdot B = A \cap B$; $A, B \in P(X))$. Let $Z_2(X)$ be the ring of all functions $f, g: X \to \{\bar{0}, \bar{1}\}$ of X into the two element field under pointwise operations $(f + g)(x_0) = f(x_0) + g(x_0)$ and $(fg)(x_0) = f(x_0)g(x_0)$. Show that $Z_2(X) = \Pi\{Z_2 \mid X\}$. Prove that $P(X) \cong Z_2(X)$ as rings. Deduce that $P(X)$ is a Boolean ring.

16. Suppose that R is a ring with $1 \in R$ and that $u \in R$ has more than one right inverse. Prove that then u has infinitely many right inverses by verifying the following steps in a proof due to Bitzer [Bitzer 63; p. 315].

   Define $S = \{x \in R \mid ux = 1\}$. Take any $s \in S$, and define $T = \{xu - 1 + s \mid x \in S\}$. Then $T \subseteq S$. The map of S onto T given by $x \to xu - 1 + s$ is one to one. If S is finite then $T = S$. In this case $s \in T$, and $xu - 1 + s = s$ for some x, or $xu = 1$. Take $t \in S$, $t \neq x$. Then $x = x(ut) = (xu)t = t$, a contradiction.

Some standard rings

17. Let K be any ring with $1 \in K$ whose center is a field and $0 \neq x$, $0 \neq y \in$ center K any elements. Let I, J, and IJ be symbols not in K. Form the set $K[I, J] = K + KI + KJ + KIJ$ of all K-linear combinations of $\{1, I, J, IJ\}$. (a) Show that $K[I, J]$ becomes an associative ring under the following multiplication rules: $I^2 = x$, $J^2 = y$, $IJ = -JI$, $cI = Ic$, $cJ = Jc$, $cIJ = IJc$ for all $c \in K$. ($K[I, J]$ is called a generalized quaternion algebra over K.)

18. For $K[I, J]$ as in 17 prove that the subring $K[I] = K + KI$ has an automorphism of order two $a = a_1 + a_2 I \to \bar{a} = a_1 - a_2 I$, a_1, $a_2 \in K$, which extends to an inner automorphism of order two of all of $K[I, J]$, where $q \to \bar{q} = J^{-1}qJ = (1/y)JqJ$ for $q = a + bJ$; $a, b \in K[I]$. Show that $\bar{q} = \bar{a} + \bar{b}J$; the latter is called the conjugate of q.

19. Prove that the ring $K[I, J]$ in 17 is isomorphic to a ring of 2×2 matrices

$$a + bJ \to \begin{vmatrix} a & by \\ b & \bar{a} \end{vmatrix} \qquad a, b \in K[I].$$

20. Now let $K = F$ be any (commutative) field in 17.
   (a) Show that $F[I, J]$ is a division ring if and only if $\forall a_1, a_2$, $b_1, b_2 \in F$
   $$a_1^2 - a_2^2 x - b_1^2 y + b_2^2 xy = 0 \Leftrightarrow a_1 = a_2 = b_1 = b_2 = 0.$$
   For any field $K = F$, in 17 define $\mathbb{H}_F = F[I, J]$, where $x = y = -1$.
   (b) If $\mathbb{R}$ are the reals, then show that the real quaternions $\mathbb{H}_\mathbb{R}$ is a division ring.
   (c) If $\mathbb{C}$ are the complexes, show that $\mathbb{H}_\mathbb{C}$ is not a division ring.

21. Let Γ be a multiplicatively written (in general noncommutative) semigroup and K any ring. Let $K[\Gamma]$ be the set of all functions α, $\beta\colon \Gamma \to K$ which are zero except for at most a finite subset of Γ. Write α and β as formal sums:

$$\alpha = \sum x\alpha(x), \quad \beta = \sum y\beta(y) \quad x, y, z \in \Gamma; \ \alpha(x), \ \beta(y) \in K.$$

    (a) Show that $K[\Gamma]$ is a ring under the following operations:

$$\alpha - \beta = \sum z[\alpha(z) - \beta(z)],$$
$$\alpha\beta = \sum \{xy\alpha(x)\beta(y) \mid (x, y) \in \Gamma \times \Gamma\}.$$

       Show that $0x = x0 = 0y = y0$ is the zero element of $K[\Gamma]$.

    (b) Show that alternatively $K[\Gamma]$ may be viewed as a function ring: $(\alpha - \beta)(z) = \alpha(z) - \beta(z)$,

$$(\alpha\beta)(z) = \sum \alpha(x)\beta(y), \quad z = xy \in \Gamma, \ (x, y) \in \Gamma \times \Gamma.$$

    (c) In addition assume that both Γ and K have identities $e \in \Gamma$ and $1 \in K$. For $x \in \Gamma$, define $1x = x1 = x$. Show that $\Gamma \subset K[\Gamma]$ and that e is the identity element of $K[\Gamma]$.

    (d) Suppose that $0_\Gamma \in \Gamma$, $0_\Gamma x = x0_\Gamma = 0_\Gamma$ for all $x \in \Gamma$. Show that 0_Γ is not the zero element of $K[\Gamma]$. (The ring $K[\Gamma]$ is called the *semigroup ring* over K. If Γ is a group it is called the *group ring*.)

22. For a field F, let $\mathbb{Z}^+$ be the additive semigroup $\mathbb{Z}^+ = 0, 1, 2, \ldots$ What is $F[\mathbb{Z}^+]$?

23. Let K be a division ring and $\theta\colon K \to K$ an isomorphism. For $0 \leqslant m \in \mathbb{Z}$, set $\theta(m) = \theta^m$, $\theta(0) = $ identity, $\theta(-m) = (\theta^{-1})^m$. For $k \in K$, write $\theta(k) = k^\theta$, and $\theta(n)(k) = k^{\theta(n)}$. Let $K((x; \theta))$ be the set of all formal power series α, β with only a finite number of negative exponent terms, and right side coefficients in K

$$\alpha = \sum_{N \leq i} x^i\alpha(i), \quad \beta = \sum_{M \leq j} x^j\beta(j) \quad N, M \in \mathbb{Z}; \ \alpha(i), \ \beta(j) \in K.$$

    (a) Show that $K((x; \theta))$ is a ring under pointwise addition, and the usual formal power series multiplication, but where scalars from K are moved past x's and x^{-1}'s according to the rule $kx = xk^\theta$ and $kx^{-1} = x^{-1}k^{\theta(-1)}$. Deduce that

$$\alpha\beta = \sum_{N+M \leq n} x^n \sum_{(i, j) \in \mathbb{Z} \times \mathbb{Z}; i+j=n} \alpha(i)^{\theta(j)}\beta(j))$$

    (b) Let $K[[x; \theta]] \subset K((x; \theta))$ be the subring of all α and β with $0 \leq M, N$. Show that $\gamma \in K[[x; \theta]]$ has an inverse

$\gamma^{-1} \in K[[x; \theta))$ if and only if $0 \neq \gamma(0) \in K$. Prove that $K((x; 0))$ is a division ring.

Definition. $K[[x; \theta]] \subset K((x; \theta))$ are called the *formal skew power series* and *Laurent series rings* over K. If $\theta = $ identity these are abbreviated as $K[[x]] \subset K((x))$.

24. Above in 23, take $K = \mathbb{C}$ the complex numbers, $\mathbb{R}$ the reals, and $\theta: \mathbb{C} \to \mathbb{C}$ complex conjugation, $\theta c = \bar{c}$. (a) Show that the center of the division ring $\mathbb{C}((x; \theta))$ is the subring $\mathbb{R}((x^2)) \subset \mathbb{C}((x; \theta))$. (b) Set $I = \sqrt{-1} \in \mathbb{C}$ and $J = x$. Show that $\mathbb{C}((x; \theta))$ is a generalized quaternion algebra over $\mathbb{R}((x^2))$. (See exercise 17).

The next sequence of exercises about generators foreshadow some ideas of Chapters II and III, and can be done either before or after reading these chapters.

Terminology. Let M be a unital right R-module over a ring R and $X \subset M$ a subset. Then

(1) X is a *free* subset if it generates a free submodule of M, i.e. $\langle X \rangle = \oplus \{xR \mid x \in X\} \leqslant M$ and $xr = 0$ only if $r = 0$, $x \in X$, $r \in R$.

(2) X is an *independent* subset if for any n any finite subset $\{x_1, \ldots, x_n\} \subseteq X$ generates a direct sum of modules $\langle x_1, \ldots, x_n \rangle = \oplus_1^n x_i R$.

(3) A module $0 \neq U$ is uniform if for any $0 \neq V_1 \leqslant U$, and $0 \neq V_2 \leqslant U$, it follows that $V_1 \cap V_2 \neq 0$. X is a *uniform independent* set if X is independent, and also for every $x \in X$, xR is a uniform module.

A submodule $L \leqslant M$ is large or *essential* if for every $0 \neq P \leqslant M$, also $L \cap P \neq 0$.

25. Show that every unital module M contains the following maximal types of sets $X, Y, Z \subseteq M$: (i) X is free; (ii) Y is independent; and (iii) Z is uniform. Show that $\langle Y \rangle$ is essential in M.

26. For $R = \mathbb{Z} = 0, \pm 1, \pm 2, \ldots$ and any $0 < n \in \mathbb{Z}$, view $Z_n = \mathbb{Z}/n\mathbb{Z}$ as a $\mathbb{Z}$-module. For $n = 6$, 8, and 12 find maximal independent sets and maximal uniform independent sets in Z_n. Which of these generate essential submodules?

27. Next, let $R = Z_n$ and view Z_n as a module over itself. For $n = 6$, 8, and 12 find maximal free sets X, maximal independent sets Y, and maximal uniform independent ones Z. Which of $\langle X \rangle$, $\langle Y \rangle$, and $\langle Z \rangle$ are essential in Z_n?

28. Repeat the previous exercise for the rings of upper triangular

matrices over the rationals $\mathbf{Q}$, the integers $\mathbf{Z}$, and the $n \times n$ ring of matrices over $\mathbf{Q}$.

29. Let $R = \mathbf{Z}$ be the ring of integers, p a prime number, and $\mathbf{Z}(p^\infty)$ the abelian group generated by generators $c_1, c_2, \ldots, c_i, \ldots$ subject to the relations $pc_{i+1} = c_i$, $i = 1, 2, \ldots$; and $pc_1 = 0$. Establish the following properties of $\mathbf{Z}(p^\infty)$.
   (a) Let $\mathbf{Z}_{(p)} = \{a/p^i \mid a \in \mathbf{Z}, (a, p) = 1\}$ $((a, p) = \gcd(a, p)$ is their greatest common divisor). Then $c_i \to 1/p_i$ induces an isomorphism $\mathbf{Z}(p^\infty) \cong \mathbf{Z}_{(p)}/\mathbf{Z}$.
   (b) $\mathbf{Z}(p^\infty)$ is *divisible* (for any $0 < n \in \mathbf{Z}$ and $0 \neq y \in \mathbf{Z}(p^\infty)$, $y = nx$ for some $x \in \mathbf{Z}(p^\infty)$).
   (c) Every nonzero element of $\mathbf{Z}(p^\infty)$ is uniquely of the form ac_i, $0 < a \in \mathbf{Z}$, $(a, p) = 1$.
   (d) For each $0 < n \in \mathbf{Z}$, $\mathbf{Z}(p^\infty)$ has a unique subgroup of order p^n, and $\mathbf{Z}(p^\infty)$ is the ascending union of all these cyclic subgroups; $\mathbf{Z}(p^\infty)$ satisfies the descending chain condition on subgroups.
   (e) $\mathbf{Z}(p^\infty)$ is isomorphic to the multiplicative group of p^i-th roots of unity for $i = 0, 1, 2, \ldots$.

The following definition and set of exercises define and develop the p-adic integers and numbers.

Definition. For a prime $p \in \mathbf{Z} = 0, \pm 1, \pm 2, \ldots$ the *p-adic integers* J_p are the set of all formal power series expressions π, ρ of the form $\pi = a_0 + a_1 p + a_2 p^2 + \cdots$, $\rho = b_0 + b_1 = p + b_2 p^2 + \cdots$ where $0 \leq a_i, b_i \leq p - 1$. (To add or multiply π and ρ first add or multiply formally as power series in an indeterminate p. Then reduce the coefficients of the p^n modulo p adding the remainders to the higher powers of p.) Define $\pi + \rho = c_0 + c_1 p + c_2 p^2 + \cdots$ where $c_0 = a_0 + b_0 - k_0 p$, $0 \leq c_0 \leq p - 1$. Subsequently all $0 \leq k_i, l_i \in \mathbf{Z}$. Then $c_1 = a_1 + b_1 + k_0 - k_1 p$, $0 \leq c_1 \leq p - 1$; $c_2 = a_2 + b_2 + k_1 - k_2 p$, $0 \leq c_2 \leq p - 1$. Similarly $\pi \rho = d_0 + d_1 p + d_2 p^2 + \cdots$, $d_0 = a_0 b_0 - l_0 p$, $d_1 = a_0 b_1 + a_1 b_0 + l_0 - l_1 p$, etc.

Let $R_p = \{p^{-n} \pi \mid \pi \in J_p, \ 0 \leq n \in \mathbf{Z}\}$. For x, $y \in R_p$ write $x = p^{-n} \pi$, $y = p^{-n} \varphi$. Define $x + y = p^{-n}(\pi + \rho)$, and $xy = p^{-2n} \pi \rho$. Then R_p is called the field of p-adic numbers, or just the *p-adic numbers*.

30. Prove that the ring J_p is associative.
31. Prove that $\pi = a_0 + a_1 p + a_2 p^2 + \cdots \in J_p$ is a unit iff $a_0 \neq 0$, and show that R_p is a field.

32. Show that every abelian p-group G (i.e. the order of every element is a power of p) is a module over J_p and that abelian group homomorphisms of p-groups are J_p-module homomorphisms.

33. For $\pi = a_n p^n + \ldots, a_n \neq 0$, show that

$$-\pi = (p - a_n)p^n + (p - a_{n+1} - 1)p^{n+1} + (p - a_{n+2} - 1)p^{n+2} + \cdots$$

34*. Let $\mathbf{Z}_{(p)} = \{a/b \mid a, b \in \mathbf{Z}, (b, p) = 1\}$. (i) Prove that every ideal of $\mathbf{Z}_{(p)}$ is of the form $p^n \mathbf{Z}_{(p)}$, $1 \leqslant n$. (ii) Show that $\mathbf{Z}_{(p)}$ is a topological ring where $\{p^n \mathbf{Z}_{(p)}\}$ are a base of neighborhoods of zero. (iii) Prove that J_p is the topological completion of $\mathbf{Z}_{(p)}$.

35*. (i) Identify $\mathbf{Z}$ as a subset of J_p.
    (ii) Give a direct proof that $\mathbf{Z}_{(p)} \subset J_p$.

CHAPTER 2

Free modules

Introduction

In Chapters 2 and 3 in our discussion of free and injective modules, it will be assumed that the ring R has an identity $1 \neq 0$ and that all modules are unital.

One of two ways to begin the study of free modules is to define them by means of a universal property (see 2-1.7). Here the other approach is used which views free modules as a generalization of vector spaces.

It will turn out that the free modules are exactly those which are isomorphic to a direct sum of the ring R viewed as a right module. And the rank of a free module is the number of copies of R needed. Thus in the special case when R is equal to a field F, all R-modules, that is vector spaces are free, and the rank is the dimension. The analogy does not stop here. Just like in the vector space case there is a concept of a basis of a free module, and that of independence generalizes linear dependence in a vector space.

If everything easily generalized from vector spaces to free modules our whole endeavour of studying modules would become trivial, we could merely treat modules as footnotes to linear algebra. The observant reader will see here already unfolding in Chapter 2 a phenomenon which will repeat many times over and over again. Namely, many familiar vector space or commutative ring concepts either generalize, or more accurately have analogues, for arbitrary noncommutative rings but many of these are in no sense trivial generalizations of familiar patterns. A simple but striking example shows that there are rings R for which $R \cong R \oplus R$ (2-2.4). However, this does not mean that we have to give up the notion of rank. By use of the facts about cardinalities of infinite sets from Appendix 0 (0-1.12), it is shown that infinite rank free modules have

19

unique rank. Then it will be shown that for commutative rings all (finite rank) free modules have unique rank. It turns out that quite a few other classes of rings have this pleasant property, in which case a ring R is said to have the IBN property (invariant basis number). For example, later we will show that rings which satisfy either the descending (Artinian) or ascending (Noetherian) chain condition likewise have the IBN property (11-3.15). Other examples of this phenomenon in this book are injective modules which generalize divisible abelian groups in Chapter III and primitive rings which are analogues of rings of linear transformations on a vector space in Chapter 6.

A sizeable amount of the overall research effort in module theory in the decade of 1960–1970 went into building a theory of torsion and torsion free modules, generalizing the familiar trivial definition of a torsion and torsion free abelian group. However, the end result which still is in the process of being reached is certainly nothing at all like the abelian group case. The interested reader will find full coverage of this advanced topic in the long book [Golan 86].

2-1   Definition of free modules

The simplest example of a free module is the case when the ring is a skew field and the module is a vector space over that skew field. Here, some of the definitions and proofs are modelled as generalizations of this case.

2-1.1   Definition. (i) A subset $X \subseteq M$ of a right R-module M *generates* M if $\langle X \rangle = M$. (ii) $X \subset M$ is *R-independent* if for any finite number n of distinct elements $x_1, \ldots, x_n \in X$ and any $r_i \in R$, if

$$x_1 r_1 + \cdots + x_n r_n = 0, \text{ then} \Rightarrow r_i = 0, \quad \text{all } i = 1, \ldots, n.$$

2-1.2   Definition. A subset $X \in M_R$ is a *basis* (R-basis, base) of M if (i) $\langle X \rangle = M$ and, (ii) X is R-independent.

Remark. It follows from the definitions that the empty subset of any module is R-independent, that $\langle \varphi \rangle = (0)$, and that φ is a basis of (0).

2-1.3   Lemma. *For a right R-module M and a subset $X \subset M$ following are all equivalent.*

(1)   *X is a basis.*

(2)   *Each $0 \neq m \in M$ is uniquely of the form $m = x_1 r_1 + \cdots + x_n r_n$, for distinct $x_i \in X$, and $0 \neq r_i \in R$.*

(3)   *X generates M. If for any finite number n of distinct $x_i \in X$, and $r_i \in R$, $x_1 r_1 + \cdots + x_n r_n = 0$, then all $r_i = 0$.*

Proof. Omitted.

In general a basis is not unique.

2-1.4   Definition. A right R-module M is *free* if it has a basis X. Then M is said to be free on X.

2-1.5   Construction. For any set T, let $R_t = R$ for $t \in T$. Then $\oplus \{R_t \mid t \in T\}$ is free with basis $\{e_t \mid t \in T\}$, where $e_t = (\ldots 0, 1, 0, \ldots)$ has all zero entries except for a "1" in the t-th coordinate. Alternatively, $e_t \colon T \to R$, $e_t(s) = 0$ if $s \neq t$ and $e_t(s) = 1$ if $s = t$. Abbreviate

$$\bigoplus_{t \in T} R_t = \bigoplus_T R = \oplus \{R \mid T\} = \bigoplus_{t \in T} e_t R.$$

It is called the *free R-module on T.*

2-1.6   Lemma. *For a right R-module F and a subset $X \subset F$, the following are equivalent.*

(1)   *F is free with basis X.*

(2)   *$F \cong \oplus_{x \in X} R_x$, $R_x = R$, $x \in X$, under $x \leftrightarrow e_x$.*

Proof. This is proved only when $X = \{x_1, \ldots, x_n\}$ is finite since the proof in the general case is the same except for slightly more complicated notation. Abbreviate $e_i = (0, \ldots, 0, 1, 0, \ldots, 0)$ with a "1" in the i-th coordinate and the rest — 0's, and $R^{(n)} \equiv \oplus \{R_x \mid x \in X\}$.

(1) $\Rightarrow$ (2). By (1), each $\alpha \in F$ is uniquely of the form $\alpha = x_1 a_1 + \cdots + x_n a_n$, $a_i \in R$; $i = 1, \ldots, n$. The map $F \to R^{(n)}$, defined by $\alpha \to (a_1, \ldots, a_n)$ is an R-isomorphism of F onto $R^{(n)}$.

(2) $\Rightarrow$ (1). By (2), without loss of generality, $F = R^{(n)}$. Each $a \in R^{(n)}$ is uniquely of the form $a = (a_1, \ldots, a_n) = e_1 a_1 + \cdots + e_n a_n$. Hence $F = R^{(n)}$ is free with basis $\{e_1, \ldots, e_n\}$.

A function of one set into another set will sometimes be called a set function, or a set map, in order to distinguish it from module homomorphisms.

2-1.7   Lemma. *For a subset $X \subset F$ of a right R-module F the following are all equivalent.*

(1) *U.P. (universal property): For any right R-module V, for any set function $\varphi: X \to V$, there exists a unique homomorphism $\bar\varphi: F \to V$ such that $\bar\varphi \,|\, X = \varphi$, i.e. there is a commutative diagram*

(2) *F is free with basis X.*

Proof. $(2) \Rightarrow (1)$. Let φ be as in (1) and let $r \in R$. For any α, $\beta \in F$ — by allowing zero coefficients $a_i = 0$ or $b_j = 0$ in R if necessary — we may assume that $\alpha = x_1 a_1 + \cdots + x_n a_n$ and $\beta = x_1 b_1 + \cdots + x_n b_n$ for the same n elements $x_1, \ldots, x_n \in X$. The uniqueness of the nonzero a_i and x_i in the expression for α guarantee that a function $\bar\varphi: F \to V$ is well defined by $\bar\varphi(\alpha) = (\varphi x_1) a_1 + \cdots + (\varphi(x_n) a_n$. Again by the uniqueness, $\bar\varphi(\alpha + \beta) = \bar\varphi\alpha + \bar\varphi\beta$ and $\bar\varphi(\alpha r) = (\bar\varphi\alpha) r$. Thus $\bar\varphi$ is a module homomorphism $\bar\varphi: F \to V$ with $\bar\varphi \,|\, X = \varphi$.

If $g: F \to V$ is any other module map extending φ with $g \,|\, X = \varphi$, then since g is a module map, $g\alpha = (gx_1) a_1 + \cdots + (gx_n) a_n = (\varphi x_1) a_1 + \cdots + (\varphi x_n) a_n = \bar\varphi\alpha$. Since $\alpha \in F$ is arbitrary, $g = \bar\varphi$ and $\bar\varphi$ is unique.

$(1) \Rightarrow (2)$. It suffices to show that X is a basis of F. Suppose that $x_1, \ldots, x_n \in X$ are any n different elements that satisfy $x_1 r_1 + \cdots + x_n r_n = 0$ for $r_i \in R$. In (1), take $V = R_R$ and for any $1 \leqslant i \leqslant n$, $\varphi: X \to R$ by $\varphi x_i = 1$ and $\varphi x = 0$ for all other $x_i \neq x \in X$, i.e. $\varphi(X \backslash \{x_i\}) = \{0\}$. By hypothesis (1), there exists a module homomorphism $\bar\varphi: F \to R$ with restriction $\bar\varphi \,|\, X = \varphi$. Thus

$$0 = \bar\varphi(0) = \bar\varphi(x_1 r_1 + \cdots + x_n r_n)$$
$$= (\varphi x_1) r_1 + \cdots + (\varphi x_i) r_i + \cdots + (\varphi x_n) r_n$$
$$= 0 + \cdots + 1 \cdot r_i + \cdots + 0,$$

and hence each $r_i = 0$. Thus X is R-independent.

Let $N \equiv \langle X \rangle = \Sigma\{xR \,|\, x \in X\}$. Form F/N. We want to show that $F/N = \{\bar0\}$, i.e. that $F = N$. Define $\varphi: X \to F/N$ by $\varphi x = N = 0 + N = \bar0$ for all x, and then extend it to module maps $\bar\varphi_1$, $\bar\varphi_2: F \to F/N$ two ways as follows. Let $\bar\varphi_1 = 0$ and $\bar\varphi_2: F \to FN$ be the natural projection, i.e. for any $\alpha \in F$, $\bar{\varphi}_1 \alpha = 0$ and $\bar{\varphi}_2 \alpha = \alpha + N$. I.e. the following diagram commutes with either $\bar{\varphi}_1$ or $\bar{\varphi}_1$.

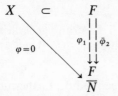

By the uniqueness hypothesis, $\bar{\varphi}_1 = \bar{\varphi}_2$, and thus $\{\bar{0}\} = \bar{\varphi}_1 F = \bar{\varphi}_2 F = F/N$ shows that $F = N = \langle X \rangle$. Hence X is a basis of F and F is free on X.

Remark. In the above proof and diagrams it is said that φ lifts to $\bar{\varphi}$, or φ lifts to $\bar{\varphi}_i$.

2-1.8   Corollary. *Let* $F = F_R$, *T a set and* $\lambda: T \to F$ *a set function such that the following holds:*
   U.P. $\forall V_R$, $\forall$ *function* $\varphi: T \to V \Rightarrow \exists!$ *R-map* $\bar{\varphi}: F \to V$ *such that* $\bar{\varphi}\lambda = \varphi$, *i.e. the following diagram commutes.*

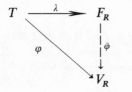

Then λ *is one to one, and* F *is free on* λT.

Proof. Suppose that $x \neq y \in T$. Take $V = U \oplus U$ where U is any nonzero right R-module. Then define $\varphi: T \to V$ by $\varphi x = (u, 0)$, $\varphi y = (0, u)$ where $0 \neq u \in U$ is arbitrary, and $\varphi(T \backslash \{x, y\}) = \{(0, 0)\}$. Let $\bar{\varphi}: F \to V$ be the extension guaranteed by the hypothesis with $\varphi = \bar{\varphi}\lambda$. Thus $\bar{\varphi}(\lambda x - \lambda y) = \bar{\varphi}(\lambda x) - \bar{\varphi}(\lambda y) = \varphi x - \varphi y = (u, -u) \neq (0, 0)$ shows that $\lambda x \neq \lambda y$.

Remarks. (1) Although we are assuming that $1 \in R$ and that all modules are unital in this chapter, nevertheless the above corollary holds also for rings without an identity, and even for the zero ring.
   (2) In the situation of the above corollary, F is also said to be free on T.

   The next lemma allows one to estimate the cardinality of the free module. In some applications (see 2-3.5 and 2-3.8) it is useful to choose the generating set T below as small as possible.

2-1.9   Lemma. Let M be any module and $T \subseteq M$ any generating set of M. Let $F = \oplus\{R \mid T\}$ be the free R-module on T (2-1.5). Then $M \cong F/K$ for some submodule $K \subseteq F$. In particular,

   (i)   every module is the quotient of a free module; furthermore
  (ii)   this free module may so be chosen that it has a base of the same cardinality as any generating set of the original module.

Proof. The set map $\lambda: T \to F = \oplus\{e_t R \mid t \in T\}$, $\lambda t = e_t$ for $t \in T$, is one-to-one, and F is free on λT and T. Consequently, the set function $\varphi: T \to M$, $\varphi t = t$, extends to a (unique) module homomorphism $\bar{\varphi}: F \to M$ with $\bar{\varphi}\lambda = \varphi$ (and $\bar{\varphi}e_t = t$). Since image $\bar{\varphi} \supseteq T$ and $\langle T \rangle = M$, im $\bar{\varphi} = M$, and $\bar{\varphi}$ is onto. Thus $M \cong F/K$ where $K \subseteq F$ is the kernel of $\bar{\varphi}$.

2-1.10   Corollary. For any module P, there is a free module F having as basis the elements of P and there is an exact sequence $F \to P \to 0$ (or epic map $F \twoheadrightarrow P$).

Proof. If in the proof of the last lemma $M = P$, $T = P$, and $\lambda: P \to F$ are as above, then $\varphi = 1: P \to P$ turns out to be actually the identity function. Its extension $\bar{\varphi}: F \to P$ gives the required epimorphism.

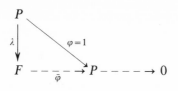

Remarks. (1) If $P = \{0\}$, then $F = \lambda(0)R \neq 0$. (2) Always above $\lambda(0) \neq 0_F$.

2-2   Bases of free modules

A base of a free module is a generalization of the notion of a vector space basis. It is shown in this section that it is possible for a finitely generated module F to be free on two bases $X \subset F$ and $Y \subset F$, where X and Y contain a different number of elements, $|X| \neq |Y|$. However as Proposition 2-2.5 will show, this pathology can occur only if X and Y are finite. Just as the dimension of a vector space determines it, so also a nonfinitely generated free module is determined uniquely up to isomorphism by the cardinality of any base.

For what kinds of rings R is the number of elements of a basis of a free module not unique?

2-2.1   Lemma. There exists a free module F with two bases $\{x\}$ and $\{y_1, y_2\}$ such that $F = xR = y_1R \oplus y_2R$, $xR \cong R$, $y_iR \cong R$, if and only if there exist four elements π_1, π_2, α_1, and $\alpha_2 \in R$ satisfying

(1) $\pi_1\alpha_1 + \pi_2\alpha_2 = 1$;
(2) $r_1, r_2 \in R$, if $\pi_1r_1 + \pi_2r_2 = 0$, then $r_1 = 0$ and $r_2 = 0$.

Proof. $\Rightarrow$: Since $y_1, y_2 \in xR$, $y_1 = x\pi_1$ and $y_2 = x\pi_2$ for some $\pi_1, \pi_2 \in R$. Also $x \in y_1R + y_2R$, and thus $x = y_1\alpha_1 + y_2\alpha_2$ for some $\alpha_1, \alpha_2 \in R$. Consequently $x = x\pi_1\alpha_1 + x\pi_2\alpha_2$ implies that (1) $\pi_1\alpha_1 + \pi_2\alpha_2 = 1$ because xR is free on $\{x\}$. For any $r_1, r_2 \in R$

$$\pi_1r_1 + \pi_2r_2 = 0 \leftrightarrow x(\pi_1r_1 + \pi_2r_2) = y_1r_1 + y_2r_2 = 0$$
$$\leftrightarrow r_1 = 0, \quad r_2 = 0.$$

$\Leftarrow$: Given that (1) and (2) hold, define $F = R_R$, $x = 1$, $y_1 = \pi_1$, and $y_2 = \pi_2$. If $z \in \pi_1R \cap \pi_2R$, then $z = \pi_1a_1 = \pi_2a_2$ with $a_1, a_2 \in R$. By (2) it follows from $\pi_1a_1 - \pi_2a_2 = 0$ that $a_1 = 0$ and $a_2 = 0$. Thus $z = \pi_2a_2 = 0$, $\pi_1R \cap \pi_2R = 0$, and the sum $\pi_1R + \pi_2R = \pi_1R \oplus \pi_2R$ is direct. Use of (1) shows that $1 = \pi_1\alpha_1 + \pi_2\alpha_2 \in \pi_1R + \pi_2R$, and hence that $R = \pi_1R \oplus \pi_2R = \langle \pi_1, \pi_2 \rangle$. By (2), the generating set $\{\pi_1, \pi_2\}$ is R-independent. Thus $\{1\}$ and $\{\pi_1, \pi_2\}$ are two bases for the free module $F = xR = y_1R \oplus y_2R$.

The next two corollaries concretely exhibit the isomorphism $R \oplus R \cong R$ whose existence is guaranteed by the previous lemma.

2-2.2   Corollary. 1. Let α_i, $\pi_j \in R$ satisfy (1) of 2-1.11. Then $(2) \Leftrightarrow (3)$, where (3) $\alpha_i\pi_j = \delta_{ij}$; $i, j = 1, 2$; where $\delta_{ij} = 1$ if $i = j$ and 0 otherwise.

Proof. $(2) \Rightarrow (3)$. By assumption, $\pi_1\alpha_1 + \pi_2\alpha_2 = 1$. Thus $(\pi_1\alpha_1 + \pi_2\alpha_2)\pi_1 = \pi_1$, and $\pi_1(\alpha_1\pi_1 - 1) + \pi_2(\alpha_2\pi_1) = 0$. Use of (2) (with $r_1 = \alpha_1\pi_1 - 1$ and $r_2 = \alpha_2\pi_1$) shows that $\alpha_1\pi_1 = 1$ and $\alpha_2\pi_1 = 0$. Similarly replacement of π_1 with π_2 above (in $(\pi_1\alpha_1 + \pi_2\alpha_2)\pi_2 = \pi_2$) shows that also $\alpha_2\pi_2 = 1$ and $\alpha_1\pi_2 = 0$.

$(3) \Rightarrow (2)$. If $\pi_1r_1 + \pi_2r_2 = 0$, then $0 = \alpha_i(\pi_1r_1 + \pi_2r_2) = r_i$ for $i = 1, 2$.

2-2.3   Corollary. 2. *Now assume conditions* (1)–(3) *of* 2-2.1 *and* 2-2.2. *Then for* $k = 1, 2$, *left multiplication by* α_k

(i) $\pi_k R \to R$, $\pi_k r \to \alpha_k \pi_k r$ *is an isomorphism* $\pi_k R \cong R$.
(ii) $R \oplus R \to R$, $\binom{a}{b} \to [\pi_1 \pi_2]\binom{a}{b} = \pi_1 a + \pi_2 b$, $a, b \in R$ *is an isomorphism*.
(iii) $(\pi_i \alpha_i)(\pi_j \alpha_j) = \delta_{ij}(\pi_j \alpha_j)$ *for* $i, j = 1, 2$.

Proof. (i) Follows from 2-2.1(2) and 2-2.1(3). (ii) By 2-2.1(2), this map is monic. For any $r \in R$, the column vector $(\alpha_1 r, \alpha_2 r)$ maps onto $(\pi_1 \alpha_1 + \pi_2 \alpha_2)r = r$ by 2-2.1(1). (iii) This is immediate from 2-2.2(3).

Remark. By iteration from $R \cong R \oplus R$ to $R \oplus R = R \oplus R \oplus R$, it follows that for any two positive integers n and m, $R^{(n)} \cong R^{(m)}$.

Above, $\alpha_1 \pi_1 = 1$ while $\pi_1 \alpha_1 \neq 1$ suggest that α_1 could be linear operator having only a one-sided inverse as in the next example of a ring R such that $R \cong R \oplus R$.

2-2.4   Example. Let F be any field; $V = \oplus_0^\infty F_n$, $F_n = F$ for $n = 0, 1, 2, \ldots$, a countable dimensional vector space, and $R = \operatorname{Hom}_F(V, V)$ the ring of all linear transformations of V into itself acting on the right. Then V is a right R-module.

Take $x = (x_0, x_1, x_2, \ldots) \in V$ arbitrary. Define $\alpha_1, \alpha_2, \pi_1, \pi_2 \in R$ by:

$$x\alpha_2 = (x_0, 0, x_1, 0, x_2, \ldots) \ldots 0 \text{ even components};$$

$$x\alpha_1 = (0, x_0, 0, x_1, 0, \ldots) \ldots 0 \text{ odd components};$$

$$x\pi_2 = (x_0, x_2, x_4, \ldots) \ldots \text{ projection onto even indexed components};$$

$$x\pi_1 = (x_1, x_3, x_5, \ldots) \ldots \text{ projection onto odd indexed components};$$

The required equations (1), (2) or (1), (3) of 2-2.1, 2.2 are now easily verified.

2-2.5   Proposition. *If* F *is a free module with* $F = \oplus\{xR \mid x \in X\} = \oplus\{yR \mid y \in Y\}$ *where* $X \subset F$ *and* $Y \subset F$ *are free bases of* F *and* X *is infinite, then both* X *and* Y *have the same cardinality* $|X| = |Y|$.

Proof. Every $x_0 \in X$ appears in the expression of some $y_0 \in Y$ as an R-linear combination of the x's with a nonzero coefficient, $y_0 = \cdots + x_0 r + \cdots$, $0 \neq r \in R$. For if this was not the case then

$\oplus\{yR \mid y \in Y\} \subseteq \oplus\{xR \mid x \in X, x \neq x_0\}$ would contradict the directness of the sum $\oplus\{xR \mid x \in X\}$. Furthermore, this also shows that Y is infinite $|Y| = \infty$. Define a function $f: X \to Y$ by $f(x_0) = y_0$. For any $y \in Y$, $f^{-1}(\{y\}) = \{x_1, \ldots, x_n\}$ is a finite set, where $y = x_1 r_1 + \cdots + x_n r_n + \Sigma x_\alpha r_\alpha$, with $x_\alpha \neq x_1, \ldots, x_n$; and all $0 \neq r_i \in R$. In fact, always $f^{-1}(y) \subseteq \{x_\beta \mid r_\beta \neq 0, y = \Sigma x_\beta r_\beta\}$. Now by 0-1.13, $|X| \leq |Y|$. From the symmetry of the hypotheses on X and Y, it follows that $|X| = |Y|$.

For a ring $\bar{R}$, free $\bar{R}$-modules are said to have a well defined rank (or unique rank) if for every (finitely generated) free $\bar{R}$-module, the number of elements of every free basis if the same.

2-2.6 Proposition. *If $\varphi: R \to \bar{R}$ is a surjective ring homomorphism of R onto another nonzero ring $\bar{R}$ over which free modules have unique rank, then so also do free R-modules.*

Proof. Let $\bar{R} = R/B$ for an ideal $B \lhd R$, and suppose that F is a free R-module $F = \oplus\{xR \mid x \in X\} = \oplus\{yR \mid y \in Y\}$ with two bases X and Y. Then $FB = \{\Sigma fb \mid f \in F, b \in B\} = \oplus\{xB \mid x \in X\}$ is an R-submodule of F. The freeness of F guarantees that $xR \cap FB = xB$, and that $xR/xB \to \bar{R} \to x(R/B)$, $xr + xB \to r + B \to x(r + B)$ are both well defined isomorphisms of right R-modules. The "x" in $x\bar{R} \cong \bar{R}$ is a dummy index. Thus

$$(xR + FB)/FB \cong xR/xB \cong x(R/B) \cong \bar{R}$$

is a homomorphism of right R-modules. Since B is the annihilator of all of these modules, —i.e. $(F/FB)B = 0$—, first, all of the above modules are in a natural way also $\bar{R}$-modules, and secondly, they are also all isomorphic as right $\bar{R}$-modules. Thus

$$\frac{F}{FB} = \bigoplus_{x \in X} \frac{xR + FB}{FB} \cong \bigoplus_{x \in X} x\bar{R} \cong \bigoplus_{y \in Y} y\bar{R}.$$

Since $\bar{R} \neq 0$ and the latter is a free $\bar{R}$-module, $|X| = |Y|$.

2-2.7 Corollary. *Free modules over a commutative ring R have unique rank.*

Proof. If R is commutative and B is an ideal maximal with respect to not containing the identity element $1 \in R$, then R/B is a field. Hence the rank of a free R/B-module is the vector space dimension over the field R/B, which is unique.

Remark. The last proposition tells us that the ring in the last example 2-2.4 does not have a commutative nonzero homomorphic image.

The next objective is to describe R-module homomorphisms between free modules in terms of matrices with entries in R. We establish a convenient easily remembered notational scheme for coping with a change of free basis.

2-2.8   Consider an R-map $\varphi: F = \oplus_{j=1}^{m} x_j R \to G = \oplus_{i=1}^{n} y_i R$ of free modules F and G with free bases $X = \{x_j\}$ and $Y = \{y_i\}$. Define an $n \times m$ matrix $\left(\begin{smallmatrix}\varphi\\XY\end{smallmatrix}\right)$ by $\left(\begin{smallmatrix}\varphi\\XY\end{smallmatrix}\right) = \|r_{ij}\|$, where $\varphi x_j = \Sigma_{i=1}^{n} y_i r_{ij}$. Define an R-module isomorphism $\left(\begin{smallmatrix}\cdot\\x\end{smallmatrix}\right): F \to R^n$ where for $\alpha = x_1 a_1 + \cdots + x_m a_m \in F$, $a_i \in R$, the image of α under the R-map $\left(\begin{smallmatrix}\cdot\\x\end{smallmatrix}\right)$ is the column vector $\left(\begin{smallmatrix}\alpha\\X\end{smallmatrix}\right) = [a_1, \ldots, a_m]^T$, where T denotes transpose of matrices. Thus exactly as in vector spaces we have:

$$\alpha = x_1 a_1 + \cdots + x_m a_m, \quad \varphi\alpha = y_1 b_1 + \cdots + y_n b_n \quad a_j, b_i \in R$$

$$\begin{pmatrix}b_1\\ \vdots \\ b_n\end{pmatrix} = \begin{pmatrix}\beta\\Y\end{pmatrix} = \begin{pmatrix}\varphi\alpha\\Y\end{pmatrix} = \begin{pmatrix}\varphi\\YX\end{pmatrix}\begin{pmatrix}\alpha\\X\end{pmatrix} = \|r_{ij}\|\begin{pmatrix}a_1\\ \vdots \\ a_m\end{pmatrix}.$$

Let $H = \oplus_{i=1}^{l} u_i R$ be free on $U = \{u_i\}$ and $\psi: G \to H$ an R-map. Then $\left(\begin{smallmatrix}\psi\varphi\\UX\end{smallmatrix}\right) = \left(\begin{smallmatrix}\psi\\UY\end{smallmatrix}\right)\left(\begin{smallmatrix}\varphi\\YX\end{smallmatrix}\right)$. Note how the repeated adjacent index Y gets cancelled or disappears in the end result.

2-2.9   Change of free bases. In the previous notation of 2.8, now let $Z = \{z_q \mid q = 1, \ldots, M\}$ and $W = \{w_p \mid p = 1, \ldots, N\}$ be other free bases of $F = \oplus_{q=1}^{M} z_q R$ and $G = \oplus_{p=1}^{N} w_p R$ for some finite M, N not necessarily equal to m and n. First concentrating our attention on F, we define an $M \times m$ matrix $\|s_{qj}\|$ by $x_j = \Sigma_{q=1}^{M} z_q s_{qj}$, and similarly an $m \times M$ matrix $\|t_{jq}\|$ by $z_q = \Sigma_{j=1}^{m} x_j t_{jq}$. As a consequence of the uniqueness of the R-coefficient of a vector of F in terms of a free basis, we conclude that $\|s_{qj}\|\|t_{jq}\| = 1_M$ and $\|t_{jq}\|\|s_{qj}\| = 1_m$ where 1_M and 1_m denote $M \times M$ and $m \times m$ identity matrices, while $1: F \to F$ will denote the identity right R-module homomorphism. Now let $\alpha = x_1 a_1 + \cdots + x_m a_m = z_1 c_1 + \cdots + z_M c_M$, $c_q \in R$. Then $[c_1, \ldots, c_M]^T = \|s_{qj}\|[a_1, \ldots, a_m]^T$ and $[a_1, \ldots, a_m]^T = \|t_{jq}\|[c_1, \ldots, c_M]^T$ become

$$\begin{pmatrix}c_1\\ \vdots \\ c_M\end{pmatrix} = \begin{pmatrix}1\alpha\\Z\end{pmatrix} = \begin{pmatrix}1\\ZX\end{pmatrix}\begin{pmatrix}\alpha\\X\end{pmatrix} = \|s_{qj}\|\begin{pmatrix}a_1\\ \vdots \\ a_m\end{pmatrix}, \quad \begin{pmatrix}\alpha\\X\end{pmatrix} = \begin{pmatrix}1\\XZ\end{pmatrix}\begin{pmatrix}\alpha\\Z\end{pmatrix}$$

$$\|s_{qj}\|\,\|t_{jq}\| = \left(\frac{1}{ZX}\right)\left(\frac{1}{XZ}\right) = \left(\frac{1}{ZZ}\right) = 1_M,$$

$$\left(\frac{1}{XZ}\right)\left(\frac{1}{ZX}\right) = \left(\frac{1}{XX}\right) = 1_m.$$

Then for $\varphi\colon F \to G$ as before in 2-2.8, $\left(\frac{\varphi}{WZ}\right) = \left(\frac{1}{WY}\right)\left(\frac{\varphi}{YX}\right)\left(\frac{1}{XZ}\right)$. In the special case when $F = G$, $X = Y$, $W = Z$, and $n = m = N = M$, the above reduces to $\left(\frac{\varphi}{ZZ}\right) = \left(\frac{1}{XZ}\right)^{-1}\left(\frac{\varphi}{XX}\right)\left(\frac{1}{XZ}\right)$, where $\left(\frac{1}{XZ}\right)^{-1} = \left(\frac{1}{ZX}\right)$.

2-2.10 Example. Under the hypotheses and notation of 2-2.1, 2.2 we have 1×2 and 2×1 matrices for which

$$[\pi_1 \pi_2]\begin{bmatrix} \alpha_1 \\ \alpha_2 \end{bmatrix} = [1] \quad \begin{bmatrix} \alpha_1 \\ \alpha_2 \end{bmatrix}[\pi_1 \pi_2] = \begin{bmatrix} 1 & 0 \\ 0 & 1 \end{bmatrix}.$$

2-3   Exercises

1. For $R = \mathbf{Z}$, prove that a subgroup of a finitely generated free group is free. (A subgroup of any free group is free, but the proof is harder.)

2*. Prove that the rational as a module over the integers do not have a minimal generating set.

3. Let D be a commutative principal ideal domain ($1 \in D$; D has no nonzero divisors of zero, and every ideal of D is of the form aD, $a \in D$). Show that my submodule $G < D^n = D \oplus \cdots \oplus D$ is (a) free, and has a free basis of n or less elements. (Hint: Use induction.)

   (b) For D as above, let $D_i = D$, $i = 0, 1, 2, \ldots$. Show that any submodule $G < \oplus_{i=1}^{\infty} D_i$ is free with a free basis that is either countable or finite. (c) Generalize (a) and (b) to arbitrary index sets. (Hint: For (c), use ordinal induction.)

4. Let $\varphi\colon F = \oplus_{j \in J} x_j R \to G = \oplus_{i \in I} y_i R$ be an R-map of free modules with free bases $X = \{x_j \mid j \in J\}$ and $Y = \{y_i \mid i \in I\}$. Define an $I \times J$ matrix $\left(\frac{\varphi}{XY}\right) = \|r_{ij}\|$ by $\varphi x_j = \Sigma_{i \in I} y_i r_{ij}$. Generalize the rest of the notation and results of 2-2.8 and 2-2.9 to this case.

CHAPTER 3

Injective modules

Introduction

Baer's Criterion (3-1.6) in order for a module to be injective is one of the more frequently used and quoted results in todays module theory. Its proof unavoidably requires the use of the axiom of choice, or some other transfinite equivalent. We already need to use this criterion in proving one of the main results of this chapter, namely that for any divisible abelian group, $\text{Hom}_{\mathbb{Z}}(R, D)$ is an injective right R-module (Theorem 3-3.3). The proof of this is intriguing, for who but an algebraist would ever think of making $\text{Hom}_{\mathbb{Z}}(R, D)$ into an R-module, and then proving it to be injective. From this, it is then an easy step to conclude that any right R-module can be embedded as a submodule of an injective module. It is precisely this latter embedability property which allows the subsequent smooth development of the whole theory of injectivity, culminating in the construction and characterization of the injective hull of a module in Theorem 3-4.13. In order to develop the theory of injectivity, the proofs have to be done in just the right order.

3-1   Properties of injectives

The reader might wish to also look at 10-1.2 for a slightly different reformulation of the next definition.

3-1.1   Definition. A module E is *injective* if for any modules $A < B$ and any homomorphism $\mu: A \to E$, there exists an extension $\bar{\mu}$ of μ to a map $\bar{\mu}: B \to E$ whose restriction to A is of course $\bar{\mu} \mid A = \mu$.

diagram $\Rightarrow$ a commutative diagram:

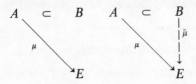

Remark. Recall that for any product $\Pi\{M_j | j \in I\}$ of modules, if $\pi_i: \Pi M_j \to M_i$ is the i-th projection, that then any $x \in \Pi M_i$ can be written as $x = (\pi_i x)_{i \in I}$.

3-1.2   Lemma. *For any indexed family of R-modules M_i, $i \in I$,*

$$\prod_{i \in I} M_i \text{ is injective } \Leftrightarrow \text{ For any } k \in I, \text{ each } M_k \text{ is injective.}$$

Proof. $\Leftarrow$: Given modules $A < B$ and a homomorphism $\mu: A \to \Pi M_j$, since M_i is injective, the map $\pi_i \mu: A \to M_i$ extends to some R-map $\mu_i: B \to M_i$ with $\mu_i | A = \pi_i \mu$. Now define $\bar{\mu}: B \to \Pi M_i$ by mapping $b \in B$ to $\bar{\mu}(b) = (\mu_i(b))_{i \in I}$. For $a \in A$, since $\mu_i a = \pi_i \mu a$, $\bar{\mu}(a) = (\mu_i a) = (\pi_i(\mu a)) = \mu a$ by the previous Remark.

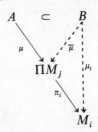

$\Rightarrow$: For any $k \in I$, given a monic $\alpha: A \to B$ and any R-map $v: A \to M_k$, it has to be shown that there is a $\bar{v}: B \to M_k$ such that $v = \bar{v} \alpha$. If $j_k: M_k \to \Pi M_i$ is the inclusion, then $j_k v: A \to \Pi M_i$, and by the injectivity of the latter there is a $\tau: B \to \Pi M_i$ with $\tau \alpha = j_k v$. Now

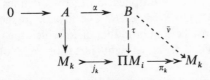

Simply define $\bar{v}: B \to M_k$ by $\bar{v} = \pi_k \tau$. Since $\pi_k j_k = 1$, $\bar{v} \alpha = \pi_k \tau \alpha = \pi_k j_k v = v$. Thus M_k is injective.

In particular, a finite direct sum of injective modules is injective. Because of frequent applications of the latter, it is worth restating.

3-1.3   Corollary. *A summand of an injective module is injective.*

The construction below does not use injective modules and is perfectly general.

3-1.4   Construction. Let $N \subset M$ and E be any modules, and $f: N \to E$ any homomorphism. Then there exist a maximal extension of f in M.

Proof. An extension of f is a pair (V, g) where $N \subseteq V \subseteq M$, where $g: V \to E$ with $g \mid N = f$. The set $\mathscr{S} = \{(V, g)\}$ of all such extensions of f is partially ordered by $(V, g) < (V_1, g_1)$ if $V \subset V_1$ and $g_1 \mid V = g$.

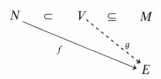

For any linearly ordered subset $\mathscr{L} = \{(V, g_V)\} \subset \mathscr{S}$, the set $W = \cup\{V \mid \exists (V, g_V) \in \mathscr{L}\}$ is a submodule $W \leqslant M$. For any $x \in W$, there is a $(V, g_V) \in \mathscr{L}$ with $x \in V$. Define $h: W \to E$ by $h(x) = g_V(x)$. If $(P, g_P) \in \mathscr{L}$ and also $x \in P$, then

$$\text{either} \quad V \subset P, \quad g_P \mid V = g_V;$$

$$\text{or} \quad P \subseteq V, \quad g_V \mid P = g_P;$$

and in either case $g_P x = g_V x = hx$. Thus $(W, h) \in S$. Moreover $(N, f) \in \mathscr{S} \neq \varnothing$. Hence by Zorn's lemma there exists a maximal element $(V, g) \in \mathscr{S}$ with $N \subseteq V \subseteq M$ and $g \mid V = f$.

3-1.5   For any right R-modules $V \subset M$, let $x \in M$ be arbitrary. View R also as a right R-module. Then left multiplication L_x by x defines a right R-homomorphism $L_x: R \to M$, $L_x r = xr$, $r \in R$. The inverse image $L_x^{-1} V$ of the submodule $V < M$ must be a submodule of R, which means that $L_x^{-1} V = x^{-1} V = \{r \in R \mid xr \in V\} \leqslant R$ is a right ideal of R. Note that $x^{-1} V = R$ if $x \in V$.

For the validity of Baer's criterion the existence of an identity element for R is vital. Note that the criterion (3) below simply says that $\varphi = L_y$.

3-1.6   Baer criterion. For a right R-module E the following are all equivalent.

(1) E is injective.

(2)   For any right ideal $B \subset R$ and for any right R-module map $\varphi: B \to E$; there exists an extension $\bar{\varphi}$ of φ to $\bar{\varphi}: R_R \to E$.

(3)   For any right ideal $B \to R$ and any right R-map $\varphi: B \to E$, there exists an element $y \in E$ such that $\varphi b = yb$ for all $b \in B$.

Proof. Obviously, $(1) \Rightarrow (2)$, and next we show the easy equivalence of $(2) \Leftrightarrow (3)$.

$(2) \Rightarrow (3)$.   Given $\varphi: B \to E$ as in (3), take any $\bar{\varphi}: R \to E$ given by (2), and define $y = \bar{\varphi}1 \in E$. Then for any $b \in B$, $\varphi b = \varphi(1b) = \bar{\varphi}(1b)$. But since $\bar{\varphi}$ is an R-map, $\bar{\varphi}(1b) = (\bar{\varphi}1)b = yb$. Thus $\varphi b = yb$.

$(3) \Rightarrow (2)$.   Given $\varphi: B \to E$ we need $\bar{\varphi}: R \to E$. Define $\bar{\varphi}$ to be left multiplication by y, $\bar{\varphi}r = yr$. If $b \in B$, then $\bar{\varphi}b = yb = \varphi b$ by hypothesis (3). Thus $(1) \Rightarrow (2) \Leftrightarrow (3)$.

$(3) \Rightarrow (1)$.   Given $N < M$ and $f: N \to E$, we want a $g: M \to E$ such that $g \mid N = f$. as in 3-1.4, let $g: V \to E$, $N \subseteq V$, $g \mid N = f$, be a maximal extension of f in M. If $V = M$, we are done. If $V \neq M$, take any $x \in M \backslash V$ and form the right ideal $B = x^{-1}V = \{b \in R \mid xb \in V\} < R$. Define $\varphi: B \to E$ by $\varphi b = g(xb)$, $b \in B$. Since $L_x B = xB \subseteq V$ and since $\varphi = gL_x$ is the composition of the two R-maps L_x and g, φ itself is a right R-homomorphism. By hypothesis (3) there exists an element $y \in E$ such that $\varphi b = yb$ for all $b \in B$. Let us try to extend g to $\tilde{g}: V + xR \to E$ by setting $\tilde{g}(v + xr) = gv + yr$. This is well defined: if $v + xr = v' + xr'$; $v, v' \in V$; r, $r' \in R$; then $v - v' = x(r - r') \in V$. Thus $r' - r \in B$, and by definition of φ

$$g(v - v') = g[x(r' - r)] = \varphi(r' - r) = y(r' - r).$$

Hence $gv + yr = gv' + yr'$, and $\tilde{g}$ is a well defined R-homomorphism. Since $\tilde{g} \mid V = g$, $V < V + xR$ contradicts the maximality of g. Thus $V = M$ and f extends to $g: M \to E$. Hence E is injective.

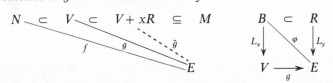

3-2   Divisibility

The motivation behind this section is the fact for modules over a commutative principal ideal domain, injectivity is the same as divisibility. Also in this section all of the specialized rings will contain an identity.

3-2.1   Definition. A module M is *generalized divisible* if for any $(v, a) \in M \times R$ such that $a^{\perp} \subseteq v^{\perp}$, there exists an $x \in M$ with $v = xa$.

3-2.2   Remarks. (1) If $v = xa$, the necessarily $a^{\perp} \subseteq v^{\perp}$.

(2) Intuitively such an x with $v = xa$ may be thought of as $x = \text{``}v(1/a)\text{''}$ provided the reader is aware that in general such an x need not be unique.

(3) The above definition says that we can solve for the unknown x the equation $v = xa$.

3-2.3   Observation. If M is injective then M is generalized divisible.

Proof. Map $\varphi: aR \to M$, $a \to v$, and hence by $\varphi(ar) = vr$ for $r \in R$. If $ar = ar_1$, then $r - r_1 \in a^{\perp} \subseteq v^{\perp}$, and so $vr = vr_1$. Thus φ is well defined. By Baer's criterion, φ extends to $\bar{\varphi}: R \to M$. Set $x = \bar{\varphi}1$. Then $v = \varphi a = \bar{\varphi}(1 \cdot a) = (\bar{\varphi}1)a$ because $\bar{\varphi}$ is an R-homomorphism. Thus $v = xa$.

3-2.4   Terminology. Recall that a ring R is a *domain* it it has no zero divisors (except trivially 0) ($0 \neq a$, $0 \neq b \in R \to ab \neq 0$). A domain need not be commutative.

A ring R is a *principal right ideal* ring if every right ideal is principal, that is, generated by a single element of R (abbreviation: *pri ring*). If R is a domain and a pri ring then it is a *principal right ideal domain* or a *prid* for short.

For any R whatever, R^* will denote the set $R^* = R \backslash \{0\}$ which of course is a multiplicative semigroup if R is a domain.

3-2.5   Observation. If R is a pri ring and $I \lhd R$ an ideal, then R/I is also a pri ring.

Proof. Every right ideal of R/I is of the form $A/I < R/I$ where $A < R$ is a right ideal. By hypothesis, $A = aR$ for some $a \in R$. Thus $A/K = (aR + K)/K = (a + K)R$ is also principal.

3-2.6   Definition. If R is a domain and M a module, then M is *divisible* if for any $(v, a) \in M \times R$, there is an $x \in M$ such that $v = xa$.

3-2.7 **Consequences.** Let M be a module over a domain R.

(1) M is divisible $\Leftrightarrow M$ is generalized divisible.
(2) Any quotient module of a divisible module is divisible.

Proof. If M is divisible, $N < M$, and $v + N \in M/N$ and $a \in R$ are arbitrary, then $v = xa$ for some $x \in M$. Thus $v + N = (x + N)a$.

(3) Assume now that for any $0 \neq m \in M$, $m^\perp = 0$. (This actually implies that R is a domain.) Then if an element $v \in M$ is at all divisible by $a \in R^*$, then it is uniquely divisible. For if $v = xa = ya$ for some $x, y \in M$, then $(x - y)a = 0$. Since $a \neq 0$, $x - y = 0$ by hypothesis.

3-2.8 **Observation.** If M is a right module over a pri ring R, then

M is injective $\Leftrightarrow M$ is generalized divisible.

Proof. $\Rightarrow$: By observation 2-2.3, even without assuming R to be a pri, M is general divisible. $\Leftarrow$. Baer's criterion will be verified. Let $B \subset R$ be any right ideal and $\varphi: B \to M$ a right R-module homomorphism. Then $B = aR$ for some $a \in R$. Set $v = \varphi a$. If $ar = 0$, then $0 = \varphi(ar) = vr$, and hence $a^\perp \subseteq v^\perp = (\varphi a)^\perp$. By the general divisibility of M, there exists an $x \in M$ such that $v = xa$. Define $\bar{\varphi}: R \to M$ by $\bar{\varphi}r = xr$. Since $\bar{\varphi} = L_x$, it is a module map. Furthermore, for any $ar \in B$, $\bar{\varphi}(ar) = xar = vr = \varphi(ar)$. So $\bar{\varphi} | B = \varphi$. Hence M is injective.

3-2.9 **Lemma.** *Suppose that (a) R is a domain and (b) a pri ring. Let $Q = Q_R$ be any divisible module. Then*

(i) *every quotient module Q/K of Q is injective, i.e. homomorphic images of divisible modules remain divisible.*
(ii) *For any index set X, $\oplus\{Q \,|\, X\}$ and $\Pi\{Q \,|\, X\}$ are also divisible, and hence in particular $\oplus\{Q \,|\, X\}$ is an injective R-module.*

Proof. (i) over a pri ring, any module Q/K is injective if and only if it is generalized divisible. But since R is a domain, generalized divisibility coincides with divisibility, and quotients of divisible modules remain divisible. Thus Q/K is injective. Conclusion (ii) can be verified componentwise.

3-2.10 **Definition.** Let D be a commutative principal ideal domain (pid) with $1 \in D$ and $M = M_D$. Then $\operatorname{tor} M = \{x \in M \mid xd = 0$ for some $0 \neq d \in D\}$ is the *torsion submodule* of M. If $\operatorname{tor} M = 0$, M is *torsion free*, whereas M is *torsion* if $M = \operatorname{tor} M$.

3-2.11   Consequences. (1) For $M = M_D$ as above, $M/\operatorname{tor} M$ is torsion free.

(2) If $\operatorname{tor} M_D = 0$ and M is divisible, then for any $(v, d) \in M \times D^*$, there is a unique $x \in M$ such that $v = xd$.

(3) Let Q be the quotient field of D. Then Q is a D-module. Since Q is divisible it is injective.

(4) A module M over a commutative pid D is injective if and only if M is divisible.

3-2.12   Lemma. *Any module M over a commutative principal ideal domain D can be embedded in a divisible D-module, and hence in an injective D-module.*

Proof. Let Q be the quotient field of D. Then Q is a divisible D-module. If $F = F_D$ is any free module, then $F = \oplus\{D \mid X\}$, and hence $F = \oplus\{D \mid X\} \subset \oplus\{Q \mid X\}$, where the latter is divisible. But any module M is the quotient $M = F/K$ of a free module F for some $K < F$. Thus

$$M = \frac{F}{K} \subset \frac{\oplus\{Q \mid X\}}{K},$$

where $\oplus\{Q \mid X\}/K$ is divisible by 3-2.7(2). So far the principal ideal property of D was not used. However now it is required to conclude that $\oplus\{Q \mid X\}/K$ is injective (see 3-2.8, or 3-2.10(4)).

3-2.13   Corollary. Every abelian group may be embedded in a divisible group.

3-3   Embeddings in injectives

First it will be shown that every module is a submodule of an injective one. This not only is the first step towards showing that every module has an injective hull, but is itself a useful device which will be used to establish various properties of injective modules.

3-3.1   Constructions. Let $\mathbb{Z} = 0,\ \pm 1,\ \pm 2, \ldots$ let D be any additive abelian group and let $a,\ b,\ r \in R$.

(i)   Then $\mathrm{Hom}_{\mathbb{Z}}(R, D)$ becomes a right R-module as follows:

$f \in \mathrm{Hom}_{\mathbb{Z}}(R, D), \ f * a \in \mathrm{Hom}_{\mathbb{Z}}(R, D);$

$f, f * a \colon R \to D, \ (f * a)(r) = f(ar);$

$\{f * (ab)\}(r) = f((ab)r),$

$\{(f * a) * b\}(r) = \{f * a\}(br) = f(a(br)) \Rightarrow (f * a) * b = f * (ab).$

(ii) Let $\varepsilon \colon \mathrm{Hom}_{\mathbb{Z}}(R, D) \to D$ be the abelian group homomorphism which maps $v \in \mathrm{Hom}_{\mathbb{Z}}(R, D)$, where $v \colon R \to D$, into $\varepsilon(v) = v(1) \in D$. That is, ε is evaluation at $1 \in R$.

(iii) Suppose that M is any right R-module and that $g \colon M \to \mathrm{Hom}_{\mathbb{Z}}(R, D)$ is an abelian group homomorphism. Thus for $m \in M$, $g[m] \colon R \to D$. To say that g is an R-homomorphism means that $g[ma] = g[m] * a$ for $a \in R$. Evaluation of the latter at $b \in R$ gives $\{g[m] * a\}(b) = \{g[m]\}(ab)$. Thus g is a right R-module homomorphism if and only if $\{g[ma]\}(b) = \{g[m]\}(ab)$ for all $a, b \in R$.

(iv) Let N be a right R-module and $\psi \colon N \to D$ just an additive abelian group homomorphism. Define $\psi^* \colon N \to \mathrm{Hom}_{\mathbb{Z}}(R, D)$, where for $n \in N$, $\psi^*[n] \colon R \to D$ is the map $\{\psi^*[n]\}(b) = \psi(nb)$. Since

$\{\psi^*[na]\}(b) = \psi((na)b),$

$\{\psi^*[n] * a\}(b) = \{\psi^*[n]\}(ab) = \psi(n(ab)), \qquad \psi^* = \psi L_n,$

it follows that ψ^* is a right R-homomorphism.

(v)   For $n \in N$, $\varepsilon(\psi^*[n]) = \{\psi^*[n]\}(1) = \psi(n)$. Thus $\varepsilon\psi^* = \psi$.

(vi) Here and in the next theorem when the value of a homomorphism is another homomorphism, square brackets are used, while braces are sometimes used when a function is evaluated at a point.

3-3.2 Lemma. *Suppose that as before $N = N_R$, D, and $\psi \colon N \to D$ are given; and that $\mathrm{Hom}_{\mathbb{Z}}(R, D)$, and*

$\varepsilon \colon \mathrm{Hom}_{\mathbb{Z}}(R, D) \to D, \ v \mapsto \varepsilon(v) = v(1)$

$\psi^* \colon N \to \mathrm{Hom}_{\mathbb{Z}}(R, D), \ \psi^*[n] \colon R \to D,$

$\psi^*[n](r) = \psi(nr) \qquad n \in N, r \in R;$

are as before. Then

(i)   *ε is an abelian group homomorphism;*

(ii) *ψ^* is a right R-module homomorphism; and*

(iii) *$\varepsilon\psi^* = \psi$, i.e. the diagram below commutes.*

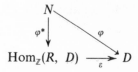

3-3.3   Theorem. *If D is a divisible abelian group, then* $\text{Hom}_{\mathbb{Z}}(R, D)$ *is an injective right R-module.*

Proof. Given any two R-modules M and N, and any R-homomorphisms $g: M \to \text{Hom}_{\mathbb{Z}}(R, D)$, $f: M \to N$ with f monic, we need an R-homomorphism $\varphi^*: N \to \text{Hom}_{\mathbb{Z}}(R, D)$ such that $\varphi^* f = g$. For ε as before, since D is divisible, the abelian group homomorphism $\varepsilon g: M \to D$ extends to an abelian group homomorphism $\varphi: N \to D$ such that $\varphi f = \varepsilon g$.

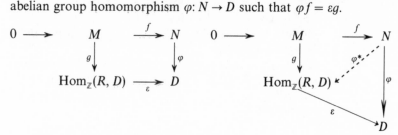

It has to be shown that for the previous right R-map φ^*: $N \to \text{Hom}_{\mathbb{Z}}(R, D)$, $\varphi^* f = g$. For any $m \in M$, $f(m) \in N$, and we have two functions, $\varphi^*[f(m)]$, $g[m]: R \to D$. It suffices to show that for any $r \in R$, $\{\varphi^*[f(m)]\}(r) = \{g[m]\}(r)$. By definition of φ and φ^*, $\{\varphi^*[f(m)]\}(r) = \varphi(f(m)r) = \varphi(f(mr)) = \{\varphi f\}(mr)$. Since $\varphi f = \varepsilon g$, $\{\varphi^*[f(m)]\}(r) = \{\varepsilon g\}(mr) = \varepsilon(g[mr]) = \{g[mr]\}(1)$. Since g is an R-homomorphism, $g[mr] = g[m] * r$ where $\{g[m] * r\}(s) = \{g[m]\}(rs)$ for any $s \in R$, and in particular for $s = 1$. Thus now $\{\varphi^*[f(m)]\}(r) = \{g[m] * r\}(1) = \{g[m]\}(r)$. Since r is arbitrary, $\varphi^*[f(m)] = g[m]$, and since m is arbitrary, $\varphi^* f = g$. Thus $\text{Hom}_{\mathbb{Z}}(R, D)$ is injective.

One of the main reasons for proving the last theorem is the next corollary.

3-3.4   Corollary 1 to theorem 3.3. *Every R-module is a submodule of an injective module.*

Proof. For any $M = M_R$, $M \subset D$ where D is a divisible abelian group. The map $l: M \to \text{Hom}_{\mathbb{Z}}(R, D)$ defined by $m \to l_m: R \to D$, where

$l_m(a) = ma \in M \subset D$ is a monic R-homomorphism:

$$l_{m+n} = l_m + l_n \qquad m, n \in M; \ a, r \in R;$$
$$\{l_m * r\}(a) = l_m(ra) = m(ra) = \{l_{mr}\}(a);$$
$$l_{mr} = l_m * r; \quad l_m(1) = m1 = m.$$

Thus $M \cong \{l_m \mid m \in M\} \subset \mathrm{Hom}_{\mathbb{Z}}(R, \ C)$ as an R-submodule.

3-3.5   Corollary 2 to theorem 3.3. *A module M is injective if and only if M is a direct summand of every module of which it is a submodule.*

Proof. $\Rightarrow$: Given $M_R \subset N_R$ with M injective, the identity function $1_M: M \to M$ extends to a homomorphism $\rho: N \to M$ such that $\rho j = 1_M$, where $j: M \to N$ is the inclusion map. Hence $N = M \oplus T$ for some submodule $T < N$.

$\Leftarrow$: By the last theorem, there exists an injective module E such that $M \subset E$. By hypothesis $E = M \oplus T$ for some $T \leqslant E$. Since any direct summand of an injective module is injective, M is injective.

3-4   Injective hulls

The basic facts about essential extensions presented below are not only used to obtain and characterize the injective hull, but have a fundamental importance in their own right for module theory as a whole.

3-4.1   Definitions. For any $M = M_R$ and $V \leqslant M$, V is an *essential* submodule of M if it intersects every nonzero submodule of M non-trivially, i.e.

$$0 \neq A < M \to V \cap A \neq 0.$$

Synonyms for essential: V is *large* in M; V is essential in M; $V \leqslant M$ is an essential extension, $V \leqslant M$ is essential; and $V \leqslant M$ is large.
   An essential extension $V < M$ is *proper* if $V \neq M$.

3-4.2   Definition. An essential extension $M \leqslant N$ is an *absolute maximal* *essential* extension of M, provided that for any extension $N < P$ with $N \neq P$ whatever, $M < P$ is *not* essential.

A complement submodule, which is defined below, should not be confused with a direct summand. Actually, the concept of the injective hull could be developed without the next definition. However, Lemma 3-4.5 helps to explain what is really happening in Lemma 4.6.

3-4.3   Definition. A submodule $K \leqslant M$ is a *complement* submodule of M if there exists some fixed submodule (in general nonunique) $B \leqslant M$ such that $K \cap B = 0$, and furthermore, such that among all the submodules of M, K is maximal with respect to intersecting B trivially.

An alternative definition is that $K \leqslant M$ is a complement if there exists $B \leqslant M$ such that (i) $K \cap B = 0$ and such that for any $K \subset A \leqslant M$, $K \neq A$, (ii) $A \cap B \neq 0$.

A complement $C < M$ is a maximal complement of M if $C \neq M$, and if for any other complement submodule $A \leqslant M$ with $C \subseteq A$, either $C = A$ or $C = M$.

Although in this section $1 \in R$, all of the properties below except (5) would also hold in the absence of an identity.

3-4.4   Properties of essential extensions. Here M is any nonzero R-module.
   (1) $0 < M$ is not essential; $M \leqslant M$ is large.
   (2)   If R is a domain and $0 \neq I \lhd R$, then $I \leqslant R$ is essential.
   (3)   If $V < M$ is essential, and $V < W < M$, then both $V < W$ and $W < M$ are essential extensions.
   (4)   Transitive: $V < W$ essential, $W < M$ essential $\Rightarrow V < M$ is essential.
   (5)   Since $1 \in R$, $M < N$ is essential $\Leftrightarrow \forall\, 0 \neq v \in N$, $vR \cap M \neq 0$.
   (6)   Let N be any fixed module containing M as a submodule $M < N$.
      (i)   Suppose that $\mathscr{L} = \{V\}$ is any linearly ordered chain of submodules of N such that for any $V \in \mathscr{L}$, $M \leqslant V$ is essential. Then $M \leqslant \cup\{V \in \mathscr{L}\}$ is an essential extension of modules.
      (ii)   Use of (i) and Zorn's Lemma now shows that any extension of M contains a maximal essential extension module of M inside N. I.e. there exists $P \leqslant N$ with $M \leqslant P$ essential and for any $P \subset Q \leqslant N$, $M < Q$ is not essential.

(7) For any $M < N$, let $T < N$ be maximal with respect to $M \cap T = 0$. Then $M \oplus T \leqslant N$ is large.

(8) Assume that $M < N$ is essential, and that $\varphi: M \to E$ is a monic homomorphism of R-modules. Then if $\bar{\varphi}: N \to E$ is any extension of φ (with $\bar{\varphi} \mid M = \varphi$) then $\bar{\varphi}$ is also monic.

Proof. Always, kernel $\varphi = M \cap \ker \bar{\varphi}$. Since $\ker \varphi = 0$ and $M < N$ is large, necessarily $\ker \bar{\varphi} = 0$.

(9) If $V < M$ and $W < M$ are both essential, then so also is $V \cap W < M$.

(10) Suppose $M \subset N'$ and $M \subset N''$ are two extension modules of M, and that $\mu: N' \to N''$ is an isomorphism such that $\mu M = M$. If $M \subset N'$ is essential then so also is $M \subset N''$.

Proof. In general, if $f: X \to Y$ is any function of sets, and $A, B \subseteq X$ are any subsets then always $f^{-1}(A \cap B) = (f^{-1}A) \cap (f^{-1}B)$, but only $f(A \cap B) \subseteq (fA) \cap (fB)$. However, if f is one to one then also $f(A \cap B) = (fA) \cap (fB)$.

If $0 \neq C < N''$, then $0 \neq \mu^{-1}C < N'$, and hence $M \cap \mu^{-1}C \neq 0$. Since μ is monic, $\mu(\mu^{-1}C) = C$, and also $0 \neq \mu(M \cap \mu^{-1}C) = (\mu M) \cap (\mu \mu^{-1}C) = M \cap C$.

(11) An extension $M \leqslant N$ is an absolute maximal essential extension of M if and only if N has no proper essential extensions.

In general, passage to quotient modules destroys largeness. The significance of a complement submodules is that modulo a proper complement, the image of a large submodule remains large. This condition actually characterizes complement submodules.

3-4.5 Proposition. *For a fixed module M, any submodule $K < M$ with $K \neq M$ satisfies the following.*

(i) $K < M$ *is a complement* $\Leftrightarrow \forall$ *large* $L \leqslant M$, $(K + L)/K \leqslant M/K$ *is also large.*

(ii) *If* $K < M$ *is a complement, then* $(K + L)/K \neq \{K\}$.

Proof. Take any $B \leqslant M$ such that K is maximal with respect to $K \cap B = 0$. Thus $B \neq 0$.

(ii) Since $L < M$ is large, $L \cap B \neq 0$, and $\bar{0} \neq [K \oplus (L \cap B)]/K \leqslant (K + L)/K$ proves that $(K + L)/K \neq \{K\}$.

(i) ⇒: Next suppose that $\bar{0} \neq P/K \leqslant M/K$ is any nonzero submodule. Since $P \neq K \subset P$, and since K was maximal with respect to $K \cap B = 0$, necessarily $P \cap B \neq 0$. Because $L < M$ is large, also $L \cap P \cap B \neq 0$. But $K + L \cap P \cap B \subseteq K + P = P$, and hence

$$\{K\} \neq \frac{K \oplus L \cap P \cap B}{K} \leqslant \frac{K + L}{K} \cap \frac{P}{K}.$$

Thus $\bar{0} \neq (K + L)/K \leqslant M/K$ is large.

(i) ⇐: If K is not a complement, then K is not maximal with respect to $K \cap B = 0$. Thus there exists $N < M$ with $K \subset N$, $K \neq N$, and $N \cap B = 0$. By the maximality of B, $K \leqslant N$ is large in N, for otherwise we could enlarge B so that it still would intersect K trivially. Thus $L = K \oplus B < N \oplus B \leqslant M$ are all essential in each other. Note that $N \cap (K \oplus B) = K + N \cap B = K$ by the modular law. Thus $N/K \neq \bar{0}$ and by hypothesis $(K + L)/K = (K \oplus B)/K \leqslant M/K$ is large. But then

$$\frac{K + L}{K} \cap \frac{N}{K} = \frac{(K \oplus B) \cap N}{K} = \frac{K}{K}$$

is a contradiction. Hence $K < M$ is a complement.

3-4.6   Lemma. *Let $0 \neq M \subset E$ be any modules. Suppose that $T \subseteq E$ is a submodule maximal with respect to $M \cap T = 0$. Then $(M \oplus T)/T \leqslant E/T$ is an essential extension.*

Proof. Suppose that $0 \neq K/T < E/T$, $T \subsetneqq K$, is any non-zero submodule. It has to be shown that $[(M + T)/T] \cap (K/T) \neq 0$. Suppose that $[(M + T)/T] \cap (K/T) = 0$. Then $K \cap (M + T) = T$. Hence

$$\left. \begin{array}{l} K \cap M \subseteq K \cap (M + T) = T \\ K \cap M \subseteq M \end{array} \right\} \Rightarrow K \cap M \subseteq T \cap M = 0.$$

But then $T \subsetneqq K$, $T \neq K$, and $K \cap M = 0$ contradicts the maximality of T. Thus $(M \oplus T)/T \leqslant E/T$ is essential.

3-4.7   Proposition. *For any nonzero module M, M is injective ⇔ M has no proper essential extensions.*

Proof. ⇒: Let M be injective, and $M \leqslant V$ essential. Then $V = M \oplus T$ by 3-3.5. Thus $M \cap T = 0$ and $M \subseteq V$ essential requires that $T = 0$. Hence $M = V$.

⇐: Now M has no proper essential extensions. Let $M \subset E$ where E is any injective module containing M as a submodule. Let $T < E$ be a submodule of E maximal with respect to $M \cap T = 0$. Then $M \cong (M \oplus T)/T \leqslant E/T$ is essential by the last lemma. Since M has no proper essential extensions, $(M \oplus T)/T = E/T$, or $M \oplus T = E$. Thus M is a direct summand of the injective module E, and hence M is injective.

The following special case of 3-4.4(8) is used sufficiently often in order to restate it as the next lemma.

3-4.8   Lemma. *For any modules M, N, and E, let $M \subset N$ be essential, E injective, and $\varphi \colon M \to E$ monic. If $\bar{\varphi} \colon N \to E$ is any extension of φ, $\bar{\varphi} \mid M = \varphi$, then also $\bar{\varphi}$ is monic.*

3-4.9   Proposition. *Every module M has an absolute maximal essential extension $M \subseteq N$.*

Uniqueness. If $M \subseteq N'$ is another absolute maximal essential extension, then the identity map of $M \to M$ extends to an isomorphism $N \to N'$.

Proof. Let $M \subseteq E$, where E is any injective module. Let N be a maximal essential extension of M in E, that is $M \subseteq N \subseteq E$, and N has no proper essential extensions inside E. We will show that N is actually an absolute maximal essential extension of M. Let $M \subset N'$ be an essential extension such that $M \subset N \subseteq N'$. Let $i \colon N \to E$ be the inclusion map and $\mu \colon N' \to E$ any extension of i with $\mu \mid N = i$. Then μ is monic since $M \subset N'$ is essential. Thus the corestriction $\mu \colon N' \to \mu N'$ is an isomorphism such that $\mu \mid M = 1_M$ is the identity map of M.

$$
\begin{array}{ccc}
N & \subseteq & N' \\
\scriptstyle i \downarrow & \nearrow \scriptstyle \mu & \\
E & &
\end{array}
\qquad
\begin{array}{ccccc}
M & \subseteq & N & \subseteq & N' \\
\| & & \| & & \downarrow \scriptstyle \mu \\
M & \subseteq & N & \subseteq & \mu N' & \subseteq & E.
\end{array}
$$

It follows from 3-4.4(10) that $M \subset \mu N'$ is also essential. But then since $M \subseteq N \subseteq \mu N' \subseteq E$ with $M \subseteq N$ a maximal essential extension of M in N, it follows that, $N = \mu N'$. In order to show that $N = N'$, take any $n' \in N'$. Since

$\mu \mid N = 1_N$ and $\mu\mu = \mu$, we have $\mu(n' - \mu n') = 0$, and hence $n' - \mu n' = 0$. Thus $n' = \mu n' \in N$, or $N' = N$. Hence N is an absolute maximal essential extension of M.

3-4.10 Corollary. *The above N has no proper essentials extensions. Hence N is injective.*

3-4.11 Corollary. *If M is any module and I is any injective module containing M, $M \subset I$, then any maximal essential extension N of M in I is (1) an absolute maximal extension of M. (2) Hence such an N is injective.*

3-4.12 Definition. A module N is a *minimal injective* extension of a module M if $M \subseteq N$, if N is injective, and if $M \subseteq K < N$ with $K \ne N$, then K is *not* injective.

3-4.13 Theorem. *For modules $M \subseteq N$ the following are all equivalent.*

(1) N is an absolute maximal essential extension of M.
(2) N is an essential extension of M and N is injective.
(3) N is a minimal injective extension of M.

Proof. (1) $\Rightarrow$ (2): Let $M \subseteq N$, N an absolute maximal essential extension of M. By 3-4.7, N is injective, and by hypothesis (1), $M \leqslant N$ is an essential extension.

(2) $\Rightarrow$ (3): Suppose that E is any injective submodule in $M \subseteq E \subseteq N$. Since E is injective and $E \subseteq N$, it follows that $N = E \oplus E'$ for some submodule $E' \subseteq N$ by 3-3.5. However, since $M \subseteq E \oplus E'$ is essential and since $M \cap E' = 0$, it follows that $E' = 0$ and $E = N$. Thus N is a minimal injective extension of M.

(3) $\Rightarrow$ (1): Let $M \subseteq N$ be a minimal extension of modules such that N is injective. By use of Zorn's lemma as in 3-4.4(6), choose $N' \leqslant N$ with $M \subseteq N'$, and with $M \subseteq N'$ a maximal essential extension of M in N. By 3-4.11(2), N' is injective. But since N is minimal, $N' = N$. Again by 3-4.11(1), $N' = N$ is an absolute maximal essential extension of M.

3-4.14 Definition. For a module M, the module N above satisfying 3-4.13, (1)–(3) is called the *injective hull* of M. (Frequently used notations for the injective hull of M are $M \subseteq E(M) = EM$, and $M \subseteq \hat{M}$.)

3-4.15   Corollary to theorem 4.13. *Let $M \subseteq E$ and $M \subseteq G$ be injective hulls of M. Then there exists an isomorphism $E \to G$ leaving M elementwise fixed.*

Proof. Both modules E and G by (1) of 3-4.13 are absolute maximal essential extensions of M. Now use of 3-4.9 shows that E and G are isomorphic $E \cong G$ under an isomorphism which leaves M pointwise fixed.

3-5   Noetherian rings

This section studies injective modules over Noetherian rings.

3-5.1   Definition. A module V is *Noetherian* if every strictly increasing chain of submodules of V is finite, in which case V is said to satisfy the ascending chain condition, abbreviated as "A.C.C.". Also, more precisely, a right R-module V is said to be right Noetherian if it satisfies the A.C.C. on right R-submodules.

A ring R is right Noetherian if it is Noetherian as a right R-module. A left Noetherian ring is defined analogously.

3-5.2   Observation. For $V = V_R$, the following are all equivalent.

(i) V satisfies the A.C.C. on submodules.
(ii)   For any submodule $W \leqslant V$, W is finitely generated.
(iii) Every nonempty set of submodules of V contains a maximal element.

Proof. (i) $\Rightarrow$ (ii): If not, for every finite number of elements $x_1, \ldots, x_{n-1} \in W$, there exists an $x_n \in W \backslash [x_1 R + \cdots + x_{n-1}R]$. Thus a strictly ascending infinite chain of submodules of W, and hence of V, can be selected of the following form:

$$0 \subsetneqq x_1 R \subsetneqq \cdots \subsetneqq x_1 R + \cdots + x_{n-1}R \subsetneqq x_1 R + \cdots + x_n R \subsetneqq \cdots W.$$

Consequently W must be finitely generated.

(ii) $\Rightarrow$ (i). It will be shown that every chain $V_0 \subseteq V_1 \subseteq \cdots \subseteq V_n \cdots \subseteq V$ of submodules of V is finite. By hypothesis the submodule $\cup \{V_n \mid n = 0, 1, \ldots\}$ is generated by some finite subset $\{y_1, \ldots, y_m\} \subset V$ as $\cup V_n = y_1 R + \cdots + y_m R$. For each i, there exists an $n(i)$ such that $y_i \in V_{n(i)}$. If $N = \text{maximum } \{n(i) \mid 1 \leqslant i \leqslant m\}$, then $\{y_1, \ldots, y_m\} \subset V_N$. Hence $\cup V_n = V_N$ and thus the chain stabilizes at N, $V_N = V_{N+1} = \ldots$.

The implications "(iii) $\Rightarrow$ (i)" and "(i) $\Rightarrow$ (iii)" are easily proved by way of contradiction.

3-5.3   Proposition. A ring R is right Noetherian if and only if for any indexed family $\{E_i | i \in I\}$ of injective right R-modules E_i for any index set I, their sum $\oplus \{E_i | i \in I\}$ is injective.

Proof. $\Rightarrow$: If R is right Noetherian and E_i, $i \in I$, are given injectives, it will be shown that $\oplus \{E_i | i \in I\}$ satisfies Baer's criterion. Suppose that any homomorphism $\varphi: B \to \oplus E_i$ of a right ideal $B \subset R$ is given. Since every right submodule of R is finitely generated, $B = x_1 R + \cdots + x_m R$ for some finite set of elements $x_1, \ldots, x_m \in R$. There exist a finite number N of indices $i(1), \ldots, i(N) \in I$ such that all $\{\varphi x_1, \ldots, \varphi x_m\} \subset \oplus \{E_{n(i)} | i = 1, \ldots, N\}$. Since $\varphi B \subset \oplus_{i=1}^{N} E_{n(i)}$, and since the latter is injective, there exists an extension $\bar{\varphi}: R \to \oplus_{i=1}^{N} E_{n(i)}$ of φ to R. Since $\bar{\varphi}$ can be viewed as a map into $\oplus E_i$, Baer's criterion is satisfied and $\oplus E_i$ is injective.

$\Leftarrow$: If R is not right Noetherian, there exists an infinite properly ascending chain of right ideals $(0) = L_0 \subset L_1 \subset \cdots \subset L_i \subset \cdots \subset R$. Then $L = \cup L_i \subseteq R$ is also a right ideal. Define $E_i = E(R/L_i)$ to be the injective hull of R/L_i. Thus $R/L_i \leqslant E_i$. Set $G = \oplus \{E_i | i = 1, \ldots\}$.

For each $0 \neq a \in L$, there exists a unique integer $n = n(a) \geqslant 1$ such that $a \in L_{n+1} \backslash L_n$. Hence $a + L_{n+k} = \bar{0} \in E_{n+k}$ for all $k = 1, 2 \ldots$. Map $\varphi: L \to G$ by $\varphi a = \{a + L_i | i = 1, \ldots\} \in G$. Since G is injective, by Baer's criterion there exists a $y \in G$ such that $\varphi b = yb$ for all $b \in L$. Since $y \in \oplus_{i=1}^{N} E_i$ for some integer N, also $\varphi L \subseteq \oplus_{i=1}^{N} E_i$. But there exists an element $c \in L_{N+2} \backslash L_{N+1}$, and thus the $N+1$ coordinate of φc in E_{N+1} is $c + L_{N+1} \neq L_{N+1}$. Hence $\varphi c \notin \oplus_{i=1}^{N} E_i$, a contradiction. Hence R is right Noetherian.

The following classical theorem is useful in constructing new Noetherian rings from old ones. This theorem is well known because it is proved in many texts ([Bhattacharya, Jain, and Nagpaul 86; p. 363, Theorem 2.14]), and therefore we merely state it.

3-5.4   Hilbert basis theorem. *Let K be any ring (with or without identity), and $K[x]$ the ring of polynomials with coefficients in K with the usual operations:*

$$kx = xk, (x^i a)x^j b = x^{i+j} ab, \quad k, a, b \in K.$$

If K is right Noetherian, then so also is $K[x]$.

A proof quite similar to that of 3-5.4, can be used to also prove the next corollary (of the proof of the Hilbert basis theorem).

3-5.5   Corollary. *For any ring K, let $K[[x]]$ be the rings of formal power series in positive powers of x with the usual operations:*

$$\alpha = \sum_{i=0}^{\infty} x^i a_i, \qquad \beta = \sum_{j=0}^{\infty} x^j b_j;$$

$$\alpha\beta = \sum_{n=0}^{\infty} x^n \sum_{k=0}^{n} a_{n-k} b_k, \qquad a_i, b_j \in K.$$

Then $K[[x]]$ is right Noetherian if and only if K is.

3-6   Examples

The class of rings defined below will be used to give examples of injective modules of the kind in 3-2.8. These rings will also be used later.

3-6.1   Definition. For a ring K with identity, and an identity preserving ring endomorphism $\theta: K \to K$, $k\theta = k^\theta$; $k \in K$, a θ-*derivation* δ is an additive group homomorphism $\delta: K \to K$ satisfying

$$(ab)^\delta = a^\delta b^\theta + ab^\delta \quad \text{for all } a, b \in K.$$

The (twisted) *skew polynomial ring* $K[x; \theta; \delta]$ with right side coefficients in K is the set of all polynomials $p = \Sigma_{i=0}^n x^i k_i$, $k_i \in K$ under pointwise addition, and subject to the multiplication rule extended linearly and distributively:

$$kx = xk^\theta + k^\delta, \quad kx^2 = x^2 k^{\theta\theta} + x(k^{\theta\delta} + k^{\delta\theta}) + k^{\delta\delta}, \text{ etc.}$$

The result is an associative ring with identity.

Associativity of $K[x; \theta; \delta]$ will be proved next. So far we have an additive abelian group and a binary composition law $K[x; \theta; \delta] \times K[x; \theta; \delta] \to K[x; \theta; \delta]$ which distributes over addition, i.e. a so-called nonassociative ring. Observe that this $K[x; \theta; \delta]$ is generated as a nonassociative ring by the subset $K \cup \{x\} \subset K[x; \theta; \delta]$. We do not need here the concept of a nonassociative ring; all that the latter means is that every element of $K[x; \theta; \delta]$ is a finite linear combination of finite products of elements of $K \cup \{x\}$. Consequently, in order to prove associativity, it suffices to show that for any three elements $g_1, g_2, g_3 \in K \cup \{x\}$ with repetitions allowed, $(g_1 g_2)g_3 = g_1(g_2 g_3)$. Letting $a, b, c \in K$, we have only eight cases to verify: $abc, axb, xab, axx, xax, xxa, xxx$, and abx. The first seven are trivial to check, while the last one that $a(bx) = (ab)x$ is established as follows:

$$(ab)^\theta = a^\theta b^\theta; \; (ab)^\delta = a^\delta b^\theta + ab^\delta;$$
$$a(bx) = a(xb^\theta + b^\delta) = xa^\theta b^\theta + a^\delta b^\theta + ab^\delta$$
$$= x(ab)^\theta + (ab)^\delta = (ab)x,$$

where the next to the last equality uses the properties of θ and δ in the top equation.

When $\theta = 1_K = 1$ is the identity automorphism of K, abbreviate $K[x; 1; \delta] = K[x; \delta]$, where now δ is an ordinary derivation of K. If $\delta = 0$, set $K[x; \theta; 0] = K[x; \theta]$; and if both $\theta = 1$ and $\delta = 0$, we get the ordinary polynomial ring $K[x]$.

Only if K is a domain, the degree of any nonzero polynomial p as above is defined to be $\deg p = n$ if $k_n \neq 0$. In the special case when K is a skew field, note that $(1/k)\theta = 1/(k\theta)$, $K\theta \subseteq K$ is a skew subfield, and that $(1\delta) = 1^\delta = 0$.

For K as above and any element $0 \neq d \in K$ the map $K \to K$, $k \to kd - dk^\theta$ is a θ-derivation called an *inner θ-derivation*. Note that for $d = x \in K[x; \theta; \delta]$, the restriction of the inner θ-derivation by x to K is $\delta: kx - xk^\theta = k^\delta$.

In ring theory, and in particular in applications of ring theory in constructing rings, the above problem of showing associativity occurs frequently. It is worthwhile to observe that the above procedure works in general. Suppose that we have an additive abelian group R with a binary composition $R \times R \to R$ which would be a ring if only it were associative. And suppose that for some subset $\{g_i \,|\, i \in \Gamma\} \subseteq R$, every element of R is a finite sum of a finite number of the g_i's, with repetitions allowed. (The last two sentences simply say that R is generated as a nonassociative ring by $\{g_i \,|\, i \in \Gamma\}$.) Then R is associative if and only $g_i(g_j g_k) = (g_i g_j)g_k$ for any $i, j, k \in \Gamma$. Unlike the case for $K[x; \theta; \delta]$, sometimes $\{g_i \,|\, i \in \Gamma\}$ is closed under the binary composition law, i.e. for any $g_1, g_2 \in \{g_i \,|\, i \in \Gamma\}$ there exists a third $g_1 g_2 = g_3 \in \{g_i \,|\, i \in \Gamma\}$. Now associativity of R can be routinely verified directly from the multiplication table of $\{g_i \,|\, i \in \Gamma\}$. As a historical digression, note that the above is strictly a semigroup procedure for $\{g_i \,|\, i \in \Gamma\}$, whose systematization and automation is called Light's associativity test. For more details, see [A. H. Clifford and G. B. Preston 16; pages 7–8].

3-6.2   Division algorithm. For a skew field K, the twisted skew polynomial ring $K[x; \theta; \delta]$ satisfies a right side *division algorithm*. For any f, $0 \neq g \in K[x; \theta; \delta]$, there exist elements h, $r \in K[x; \theta; \delta]$ such that

$$f = gh + r, \quad \text{where either } r = 0, \text{ or } \deg r < \deg g.$$

3-6.3   Ore quotient rings of domains. A *domain* K satisfies the *right Ore condition* if every pair a, $b \in K \backslash \{0\}$ of nonzero elements of K have a common right multiple $0 \neq at = bs \in aK \cap bK \neq 0$. Define an equivalence relation $\sim$ on the set $K \times (K \backslash \{0\})$ by $(a, s) \sim (b, t)$ if $ax = by$ and also $sx = ty$ for some x, $y \in K \backslash \{0\}$. The equivalence class of (a, s) is denoted by a/s and also as $a/s = as^{-1}$, but not as $s^{-1}a$. Define $-(a/s) = (-a)/s$. Addition or subtraction and multiplication of equivalence classes is defined as follows:

$$a/s - b/t = (ax - by)/sx \quad \text{where } 0 \neq sx = ty$$
$$(a/s)(b/t) = ax/ty \quad \text{where } 0 \neq sx = by; x, y \in K \backslash \{0\}.$$

A long, but straightforward verification shows the above operations are (1) independent of the particular choice of x and y; and (2) are independent of the representatives (a, s) and (b, t) of the equivalence classes a/s and b/t. We do this for the second equation, leaving the first to the reader.

We work in $K \backslash \{0\}$. (1) Let also $sx_1 = by_1$, we will show that $ax/ty = ax_1/ty_1$. First, choose u, $w \in K \backslash \{0\}$ with $xu = x_1 w$. From $sx = by$ we get that $sx_1 w = sxu = byu$. But $sx_1 = by_1$, hence $byu = by_1 w$. Then $yu = y_1 w$ by cancelling $b \neq 0$. But then $tyu = ty_1 w$. Since $xu = x_1 w$, we get that $axu = ax_1 w$. Thus $(ax, ty) \sim (axu, tyu) = (ax_1 w, ty_1 w) \sim (ax_1, ty_1)$, or $ax/ty = ax_1/ty_1$ as required.

(2) Let $a/s = a_1/s_1$ and $b/t = b_1/t_1$, we show that $(a_1/s_1)(b_1/t_1) = (a/s)(b/t)$, where each side is to be calculated by the above prescription. There exist x_0, x_1, y_0, y_1, $\bar{x}$, and $\bar{y}$ such that

$$ax_0 = a_1 x_1, \quad by_0 = b_1 y_1$$
$$sx_0 = s_1 x_1, \quad ty_0 = t_1 y_1, \quad 0 \neq s_1 x_1 \bar{x} = b_1 y_1 \bar{y}.$$

In view of these five equations also $0 \neq sx_0 \bar{x} = by_0 \bar{y}$. It follows, first from the latter, and secondly the above five equations that

$$(a/s)(b/t) = ax_0 \bar{x}/ty_0 \bar{y}$$
$$a_1 x_1 \bar{x} = ax_0 \bar{x}, \quad \text{and} \quad b_1 y_1 \bar{y} = by_0 \bar{y},$$
$$(a/s)(b/t) = a_1 x_1 \bar{x}/t_1 y_1 \bar{y},$$

where $ty_0 = t_1 y_1$ was used. But use of $s_1 x_1 \bar{x} = b_1 y_1 \bar{y}$ shows that

$(a_1/s_1)(b_1/t_1) = a_1 x_1 \bar{x}/t_1 y_1 \bar{y} = (a/s)(b/t)$. For alternate ways of constructing the ring of quotients see [Bhattacharya, Jain, and Nagpaul 86; p. 223, Theorem 2.4] or [Koh 74], or [Jacobson 64; pages 142–163].

The set $Q(K)$ of all equivalence classes under the above operations forms an associative division ring with identity element x/x for any $0 \neq x \in K$. The ring $Q(K)$ is called the *right Ore quotient* division *ring* of K. Note that $K < Q(K)$ is essential as a right K-module.

Now we have all the ingredients to obtain a mechanism for constructing injective modules over noncommutative rings.

3-6.4   Let K be a skew field. It is a consequence of the division algorithm that $K[x; \theta; \delta]$ is a prid. But any prid is right Ore. Thus $K[x; \theta; \delta]$ has a right Ore quotient division ring denoted by $K(x; \theta; \delta)$. Hence $K(x; \theta; \delta)$ viewed as a right $K[x; \theta; \delta]$-module can be shown to be injective by Baer's criterion. Moreover, the injective hull of $K[x; \theta; \delta]$ viewed as a right module over itself is $K(x; \theta; \delta)$.

3-6.5   Example. For the complex numbers $\mathbf{C}$, let $\theta \colon \mathbf{C} \to \mathbf{C}$ be complex conjugation, $c\theta = \bar{c}$, $c \in \mathbf{C}$. The twisted polynomial ring $\mathbf{C}[x; \theta]$, where $cx = x\bar{c}$, is right Ore, and its right Ore division ring $\mathbf{C}(x; \theta)$ is an injective right $\mathbf{C}[x; \theta]$-module.

3-6.6   Example. For a commutative field F, let $K = F(y)$ be the rational function field in an indeterminate y, and let $\delta \colon K \to K$ where $\delta k = -(d/dy)k$, $k \in K$, is minus one times the derivative. Then $K[x; \delta] = F(y)[x; \delta]$ is right Ore. Hence $F(y)(x; \delta)$ is an injective right $F(y)[x; \delta]$-module.

Note that $K[x; \delta]$ contains the so called *Weyl algebra*, that is the ring of all non-commutative polynomials $F[x, y]$ in indeterminates x and y subject to the sole relation $xy - yx = 1$. I.e. $F[x, y] = F[y][x; \delta] \subset F(y)(x; \delta) \subset F(y)(x; \delta)$.

In the next example some injective modules are described without proofs.

3-6.7   Example. For a right vector space $V = V_D$ over a division ring D, let $\mathrm{End}_D V$ be the ring of all D-linear transformations of V written on the left side of V. Then $\mathrm{End}_D V$ is injective as a right module over itself. It is left self injective if and only if V is finite dimensional over D. (See [Goodearl 76; pp. 52–53] for proofs.)

Now assume in addition that D is commutative. For the identity transformation 1_V and a fixed $d \in D$, the map $d1_V: V \to V$, $d1_V(v) = vd$, $v \in V$, is D-linear. Thus $D1_V \subset \mathrm{End}_D V$ is a subring.

Now let $S \lhd \mathrm{End}_D V$ be the ideal of linear transformations whose range is finite dimensional, and let $R = S + D1_V$.

For any $0 \neq v \in V$, write $V = vD \oplus W$ and let $\pi: V \to vD$ be the corresponding projection. Then πR is an injective module. The right R-module $R/S \cong D$ is also injective. (Proofs can be found in [Cozzens and Faith 75; p. 88].)

3-7   Exercises

The following definitions and the immediately following sequence of exercise develop the Goldie dimension.

Definition. A right R-module M has finite uniform dimension if every direct sum of nonzero submodules of M contains only a finite number of direct summands.

The uniform or *Goldie dimension* of M is the largest integer m such that M contains a direct sum of m nonzero submodules. (Hence it does not contain a direct sum of $m + 1$ nonzero submodules.) If no such m exists, the Goldie dimension of M is said to be infinite. The zero module has zero Goldie dimension. Abbreviate the Goldie dimension of M as $\mathrm{Gd}\, M = m$.

1. For a module M, prove that the following are all equivalent:
   (i) M has finite uniform dimension.
   (ii) M satisfies the descending chain condition on complement submodules.
   (iii) M satisfies the ascending chain condition on complement submodules.
2. Prove that if the uniform dimension of M is finite that then every nonzero submodule of M contains a uniform submodule. (Hint. Let $0 \neq U \leqslant M$ be a minimal complement submodule of M.)
3. Suppose that a module M contains as an essential submodule $V_1 \oplus V_2 \oplus \cdots \oplus V_n \leqslant M$.
   (i) Prove that $E(M) = E(V_1) \oplus E(V_2) \oplus \cdots \oplus E(V_n)$.
   (ii) Show that an injective module is directly indecomposable if and only if it is uniform.
4. Suppose that M has finite uniform dimension and that $\{V_1, \ldots, V_m\}$ is a maximal set of uniform submodules whose sum is direct. Prove that $V_1 \oplus \cdots V_m \leqslant M$ is essential.

5. Suppose that $V_1 \oplus \cdots \oplus V_m \leqslant M$ is essential where the V_i are uniform. Prove that any nontrivial direct sum of submodules of M contains at most m nonzero summands.

6. Use 5 to deduce that M has finite uniform dimension if and only if M has Goldie dimension $m < \infty$.

7. If $K < M$ is a complement submodule show that $\operatorname{Gd} M = \operatorname{Gd} K + \operatorname{Gd} M/K$. Give a counterexample to show that this fails if K is not a complement submodule.

8. If R is a right Ore domain with right Ore quotient ring Q, prove that $\operatorname{Gd} R_R = \operatorname{Gd} Q_Q$.

9. Let $\oplus\{V_i \mid i \in I\} \leqslant M$ be essential. If $W_i \leqslant V_i$ is large for all i, show that $\oplus_i W_i \leqslant M$ is large. Give a counterexample to show that the corresponding property for infinite products is false.

10. Find modules $\{V_n\}_{n=1,2,\ldots}$ such that $\oplus_1^\infty V_n < \prod_1^\infty V_n$ is not essential.

11. Show that any principal right ideal ring R is right Noetherian and right Ore.

12. Prove that a domain R with a finite Goldie dimension as a right R-module is right Ore.

13. Prove the division algorithm 3-6.2.

14. Suppose that K is a right Ore domain. Prove that the ring $K[x; \theta; \delta]$ (see 3-6.1) is right Ore. What is its right Ore quotient ring?

15. Prove that for a skew field K, the ring $K[x; \theta; \delta]$ is left Ore if and only if θ is an automorphism.

CHAPTER 4

Tensor products

Introduction

Tensor products of algebras and modules are constructed and their universal mapping properties investigated. The tensor products of modules and those of algebras are not only treated separately, but are investigated from different perspectives. The tensor product of modules is defined externally in terms of a universal property. Whereas the tensor product C of two algebras A and B is defined internally with reference only to A, $B \subset C$ in terms of C only. The latter approach is well suited to answer questions of the following type. Given an algebra C containing subalgebras $A \subset C$ and $B \subset C$, when does C have the structure of a tensor product $C \cong A \otimes_F B$? Aside from the obvious condition that elements of A and B commute and that C is "generated" by A and B, intuitively, C is a tensor product of A and B if there are "no algebraic relations between A and B" in C. The latter vague notion is made precise by defining and using the concept of linear disjointness. Both for modules and also algebras several concrete constructions of tensor products are given.

One of the more interesting aspects of this chapter is the Universal Mapping Theorem. It gives a necessary and sufficient condition on an algebra C to be an algebra tensor product of A and B, but this condition is expressed entirely in terms of algebra homomorphisms.

4-1   Tensor products of modules

Before even defining the tensor product of two modules, an abelian group is constructed which later will actually turn out to be their tensor product.

4-1.1   Notation. In the first section, R can be any ring with or without an identity element; $A = A_R$ will be a right and $B = {}_R B$ a left R-module. Let S be the free $\mathbb{Z} = 0,\ \pm 1,\ \pm 2,\ldots$-module on the set $A \times B$, $S = \oplus \{\mathbb{Z}(a, b) \,|\, (a, b) \in A \times B\}$. The usual abbreviations and pointwise operations are used for $p = \{p_{(a,b)} \,|\, (a, b) \in A \times B\}$, $q = (q_{(a,b)})_{(a,b) \in A \times B} = (q_{(a,b)})$; $p_{(a,b)}$, $q_{(a,b)}$, $j \in \mathbb{Z}$; $jp = pj = (jp_{(a,b)}) \in S$, $p - q = (p_{(a,b)} - q_{(a,b)}) \in S$.

There are three other ways of visualizing elements of S. First, each element of S can be written uniquely as a finite sum

$$\sum_{i=1}^{n} p_i(a_i, b_i) \qquad p_1, \ldots, p_n \in \mathbb{Z} \setminus \{0\};$$

$$(a_j, b_j) \neq (a_i, b_i) \in A \times B \quad \text{if } j \neq i.$$

Secondly, each element of S is a finite sum of the form $\Sigma_i \varepsilon(i)(a_i, b_i)$ where $\varepsilon(i) = \pm 1$, and where repetitions in $(a_i, b_i) \in A \times B$ are allowed.

Thirdly, an element of S can be represented as formal sum $p = \Sigma p_{(a,b)}(a, b)$ where (a, b) ranges over all of $A \times B$, but where only a finite number of the integers $p_{(a,b)} \in \mathbb{Z}$ are allowed to be nonzero. The algebraic operations of addition of two elements $p, q \in S$ and multiplication of an element $p \in S$ by an integer $j \in \mathbb{Z}$ are easily carried out in this representation

$$\Sigma p_{(a,b)}(a, b) + \Sigma q_{(a,b)}(a, b) = \Sigma (p_{(a,b)} + q_{(a,b)})(a, b);$$

$$j \Sigma p_{(a,b)}(a, b) = \Sigma j p_{(a,b)}(a, b).$$

The empty sum is always the zero element $0_S \in S$. Note that there are four other zero elements $0_A \in A$, $0_B \in B$, $0_R \in R$, and $0 = 0_{\mathbb{Z}} \in \mathbb{Z}$ any one of which is usually simply written as 0. Define $(a, b) = 1(a, b) = (a, b)1 \in S$ for $(a, b) \in A \times B$. Note that $0_S \neq (0_A, 0_B) = 1(0_A, 0_B) \in S$.

4-1.2   Construction. Let H be the additive subgroup of S generated by all $\mathbb{Z}$-linear combinations of all possible elements of the type y_1, y_2, and y_3 for varying $a, a' \in A$; $b, b' \in B$, and $r \in R$:

(1) $(a + a', b) - (a, b) - (a', b) = y_1$,
(2) $(a, b + b') - (a, b) - (a, b') = y_2$,
(3) $(ar, b) - (a, rb) = y_3$.

Define "$A \otimes_R B$" as the abelian quotient group $A \oplus_R B = S/H$. If R is fixed and understood, the abbreviation $A \otimes B = A \otimes_R B$ will be used. For $(a, b) \in S$, define $a \otimes b = (a, b) + H = 1(a, b) + H \in A \otimes B$. Let $\rho: S \to S/H$ be the natural quotient map $\rho s = s + H$ and let π be the restriction $\pi = \rho \,|\, A \times B$ of ρ to the subset $A \times B \subset S$. Thus $\pi(a, b) = a \otimes b$.

First, in order to gain familiarity with $A \otimes B$, a few computations are performed in S/H below.

4-1.3   Remarks. Recall that in general for any element of an abelian group like $a \in A$ (or $a \otimes b \in A \otimes B$, or $b \in B$) by definition for $0 \leqslant n$, na is $na = a + \cdots + a$, a added n-times, while for $n \leqslant -1$, na is $na = (-a) + \cdots + (-a)$, $-a$ taken n-times. Let $a \in A$ and $b \in B$ be arbitrary.

   (i)   If $1 \notin R$, then the element $0_S = (a, b) - (a, b) \in H$ is not of the
         form y_3.
   (ii) $(0_A, 0_B) \in H$.

Proof. $(0_A, 0_B) + H = (0_A, 0_B) + [(0_A + 0_A, 0_B) - (0_A, 0_B) - (0_A, 0_B)] + H = 2(0_A, 0_B) + (-2)(0_A, 0_B) + H = 0(0_A, 0_B) + H = H$.

   (iii) $(0_A, b), (a, 0_B) \in H$.

Proof. Since $(0_A, b) + H = (0_A + 0_A, b) - [(0_A + 0_A, b) - (0_A, b) - (0_A, b)] + H = 2(0_A, b) + H$, it follows that $(0_A, b) + H = H$. Similarly, also $(a, 0_B) \in H$.

   (iv)   For $1 \leqslant n \in \mathbb{Z}^+$, $(na) \otimes b = a \otimes (nb) = n(a \otimes b)$.

Proof. Assume by induction that $((n - 1)a) \otimes b = a \otimes ((n - 1)b) = (n - 1)(a \otimes b)$. Then $(na) \otimes b = (na, b) - [(a + (n - 1)a, b) - (a, b) - ((n - 1)a, b)] + H = -[(-a, b) - (n - 1)(a, b)] + H = +n(a \otimes b)$. Analogously, $a \otimes (nb) = n(a \otimes b)$.

   (v) $(-a) \otimes b = -(a \otimes b) = a \otimes (-b)$.

Proof. By 4-1.3(iii) and 4-1.3(iv), $0_{S/H} = (0_A, b) + H = (a + (-a), b) + H = a \otimes b + (-a) \otimes b$. Thus $(-a) \otimes b = -(a \otimes b)$. The same way also $a \otimes (-b) = -(a \otimes b)$.

   (vi)   For $1 \leqslant n$, $(-na) \otimes b = -n(a \otimes b) = a \otimes (-nb)$.

Proof. Apply 4-1.3(iv) with "a" replaced by "$(-a)$" and then 4-1.3(v) to conclude that $(-na) \otimes b = (n(-a)) \otimes b = n((-a) \otimes b) = -n(a \otimes b)$. Also the same way $a \otimes (-nb) = -n(a \otimes b)$.

   (viii)   So far, the relation 4-1.2 (3) has not been used.

4-1.4   Definition. For an additive abelian group C, a function

$\varphi: A \times B \to C$ is R-bilinear, or just *bilinear* if R is fixed, if for all $a, a' \in A$; $b, b' \in B$; and $r \in R$ the following hold:

(1) $\varphi(a + a', b) = \varphi(a, b) + \varphi(a', b)$,
(2) $\varphi(a, b + b') = \varphi(a, b) + \varphi(a, b')$,
(3) $\varphi(ar, b) = \varphi(a, rb)$.

An equivalent definition is that $\varphi: A \times B \to C$ is bilinear, provided (3) above holds, and with one variable $b \in B$ or $a \in A$ fixed, the induced maps $\varphi(\ , b): A \to C$ and $\varphi(a, \ ): B \to C$ are abelian group homomorphisms.

The following remarks are intended to free the reader from any attempt to draw some erroneous vague connection between $A \oplus B$ and either S, or $A \otimes B$, or bilinear maps. There simply is not any.

4-1.5   Remarks. (1) The set $A \times B$ can be viewed as the group $A \times B = A \oplus B$, and as a subset in $A \times B \subset S$, but it is not a subgroup of S because $(a + a', 0_A) - (a, 0_A) - (a', 0_A)$ equals zero in $A \oplus B$ but never in S.

(2) A bilinear map $\varphi: A \times B \to C$ can in no way be constructed as a group homomorphism $A \oplus B \to C$. First of all, "$\varphi[(a, b) + (a', b')]$" is undefined because "$(a, b) + (a', b')$" $\notin A \times B$. Next, let us suppose that we define $(a, b) + (a', b') \equiv (a + a', b + b')$ as dictated by the identification $A \times B \leftrightarrow A \oplus B$. Even then, $\varphi[(a, b) + (a', b')] \equiv \varphi[(a + a', b + b')] = \varphi(a, b) + \varphi(a, b') + \varphi(a', b) + \varphi(a', b') \neq \varphi(a, b) + \varphi(a', b')$ in general, because in general $\varphi(a, b') + \varphi(a', b) \neq 0$.

(3) It can happen that $\pi(a, b) = a \otimes b = 0$ for $0 \neq a \in A$, $0 \neq b \in B$.

(4) The subset $\pi(A \times B) \subset A \otimes B$ does not in any way inherit the group structure from $A \oplus B$. If $A \otimes B \neq 0$, then there exists a $0 \neq a \otimes b \in A \otimes B$. Then $(a, b) = (a, 0) + (0, b)$ in $A \oplus B$, but modulo H, $0 \neq a \otimes b = (a, b) + H \neq (a, 0) + (0, b) + H = H$.

(5) Assume that $\bar{\varphi}: A \otimes B \to C$ is some abelian group homomorphism such that $\bar{\varphi} \,|\, \pi(A \times B) = \varphi$, i.e. such that $\bar{\varphi}\pi = \varphi$. Even if π is monic and we endow the subset $A \oplus B = A \times B \leftrightarrow \pi(A \times B)$ with the group structure inherited from $A \oplus B$ by way of π, then any such $\bar{\varphi}: \pi(A \times B) \to C$ in general cannot be a group homomorphism. If $\bar{\varphi} \neq 0$, then there exists $a \otimes b \in A \otimes B$ such that $0 \neq \bar{\varphi}(a \otimes b) = \bar{\varphi}(\pi(a, b)) \neq \bar{\varphi}[\pi(a, 0) + \pi(0, b)] = \bar{\varphi}(0) = 0$.

4-1.6   Observations. (1) Since S is a free $\mathbb{Z}$-module, for any bilinear map $\varphi: A \times B \to C$, there exists a unique group homomorphism $\varphi^{\#}: S \to C$ such that its restriction to $A \times B$ is $\varphi^{\#} \,|\, A \times B = \varphi$.

(2) As a consequence of (1)–(3) in both 4-1.2 and 4-1.4, $\varphi^\#(y_i) = 0$ for all y_i, and thus $\varphi^\#$ vanished on H, $\varphi^\#(H) = 0$. Hence there exists a unique group homomorphism $\bar\varphi : A \otimes B \to C$ such that $\bar\varphi\rho = \varphi^\#$. Since $\pi = \rho \,|\, A \times B$, $\varphi = \varphi^\# \,|\, A \times B = \varphi^\# \rho \,|\, A \times B = \bar\varphi\pi$. The diagram below commutes.

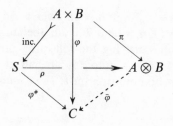

(3) By 4-1.1 and 4-1.3(v), every element $\xi \in A \otimes B$ is a finite sum $\xi = \Sigma_{i=1}^{n} a_i \otimes b_i \equiv \Sigma a \otimes b$ where the latter means a finite sum of such terms and the n and index i are suppressed. Then $\xi = \Sigma\pi(a, b)$ and $\bar\varphi(\xi) = \Sigma\bar\varphi\pi(a, b) = \Sigma\varphi(a, b)$.

(4) Modulo H in S/H the identities 4-1.2(1), (2), and (3) become:

$$(a + a') \otimes b - a \otimes b - a' \otimes b = 0,$$
$$a \otimes (b + b') - a \otimes b - a \otimes b' = 0,$$
$$ar \otimes b \qquad\quad - a \otimes rb = 0.$$

Therefore the map $\pi : A \times B \to A \otimes B$ is bilinear, that is π satisfies 4-1.4(1), (2), and (3) with $C = A \otimes B$.

In the next proposition, the ring R need not have an identity. Even if it does, the modules need not be unital.

4-1.7   Proposition. For any ring R and any right and left R-modules $A = A_R$ and $B = {}_R B$, let the abelian group $A \otimes_R B = S/H$, and the bilinear map $\pi : A \times B \to A \otimes_R B$, $\pi(a, b) = a \otimes b$, $a \in A$, $b \in B$, be as before (in 4-1.2). Then for any bilinear map $\varphi : A \times B \to C$ into any abelian group C, there exists a unique abelian group homomorphism $\bar\varphi : A \otimes_R B \to C$ such that $\bar\varphi\pi = \varphi$.

∀ bilinear φ ∃! $\bar\varphi$ and a commutative diagram

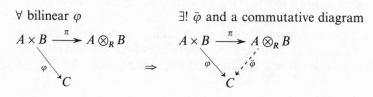

Proof. By 4-1.6(1), (2), and (4), everything follows except the uniqueness. If $f: A \otimes B \to C$ is any abelian group homomorphism such that $\varphi = f\pi$, then use of 4-1.6(3) gives that for any $\xi = \Sigma a \otimes b \in a \otimes B$, $f(\xi) = \Sigma f\pi(a, b) = \Sigma \varphi(a, b) = \bar{\varphi}(\xi)$. Thus $f = \bar{\varphi}$.

4-1.8   Definition. For modules $A = A_R$ and $B = {}_R B$ over a ring R, a *tensor product* of A and B is a pair (T, η) where T is an abelian group, and $\eta: A \times B \to T$ a bilinear function satisfying the following universal property. For any bilinear map $f: A \times B \to C$ into any abelian group C, there exists a unique abelian group homomorphism $\bar{f}: T \to C$ such that $f = \bar{f}\eta$.

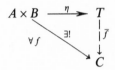

4-1.9   Proposition. For modules $A = A_R$ and $B = {}_R B$ over a ring R suppose that $(A \otimes B, \pi)$ is as before and (T, η) is any tensor product of A and B. Then

   (i) $(A \otimes B, \pi)$ is a tensor product of A and B.
  (ii) UNIQUENESS. There exists an abelian group isomorphism $g: A \otimes B \to T$ such that $\eta = g\pi$.

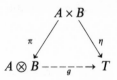

Proof. (i) This follows from 4-1.7. (ii) By the universal property for $A \otimes B$, the bilinear map η lifts to $g: A \otimes B \to T$ such that $\eta = g\pi$. On the other hand, by the universal property of T, the bilinear map π factors through $f: T \to A \otimes B$ with $\pi = f\eta$. Since $gf\eta = g\pi = \eta$ and also $1_T\eta = \eta$ where 1_T is the identity map of T, the uniqueness hypotheses for T implies that $gf = 1_T$. Similarly $fg = 1_{A \otimes B}$ is the identity on $A \otimes B$ and hence g is an isomorphism. The above arguments can be expressed in terms of the following four commutative diagrams, where the universal property or the uniqueness is always applied to the top group.

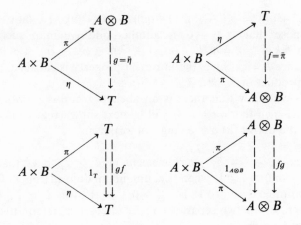

4-1.10 Example. For any $2 \leqslant n \in \mathbb{Z}$, for $(n) = \mathbb{Z}n \lhd \mathbb{Z}$ and $\mathbb{Q}$ the rationals, $\mathbb{Q} \otimes_{\mathbb{Z}} \mathbb{Z}/(n) = 0$. Every element $v \in \mathbb{Q}$ is of the form $v = nq$ for some $q \in Q$. Thus $v \otimes 1 + (n) = qn \otimes 1 + (n) = q \otimes n(a + (n)) = q \otimes \bar{0} = 0$.

4-1.11 Example. If $p \neq q \in \mathbb{Z}$ are two unequal prime numbers, then $\mathbb{Z}/(p) \otimes_{\mathbb{Z}} \mathbb{Z}/(q) = 0$. Since $sp + tq = 1$ for some $s, t \in \mathbb{Z}$,

$$(1 + (p)) \otimes (1 + (q)) = (sp + tq + (p)) \otimes (1 + (q))$$
$$= ((t + (p))q) \otimes (1 + (q)) = 0.$$

4-1.12 Example. For $1 \leqslant n, m \in \mathbb{Z}$, their greatest common divisor $1 \leqslant d$ is the generator of the ideal $(n) + (m) = \mathbb{Z}n + \mathbb{Z}m = \mathbb{Z}d = (d)$. Set $\mathbb{Z}_n = \mathbb{Z}/(n)$, and the same for $\mathbb{Z}_m$ and $\mathbb{Z}_d$. As before $\pi: \mathbb{Z}_n \times \mathbb{Z}_m \to \mathbb{Z}_n \otimes \mathbb{Z}_m = \mathbb{Z}_n \otimes_{\mathbb{Z}} \mathbb{Z}_m = S/H$. Then as $\mathbb{Z}$-modules, $\mathbb{Z}_n \otimes \mathbb{Z}_m \cong \mathbb{Z}_d$.

Proof. define $\eta: \mathbb{Z}_n \times \mathbb{Z}_m \to \mathbb{Z}_d$ to be the bilinear map $\eta(i + (n), j + (m)) = ij + (d)$ for $i, j \in \mathbb{Z}$. The fact that this map is well defined independent of choice of coset representatives i and j requires d to be any common divisor of n and m whatsoever. Here d is the greatest common divisor of n and m.

If $\varphi: \mathbb{Z}_n \times \mathbb{Z}_m \to C$ is any $\mathbb{Z}$-bilinear map into some abelian group C, define an abelian group homomorphism $\bar{\varphi}: \mathbb{Z}_d \to C$ by $\bar{\varphi}(k + (d)) = \varphi(k + (n), 1 + (m))$ for $k \in \mathbb{Z}$. The fact that this is well defined independent of the coset representative k of $k + (d)$ requires not only that the integer d is of the form $d = in + jm$, but also that φ is $\mathbb{Z}$-bilinear. Then $\bar{\varphi}\eta(i + (n),$

$j + (m)) = \bar{\varphi}(ij + (d)) = \varphi(ij + (n), \ 1 + (m)) = \varphi(i + (n), \ j + (m))$. Thus $\varphi = \bar{\varphi}\eta$. Suppose that $h: \mathbb{Z}_d \to C$ is another additive map such that $\varphi = h\eta$. Then $h(k + (d)) = h\eta(k + (n), \ 1 + (m)) = \varphi(k + (n), \ 1 + (m)) = \bar{\varphi}(k + (d))$. Hence $h = \bar{\varphi}$. Since the universal property is satisfied, $(\mathbb{Z}_d, \eta)$ is the tensor product of $\mathbb{Z}_n$ and $\mathbb{Z}_m$.

From 4-1.9 we know that there is an abelian group is isomorphism $g: \mathbb{Z}_n \otimes \mathbb{Z}_m \to \mathbb{Z}_d$ such that $g\pi = \eta$. It will be next shown that (i) $\mathbb{Z}_n \otimes \mathbb{Z}_m$ is a ring, and the above (ii) g is a ring isomorphism.

Proof. (i) Since $\mathbb{Z}_n \otimes \mathbb{Z} \cong \mathbb{Z}_d$, every element of $\mathbb{Z}_n \otimes \mathbb{Z}_m$ is of the form $(i + (n)) \otimes (1 + (m)) \equiv \bar{i} \otimes \bar{1}$, where i is unique modulo (d) only. This is enough to guarantee that $(\bar{i} \otimes \bar{1})(\bar{j} \otimes \bar{1}) = (ij)^- \otimes \bar{1}$ turns $\mathbb{Z}_n \otimes \mathbb{Z}_m$ into a ring independent of the two separate choices of coset representatives i of $i + (n)$ and j of $j + (m)$.

(ii) First of all, by definition of π, $\pi(i + (n), \ j + (m)) = \bar{i} \otimes \bar{j} = \bar{i} \otimes j(\bar{1}) = (ij)^- \otimes \bar{1}$. However, also $(\bar{i} \otimes \bar{1})(\bar{j} \otimes \bar{1}) = (\overline{ij}) \otimes \bar{1} = \pi(ij + (n), \ 1 + (m))$. Since $g\pi = \eta$, a typical element $\bar{k} \otimes \bar{1}$ is mapped into $g(\bar{k} \otimes \bar{1}) = g\pi(k + (n), \ 1 + (m)) = \eta(k + (n), \ 1 + (m)) = k + (d)$. Thus

$$g[(\bar{i} \otimes \bar{1})(\bar{j} \otimes \bar{1})] = g\pi(ij + (n), \ 1 + (m)) = ij + (d)$$
$$= [i + (d)][j + (d)] = (g(\bar{i} \otimes \bar{1}))(g(\bar{j} \otimes \bar{1}))].$$

Hence g is actually a ring isomorphism.

In the special case when n and m are relatively prime, $d = 1$, and $\mathbb{Z}_m \otimes \mathbb{Z}_m = 0$.

4-2   Definitions for algebras

For the sake of later applications, algebras in this section need not necessarily contain identity elements. However, tacitly they will be assumed to be nonzero. Division ring, skew field, or field refers to a not necessarily commutative ring in which every nonzero element has a multiplicative inverse, while the term "field" in this book will always mean a commutative field.

Unless otherwise stated, in a sum involving various indices but with an unspecified summation index, all double indices are to be summed over.

In this and the next section by an algebra will be meant an algebra over a field F unless some other field is explicitly specified.

4-2.1   Definition. For a field F a ring A is an *algebra* over F, or an F-algebra, if A is a right vector space over F, and $(ab)k = (ak)b = a(bk)$ holds for all $a, b \in A$ and $k \in F$. Define ka to be $ka = ak$.

4-2.2   Consequences. Let A be an F-algebra and $a, b \in A$, $K \in F$.

(1) Since A is an F-vector space, $a1_F = a$ for all $a \in A$.

(2) If there exists $1_A = e \in A$, then $F \to A$, $k \to ek$ is a ring homomor-
     phism whose kernel consequently must either be (0) or F. Since
     $A \neq 0$, also $e \neq 0$, and hence $e = e1_F \neq 0$. Thus $F \cong Fe$. In this
     case, identity $F \cong Fe \subset A$ and view $F \subset A$ as a subring. Note
     that $1_F = e1_F = e$ are identified also.

(3) Since $(ak)b = a(kb) = akb$, there is no distinction between right
     and left vector subspaces of A.

(4) Every algebra A over F with $1 \in A$ of finite dimension $= n \geqslant 2$
     is isomorphic to a proper subalgebra of F_n, the $n \times n$ matrices
     over F.

Proof. For $x \in A$, let L_x be the F-linear transformation $L_x: A \to A$,
$L_x a = xa$ for $a \in A$. There is a monic F-algebra map $A \to F_n$, where x is
mapped to the matrix of L_x with respect to some fixed F-vector space
basis of A.

Next are described two frameworks – also called categories – within
which one can work: arbitrary algebras as in this section, or algebras with
identity and identity preserving morphisms as in the next one.

4-2.3   Terminology. In the literature, F is sometimes referred to as the
ground field (over which everything takes place). A subring $C \subset A$ of an
F-algebra A is a *subalgebra* if C is also a vector subspace of A. An algebra
homomorphism $f: A \to B$, also called an F-algebra map, is a ring
homomorphism that is also an F-vector space homomorphism. The same
applies to algebra isomorphism, monomorphisms, and epimorphism. (In
the category of algebras with identity, in addition it is required that $1_A \in C$
and that f preserves identities, i.e., $f1_A = 1_B$. Not only is a subalgebra C
required to have an identity element, but it must necessarily contain the
same identity 1_A as the larger algebra A.)

A right module V over A is an *algebra module* if in addition also

(i) V is an F-vector space, and
(ii) $(va)k = (vk)a$ for all $a \in A$, $k \in F$.

(In the category of algebras with identity, in addition, all algebra modules
are to be taken to be unital modules.)

In any category of algebras, submodules of an F-algebra module have
to be vector subspaces. Hence in particular, algebra right ideals, algebra
left ideals, and algebra ideals must also be vector subspaces. The basic
properties and definitions of modules, right ideals, and the four isomor-

phism theorems carry over to both categories of algebras. In dealing with an F-algebra A, it is convenient to have the terms "ring right ideal", "ring ideal" etc. refer to A purely as a ring without any regard for F. For any ring right ideal $L \subset A$, $FL = \{\Sigma f_i e_i \mid f_i \in F, e_i \in L\}$ is an algebra right ideal.

In the literature and also here, after a while one gets tired of constantly quantifying everything by the word algebra, or F-algebra, and one simply uses the usual ring terms and symbols like "ideal", "homomorphism", "$\leqslant$", "$\lhd$" etc. to refer to F-algebra concepts when it is clear from the context what is meant.

A method is given for constructing associative algebras from a vector space, and then a test is devised for verifying that a linear map preserves multiplication.

4-2.4   Let A be a vector space over F with a vector space basis $\{u_i \mid i \in I\}$.

(1) Let $c_{ijk} \in F$; $i, j, k \in I$; be the so-called multiplication constants to be determined later. First, define $u_i u_j = \Sigma c_{ijk} u_k$, and then extend this definition by linearity and distributivity which can only be done in one way. Every $a, b \in A$ is uniquely of the form $a = \Sigma a_i u_i$ and $b = \Sigma b_j u_j$; $a_i, b_j \in F$. Then $A \times A \to A$, $(a, b) \to ab = \Sigma a_i b_j c_{ijk} u_k$ defines a binary composition law on A; A is associative $\Leftrightarrow \forall i, j, k \in I$, $(u_i u_j) u_k = u_i(u_j u_k) \Leftrightarrow \forall i, j, k$, $\Sigma c_{ij\alpha} c_{\alpha k \beta} = \Sigma c_{i\alpha\beta} c_{jk\alpha}$.

(2) Assume that A and B are algebras

 (i) Of course for any choice of $\{b_i\} \subset B$, $u_i \to b_i$ defines a unique F-linear transformation $\varphi \colon A \to B$, $\varphi(\Sigma c_i u_i) = \Sigma c_i b_i$, $c_i \in F$.

 (ii) φ is an algebra homomorphism if and only if $\varphi(u_i u_j) = (\varphi u_i)(\varphi u_j)$ for all i, j.

Next, two ways of constructing new algebras from old ones are described; one is by extending the ground field F to a larger field, and the other is by constructing a quotient algebra.

4-2.5   (1) Suppose that A is an F-algebra with (associative) multiplication constants $c_{ijk} \in F$ obtained from any F-vector space basis $\{u_i \mid i \in I\}$ of A by $u_i u_j = \Sigma c_{ijk} u_k$. (See 2-2.4(1).) Also suppose that $F \subset K$ is any field extension of F. Let A^K be the K-vector space with the same basis $\{u_i \mid i \in I\}$ as that of A, i.e. $A^K = \{\Sigma k_i u_i \mid k_i \in K\}$. Since $c_{ijk} \in F \subset K$, they can be viewed as multiplication constants in K, and exactly as before in A, used to define an associative multiplication on A^K. Then

(i) A^K is a K-algebra.
(ii) A^K is an F-algebra, and $A \subset A^K$ is an F-subalgebra. If $F \neq K$, then $A \neq A^K$.

(2) Suppose that A and B are F-algebras and that $Y \subset A$ is a subset. Let I be the algebra ideal generated by Y, $I = \cap\{J \mid Y \subseteq J \lhd A, JF \subseteq J\}$. Then I consists of all finite sums $\Sigma a_0 t_1 a_1 t_2 a_2 \cdots t_n a_n$ where $1 \leqslant n$, $t_i \in Y$ and $a_0, \ldots, a_n \in A \cup F$ are arbitrary. Then $\pi: A \to A/I$ is an F-algebra epimorphism.

(3) Suppose that $\psi: A \to B$ is an algebra homomorphism such that $\psi Y = 0$. Then by (2) above, $\psi I = 0$, and hence there exists an induced monic algebra homomorphism $\bar{\psi}: A/I \to B$ such that $\psi = \bar{\psi}\pi$.

4-3 Tensor products of algebras

This section takes place in the category of F-algebras with identity and identity preserving maps (see 4-2.3).

4-3.1 Definition. For an algebra C over F subspaces P, $Q \subset C$ are *linearly disjoint* if for any p and q

$$\left.\begin{array}{l} \forall\, x_1, \ldots, x_p \in P \text{ linearly independent} \\ \forall\, y_1, \ldots, y_q \in Q \text{ linearly independent} \end{array}\right\} \Rightarrow \begin{array}{l} \{x_i y_j\} \text{ are} \\ \text{linearly independent.} \end{array}$$

The next lemma greatly simplifies the task of proving that two subspaces are linearly disjoint.

4-3.2 Lemma. Let $\{u_i\}$, $\{w_j\}$ be any arbitrary bases of the subspaces P, $Q \subset C$, where C is an algebra over F. Then

$$P, Q \text{ are linearly disjoint} \Leftrightarrow \{u_i w_j\} \text{ are linearly independent} \atop \text{over } F.$$

Proof. $\Rightarrow$: Obvious. $\Leftarrow$: Given are linearly independent $x_1, \ldots, x_p \in P$; and linearly independent $y_1, \ldots, y_q \in Q$.

First assume that dimension $P = s < \infty$ and $\dim Q = t < \infty$. Then $\dim PQ = st$ because by hypothesis $\{u_i w_j\}$ span PQ and are linearly independent. Linearly independent subsets may be extended to basis. Thus extend $x_1, \ldots, x_p \to$ to $x_1, \ldots, x_p, x_{p+1}, \ldots, x_s$ a basis of P; extend

$y_1, \ldots, y_q \to$ to $y_1, \ldots, y_q, y_{q+1}, \ldots, y_t$ a basis of Q. The set $\{x_i y_j \mid i \leq s, j \leq t\}$ spans PQ and $|\{u_i w_j\}| = st$; hence $\{x_i y_j\}$ is linearly independent. We are done if $\dim P < \infty$ as well as $\dim Q < \infty$.

In the general case, there exist finite number of $\{u_1, \ldots, u_s\} \subset \{u_i\}$ such that $x_1, \ldots, x_p \in P_0$ where P_0 is the finite dimensioned subspace spanned by $\{u_1, \ldots, u_s\}$. The analogous holds for $\{w_1, \ldots, w_t\} \subset \{y_j\}$ and Q_0. Now the first case when applied to P_0 and Q_0 shows that $\{x_i y_j\}$ are linearly independent again.

4-3.3   Definition. A *tensor product of algebras* A and B over F is a triple $\langle C, \alpha, \beta \rangle$, where

   (i) $\alpha: A \to C$ is a monic algebra map, $\beta: B \to C$ is a monomorphism, and $C = (\alpha A)(\beta B)$.

   (ii) $(\alpha a)(\beta b) = (\beta b)(\alpha a)$ for all $a \in A$, $b \in B$.

   (iii) Subspaces αA and βB are linearly disjoint.

Remark. Above one may identify $A \equiv \alpha A \subset C \supset \beta B \equiv B$ above; $C = AB = BA$; $1_C \in A \cap B$.

4.3.4   Construction of a tensor product of A and B.

Proof. Let $\{x_i\} \subset A$ and $\{y_j\} \subset B$ be bases of A and B. Form the vector space C with basis $\{(x_i, y_j)\}$, $C = \{t = \Sigma c_{ij}(x_i, y_j) \mid c_{ij} \in F\}$. Thus $t = 0$ if and only if all $c_{ij} = 0$, and $1_F(x_i, y_j) = (x_i, y_j)$.

Let $x = \Sigma c_i x_i \in A$, $y = \Sigma d_j y_j \in B$ where $c_i, d_j \in F$ are arbitrary. Define $x \otimes y = \Sigma c_i d_j (x_i, y_j)$. It follows from this definition that $x_i \otimes y_j = (x_i, y_j)$ and hence that $x \otimes y = \Sigma c_i d_j x_i \otimes y_j$. Thus the set of all finite linear combinations of $x \otimes y$ is exactly C.

For any $x, x' \in A$; $y, y' \in B$, and $c \in F$ the following hold.

   (0) $x \otimes y = 0 \Leftrightarrow$ all $c_i d_j = 0 \Leftrightarrow$ either $x = 0$, or $y = 0$.

   (1) $(x + x') \otimes y = x \otimes y + x' \otimes y$.

   (2) $x \otimes (y + y') = x \otimes y + x \otimes y'$.

   (3) $c(x \otimes y) = (cx) \otimes y = x \otimes cy$.

Now define multiplication in C by $(x_i \otimes y_j)(x_k \otimes y_p) = x_i x_k \otimes y_j y_p$ and extend by linearity and distributivity (4-2.4(1)), and verify associativity (4-2.4(1)). Define $\alpha: A \to C$, $\alpha x = x \otimes 1$ and $\beta: B \to C$, $y = 1 \otimes y$; both are monic algebra homomorphisms and $(\alpha x)(\beta y) = (\beta y)(\alpha x)$.

We have the following bases:

$$\{x_i \otimes 1\} \quad \text{for } A \otimes 1 = \alpha A \subset C;$$

$\{1 \otimes y_j\}$ for $1 \otimes B = \beta B \subset C$; and

$\{(x_i, y_j)\} = \{x_i \otimes y_j\} = \{(x_i \otimes 1)(1 \otimes y_j)\}$ for C.

The latter implies first that $C = \alpha A \beta B$, and secondly with the aid of Lemma 4-3.2 that $\alpha A = A \otimes 1$ and $\beta B = 1 \otimes B$ are linearly disjoint. Thus $\langle C, \alpha, \beta \rangle$ satisfies Definition 4-3.3, which is summarized below in the "$\_\_ \otimes \_\_$" notation

(i) $A \simeq \alpha A = A \otimes 1 \subset C \supset 1 \otimes B = \beta B \simeq B$;
      $C = (A \otimes 1)(1 \otimes B)$.
(ii) $(x \otimes 1)(1 \otimes y) = x \otimes y = (1 \otimes y)(x \otimes 1)$, $x \in A$, $y \in B$;
(iii) $A \otimes 1$ and $1 \otimes B$ are linearly disjoint in C.

Define $A \otimes_F B$ to be this particular algebra C written in this particular notation, and abbreviate it as $A \otimes_F B = A \otimes B = C$ if F is fixed. The following computational rule is useful. For any a_i, $a'_j \in A$ and b_i, $b'_j \in B$, $(\Sigma a_i \otimes b_i)(\Sigma a'_j \otimes b'_j) = \Sigma a_i a'_j \otimes b_i b'_j$.

4-3.5   Universal mapping theorem. For F-algebras A, B, C, and D and for arbitrary $a \in A$, $b \in B$, let

(a) $\alpha: A \to C$, $\beta: B \to C$ be algebra monomorphisms such that $(\alpha a)\beta b = \beta b(\alpha a)$ and $C = \alpha A \beta B$.
(b) Let $g: A \to D$, $h: B \to D$ be algebra homomorphisms such that $(ga)hb = (hb)ga$.

Then $\langle C, \alpha, \beta \rangle$ is a tensor product of A and $B \Leftrightarrow \forall D$, $\forall g$, h as above

$\exists!\ f: C \to D$ algebra homomorphism such that $f\alpha = g$, $f\beta = h$

$\Leftrightarrow \forall$ diagram: $\Rightarrow \exists!$ f and a, commutative diagram:

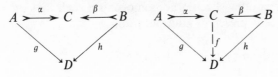

Proof. $\Rightarrow$: Assume that $\langle C, \alpha, \beta \rangle$ satisfies Definition 4-3.3, and let D, g, and h as above be given. Let $\xi = \Sigma(\alpha a_k)\beta b_k$, $\xi' = \Sigma(\alpha a'_k)\beta b'_k \in C$ be arbitrary. Define $f(\xi) = \Sigma(ga_k)hb_k$. Hence $f(\xi - \xi') = f(\xi) - f(\xi')$ and $f(k\xi) = kf(\xi)$ for $k \in F$. The expression for ξ as a sum of terms of the form $(\alpha a)\beta b$ need not be unique, and to show that f is well defined it suffices to show that if $\xi = 0$, that then also $f\xi = 0$.

Let the subspaces $P = \Sigma F a_k$ and $Q = \Sigma F b_k$ have bases $\{x_i\}$ and $\{y_j\}$; write $a_k = \Sigma c_{ki} x_i$ and $b_k = \Sigma d_{kj} y_j$. Since $\{\alpha x_i\} \subset C$ and $\{\beta y_j\} \subset C$ are linearly independent, Definition 4-3.3(iii) shows that $\{(\alpha x_i)\beta y_j\}$ are also linearly independent. For any ξ, zero or otherwise,

$$\xi = \sum_k (\alpha a_k)\beta b_k = \sum_{i,j} \left(\sum_k c_{ki} d_{kj} \right) \alpha(x_i)\beta(y_j).$$

Hence

$$\xi = 0 \to \sum_k c_{ki} d_{kj} = 0 \quad \text{for all } i, j.$$

Use of the latter now yield that

$$\sum_k (ga_k)hb_k = \sum_{i,j} \left(\sum_k c_{ki} d_{kj} \right) g(x_i)h(y_j) = 0.$$

Hence $\xi = 0$ implies that $f\xi = 0$ and f is a well defined linear map.

In view of the pointwise commutativity hypothesis (a) (which is the same as Definition 4-3.3(ii))

$$\xi\xi' = \sum_k (\alpha a_k)\beta b_k \sum_m (\alpha a'_m)\beta b'_m = \sum_{k,m} \alpha(a_k a'_m)\beta(b_k b'_m)$$

Thus by definition of f and hypothesis (b)

$$f(\xi\xi') = \sum_{k,m} g(a_k a'_m)h(b_k b'_m)$$

$$= \sum_k (ga_k)(hb_k) \sum_m (ga'_k)(hb'_m) = (f\xi)(f\xi').$$

Thus $f(\xi\xi') = f(\xi)f(\xi')$ and f is an algebra homomorphism.

Since $\alpha 1_A = 1_C$, $\beta 1_B = 1_C$, and $h 1_B = 1_D$, $f\alpha a = f[(\alpha a)(\beta 1_B)] = (ga)h 1_B = ga$, and $f\alpha = g$. Similarly, $f\beta = h$.

If $\varphi: C \to D$ is another algebra homomorphism such that $\varphi\alpha = g$ and $\varphi\beta = h$, then since φ preserves multiplication

$$\varphi\xi = \varphi \left(\sum_k (\alpha a_k)\beta b_k \right) = \sum_k (\varphi\alpha a_k)\varphi\beta b_k$$

$$= \sum_k (ga_k)hb_k = f\xi.$$

Thus $\varphi = f$ and f is unique.

$\Leftarrow$: The first part of the proof is used in the second part of the proof. The first part implies that if $\langle T, \gamma, \delta \rangle$ is ANY tensor product of A and B, that then it has the above map extension property, that is the universal property.

So we are given that $\langle C, \alpha, \beta \rangle$ satisfies the universal property. Next let $\langle T, \gamma, \delta \rangle$ be a fixed tensor product of A and B. (There exist such, e.g. take $T = A \otimes B$ as in 4-3.4.)

(1) By the universal property for C (with $D = T$, $g = \gamma$, and $h = \delta$), there is a (unique) F-algebra map $f : C \to T$ such that $f\alpha = \gamma$ and $f\beta = \delta$.

(2) On the other hand by the universal property for T (with $D = C$, $g = \alpha$, and $h = \beta$), there is a (unique) map $g : T \to C$ with $g\gamma = \alpha$ and $h\delta = \beta$. In all of the four diagrams the universal property is used for the top algebra

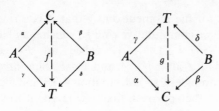

Diagram 1   Diagram 2

Diagram 3 commutes with either the identity map $\mathbb{1}_C$ of C – obviously $\alpha = \mathbb{1}_C\alpha$, $\beta = \mathbb{1}_C\beta$ –, as well as with gf, because $f\alpha = \gamma$, $g\gamma = \alpha$, and hence $gf\alpha = g\gamma = \alpha$. Similarly also $gf\beta = \beta$.

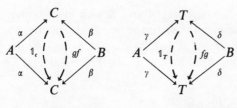

Diagram 3   Diagram 4

Thus by the uniqueness in the universal property for the top algebra C, $gf = \mathbb{1}_C$. Similarly, from Diagram 4, $fg = \mathbb{1}_T$. Hence g and f are isomorphisms and inverses.

Next, it will be shown that if $\langle T, \gamma, \delta \rangle$ is a tensor product of A and B and if there is an isomorphism $g : T \to C$ and a commutative diagram 2, that then necessarily also $\langle C, \alpha, \beta \rangle$ is a tensor product of A and B.

(i) Since $\alpha = g\gamma$ with g and γ monic, also α and similarly β are monic. From $T = \gamma A \delta B$, one obtains that $C = gT = (g\gamma A)(g\delta B) = (\alpha A)(\beta B)$.

(ii) From $(\gamma a)\delta b = \delta b(\gamma a)$ and that g is multiplicative, it follows that $(\alpha a)\beta b = (g\gamma a)(g\delta h) = g[(\delta b)\gamma a] = (\beta b)\alpha a$.

(iii) If $\{x_i\} \subset A$ and $\{y_j\} \subset B$ are linearly independent, then so are $\{(\gamma x_i)\delta y_j\}$ because γA and δB are linearly disjoint. Since g is an isomorphism so are also $\{(g\gamma x_i)(g\delta y_j)\} = \{(\alpha x_i)\beta y_j\}$. Hence $\alpha A \subset C$ and $\beta B \subset C$ are linearly disjoint. Thus $\langle C, \alpha, \beta \rangle$ is a tensor product of A and B.

Our next objective is to give an alternate construction of the algebra tensor product by the procedure of section 4-1 except that $\mathbb{Z}$ is now replaced by F. The following remark will not be used in the sequel.

4-3.6 Remark. Suppose for the moment only – that A is a ring and a unital right module over some commutative ring F (which may have zero divisors). Then just as in 4-2.1, A is called an F-algebra if $(ab)k = (ak)b = a(bk)$ holds for all $a, b \in A$ and $k \in F$. Sine now linear independence over F is meaningless, so is Definition 4-3.3 and Construction 4-3.4. The advantage of the approach that we are about to follow is that it never uses the fact that F is a field, and would produce an algebra which is the correct tensor product in this more general setting – namely an algebra which possesses the universal map extension property of 4-3.5. However, from now on again, F will be a field.

4-3.7 Construction. For algebras A and B with identity over the field F, let $S = \oplus\{F(a, b) \mid (a, b) \in A \times B\}$ be the free F-module on $A \times B$, i.e. an F-vector space with basis $A \times B$. Let $Y = \{y_1, y_2, y_3, y_4\}$ be the set of all elements of the following form for any $a, a' \in A$; $b, b' \in B$; and $k \in F$:

(1) $(a + a', b) - (a, b) - (a', b) = y_1$;
(2) $(a, b + b') - (a, b) - (a, b') = y_2$;
(3) $(ka, b) - k(a, b) = y_3$;
(4) $(a, kb) - k(a, b) = y_4$.

Note that $y_3 - y_4 = (ak, b) - (a, kb)$ as before in 4-1.1. Next let H be the vector subspace of S spanned by Y, that is H consists of all finite F-linear combinations of the y_i.

Define $(a, b)(a', b') = (aa', bb')$ and extend by linearity and distributivity to turn S into an (associative) F-algebra with identity element $1_S = (1_A, 1_B)$.

For any $y \in Y$ and any $(c, d) \in A \times B$, $(c, d)y \in Y$ and $y(c, d) \in Y$. Hence $H \lhd S$ is an algebra ideal. Thus H is the algebra ideal generated by Y as in 4-2.5(2).

Define T to be the F-algebra $T = S/H$; the projection map $\rho: S \to T$ is an algebra homomorphism. Map $\alpha: A \to$ by $\alpha a = (a, 1) + H$ and $\beta: B \to T$.

$\alpha b = (1, b) + H$; α and β are monic. Define $a \otimes b = (a, b) + H$. Thus $\alpha a = a \otimes 1$ and $\beta b = 1 \otimes b$.

If $i: A \to S$, $ia = (a, 1)$ and $j: B \to S$, $jb = (1, b)$ are the natural inclusion maps, then $\alpha = \rho i$ and $\beta = \rho j$.

4-3.8   Proposition. The above $\langle T, \alpha, \beta \rangle$ is a tensor product of algebras A and B.

Proof. It suffices to show that T has the universal map extension property. Let $g: A \to D$ and $h: B \to D$ such that $(ga)hb = hb(ga)$ be given as in 4-3.5. Each $s \in S$ is uniquely of the form $s = \Sigma k_i(a_i, b_i)$ for $0 \neq k_i \in F$. Then $\sigma s = \Sigma k_i(ga_i)(hb_i)$ defines an algebra map $\sigma: S \to D$. A computation of the type used in 4-3.5 easily shows that for $s, t \in S$, $\sigma(st) = (\sigma s)\sigma t$ For all y_i, $\sigma y_i = 0$. For example, $\sigma y_3 = g(ka)h(b) - k(ga)(hb) = 0$, etc. Since kernel $\sigma \supset Y$, also kernel $\sigma \supseteq H$. Hence as in 4-2.5(3) there is an algebra homomorphism $\tau: T \to D$ such that $\sigma = \tau \rho$ which is defined by $\tau(a \otimes b) = \sigma((a, b)) = (ha)gb$.

First, $\sigma ia = \sigma(a, 1) = ga$, or $\sigma i = g$; similarly $\sigma j = h$. Secondly $\tau \alpha a = \tau(a \otimes 1_B) = (ga)h1_B = (ga)1_D = ga$. Thus $\tau \alpha = q$ and $\tau \beta = h$ the same way. Thus there are commutative diagrams of algebra morphisms.

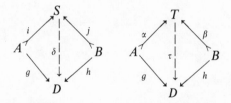

Diagram 1   Diagram 2

(Note that no claim is made that σ is a unique map giving the commutative first diagram.) Suppose that $\tau': T \to D$ is another map such that $\tau' \alpha = g$ and $\tau' \beta = h$. Then $\tau'(a \otimes 1) = \tau' \alpha a = ga = \tau \alpha a = \tau(a \otimes 1)$, and similarly, $\tau'(1 \otimes b) = \tau(1 \otimes b)$. Since $T = (A \otimes 1)(1 \otimes B)$ and both τ, τ' preserve multiplication, $\tau = \tau'$. Or more explicitly, $\tau'(a \otimes b) = \tau' [(a \otimes 1)(1 \otimes b)] = \tau'(a \otimes 1)\tau'(1 \otimes b) = \tau(a \otimes b)$. Thus $\langle T, \alpha, \beta \rangle$ is the tensor product of A and B.

4-4   Exercises

In the following exercises V and W are vector spaces over a field F.

1. If $v \otimes w = v_1 \otimes w_1 + \cdots + v_n \otimes w_n \in V \otimes_F W$, for v, $v_i \in V$ and w, $w_i \in W$ show that v is a linear combination $v = c_1 v_1 + \cdots + c_n v_n$, $c_i \in F$, of the v_i.

2. If $\dim_F V = \dim_F W = n$, show that every element of $V \otimes_F W$ can be written as a sum of no more than n elements of the form $v \otimes w$.

3. Let $\dim_F V = \dim_F W = 3$. Show that there exist elements $t \in V \otimes_F W$ such that $t \neq v_1 \otimes w_1 + v_2 \otimes w_2$ for any choice of $v_i \in V$, $w_i \in W$. Generalize this.

4. Let A and B be modules over a commutative ring C. Define $A^* = \mathrm{Hom}_C(A, C)$. (i) Show that A^* and $A \otimes_C B$ are C-modules. (ii) Define a natural group homomorphism $\rho: A^* \otimes_C B^* \to (A \otimes_C B)^*$ such that $\rho(\alpha \otimes \beta)(a \otimes b) = \alpha(a)\beta(b)$ for $\alpha \in A^*$, $\beta \in B^*$, $a \in A$, $b \in B$. (iii) Show that ρ is an isomorphism if A and B are free modules both of finite rank over C. (iv) Give examples to show that in general ρ need not be monic, or surjective. (Nonmonic: $C = Z_4 = \mathbf{Z}/4\mathbf{Z}$, $A = B = Z_4/2Z_4$. Not surjective: A and B infinite dimensional vector spaces over a field C.)

5. For rings R and S, $B = {}_R B_S$ an $R - S$ bimodule and $A = A_R$ show that $A \otimes_R B$ is a right S-module.

CHAPTER 5

Certain important algebras

Introduction

The free algebra on a set, and the tensor and exterior algebras of a vector space are constructed. Emphasis is placed on the exterior algebra of a finite dimensional vector space and its applications to determinants. Afterwards an alternate independent development of the exterior algebra is given as the quotient of a tensor algebra.

Why is the exterior and to a lesser extent the tensor algebra given such prominence in this chapter as opposed to other algebras? The exterior algebra appears in differential geometry not only as the algebra of differential forms but also from skew symmetric tensors. There seems to be an abundant supply of the latter perhaps because the Lie product $[x, y] = xy - yx$ is skew symmetric in x and y, and various Lie operations (e.g. the Lie derivative) and Lie algebras appear unavoidably in differential geometry and in physics as well. Some constructions and concepts in physics can be formulated more precisely, more easily, and be better understood if one has the tensor and exterior algebras at one's disposal. There is a lot of interesting and useful material here that is not covered in this chapter, which hopefully will serve as an introduction and invitation to further study.

This chapter takes place in the category of algebras with identity over a fixed ground field F and identity preserving homomorphisms. The identity element of a subalgebra is required to be the identity of the big algebra containing it. Here the prefix "F" will indicate an F-linear transformation or an F-subspace, and "algebra" means F-algebra.

Some additional properties and applications of the finite dimensional exterior algebra are given in the Appendix 5 to Chapter 4. There also is given a more general alternative treatment of the tensor, symmetric, and

71

exterior algebras with the underlying vector space no longer being restricted to be finite dimensional.

5-1   Free and tensor algebras

Several useful algebras can be obtained as a quotient algebra of a free or equivalently tensor algebra.

5-1.1   Notation. Let as before F be a field, $V = V_F$ an F-vector space, and $\{u_i \mid i \in I\} \subset V$ some F-vector space basis of V indexed by some index set I.

5-1.2.   Definition. For any nonempty set X, a *free algebra* on X is a pair (T, σ) where T is an F-algebra, $\sigma: X \to T$ is a one-to-one set function and where T has the following universal property (abbreviation U.P.):

> $\forall$ F-algebra A, $\forall$ set function $\varphi: X \to A \Rightarrow \exists!$ (unique) F-algebra homomorphism $\bar{\varphi}: T \to A$ such that $\bar{\varphi}\sigma = \varphi$.

It turns out that a free algebra for all purposes is the same as a tensor algebra.

5-1.3   Definition. A *tensor algebra* on V is a pair (T, ρ) where T is an F-algebra, $\rho: V \to T$ is a monic F-vector space map such that the following universal property holds.

U.P. For any F-algebra A, and for any F-linear $\varphi: V \to A$, there exists a *unique* F-algebra homomorphism $\bar{\varphi}: T \to A$ such that $\varphi = \bar{\varphi}\rho$.

The same type of proof can be used to show the uniqueness of free and exterior algebras and will not be repeated.

5-1.4   Uniqueness. If $\langle T_i, \rho_i \rangle$ are tensor algebras on V; $i = 1, 2$; then there exists a unique F-algebra isomorphism $f: T_1 \to T_2$ such that $\rho_2 = f\rho_1$. (Alternatively $f^{-1}: T_2 \to T_1$ is an F-algebra homomorphism such that $\rho_1 = f^{-1}\rho_2$.)

Proof. The universal property will always be applied at the algebra at the top of the four diagrams. In Diagram 1 with $\varphi = \rho_2$ and $A = T_2$ the universal property of T_1 gives a unique algebra homomorphism $f: T_1 \to T_2$ such that $\rho_2 = f\rho_1$. Similarly Diagram 2 gives $g: T_2 \to T_1$ with $\rho_1 = g\rho_2$.

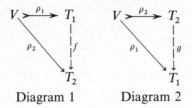

Diagram 1   Diagram 2

Use of $\rho_2 = f\rho_1$ and then $\rho_1 = g\rho_2$ gives that $gf\rho_1 = g\rho_2 = \rho_1 = \mathbb{1}\rho_1$ where $\mathbb{1}: T_1 \to T_1$ is the identity map. Diagram 3 commutes with either gf or $\mathbb{1}$. Thus the uniqueness part of the universal property of T_1 (with $\varphi = \rho_1$ and $A = T_1$) requires that $gf = \mathbb{1}$ is the identity on T_1. Either an entirely similar argument, or alternatively the symmetry of hypotheses on subscripts 1 and 2, similarly shows that $fg = \mathbb{1}$ is the identity on T_2. Thus $f = g^{-1}$ is an isomorphism.

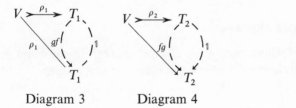

Diagram 3   Diagram 4

Remark. If T_1 and T_2 are any algebraic objects both containing a third object like V in common, then occasionally it is said that T_1 and T_2 are isomorphic over V – provided there exists an isomorphism $T_1 \cong T_2$ leaving V elementwise fixed.

5-1.5   Free algebra on X. For I as above, let I^* be the set of all finite ordered nonempty subsets of I. Thus I^* is the set of all α of the form:

$$n = 1, 2, \ldots,; \quad \alpha = (i(1), \ i(2), \ \ldots, i(n)); \quad i(k) \in I; \quad k = 1, \ldots, n;$$

where repetitions are allowed. Let $\beta = (j(1), \ldots, j(m)) \in I^*$, and $\gamma \in I^*$. Define $\alpha\beta = (i(1), \ldots, i(n), j(1), \ldots, j(m))$. Thus $(\alpha\beta)\gamma = \alpha(\beta\gamma)$.

For all $i \in I$, let "X_i" be different symbols and let "e" be another symbol distinct from these. Then define X_α as $X_\alpha = X_{i(1)}X_{i(2)} \cdots X_{i(n)}$, $X_\alpha \neq X_\beta \Leftrightarrow \alpha \neq \beta$. Define $F\{X\}$ as the F-vector space with basis $\{X_\alpha | \alpha \in I^*\} \cup \{e\}$. Thus $F\{X\} = \{c_0 e + \Sigma c_\alpha X_\alpha | \alpha \in I^*;\ c_0,\ c_\alpha \in F\}$. Abbreviate $1_F X_\alpha = X_\alpha$ and $c_0 e = c_0$. Under the definitions $X_\alpha X_\beta = X_{\alpha\beta}$; $X_\alpha e = eX_\alpha = X_\alpha$, $F\{X\}$ becomes an F-algebra which is associative because $(X_\alpha X_\beta)X_\gamma = X_\alpha(X_\beta X_\gamma)$. Note that $F\{X\}$ is a direct F-vector space sum $F\{X\} = F \oplus [\oplus\{FX_\alpha | \alpha \in I^*\}]$. Let $\sigma: X \to F\{X\}$ be the natural inclusion map $\sigma X_i = X_i \in F\{X\}$.

It is straightforward now to verify the next lemma.

5-1.6   Lemma. $F\{X\}$ *is a free algebra on the set X. If Y is any other set with $|X| = |Y|$, then any bijection $X \to Y$ extends to an algebra isomorphism $F\{X\} \to F\{Y\}$.*

5-1.7   Tensor algebra on V. For a vector space basis $\{u_i | i \in I\}$ of V, as before, form the free algebra $F\{X\}$ on the set $X = \{X_i | i \in I\}$. Define $\rho u_i = X_i$ and extend linearly to a monic vector spae map $\rho: V \to F\{X\}$. Then $(F\{X\}, \rho)$ is the tensor algebra on V, which by 5-1.4 is unique up to an isomorphism over V, and is independent of the choice of basis of V.

5-2   Exterior algebras

The previous framework of algebras with identity and identity preserving morphisms is continued throughout this chapter.

5-2.1   Definition. For a (finite or infinite dimensional) vector space V over a field an *exterior algebra* on V is a pair $(E(V), \eta)$ where $E(V)$ is an F-algebra satisfying the following

(i) η is monic;
(ii)   For any $v \in V$, $(\eta v)^2 = 0$ in $E(V)$.

Furthermore, $(E(V), \eta)$ satisfies the following universal property.

(U.P.) For any F-algebra A, and for any F-linear vector space map $\varphi: V \to A$ such that $(\varphi v)^2 = 0$ for all $v \in V$, there exists a unique F-algebra homomorphism $\hat{\varphi}: E(V) \to A$ such that $\varphi = \hat{\varphi}\eta$.

The abbreviation $EV = E(V)$ will be used, and if V is fixed and understood, $E = EV$.

The isomorphism over V of the two exterior algebras on V is obtained exactly as in 5-1.4.

5-2.2   Uniqueness. If (E_i, η_i); $i = 1, 2$; are exterior algebras on V, then there exists an F-algebra isomorphism $f: E_1 \to E_2$ such that $\eta_2 = f\eta_1$ (or $\eta_1 = f^{-1}\eta_2$).

From now on through the remainder of this section the vector space V will be finite dimensional. The exterior algebra will be constructed by first specifying a basis for a vector space, and then defining the product of two vector space basis elements.

5-2.3   Notation and definitions. Let V be a vector space of finite dimension $\dim_F V = N$ over the field F; let X be the set $X = \{1, 2, \ldots, N\}$ and $\mathscr{P}(X) = \{S \mid S \subseteq X\} = 2^X$ the set of all subsets of X, which is also called the power set of X. It contains $|\mathscr{P}(X)| = \Sigma_0^N \binom{N}{i} = 2^N$ elements. For any R, S, $T \in \mathscr{P}(X)$, introduce symbols u_R, u_S, u_T; since the void set $\varnothing \in \mathscr{P}(X)$, there also is $u_\varnothing$. We also have u_X for $X \in \mathscr{P}(X)$.

5-2.4   Definition. Define functions $\varepsilon_0: X \times X \to \{0, \pm 1\}$ and $\varepsilon: \mathscr{P}(X) \times \mathscr{P}(X) \to \{0, \pm 1\}$ by

$$\varepsilon_0(i, j) = \begin{cases} 1 & i < j \\ 0 & i = j \\ -1 & i > j \end{cases} \quad \begin{array}{l} \varepsilon(S, \varphi) = \varepsilon(\varphi, T) = 1 \\ \\ \varepsilon(S, T) = \prod_{(s,t) \in S \times T} \varepsilon_0(s, t) \end{array} ;$$

For singletons the abbreviation $\varepsilon(i, j) = \varepsilon(\{i\}, \{j\}) = \varepsilon_0(i, j)$ will sometimes be used.

5-2.5   Construction. Define $E(V)$ to be the F-vector space of dimension $\dim_F E(V) = 2^N$ having the symbols $\{u_S \mid S \in \mathscr{P}(X)\} = \{u_T, e_\varnothing \mid \varnothing \neq T \subseteq X\}$ as a vector space basis. The abbreviation $EV = E(V)$ will sometimes be used. Consequently $1_F u_S = u_S 1_F = u_S$. Elements of EV are finite sums $\Sigma c_S u_S$, $c_S \in F$. Define $u_R u_S = \varepsilon(R, S) u_{R \cup S}$. Then this multiplication can be extended to all of EV by distributivity and linearity. It will be associative, provided $(u_R u_S) u_T = u_R (u_S u_T)$.

5-2.6   Remark. It will be important for later purposes to note that in the proofs of the forthcoming next two lemmas 5-2.7 and 5-2.9, nowhere is the linear independence of the set $\{u_S \mid S \subseteq X\}$ used.

The next lemma will give us an associative algebra $E(V)$ with identity element u_φ.

5-2.7   Lemma. *For any $R, S, T \in \mathscr{P}(X)$ the following holds.*

    (i) $u_\varphi u_R = u_R u_\varphi = u_R$.
    (ii) *If $R \cap S = \varphi$, then $\Rightarrow \varepsilon(R \cup S, T) = \varepsilon(R, T) \in (S, T)$. If $S \cap T = \varphi$,*
        then $\Rightarrow \varepsilon(R, S \cup T) = \varepsilon(R, S) \in (R, T)$.
    (iii) $(u_R u_S) u_T = u_R (u_S u_T)$.

Proof. (i) By 5-2.4, $u_\varnothing u_R = \varepsilon(\varnothing, R) u_{\varnothing \cup R} = u_R$ and similarly also $u_R u_\varnothing = u_R$.
    (ii) Rearrangement of the terms in the expression in 5-2.4 for $\varepsilon(R \cup S, T)$ gives

$$\varepsilon(R \cup S, T) = \prod_{q \in R \cup S, t \in T} \varepsilon_0(q, t)$$

$$= \prod_{(r,t) \in R \times T} \varepsilon_0(r, t) \prod_{(s,t) \in S \times T} \varepsilon_0(s, t)$$

$$= \varepsilon(R, T) \varepsilon(S, T).$$

Similarly $\varepsilon(R, S \cup T) = \varepsilon(R, S) \varepsilon(R, T)$.
    (iii) Each side separately is

$$\text{left} = (u_R u_S) u_T = \varepsilon(R, S) \varepsilon(R \cup S, T) u_{R \cup S \cup T},$$

$$\text{right} = u_R (u_S u_T) = \varepsilon(S, T) \varepsilon(R, S \cup T) u_{R \cup S \cup T}.$$

Case 1. $R \cap S \neq \varnothing$. Then $\varepsilon(R, S) = 0$, and $u_R u_S = 0$. Also $\varepsilon(R, S \cup T) = 0$, and $u_R u_{S \cup T} = 0$. In Case 1, $(u_R u_S) u_T = u_R (u_S u_T) = 0$.

Case 2. $S \cap T \neq \varnothing$. On the left, $\varepsilon(R \cup S, T) = 0$, and on the right, $\varepsilon(S, T) = 0$. Again $(u_R u_S) u_T = u_R (u_S u_T) = 0$.

Case 3. $R \cap S = \varnothing$ and $S \cap T = \varnothing$. Use of (i) shows that $(u_R u_S)u_T = u_R(u_S u_T)$.

Except for the use of the integer N, so far V has not been used, but now it must appear.

5-2.8   Definition. Let $\{u_1, u_2, \ldots, u_N\}$ be any F-basis of V. Set $\eta u_i = u_{\{i\}}$ and extend η to a monic vector space map $\eta: V \to EV$. Identify $V = \eta V \subset EV$. Therefore $\eta u_i = u_i = u_{\{i\}}$. Write $u_\varnothing = 1$. Identify $Fu_\varnothing = F \subset EV$. Thus $F, V \subset EV$.

A standard monomial of degree k is an element of EV of the form $u_{i(1)}u_{i(2)} \cdots u_{i(k)}$ where $1 \leqslant i(1) < i(2) < \cdots < i(k) \leqslant N$ are k distinct integers in increasing order; $u_\varnothing = 1$ is considered to be a standard monomial of degree zero.

5-2.9   Lemma. *Take any v, $w \in V$. For an arbitrary m and $S = \{i(1) < i(2) < \cdots < i(m)\} \subset X$, let σ be any permutation of S and sgn σ the signature of σ. (I.e. sgn $\sigma = 1$ or -1 depending upon whether σ is a product of an even or an odd number of transpositions respectively.) Then*

(i) $v^2 = 0$;

(ii) $vw = -wv$;

(iii) $u_{i(1)}u_{i(2)} \cdots u_{i(k)} = u_S$;

(iv) $u_{i(1)\sigma}u_{i(2)\sigma} \cdots u_{i(m)\sigma} = (\text{sgn } \sigma)u_S$.

Proof. (i) If $v = \Sigma c_i u_i$, $c_i \in F$, then

$$i < j: u_i u_j = \varepsilon_0(i, j)u_{\{i,j\}} = u_{\{i,j\}};$$
$$u_j u_i = \varepsilon_0(j, i)u_{\{i,j\}} = -u_{\{i,j\}};$$
$$u_i^2 = \varepsilon_0(i, i)u_i = 0;$$
$$v^2 = \Sigma c_i u_i^2 + \sum_{i<j} c_i c_j(u_i u_j + u_j u_i) = 0.$$

(ii) Since the square of every vector in EV is zero, $0 = (v + w)^2 = v^2 + vw + wv + w^2$; and hence $vw + wv = 0$.

(iii) If $i < j < k$, then as seen above in 5-2.9(i), $u_i u_j = u_{\{i,j\}}$ and $u_i u_j u_k = u_{\{i,j\}}u_k = \varepsilon(\{i,j\}, \{k\})u_{\{i,j,k\}} = \varepsilon(i, k)\varepsilon(j, k)u_{\{i,j,k\}} = u_{\{i,j,k\}}$. The rest follows by repetition or induction.

(iv) For any i and j whatever, $u_j u_i = -u_i u_j$. By successive interchanges of two adjacent terms, $u_{i(1)\sigma}u_{i(2)\sigma} \cdots u_{i(m)\sigma}$ may be transformed into plus or minus u_S. The parity of the total number of such interchanges is exactly the parity of σ.

5-2.10   Corollary to 5-2.7 and 5-2.9. *Suppose that A is an algebra and $\mathscr{S} \subset A$ a subset satisfying*

(a) $\mathscr{S} = \{\bar{u}_R \mid R \subseteq X\}$;
(b) $A = \Sigma\{F\bar{u}_R \mid R \subseteq X\}$ *($\mathscr{S}$ is not assumed to be linearly independent.)*;
(c) *If* $R = \{1 \leqslant i(1) < \cdots < i(r) \leqslant N\}$, *then* $\bar{u}_R = \bar{u}_{i(1)}\bar{u}_{i(2)}) \cdots \bar{u}_{i(r)}$;
(d) $\bar{u}_i\bar{u}_j = -\bar{u}_j\bar{u}_i$ *for all* $1 \leqslant i, j \leqslant N$.

Then

(i) $\bar{u}_R\bar{u}_S = \varepsilon(R, S)\bar{u}_{R \cup S}$;
(ii) *Lemmas 5-2.7 and 5-2.9 hold in A, that is, if u_R's are replaced by $\bar{u}_R$'s.*

Proof. First note that the condition (d) $\bar{u}_i\bar{u}_j = -\bar{u}_j\bar{u}_i$ is equivalent to the condition (d') $\bar{u}_i\bar{u}_j = \varepsilon_0(i, j)\bar{u}_{\{i,j\}}$. (i) If $R \cap S \neq \varphi$, then both sides are zero. So let $R \cap S = \varnothing$. If $R = \{i(1) < \cdots < i(r)\}$ and $S = j(1) < \cdots < j(s)$, then we will arrange the $r + s$ subscripts on $\bar{u}_R\bar{u}_S = \bar{u}_{i(1)} \cdots \bar{u}_{i(r)}\bar{u}_{j(1)} \cdots \bar{u}_{j(s)}$ in increasing order by moving $\bar{u}_{j(1)}$ to its correct place and keeping track of minus signs. If $i(k) < j(1) < i(k + 1)$ for $0 \leqslant k \leqslant r$, then k does not appear on the right side of

$$\bar{u}_{i(1)} \cdots \bar{u}_{i(k)}\bar{u}_{j(1)}\bar{u}_{i(k+1)} \cdots \bar{u}_{j(s)} = \prod_{\mu=1}^{r} \varepsilon(i(\mu), \; j(1))\bar{u}_R\bar{u}_S.$$

Next, the result of moving $\bar{u}_{j(2)}$ to its correct place will equal

$$\prod_{v=1}^{2} \prod_{\mu=1}^{r} \varepsilon(i(\mu), j(v))\bar{u}_R\bar{u}_S,$$

until finally after s such operations $\bar{u}_R\bar{u}_S = \varepsilon(R, S)\bar{u}_{R \cup S}$.
(ii) This now follows from 5-2.6, 2.7, and 2.9.

5-2.11   Theorem. *The algebra $V \subset EV$ is an exterior algebra on V.*

Proof. So far, we have an associative algebra EV with identity $u_\varnothing = 1_F = 1 \in F \subset EV(\text{5-2.5}, \;\; 2.7)$, and a vector space embedding $\eta: V \rightarrowtail \eta V = V \subset EV$ such that $(\eta v)^2 = v^2 = 0$ for all $v \in V$ (5-2.8, 2.9).

In order to show that EV satisfies the universal property for an exterior algebra in 5-2.1, take any algebra A, and any vector space map $\varphi: V \rightarrow A$ such that $(\varphi v)^2 = 0$ for all $v \in V$. Set $\bar{v} = \varphi v$, and for $S = \{i(1) < \cdots < i(k)\}$, set $\bar{u}_S = (\varphi u_{i(1)})(\varphi u_{i(2)}) \cdots (\varphi u_{i(k)}) = \bar{u}_{i(1)}\bar{u}_{i(2)} \cdots \bar{u}_{i(k)}$. If a map $\hat{\varphi}: EV \rightarrow A$ is to be an algebra homomorphism, then it must necessarily map $\hat{\varphi}u_S = \bar{u}_S$. Set $\hat{\varphi}(u_\varnothing) = 1_A \in A$, and extend $\hat{\varphi}$ by linearity to $\hat{\varphi}: EV \rightarrow A$.

For any $v, w \in V$, it follows from $0 = (\bar{v} + \bar{w})^2 = \bar{v}\bar{w} + \bar{w}v + \bar{w}^2$ that $\bar{v}\bar{w} = -\bar{w}\bar{v}$. In particular, $\bar{u}_i\bar{u}_j = -\bar{u}_j\bar{u}_i$, and hence $\bar{u}_i\bar{u}_j = \varepsilon(i, j)\bar{u}_{\{i,j\}}$ holds for all i and j. Thus the subalgebra of A generated by φV satisfies the conditions (a)–(b) of 5-2.10. Consequently

$$(\hat{\varphi} u_R)(\hat{\varphi} u_S) = \varepsilon(R, S) u_{R \cup S} = \hat{\varphi}(u_R u_S),$$

and $\hat{\varphi}$ is an algebra homomorphism. It follows from the definition of $\hat{\varphi}$ that $\hat{\varphi} v = v$ for any $v \in V$, i.e. $\hat{\varphi} \mid V = \varphi$.

If $f : EV \to A$ is another algebra homomorphism such that $f \mid V = \varphi$, then

$$fu_S = f[u_{i(1)} \cdots u_{i(k)}] = (fu_{i(1)}) \cdots (fu_{i(k)}) = (\varphi u_{i(1)}) \cdots (\varphi u_{i(k)})$$
$$= \hat{\varphi}[u_{i(1)} \cdots u_{i(k)}] = \hat{\varphi} u_S.$$

Thus $f = \hat{\varphi}$. Hence $V \subset EV$ is the exterior algebra on V.

From now on EV will denote the above algebra originally constructed from some fixed given basis $\{u_1, \ldots, u_N\}$ of N.

5-2.12 Corollary 1 to theorem 5-2.11. *If $V \subset E_1$ and $V \subset E_2$ are any two exterior algebras on V, then $E_1 = E_2$. Hence in particular, $E_1 = E_2 = EV$.*

Proof. Since $V \subset E_1$, the standard monomials in the u_i are in E_1. Since as a vector space the standard monomials span EV, also $EV \subseteq E_1$. By 5-2.2, $EV \cong E_1$. Hence $\dim_F E_1 = \dim_F EV = 2^N$ is finite, and hence $EV = E_1$. Thus $E_1 = E_2 = EV$.

In almost all of our proofs of the uniqueness of maps in universal properties (including the last theorem), the following general principle is used.

5-2.13 Uniqueness. Suppose that every element of an algebra A is an F-linear combination of arbitrary finite products of elements of a given fixed subset $Y \subset A$. In this case if $f, g : A \to B$ are two algebra homomorphisms, then $f = g$ if and only if $f \mid Y = g \mid Y$.

5-2.14 Corollary 2 to theorem 5-2.11. *For any subspace $U \subset V$, $E(U) \subset E(V)$ is a subalgebra.*

Proof. Two proofs will be given. But before that, without loss of generality $u_1, \ldots, u_m$ is a basis of U which is extended to a basis $u_1, \ldots, u_m, u_{m+1}, \ldots, u_N$ of V. In both proofs, A is defined as the unique smallest subalgebra of EV containing U, i.e. A is the intersection of all subalgebras of EV which contain U.

First proof. The standard monomials formed from 1, $u_1, \ldots, u_m$ only are not only a vector space basis of EU, but are also contained in A. Since $U \subset EU \subseteq A$, necessarily $EU = A$.

Second proof. Let $\eta_U : U \to E[U]$ be an abstractly given exterior algebra on U about which we will use nothing else except that it satisfies the definition and universal property. View $E[U]$ as not embedded in EV. By the universal property of $E[U]$, the identity map $1_U : U \rightarrowtail U \subset V \subset EV$ on U extends to an algebra homomorphism $\hat{1}_U : E[U] \to EV$ such that $\hat{1}_U \eta_U = 1_U$. Consequently $U = 1_U U = \hat{1}_U \eta_U U \subseteq \hat{1}_U (E[U])$. Since $\hat{1}_U(E[U])$ is a subalgebra of EV containing U, by the definition of A, $\hat{1}(E[U]) \supseteq A$. It follows from 5-2.2 and 5-2.12 that $E[U]$ is generated (in the sense of 5-2.13) by $\eta_U U$. Consequently $\hat{1}(E[U])$ is generated by the images of $\eta_U U$ in EV, i.e. $\hat{1}(E[U])$ is generated by U. So is also A. Hence $A = \hat{1}(E[U])$. Since the standard monomials in 1 and $u_1, \ldots, u_m$ are linearly independent in EV, $2^m \leqslant \dim_F A$. By 5-2.2 and 5-2.12, $\dim_F E[U] = 2^m$. Thus $2^m \leqslant \dim_F A = \dim \hat{1}_U (E[u]) \leqslant \dim E[U] = 2^m$ shows that $\hat{1}_U$ is an isomorphism. Hence $U \subset \hat{1}_U(E[U]) = A \subset EV$ is an exterior algebra on U.

5-2.15 Let U, V, and W be F-vector spaces and $\eta_U : U \to EU$, $\eta_V : V \to EV$, and (EW, η_W) their exterior algebras. Suppose that $f : U \to V$ and $g : V \to W$ are any linear transformations. Then there exist algebra homomorphisms $Ef : EU \to EV$, $Eg : EV \to EW$, and $E(gf) : EU \to EW$ such that the following hold

(i) $\eta_V f = (Ef)\eta_U$, $\eta_W g = (Eg)\eta_V$, $\eta_W gf = E(gf)\eta_U$; and Ef, Eg, and $E(gf)$ are unique with respect to this property.
(ii) $E(gf) = (Eg)(Ef)$.

The above says that the diagram below can be completed with the dashed arrows to yield a commutative diagram.

Proof. In the four diagrams below the uniserval property is always applied at the algebra at the top of the diagram and the slanted arrows represent vector space maps such that the square of the image is zero. In the first three diagrams the dashed vertical arrows are the unique algebra homomorphisms given by the universal property making the diagrams commute. It follows from $(Ef)\eta_U = \eta_V f$ and then from $(E_g)\eta_V = \eta_W g$ that $(Eg)(E f)\eta_U = (Eg)\eta_V f = \eta_W gf$. Since the last diagram commutes with either $E(gf)$ or $(Eg)(E f)$, by the uniqueness part of the universal property, $E(gf) = (Eg)(E f)$.

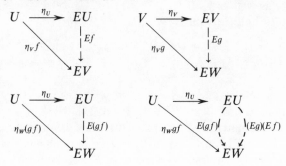

5-2.16 When $U = V$ and $f = 1_V : V \to V$ is the identity operator, $E1_V = 1_{EV} : EV \to EV$ is the identity map. The latter and 5-2.15 says that E is a functor. Write $1_V = E1_V$.

5-2.17 Let L, $L_i : V \to V$ be F-linear transformations for $i = 1$, 2. In order to distinguish the general case of $U \to V$ from the special case $U = V$, the notation $\hat{L} = EL$ will be used for the unique F-algebra homomorphism $\hat{L} : EV \to EV$ such that $\hat{L} \mid V = L$. Similarly, $\hat{L}_i : EV \to EV$ with $\hat{L}_i \mid V = L_i$, $i = 1$, 2. By letting $L = L_1 L_2$, we also get $(L_1 L_2)^\wedge : EV \to EV$. For $1_V : V \to V$, we get the identity map $\hat{1}_V : EV \to EV$. The following hold.

    (i) $(L_1 L_2)^\wedge = \hat{L}_1 \hat{L}_2$.

    (ii) $\hat{1}_V = 1_{EV}$.

    (iii) If $\exists L^{-1}$, then $\Rightarrow \exists$ also $\hat{L}^{-1}$, and furthermore, $\hat{L}^{-1} = (L^{-1})^\wedge$.

Proof. (i) This follows from 5-2.15(ii) with $U = V = W$ and $f = L_2$, $g = L_1$. It is instructive to give short alternative proofs of the above properties. The U.P. of EV (with $A = EV$ and $\varphi = L_1$, L_2, $L_1 L_2 : V \to A$ such that $(\varphi v)^2 = 0$ for all $v \in V$) gives three homomorphisms $\hat{L}_i$, $(L_1 L_2)^\wedge : EV \to EV$ such that $\hat{L}_i \mid V = L_i$ and $(L_1 L_2)^\wedge \mid V = L_1 L_2$. We ourselves can manufacture a fourth one $\hat{L}_1 \hat{L}_2 : EV \to EV$. For $v \in V$, $(\hat{L}_1 \hat{L}_2)v = \hat{L}_1 (\hat{L}_2 v) = \hat{L}_1 (L_2 v)$. Since $L_2 v \in V$, and $\hat{L}_1 \mid V = L_1$, $(\hat{L}_1 \hat{L}_2)v = L_1 L_2 v$. Thus the diagram below commutes with $\hat{L}_1 \hat{L}_2$, while by definition of $(L_1 L_2)^\wedge$ it also commutes with $(L_1 L_2)^\wedge$.

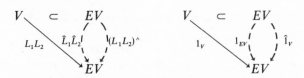

The uniqueness applied at the top algebra in the diagrams gives (i) $(L_1 L_2)^\wedge = \hat{L}_1 \hat{L}_2$ as well as $\hat{1}_V = 1_{EV}$.

(iii) Use 5-2.7(i) and (ii) with $L_1 = L$, $L_2 = L^{-1}$.

$$\left.\begin{array}{l} 1_{EV} = \hat{1}_V = (LL^{-1})^\wedge = \hat{L}(L^{-1})^\wedge \\ 1_{EV} = \hat{1}_V = (L^{-1}L)^\wedge = (L^{-1})^\wedge \hat{L} \end{array}\right\} \Rightarrow \begin{array}{l} \exists\, \hat{L}^{-1} \text{ and} \\ \hat{L}^{-1} = (L^{-1})^\wedge. \end{array}$$

5-2.18 Definition. Set $V^0 = F$, $V^1 = V$, $V^2 = \{\Sigma vw \,|\, v, w \in V\}, \ldots, V^i = V^{i-1}V$ for all i. Alternatively, V^i is the vector subspace of EV spanned by $\{u_S \,|\, S \subseteq X, |S| = i\}$. Then $V^N = Fu_1 u_2 \cdots u_N$ is one dimensional, while $0 = V^{N+1} = V^{N+2} = \cdots$. A nonzero element of EV of the form $v_1 v_2 \cdots v_i$, $v_j \in V$, in the literature is sometimes referred to as a monomial; any arbitrary element of V^i is a finite sum of monomials of degree i and is called homogeneous of degree i. Thus EV is a vector space direct sum $EV = \oplus \{V^i \,|\, i = 0, \ldots, N\} = F \oplus V \oplus V^2 \oplus \cdots \oplus V^N$, where $V^i V^j = V^{i+j}$ for all $0 \leqslant i$, j. For any F-linear map $f: V \to W$ of vector spaces, $(Ef)V^i \subseteq W^i$ for all i.

Next some applications of exterior algebras to determinants and to linear algebra are given.

5-2.19 For the vector space V with basis $u_1, \ldots, u_N$ and a linear transformation $L: V \to V$, if $Lu_j = \Sigma a_{ij} u_i$, $a_{ij} \in F$, then the determinant of L is $\det L = \det \|a_{ij}\|$. The homomorphism $\hat{L}: EV \to EV$ preserves $\hat{L}V^i \subseteq V^i$ degrees for all $i = 0, 1, 2, \ldots, N$; because $\hat{L}V = LV \subseteq V$. Define the scalar $d \in F$ by $\hat{L}(u_1 u_2 \cdots u_N) = du_1 u_2 \cdots u_N$. Then

$$\begin{aligned} du_1 u_2 \cdots u_N &= (Lu_1)(Lu_2) \cdots (Lu_N) \\ &= \Sigma a_{i(1)1} a_{i(2)2} \cdots a_{i(N)N} u_{i(1)} u_{i(2)} \cdots u_{i(N)} \\ &\quad (i(1), \ldots, i(N)); \ 1 \leqslant i(k) \leqslant N, \ 1 \leqslant k \leqslant N. \\ &= \sum_{(i(1),\ldots,i(N))} (\mathrm{sgn}\,\sigma) a_{i(1)1} \cdots a_{i(N)N} u_1 u_2 \cdots u_N \\ &= (\det L) u_1 u_2 \cdots u_N, \\ \sigma &= \begin{pmatrix} 1 & 2 & \cdots & N \\ i(1) & i(2) & & i(N) \end{pmatrix}, \end{aligned}$$

where σ is the permutation $k\sigma = i(k)$ and $\operatorname{sgn}\sigma$ is its signature. Thus $d = \det L$, and $\hat{L}(u_1 u_2 \cdots u_N) = (\det L)u_1 u_2 \cdots u_N$.

5-2.20 If L_1, $L_2: V \to V$ are linear transformations, then $(L_1 L_2)^\wedge$ $(u_1 u_2 \cdots u_N) = \det(L_1 L_2)u_1 u_2 \cdots u_N$. Since $(L_1 L_2)^\wedge = \hat{L}_1 \hat{L}_2$, also $(L_1 L_2)^\wedge$ $(u_1 u_2 \cdots u_N) = \hat{L}_1(\hat{L}_2(u_1 u_2 \cdots u_N)) = \hat{L}_1((\det L_2)u_1 u_2 \cdots u_N)$. Since $\hat{L}_1$ is F-linear, the latter equals

$$(\det L_2)\hat{L}_1(u_1 u_2 \cdots u_N) = (\det L_2)(\det L_1)u_1 u_2 \cdots u_N.$$

This proves that $\det(L_1 L_2) = (\det L_1)(\det L_2)$.

Two additional applications of finite dimensional exterior algebras to determinants and to linear algebra are given in Appendix V. There also is developed a more general unified approach to all the important algebras via the tensor algebra.

5-3   Exercises

In the following connected sequence of exercises, V is a vector space over a field F of finite dimension $\dim_F V = N$, $V^* = \operatorname{Hom}_F(V, F)$, $TV^* = \oplus_0^\infty T^n V^*$ is the tensor algebra on V^*, and for any integer $1 \leqslant m$, $B^m = B^m(V)$ is the set of all m-multilinear maps $\alpha: V \times \cdots \times V = V^m \to F$ (i.e. α is F-linear in each argument separately).

1. For α, $\beta \in B^m$ and $k \in F$ show that $\alpha + \beta$, $k\alpha \in B^m$ where $(\alpha + \beta)(v_1, \ldots, v_m) = \alpha(v_1, \ldots, v_m) + \beta(v_1, \ldots, v_m)$ and $(k\alpha)(v_1, \ldots, v_m) = k\alpha(v_1, \ldots, v_m)$ for $v_i \in V$.

2. Let $\alpha \in B^p$; β, $\gamma \in B^q$, and $\delta \in B^s$. Define $\alpha \otimes \beta \in B^{p+q}$ by $(\alpha \otimes \beta)$ $(v_1, \ldots, v_p, v_{p+1}, \ldots, v_{p+q}) = \alpha(v_1, \ldots, v_p)\beta(v_{p+1}, \ldots, v_q)$. Show that $\alpha \otimes (\beta + \gamma) = \alpha \otimes \beta + \alpha \otimes \gamma$ and that $(\alpha \otimes \beta) \otimes \delta = \alpha \otimes (\beta \otimes \delta)$.

3. Show that addition "$+$" and multiplication "$\otimes$" and scalar multiplication can be extended by linearity and distributivity to make $\oplus_1^\infty B^m$ into an F-algebra.

4. Show how each $\alpha \in B^m$ induces a unique linear functional also denoted by α, mapping $\alpha: V \otimes \cdots \otimes V = \otimes_1^m V \to F$ and conversely, that each element of $(\otimes_1^m V)^*$ induces a unique element in B^m.

5. Prove that there are natural vector space isomorphisms $B^m \cong (\otimes_1^m V)^* \cong \otimes_1^m V^* = T^m V^*$.

6. Let $F \times (\otimes_1^\infty B^m)$ be the algebra in exercise 3 with an identity adjoined (as in 1-3.1). Prove that there is an algebra isomorphism $F \times (\otimes_1^\infty B^m) \cong TV^*$ which maps $B^m \cong T^m V^*$ and is the identity on $B^1 = V^*$.

7. Which of the above remain valid if $\dim_F V = \infty$?

In the following sequence of exercises continuing 1–7 let $\Lambda^m = \Lambda^m(V) \subset B^m$ be the subspace of all alternating or skew m-multilinear functions, i.e. all $\gamma: V^m \to F$, $\gamma(\cdots v_i \cdots v_j \cdots) = -\gamma(\cdots v_j \cdots v_i \cdots)$ if v_i and v_j are switched. Let $\operatorname{sgn}\sigma$ denote the signature of a permutation $\sigma \in S_m$ on $\{1, \dots, m\}$. For $\alpha \in B^m$, define $\operatorname{Alt}\alpha: V^m \to F$ by

$$(\operatorname{Alt}\alpha)(v_1, \dots, v_m) = \frac{1}{m!} \sum_{\sigma \in S_m} \operatorname{sgn}\sigma\, \alpha(v_{1\sigma}, \dots, v_{m\sigma}).$$

8. (a) Prove that $\operatorname{Alt}: B^m \to \Lambda^m$ is a linear transformation which is the projection of B^m onto the subspace Λ^m. I.e. that for $\alpha \in B^m$, (a) $\operatorname{Alt}\alpha \in \Lambda^m$; (b) For $\alpha \in \Lambda^m$, $\operatorname{Alt}\alpha = \alpha$; and (c) $\operatorname{Alt}(\operatorname{Alt}\alpha) = \operatorname{Alt}\alpha$. (iii) Describe the kernel of Alt.
9. Show that $\Lambda^m = 0$ if $m \geqslant N + 1$.
10. For $\alpha \in \Lambda^p$, $\beta \in \Lambda^q$ define $\alpha \wedge \beta \in \Lambda^{p+q}$ by

$$\alpha \wedge \beta = \frac{(p + q)!}{p!\,q!} \operatorname{Alt}(\alpha \otimes \beta).$$

Assume that $\alpha \wedge (\beta \wedge \gamma) = (\alpha \wedge \beta) \wedge \gamma$ for all $\alpha \in \Lambda^p$, $\beta \in \Lambda^q$, $\gamma \in \Lambda^s$ for all $1 \leqslant p, q, s$. (See exercise 13*.)

Prove that the above assumption is enough to extend addition "+" and multiplication "$\wedge$" by linearity and distributivity to yield an associative algebra $\oplus_{m=1}^N \Lambda^m$.

11. Let $F \times (\oplus_1^N \Lambda^m) = \oplus_0^N \Lambda^m$, $\Lambda^0 = F$ be the previous algebra with an identity adjoined (1-3.1). Let EV^* be the exterior algebra on V^* and let $(V^*)^m$ be the subspace spanned by all products of m-elements of V^* in EV^*. Prove that there is a ring isomorphism $\oplus_1^N \Lambda^m \cong EV^*$ which is the identity on $\Lambda^1 = V^* \subset EV^*$ and which maps $\Lambda^m \cong (V^*)^m$.
12. Show that $\oplus_0^N \Lambda^m \subset \oplus_0^N B^m$ as a vector subspace. Is $\oplus_0^N \Lambda^m$ a subring of $\oplus_0^\infty B^m$?
13*. Prove that $\alpha \wedge (\beta \wedge \gamma) = (\alpha \wedge \beta) \wedge \gamma$ for $\alpha \in \Lambda^p$, $\beta \in \Lambda^q$, $\gamma \in \Lambda^s$ for $1 \leqslant p, q, s$. (Hint: Establish for $f \in B^p$, $g \in B^q$, $h \in B^s$ that (a) If $\operatorname{Alt}g = 0$, then $\operatorname{Alt}(f \otimes g) = \operatorname{Alt}(g \otimes f) = 0$. (b) Use (a) to show that $\operatorname{Alt}[\operatorname{Alt}(f \otimes g) \otimes h] = \operatorname{Alt}[f \otimes \operatorname{Alt}(g \otimes h)]$. (c) Show that $\operatorname{Alt}[\alpha \otimes [\operatorname{Alt}(\beta \otimes \gamma) - \beta \otimes \gamma]] = 0$ by (a).) (See [Spivak 65; p. 81].)
14. Let $f: V \to W$ be a linear transformation and let $f^*: W^* \to V^*$. Show by explicit formulas involving f and referring to f^* how f induces an algebra homomorphism (i) $Tf: \oplus_0^\infty B^m(W) \to \oplus_0^\infty B^m(V)$ such that $Tf \mid B^1(W) = f^*$. (ii) Show that the restric-

tion $Tf \mid \bigoplus_0^N \Lambda^m(W)$ induces an algebra homomorphism $\bigoplus_0^N \Lambda^m(W) \to \bigoplus_0^N \Lambda^m(V)$. (iii) Conclude that under our vector space embeddings "$EV^* \subset TV^*$", the same function Tf preserves the two different multiplications "$\otimes$" and "$\wedge$".

15. Let V and W be nonzero vector spaces over F (finite or infinite dimensional), and EV, EW, $E(V \oplus W)$, $E(V \otimes W)$ exterior algebras. Show that $E(V \otimes W) \ncong (EV) \otimes (EW)$ and that $E(V \otimes W) \ncong E(V \oplus W)$. Show that $\dim_F E(V \oplus W) = \dim_F (EV) \otimes (EW)$.

CHAPTER 6

Simple modules and primitive rings

Introduction

Unlike Chapters 2, 3, 4, and 5, in the next few chapters the ring need not have an identity. (i) Once an adequate notation and terminology is established, it is just as easy to prove everything for a general ring. (ii) If all our theorems were proved only for rings with an identity, they could not be applied to subrings or ideals or one sided ideals of a ring. However, some of the theorems proved here are useful tools for proving other theorems, for example by applying them to various subrings of a ring when such arise naturally in the course of some proof. (iii) Many interesting rings that occur naturally cannot have an identity as a matter of principle, e.g. nilrings, or rings of linear transformations on an infinite dimensional vector space having finite dimensional images. (iv) Some of the more influential previous authors have always traditionally treated these topics for rings without an identity. (v) Modules of the form R/L where L is a right ideal occur frequently in ring theory. To insist that R have an identity forces R/L to be cyclic, $R/L = (1 + L)R$, and the objects of study are trivialized. In the case when R is the ring of integers, R/L becomes a cyclic abelian group. To insist that the ring R should have an identity in general is the same as for an abelian group theorist to restrict his attention only to cyclic abelian groups. (vi) Some readers view ring and module theory as being a (central) part of the wider discipline of nonassociative rings. The latter, again sometimes as a matter of logic sometimes simply cannot contain multiplicative identities.

Unquestionably, the Jacobson Density Theorem (6-5.7) is one of the key theorems in this chapter if not this book. Like the Wedderburn Theorem, the Density Theorem gives conditions which guarantee that the ring is almost equal to the full ring of linear transformations of a vector space over a division ring, and in fact under some additional finiteness conditions is a full matrix ring (6-5.8, 6-5.10). It underscores the significant role played by the simple modules and primitive rings. We will use it later to give one of our several proofs of the Wedderburn Theorem. To this day the Density Theorem has spawned new versions of two types. The simple modules can be replaced by generalized simple modules and the division ring by a domain, particularly a right or left Ore domain.

6-1   Preliminaries

The notation, definitions, and facts contained in all of 1-3, and in 4-2.3 are presupposed and used.

6-1.1   Terminology. Here R is an arbitrary ring, and as before $R^1 = R$ if $1 \in R$ and $R^1 = \mathbb{Z} \times R$ otherwise.

A right R-module V is *cyclic* if there exists some $x \in V$ such that $V = xR$. Any $y \in V$ such that $V = yR$ is a *cyclic vector*.

6-1.2   Remarks. If $M = M_R$ is an R-module and $x \in M$ is arbitrary, then

(1) $xR = \{xr \mid r \in R\} \subseteq xR^1 = \{nx + xr \mid r \in R, \ n = 0, \ \pm 1, \ \pm 2, \dots\}$.
    Perhaps $xR \neq xR^1$. Moreover, $xR \neq xR^1 \Leftrightarrow x \notin xR$.

(2)   If $V = xR$ for some $x \in V$, then there exists an $e \in R$ such that $x = xe$. Also, $e^2 - e \in x^\perp$.

6-1.3   Definition. For any right ideal $L \leqslant R$, it will be convenient to introduce the notation $R : L = \{r \in R \mid Rr \subseteq L\}$, i.e. $R : L = (R/L)^\perp \lhd R$. The latter is mostly used when the module R/L is considered, whereas the former when the focus is on the right ideal L alone. Also for any $L \leqslant R$, in this chapter $\tilde{L}$ will denote $\tilde{L} = \{r \in R \mid rR \subseteq L\}$. Then $L \subseteq \tilde{L}$, $\tilde{L}R \subseteq L$, and $\tilde{L} \leqslant R$ is a right ideal. A right ideal L of R is *regular* if there exists some element $e \in R$ such that $er - r \in L$ for all $r \in R$. Then e is called a relative *left identity modulo* (the right ideal) L, or just a relative left identity. (Thus $L \leqslant R$ is regular if and only if $(e - 1)R \subseteq L$, where $e - 1 \in R^1$ and where $(e - 1)R$ is an expression written in terms of elements from R^1.

A right ideal $M < R$ is a *maximal right ideal* if $M \neq R$, and for any $K \leqslant R$ with $M \subseteq K$, either $K = M$, or $K = R$.

For any $M < R$, there is of course a one-to-one correspondence

$$\{K \mid M \subseteq K \leqslant R\} \leftrightarrow \text{Submodules of } R/M$$
$$K \leftrightarrow K/M.$$

Hence trivially, $M < R$ is maximal if and only if $R/M \neq 0$ has only two submodules (0) and R/M.

6-1.4   Terminology. For two right R-modules $A \leqslant B$.

A is *proper* iff $A \neq B$, i.e. if A is properly contained in B.

A is *maximal* (in B) iff $A \neq B$ and B/A has only the *trivial* submodules (0) and B/A.

A is *minimal* if $A \neq 0$, and A has only two submodules (0) and A.

Paradoxically a maximal submodule is always proper whereas a minimal one never zero. The above four terms apply to right ideals of a right R by viewing R as a right module over itself. (Some authors misuse proper to mean either that $A \neq 0$, or even that both $A \neq 0$ and $A \neq B$).

For any module V, in this chapter sometimes $V^0 \leqslant V$ will denote the submodule $V^0 = \{v \in V \mid vR = 0\}$. In particular when $V = R/L$, then $V^0 = \tilde{L}/L$.

If $1 \in R$, then every right ideal is regular; simply take $e = 1$.

6-1.5   Properties. Let $L \leqslant R$ be any regular right ideal and $e \in R$ a completely arbitrary left identity modulo L.

    (1) $\forall a \in L$, $e + a$ is a left identity modulo L.

    (2) e^2 is a left identity modulo L.

Proof. Since $e^2 - e \in L$, also $(e^2 - e)r \in L$ and hence $e^2r - r = (e^2 - e)r + er - r \in L$.

    (3) $e \in L \Leftrightarrow L = R$. Thus $L = R \Leftrightarrow e^2 \in L$.

    (4) $R^2 \nsubseteq L \Leftrightarrow e^2 \notin L$.

    (5) $\forall x \in R$, $x + L = ex + L$ in R/L. In particular, $x \in L \Leftrightarrow ex \in L$.

Proof. This follows from $x = ex + x - ex$, where always $x - ex \in L$.

    (6) $(e + L)^{\perp} = L$.

Proof. For all $r \in R$, $(e + L)(r = r + L$.

    (7) If $L \subset K \leqslant R$, then K is regular with the same left identity e as that of L.

    (8) $L \neq R \Leftrightarrow \tilde{L} \neq R$.

Preliminaries   89

Proof. If $L \neq R$, then $e^2 \notin L$. Hence $eR \nsubseteq L$, and $e \notin \tilde{L}$.

(9) In particular, if L is a regular maximal right ideal, then $L = \tilde{L}$.

(10) Every proper regular right ideal is contained in a maximal right ideal (which by (7) is automatically also regular).

Proof. Suppose that $e \notin L \neq R$. Under set inclusion, $\mathscr{S} = \{K \mid L \subseteq K < R,$ $e \notin K\}$ is a partially ordered set with $L \in \mathscr{S} \neq \varnothing$. If $\mathscr{C} \subseteq \mathscr{S}$ is a linearly ordered subset (= a chain), then $\cup \mathscr{C} = \cup \{K \mid K \in \mathscr{C}\} \in \mathscr{S}$. By Zorn's lemma there exists a maximal element $M \in \mathscr{S}$. If now $M \subset K \leqslant R$, then $K \notin \mathscr{S}$. Hence $e \in K$, and $K = R$. Thus $L \subseteq M$ is a maximal right ideal, which also is regular.

(11) A right ideal of R is a regular maximal right ideal if and only if it is a maximal regular right ideal.

Proof. Using obvious abbreviations for these two sets, always

$$\text{reg. (max. rt. ids.)} \subseteq \text{max. (reg. rt. ids.),}$$

while (10) gives the converse.

(12) The set of all relative left identities modulo L is exactly the coset $e + \tilde{L}$.

Proof. Suppose that $fr - r \in L$ for all $r \in R$. Then $(f - e)r = fr - r - er + r \in L$ shows that $f - e \in \tilde{L}$. Hence $f = e + (f - e) \in e + \tilde{L}$.

6-1.6 Basic Lemma. *For a regular right ideal $L < R$, if as before $R:L = \{r \in R \mid Rr \subseteq L\} \lhd R$, then*

(i) $R:L \subseteq L$.

(ii) $R:L$ *is the UNIQUE biggest ideal of R contained inside L.*

Proof. First, take any $e \in R$ such that $er - r \in L$ for all $r \in R$.

(i) If now also in addition $Rr \subseteq L$, then $er \in L$, and hence $r = r - er + er \in L$ too.

(ii) If $I \lhd R$ and $I \subseteq L$, then $RI \subseteq I \subseteq L$, and hence $I \subseteq R:L$. Thus $R:L \lhd R$ contains all ideals of R which are inside L.

The next corollary is an immediate consequence of the uniqueness above.

6-1.7   Corollary. *If* $L < R$ *is a regular right ideal, then*

$$L \lhd R \Leftrightarrow L = R : L.$$

6-2   Cyclic modules

6-2.1   Lemma. (i) *If* L *is a regular right ideal, then* R/L *is a cyclic R-module.*

(ii) *Conversely every cyclic R-module* V *is of the form* $V \cong R/L$ *for some regular right ideal.*

(iii) *Moreover, if* $V = xR$, $x = xe$, $e \in R$, *then* $V \cong R/x^{\perp}$ *where* $x^{\perp}$ *is regular with relative left identity* e.

Proof. (i) If $er - r \in L$ for all $r \in R$, then $(e + L)r = er + L = r + L$ shows that $e + L$ is a cyclic vector for the right R-module $R/L = (e + L)R$.

(ii) and (iii). If $V = xR$ for some $x \in V$, set $L = x^{\perp} < R$. Since $x \in V = xR$, there exists an $e \in R$ such that $x = xe$. Hence for any $r \in R$, $x(er - r) = 0$, and $er - r \in L$. The map $R_R \to V$, $r \to xr$ is a homomorphism of right R-modules whose kernel is L. Hence $R/L \cong xR = V$.

6-2.2   Lemma. *Cyclic modules* V_1 *and* V_2 *are isomorphic if and only if there exist cyclic vectors* $v_i \in V_i$ *for* $i = 1, 2$ *such that* $v_1^{\perp} = v_2^{\perp}$.

Proof. $\Leftarrow$: If $v_1^{\perp} = v_2^{\perp}$, then

$$V_1 \cong \frac{R}{v_1^{\perp}} = \frac{R}{v_2^{\perp}} \cong V_2.$$

$\Rightarrow$: Suppose that $\varphi : V_1 \to V_2$ is an isomorphism. If $V_1 = xR$, set $y = \varphi x$. Since φ is onto, $\varphi xR = yR = V_2$, and y is a cyclic vector for V_2. Since φ is one to one, $\varphi(xr) = 0$ if and only if $xr = 0$ for any $r \in R$. Then $\varphi(xr) = (\varphi x)r = yr$ shows that $x^{\perp} = y^{\perp}$.

6-2.3   Definition. For any right ideal $L < R$, the *idealizer* of L is the subring $N(L) = \{n \in R \mid nL \subseteq L\}$. Abbreviate $N = N(L)$ if L is fixed. Thus $N = \{n \mid L \subseteq n^{-1}L\}$ is the largest subring of R containing L as an ideal, $L \lhd N$.

In view of 6-2.1, the next theorem tells us what the endomorphism ring is for any cyclic module.

6-2.3 **Theorem.** *Let $L < R$ be a regular right ideal with relative left identity e. Let $L \subseteq \tilde{L} = \{r \mid rR \subseteq L\} \leqslant R$, and $\tilde{L} \subseteq N(L) \equiv N$ be as before. Define C to be the ring $C = \text{Hom}_R(R/L, R/L) = \text{End}_R(R/L)$. Then*

   (i) $\tilde{L} \lhd N$;

  (ii) $e + \tilde{L} = \mathbb{1} \in N/\tilde{L}$.

 (iii) There is a natural ring isomorphism $C \cong N/\tilde{L}$.

Proof. (i) Since $\tilde{L}R \subseteq L$, above $N\tilde{L}R \subseteq NL \subseteq L$. Thus $N\tilde{L} \subseteq \tilde{L}$, and $\tilde{L}$ is an $N - R$ bimodule of R.

(ii) For any $a \in L$, $ea = ea - a + a \in L$. So $e \in N$. For $n \in N$, in the ring N/L, $(e + L)(n + L) = en + L = n + L$ by 6-1.4(5). The exact same argument also works for $N/\tilde{L}$. Alternatively, any ring epimorphism such as $N/L \to N/\tilde{L}$, $n + L \to n + \tilde{L}$ maps left identities to left identities.

For any $r \in R$ and $n \in N$, $(n - ne)r = n(r - er) \in nL \subseteq L$. Thus $n - ne \in \tilde{L}$, or $n + \tilde{L} = ne + \tilde{L} = (n + \tilde{L})(e + \tilde{L})$ in the ring $N/\tilde{L}$.

(iii) Each $n \in N$ defines a unique and natural right R-homomorphism $T_n : R/L \to R/L$ simply by left multiplication of any coset representative $T_n(r + L) = nr + L$. Next, map $T : N \to C$ by $n \to T_n$; for n, $m \in N$, $T_{(n-m)} = T_n - T_m$ and $T_{nm} = T_n T_m$ because endomorphisms act on the opposite side than R. (Also $T_e(r + L) = er + L = r + L$, and $T_e = \mathbb{1} \in C$.) Then kernel $T = \{n \in N \mid nR \subseteq L\} = \tilde{L}$. Thus so far $N/\tilde{L} \cong T(N) \subseteq C$ as rings.

For any $\beta \in C$, select any element $b \in R$ (unique only up to L) satisfying $\beta(e + L) = b + L \in R/L$. It is claimed that $b \in N$. For any $\lambda \in L$,

$$[\beta(e + L)]\lambda = (b + L)\lambda = b\lambda + L, \text{ and also}$$

$$[\beta(e + L)]\lambda = \beta[(e + L)\lambda] = \beta(e\lambda + L) = \beta(\lambda + L) = \beta(L) = 0.$$

Thus $b\lambda \in L$ and $bL \subseteq L$. Next, it is asserted that $T_b = \beta$. For any $r \in R$, since $r + L = er + L = (e + L)r$, it follows that $\beta(r + L) = \beta[(e + L)r] = [\beta(e + L)]r = br + L$. Thus $\beta = T_b$. Thus $N/\tilde{L} \cong T(N) = C$.

6-3   Simple modules

6-3.1 **Definition.** An R-module V is *simple* if $VR \neq 0$, and if (0) and V are the only submodules of V.

6-3.2 **Consequences.** Suppose that V is any simple module.

  (1) Since $0 \neq VR \leqslant V$, $VR = V$.

(2) If $V^0 = \{z \in V \mid zR = 0\} \neq 0$, then $V^0 = V$, and hence $VR = 0$, a
contradiction. Thus $V^0 = 0$. (3) For any $0 \neq x \in V$, since $x \notin V^0$,
$0 \neq xR \leqslant V$. Hence $V = xR$.

6-3.3   Lemma. *Let V be a nonzero right R-module and let $L < R$ be a*
right ideal. Then

(i) V *is simple* $\Leftrightarrow \forall\, 0 \neq x \in V$, x *is a cyclic vector.*
(ii) *Assume for* (ii) *that V is simple. Then for any $0 \neq x \in V$, $x^{\perp}$ is a*
regular maximal right ideal and $V \cong R/x^{\perp}$.
(iii) *If $M < R$ is a regular maximal right ideal, then R/M is simple.*
(iv) *R/L is simple* $\Leftrightarrow L$ *is a maximal right ideal and $R^2 \nsubseteq L$.*

Proof. Remark. In (iv), L need not be regular. (i) $\Leftarrow$: The fact that for
any $0 \neq x \in V$, $0 \neq x \in xR = V$ shows that every nonzero submodule of V
is all of V. This also shows that $xR \subseteq VR \neq 0$, and hence that V is simple.

(ii) By (i), for any $0 \neq x \in V$, $V = xR$. The kernel of $R \to xR$, $r \to xr$, is
the regular right ideal $x^{\perp} < R$, by 6-2.11(iii). Since $V \cong R/x^{\perp}$ has no
nontrivial submodules, $x^{\perp}$ is maximal.

(iii) By 6-1.4, $0 \neq e^2 + M = (e + M)e \in (R/M)R$. Thus RM is simple.

(iv) The module $V = R/L$ has only the two trivial submodules if and
only if $L < R$ is maximal. Since $VR = (R/L)R = (R^2 + L)/L$, $VR \neq 0$ if
and only if $R^2 \nsubseteq L$.

6-3.4   Schur's lemma. *The endomorphism ring* $\mathrm{End}_R V$ *of a simple mod-*
ule V is a division ring.

Proof. It suffices to show that every nonzero R-homomorphism $\alpha: V \to V$
has a left inverse α^{-1} such that $\alpha^{-1}(vr) = (\alpha^{-1}v)r$ for all $v \in V, r \in R$. Since
$\alpha \neq 0$, $\alpha V \neq 0$. But a nonzero submodule $0 \neq \alpha V \leqslant V$ of a simple module
V is all of $V = \alpha V$. Again, since $\alpha V \neq 0$, kernel $\alpha = \alpha^{-1}0 \neq V$ is a proper
submodule, and hence $\alpha^{-1}0 = 0$.

The argument from now on below holds more generally for any
module V and any module map $\alpha: V \to V$ which is one to one and onto.
The left inverse α^{-1}, defined by $\alpha^{-1}v = w$ if $\alpha w = v$, is an additive abelian
group homomorphism. For $v = \alpha w \in V$, $\alpha^{-1}[(\alpha w)r] = \alpha^{-1}[\alpha(wr)]$ because
α is an R-map. Thus $\alpha^{-1}(vr) = wr = (\alpha^{-1}v)r$. Hence $\alpha^{-1} \in \mathrm{End}_R V$.

6-3.5   Lemma. *Suppose that $M \lhd R$ and that R/M is a simple module.*
Then

(i) $R: M = M$; *and if*

(ii) $L < R$ such that $R: L \subseteq L$, and $R/L \cong R/M$, then $L = M$.

Proof. (i) Since $M \lhd R$, $M \subseteq R: M$. If $x \in (R: M)\backslash M$, then $x + M \in R/M$ is a cyclic vector. Thus $R/M = (x + M)R = xR + M = R$. But then $R^2 = R(xR + M) \subseteq M$, a contradiction. Thus $R: M = M$.

(ii) Since isomorphic modules have equal annihilators, by (i), $M = (R/M)^\perp = (R/L)^\perp = R: L$. Since by hypothesis $M = R: L \subseteq L < R$, and since both M and L are maximal, $M = L$.

6-3.6   Proposition. *For a fixed simple module, let $\mathscr{S}_1 \subseteq \mathscr{S}_2$ be the nonempty sets*

$$\mathscr{S}_1 = \{L < R \mid R/L \cong V, L \text{ is regular}\} \quad \text{and}$$
$$\mathscr{S}_2 = \{L < R \mid R/L \cong V, R: L \subseteq L\}.$$

Then the following hold.

(i) $\exists I \in \mathscr{S}_2, I \lhd R \Rightarrow \mathscr{S}_1 = \mathscr{S}_2 = \{I\}$ is a singleton.
(ii) $\mathscr{S}_1 = \{I\}$ is a singleton $\Rightarrow \mathscr{S}_1 = \mathscr{S}_2 = \{I\}$ and $I \lhd R$.

Proof. (i) For any $L \in \mathscr{S}_2$, since $R/I \cong R/L$, $I = (R/I)^\perp$ by the last lemma. Thus $I = (R/L)^\perp \subseteq L$ are two maximal right ideals of R, and hence $I = L$.

(ii) For any $x \in R\backslash I$, since $x^{-1}I \in \mathscr{S}_1$, $x^{-1}I = I$. Thus $xI \subseteq I$ for all $x \in R$, and $I \lhd R$. By (i), $\mathscr{S}_1 = \mathscr{S}_2$.

6-3.7   Corollary 1. *For V and $\mathscr{S}_i$ as above for either $i = 1$ or 2:*

$\mathscr{S}_i$ *contains an ideal* $\Leftrightarrow \mathscr{S}_i$ *is a singleton.*

Proof. If $|\mathscr{S}_2| = 1$, then $\varnothing \neq \mathscr{S}_1 \subseteq \mathscr{S}_2$ and $|\mathscr{S}_1| = 1$. The rest is clear.

6-3.8   Corollary 2. *Let D be a commutative ring and $L \lhd D$, $K \lhd D$ ideals.*

(i)   *If both L and K are regular, then*

$$D/L \cong D/K \Leftrightarrow L = K$$

(ii) *If both D/L and D/K are simple D-modules, then*

$$D/L \cong D/K \Leftrightarrow L = K.$$

Proof. (i) If $e \in D$ is a relative left identity of L, then by the commutativity of D we see from $D/L = (e + L)D$ that $L = (e + L)^{\perp} = (D/L)^{\perp}$. Thus $L = (D/L)^{\perp} = (D/K)^{\perp} = K$.

(ii) By Lemma 6-3.5, $(D/L)^{\perp} = L$ and $(D/K)^{\perp} = K$. Hence again if $D/L \cong D/K$, then $L = K$.

6-4   Examples

A generally useful notation for matrix ring examples and counterexamples is established.

6-4.1   Notation. The usual matrix units with all zeros except for a one in row i and column j will be denoted by e_{ij} for $n \times n$ matrices for whatever n is being used. If E, F, G, H are abelian groups or modules, then the additive group of all 2×2 matrices with first row in E, F and second row in G, H respectively will sometimes be denoted by $[E, F; G, H]$. Elements will be written as $[e, f; g, h] = e_{11}e + e_{12}f + e_{21}g + e_{22}h$, where $e \in E$, $f \in F$, $g \in G$, and $h \in H$. If $A \leqslant E$, $B \leqslant F$, $C \leqslant G$, and $D \leqslant H$ are subgroups or submodules, then

$$[E, F; G, H]/[A, B; C, D] \to [E/A, F/B; G/C, H/D],$$

$$[d, f; g, h] + [A, B; C, D] \to [d + A, f + B; g + C, h + D]$$

is an isomorphism. All of this extends in an obvious way to $n \times n$ matrices.

The converse of Schur's Lemma is false as the following examples show.

6-4.2   The additive rationals $V = \mathbb{Q}$ are a nonsimple $R = \mathbb{Z} = 0, \pm 1, \pm 2, \ldots$ module, yet $\mathrm{End}_R V = \mathbb{Q}$ is a field. Although $\mathbb{Q}$ is not finitely generated as a $\mathbb{Z}$-module, in the next example the module is cyclic.

6-4.3   Example. For any commutative field F, let R be the lower triangular matrix ring $R = [F, 0; F, F]$. Let $L < K < R$ be the right ideals $L = [F, 0; 0, 0] \subset K = [F, 0; F, 0]$. Define V to be the right R-module $V = [0, 0; F, F]$ where R acts on the right of V simply by matrix multiplication. Then $R/L \cong V$ as R-modules. The idealizer of L is $N(L) = [F, 0; 0, F]$. Since $1 \in R$, $L = \tilde{L}$, and $\mathrm{End}_R R/L \cong N(L)/L \cong [0, 0; 0, F]$. The latter acts on V not by componentwise multiplication by an element of F but by left matrix multiplication. Thus $\mathrm{End}_R V \cong F$ is a field, yet $V = [0, 0; 0, 1]R$ is a nonsimple cyclic module with a submodule $K/L \cong [0, 0; F, 0] < V$.

The reader may alternatively wish to visualize this example as follows:

$$R = \begin{vmatrix} F & 0 \\ F & F \end{vmatrix} > K = \begin{vmatrix} F & 0 \\ F & 0 \end{vmatrix} > L = \begin{vmatrix} F & 0 \\ 0 & 0 \end{vmatrix}; \quad N(L) = \begin{vmatrix} F & 0 \\ 0 & F \end{vmatrix};$$

$$\frac{R}{L} \cong V = \begin{vmatrix} 0 & 0 \\ F & F \end{vmatrix}; \quad \frac{K}{L} \cong W = \begin{vmatrix} 0 & 0 \\ F & 0 \end{vmatrix} < V;$$

$$\frac{N(L)}{L} \cong \mathrm{End}_R V = \begin{vmatrix} 0 & 0 \\ 0 & F \end{vmatrix};$$

$$VR \subseteq V, \quad WR \subseteq W, \quad (\mathrm{End}_R V)V \subseteq V, \quad N(L)V \subseteq V.$$

In the next example R/L is simple where $L < R$ is not regular.

6-4.4   Example. (a) Set $\mathbb{Z}_4 = \mathbb{Z}/4\mathbb{Z}$. Take

$$L = \begin{vmatrix} \mathbb{Z}_4 & 2\mathbb{Z}_4 \\ 2\mathbb{Z}_4 & 2\mathbb{Z}_4 \end{vmatrix} < R = \begin{vmatrix} \mathbb{Z}_4 & 2\mathbb{Z}_4 \\ \mathbb{Z}_4 & 2\mathbb{Z}_4 \end{vmatrix}.$$

The two element module $R/L = \{L, \ e_{21} + L\}$ is simple, because $0 \neq e_{21} + L = (e_{21} + L)e_{11} \in (R/L)R$. Since $R \backslash L = \{e_{21}\}$ but $e_{21}e_{11} - e_{11} \notin L$, L is not regular.

(b) We know that $R/L \cong R/M$, where $M = (e_{21} + L)^{\perp} = [2\mathbb{Z}_4, 2\mathbb{Z}_4; \mathbb{Z}_4, 2\mathbb{Z}_4]$ is a regular maximal right ideal with relative left identity e_{11}, $(e_{11} - 1)R \subseteq M$. Since $M \lhd R$, $N(M) = R$, and $N(M)/M \cong \mathbb{Z}_4/2\mathbb{Z}_4$ is a field. Concretely, the R-isomorphism $R/L \to R/M = \{M, e_{11} + M\}$ is $e_{21} + L \to e_{11} + M$. Alternatively, the two cyclic modules $R/L = (e_{21} + L)R$ and $R/M = (e_{11} + M)R$ are isomorphic precisely because $(e_{21} + L)^{\perp} = (e_{11} + M)^{\perp} = M$.

(c) Since $R/L \cong R/M$, $(R/L)^{\perp} = (R/M)^{\perp} = M$, because $M \lhd R$ (see 6-1.7). Thus $R : L \nsubseteq L$.

6-5   Density

First, a short and simple proof of Jacobson's density theorem is given, and some corollaries are derived.

6-5.1   Notation. For any right R-module V, set $D = \mathrm{End}_R V$ and view

V as a left D right R-bimodule $V = {}_DV_R$. Let $X \subseteq V$ and $Y \subseteq R$ be any subsets. As before, $X^\perp = \{r \in R \mid xr = 0, \text{ all } x \in X\} \leqslant R$ is the annihilator right ideal of X. The kernel of Y is the additive subgroup $Y^\# = \{v \in V \mid vy = 0, \text{ all } y \in Y\}$. If $\bar{X}$ and $\bar{Y}$ are the additive subgroups generated by X and Y, then $\bar{X}^\perp = X^\perp$ and $\bar{Y}^\# = Y^\#$. If X is a submodule, then $X^\perp \lhd R$ is an ideal. If Y is a left ideal of R then $Y^\# \leqslant V$ is a submodule.

The two maps "$\perp$" and "$\#$" are

(1) inclusion reversing,
(2) closure operations in the sense that
     a) $X \subseteq X^{\perp\#}$, $Y \subseteq Y^{\#\perp}$ and
     b) $X^\perp = X^{\perp\#\perp}$, $Y^\# = Y^{\#\perp\#}$

Proof. From the descriptions $X^{\perp\#} = \{v \in V \mid \forall r \in R, \; Xr = 0 \to vr = 0\}$, and $Y^{\#\perp} = \{r \in R \mid \forall v \in V, vY = 0 \to vr = 0\}$ first conclude that a) $X \subseteq X^{\perp\#}$ and b) $Y \subseteq Y^{\#\perp}$. Application of (1) to the latter gives $X^\perp \supseteq X^{\perp\#\perp}$ and $Y^\# \supseteq Y^{\#\perp\#}$. Next substitution

$$X = Y^\# \text{ in } X \subseteq X^{\perp\#} \text{ gives } Y^\# \subseteq Y^{\#\perp\#}, \text{ and}$$

$$Y = X^\perp \text{ in } Y \subseteq Y^{\#\perp} \text{ that } X^\perp \subseteq X^{\perp\#\perp}.$$

(3) $(DX)^\perp = X^\perp$, in particular for any $x \in V$, $(Dx)^\perp = x^\perp$.
(4) For any additive subgroups $\{W_i\}$ of V, $(\Sigma W_i)^\perp = \cap(W_i^\perp)$.

6-5.2   Key Lemma. *For any simple R-module $V = V_R$, let $x \in V$ and $W \subset V$ be any $D = \mathrm{End}_R V$ – subspace of V of D-dimension* $\dim_D W$. *Then*

$$\text{if } \dim_D W \neq \infty, \quad x \notin W \Rightarrow xW^\perp = V.$$

Proof by contradiction. For a fixed V select a subspace W for which the lemma fails such that $\dim_D W = n$ is a minimum. This means that there exists some $x \in V$, $x \notin W$ such that $xW^\perp \neq V$. Since $xW^\perp \subset V$ is a submodule, $xW^\perp = 0$.

If $W = (0)$, then $W^\perp = R$, $0 \neq x \notin W$, and $V = xR = xW^\perp$ because V is simple. Therefore $1 \leqslant n = \dim_D W$, and hence there exists $0 \neq y \in W$. Write $W = Dy \overset{\circ}{+} U$, where $U \subset W$ is some D vector subspace, and where "$\overset{\circ}{+}$" denotes a D-vector space direct sum only. (I.e. $DU \subseteq U$, but perhaps $UR \nsubseteq U$, and $yR \nsubseteq Dy$). Since $\dim_D U = n - 1$, the lemma is true for U, by the minimality of n. Since $y \notin U$, $yU^\perp = V$.

Any $v \in V = yU^\perp$ is of the form $v = y\lambda$ for some $\lambda \in U^\perp$. Define a function $d: V \to V$ by $dv = d(y\lambda) = x\lambda$. If also $v = y\lambda = y\lambda_1$, $\lambda_1 \in U^\perp$, then $y(\lambda - \lambda_1) = 0$. In this case $\lambda - \lambda_1 \in y^\perp \cap U^\perp = (Dy)^\perp \cap U^\perp = (Dy + U)^\perp = W^\perp$. But $xW^\perp = 0$, and hence $x(\lambda - \lambda_1) = 0$, or $d(y\lambda_1) = x\lambda_1 = x\lambda = d(y\lambda)$. So far d is a well defined abelian group homomorphism. If $r \in R$, then also $\lambda r \in U^\perp < R$, and hence $d(vr) = d[(y\lambda)r] = d[y(\lambda r)] = x\lambda r = (dv)r$. Thus $d \in \mathrm{End}_R V$.

(Note that $y = ya$ for some $a \in U^\perp$, and that $dy = xa$.) For $\lambda \in U^\perp$ arbitrary, $(x - dy)\lambda = x\lambda - (dy)\lambda = x\lambda - d(y\lambda) = x\lambda - x\lambda = 0$. Thus $(x - dy)U^\perp = 0$. If $x - dy \notin U$, then since U satisfies the lemma, we would have to have $(x - dy)U^\perp = V$. Thus $x - dy \in U$, or $x \in dy + U \subset Dy + U = W$, a contradiction, because $x \notin W$. Thus W does not exist and the lemma holds.

6-5.3   Definition. A ring R is right *primitive* if there exists a right R-module V which is (i) simple and (ii) faithful. For simplicity here 'right' will be omitted and the term 'primitive' will be understood to mean right primitive.

6-5.4   Remark. There do exist right but not left primitive rings, see [Bergman 64].

6-5.5   Observation. A commutative ring C is primitive if and only if C is a field.

Proof. $\Leftarrow$: In general, if D is a division ring, then D is a simple faithful right D-module.

$\Rightarrow$: Any faithful C-module is isomorphic to one of the form $V = C/P$ for some regular $P \lhd C$ with $P \subseteq (C/P)^\perp = (0)$. If e is the relative identity of P, then $ea - a \in P = (0)$ for all $a \in C$. This shows that $ae = ea = a$, or that $e = 1 \in C$. Identify $V = C/(0) = C$. For any $0 \neq a \in V$, $aC = V$ implies that $ab = 1$ for some $b \in C$. Thus C is a field.

6-5.6   Definition. Let V be a left D right R-bimodule where D is some division ring. Then R is *n-transitive* with respect to D for some integer n if

$$
\left.
\begin{array}{l}
\forall\ y_1, \ldots, y_n \in V \text{ and} \\
\forall\ D\text{-linearly independent} \\
x_1, \ldots, x_n \in V
\end{array}
\right\}
\Rightarrow
\begin{cases}
\exists\, r \in R \text{ such that } x_i r = y_i \\
\text{for all } 1 \leqslant i \leqslant n.
\end{cases}
$$

The ring R is *dense* if it is n-transitive for all n.

The Density Theorem is an example of a class of results which could be classified as "characterization theorems". They take some nebulous, abstract property like 'primitive' and reduce it to something very familiar like a matrix ring.

6-5.7   Density theorem (N. Jacobson). *For any ring R and any simple right R-module V, view V as a left vector space over the division ring $D = \mathrm{End}_R V$. Then*

(i)   *R is dense on V with respect to D.*

(ii) *$R/V^\perp$ is isomorphic to a dense ring of D-linear transformations on V.*

(iii) *In particular, if R is primitive and V is a faithful simple right R-module, then R is isomorphic to a dense ring of D-linear transformations on V.*

Proof. Given n, x_i, and y_i as in the prevfous definition, form the D-vector space direct sums $W_i = \Sigma_{j=1}^{n} \overset{\circ}{+} Dx_j = Dx_1 + \cdots + \widehat{Dx_i} + \cdots + Dx_n$, where "$\widehat{}$" means the term below is missing. Since $x_i \notin W_i$, the Key Lemma shows that $x_i W_i^\perp = V$. Since $y_i \in V = x_i W_i^\perp$, there exists $r_i \in W_i^\perp$ such that $x_i r_i = y_i$ and $x_j r_i = \delta_{ji} x_i$ where $\delta_{ji} = 1$ if $j = i$ and zero otherwise. Then $r = r_1 + \cdots + r_n$ is the required element, because $x_i r = x_i r_1 + \cdots + x_i r_i + \cdots + x_n r_i = 0 + \cdots + y_i + \cdots + 0$. Hence R is dense on V.

6-5.8   Corollary 1 to the density theorem. *Assume that the simple faithful R-module V above is of finite dimension $\dim_D V = n \not\leqq \infty$ over D. Then $R \cong \mathrm{End}_D V \cong D_n$, the $n \times n$ matrix ring over D.*

6-5.9   Lemma. *For an algebra A over a field F suppose that V is a right A-algebra module, i.e. V is a right module over the ring A and V is a right F-vector space such that $(vc)a = v(ca)$ for all $v \in V$, $c \in F$, $a \in A$. Set $E = \mathrm{End}_A V$ where A is viewed as a ring without any regard to F. If $VA = V$, then $F \subseteq \mathrm{center}\, D$.*

Proof. Each $c \in F$ defines an invertible element $c \in D$ by the definition $c \colon V \to V$, $v \to vc$. Thus $F \subset D$.

Next, suppose that $d \in D$, $c \in F$ are arbitrary, but that $y \in V$ is of the special form $y = xa$ for some $x \in V$, $a \in A$. Then by our definition of $c \in D$, $cdy = (dy)c$ while $dcy = d(yc)$. Then

$$[dxa]c = (dx)(ac) = d[x(ac)] = (dc)(xa),$$

or $cdy = dcy$. The same holds for an arbitrary $y = \Sigma\, x_i a_i \in V$; $x_i \in V$, $a_i \in A$. Thus $cd = dc$.

6-5.10 Corollary 2 to the density theorem. Suppose that the ring R is an algebra $R = A$ over a field F, and that V is a simple faithful right module over the ring A, and as before that $D = \mathrm{End}_A V$. In addition also assume that

(a) V is an A-algebra module.

If

(b) $\dim_F V = m < \infty$, and
(c) F is algebraically closed,

then

(i) $D = F$, and
(ii) $A \cong F_m$.

Proof. It follows from $\dim_F V = (\dim_D V)(\dim_F D)$ that $\dim_F D \leqslant m$ is finite. Consequently for any $\theta \in D$, the commutative ring of polynomials $F[\theta]$ in powers of θ with coefficients in F is finite dimensional over F. Hence $F \subseteq F[\theta]$ is a finite algebraic field extension of F. Thus $\theta \in F$ and $D = F$. But then $\dim_D V = m$ implies that $A \cong D_m = F_m$ by the last corollary.

6-6   More on density and simples

In the next proposition, an alternate independent and self contained proof of the Key Lemma (6-6.6) and hence indirectly its consequence, the Density Theorem is given. The somewhat greater length of the subsequent proof is compensated not only by the greater generality of its conclusions, but also by the new ideas in its proof.

6-6.1   Definition. A *semi-endomorphism* of a module V is any homomorphism of a submodule of V back into V itself. A module V is called *quasi-injective* if every semi-endomorphism of V extends to an endomorphism of V. I.e. pictorially

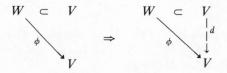

Clearly, every injective module and every simple module is quasi-injective.

In any unparenthesized formula involving the symbol $\perp$ it will be fUnderstood that $\perp$ governs as little as possible, e.g. $(M \cap Dx^\perp)^\# = [M \cap (Dx)^\perp]^\#$, or $M \cap Dx^\perp = M \cap (Dx^\perp)$.

6-6.2   Lemma. *Suppose that V is any module, D its endomorphism ring, that $M \leqslant R$ is any right ideal, and $0 \neq x \in V$ an element. Then*

(i) $M^\# + Dx \subseteq (M \cap Dx^\perp)^\#$.

(ii)   *If $xM = 0$, then $M^\# + Dx = (M \cap Dx^\perp)^\#$.*

Proof. (i) Since $M \cap Dx^\perp$ is a subset of both M and $Dx^\perp$, and since the operation $\#$ is inclusion reversing, it follows that $M^\# \subseteq (M \cap Dx^\perp)^\#$, and $(Dx^\perp)^\# \subseteq (M \cap Dx^\perp)^\#$. The latter is a subgroup of V, and thus also $M^\# + Dx \subseteq M^\# + (Dx^\perp)^\# \subseteq (M \cap Dx^\perp)^\#$.

(ii) The relation that $DxM = 0$ can be read two ways. First, $M \subseteq Dx^\perp$, and hence $M = M \cap Dx^\perp$. Thus $M^\# = (M \cap Dx^\perp)^\#$. Secondly, $Dx \subseteq M^\#$, and thus $Dx + M^\# = M^\# = (M \cap Dx^\perp)^\#$.

6-6.3   Proposition. *Consider any quasi-injective right R-module V and its endomorphism ring $D = \mathrm{End}_R V$. Then*

$$\forall\, M \leqslant R, \quad \forall\, 0 \neq x \in V \Rightarrow (M \cap Dx^\perp)^\# = M^\# + Dx.$$

Proof. By the last lemma, $M^\# + Dx \subset (M \cap Dx^\perp)^\#$ as well as $0 \neq xM \leqslant V$. Let $z \in (M \cap Dx^\perp)^\#$ be any arbitrary element. Any $v \in xM$ is of the form $v = x\lambda$, $\lambda \in M$. Let also $v = x\lambda_1$, $\lambda_1 \in M$. Define a function $\varphi: xM \to V$ by $\varphi v = z\lambda$. Then $x(\lambda - \lambda_1) = 0$, $\lambda - \lambda_1 \in M \cap x^\perp = M \cap Dx^\perp$. But $z \in (M \cap Dx^\perp)^\#$ means that $z(\lambda - \lambda_1) = 0$. Thus $\varphi(x\lambda_1) = z\lambda_1 = z\lambda$ is a well defined semi-endomorphism which by the quasi-injectivity of V extends to an endomorphism $d: V \to V$ whose restriction $d\,|\,xM$ to xM is $d\,|\,xM = \varphi$; i.e. pictorially

$$
\begin{array}{ccc}
xM & \xrightarrow{\;x\lambda \to z\lambda\;} & zM \subset M^\# \\
\big\downarrow & & \big\downarrow \\
V & \dashrightarrow{\;d\;} & V
\end{array}
$$

It is asserted that $z - dx \in M^\#$. For if $\lambda \in M$, then $(z - dx)\lambda = z\lambda - d(x\lambda) = 0$. Thus $z \in dx + M^\# \subseteq M^\# + Dx$, or $(M \cap Dx^\perp)^\# \subseteq M^\# + Dx$. Hence $(M \cap Dx^\perp)^\# = M^\# + Dx$.

The next corollary says among other things that any finite dimensional extension like $W^{\perp \#} + Dx$ of the closed left D-submodule $W^{\perp \#}$ remains closed.

6-6.4   Corollary 1. *For any quasi injective R-module and any left D-submodule $W \subseteq V$, where $D = \mathrm{End}_R V$, and any $x \in V$,*

$$(W + Dx)^{\perp \#} = W^{\perp \#} + Dx.$$

Proof. In the proposition, take $M = W^{\perp}$ (Since $WR \nsubseteq W$ perhaps, W may not be an R-submodule, and then M need not be an ideal.) Then by 6-5.1(4), $(W + Dx)^{\perp} = W^{\perp} \cap (Dx)^{\perp}$. Hence with $M = W^{\perp}$, $(W + Dx)^{\perp \#} = [(W^{\perp}) \cap Dx^{\perp}]^{\#} = W^{\perp \#} + Dx$.

6-6.5   Corollary 2. *For $V = {}_DV_R$ as in the last proposition, set $V^{\upsilon} = R^{\#} = \{v \in V \mid vR = 0\} \leqslant V$. Then for any $0 \leqslant n$, and any $x_1, \dots, x_n \in V$,*

(i) $(Dx_1 + \cdots + Dx_n)^{\perp \#} = V^0 + Dx_1 + \cdots + Dx_n$.

(ii)   *In particular, if V is a faithful R-module, then $(Dx_1 + \cdots + Dx_n)^{\perp \#} = Dx_1 + \cdots + Dx_n$.*

Proof. (i) When $x_1 = \cdots = x_n = 0$, or $n = 0$, $(0)^{\perp} = R$ and $(0)^{\perp \#} = R^{\#} = V^0$. Also, $V^{0 \perp \#} = R^{\# \perp \#} = R^{\#} = V^0$.

Next, in the last corollary, take $W = V^0 = W^{\perp \#}$. Then $(V^0 + Dx_1)^{\perp \#} = W^{\perp \#} + Dx_1 = V^0 + Dx_1$. Next, repeat with $W = V^0 + Dx_1 = W^{\perp \#}$ to conclude that $(V^0 + Dx_1 + Dx_2)^{\perp \#} = W^{\perp \#} + Dx_2 = V^0 + Dx_1 + Dx_2$. The proof now follows this same way by induction.

6-6.6   Corollary 3. *If $VR \neq 0$ and V is quasi-injective over R, and ${}_DW \subset V$ is a finitely generated left $D = \mathrm{End}_R V$-module, then for any $x \in V$,*

$$x \notin W \Rightarrow xW^{\perp} \neq 0.$$

Proof. By the previous corollary, $x \notin W = W^{\perp \#}$, i.e. $xW^{\perp} \neq 0$.

The notation "$\frown$" below means that the corresponding term underneath this symbol is missing. For the case of a simple module, the left side of the condition in the next corollary for all $j = 1, \dots, n$ merely says that $x_1, \dots, x_n$ are independent over the skew field D.

6-6.7   Corollary 4. *For a quasi-injective right R-module V with $VR \neq 0$,* *suppose that* $x_1, \ldots, x_n \in V$ *is any finite set of elements. Then for any* j

$$x_j \in Dx_1 + \cdots + \widehat{Dx_j} + \cdots + Dx_n \Leftrightarrow \bigcap_{\substack{i=1 \\ i \neq j}}^{n} x_i^{\perp} \subseteq x_j^{\perp}.$$

Proof. $\Leftarrow$; By hypothesis, there exists an $r \in \bigcap_{i \neq j} x_i^{\perp} \setminus x_j^{\perp} \neq \varnothing$ with $x_j r \neq 0$. If $x_j = \Sigma_{i \neq j} d_i x_i$ for some $d_i \in D$, then $0 \neq x_j r = \Sigma_{i \neq j} d_i x_i r = 0$ is a contradiction.

$\Rightarrow$: First note that $x_j \in Dx_j = (Dx_j)^{\perp \#} = x_j^{\perp \#}$, and that $\cap_{i \neq j} x_i^{\perp} = \cap_{i \neq j} Dx_i^{\perp} = (Dx_1 + \cdots + \widehat{Dx_j} + \cdots + Dx_n)^{\perp}$, and hence that $(\cap_{i \neq j} x_i^{\perp})^{\#} = Dx_1 + \cdots + \widehat{Dx_j} + \cdots + Dx_n$. Suppose that $\cap_{i \neq j} x_i^{\perp} \subseteq x_j^{\perp}$. The inclusion reversing property of " $\#$ " now gives the contradiction that $Dx_j = x_j^{\perp \#} \subseteq (\cap_{i \neq j} x_i^{\perp})^{\#} = Dx_1 + \cdots + \widehat{Dx_j} + \cdots + Dx_n$.

6-6.8   Proposition. *Let* V *be a simple right* R-*module and* $D = \mathrm{Hom}_R(V, V)$. *Suppose that for* $1 \leqslant n$, $x_1, \ldots, x_n \in V$ *are* D-*linearly independent. Then*

$$\mathrm{Hom}_R \left[\frac{R}{\bigcap\limits_{i=1}^{n} x_i^{\perp}}, \frac{R}{\bigcap\limits_{i=1}^{n} x_i^{\perp}} \right] \cong D_n.$$

Proof. We first define a right ideal $L < R$, and two subrings $N(L)$, $N(W) \subseteq R$:

$$L = \bigcap_{i=1}^{n} x_i^{\perp}; \quad N(L) = \{n \in R \mid nL \subseteq L\};$$

$$W = \sum_{i=1}^{n} Dx_i; \quad N(W) = \{r \in R \mid Wr \subseteq W\}.$$

By the density theorem, there exists an element $e \in R$ such that $x_i e = x_i$ for all i. Thus $er - r \in L$ for all $r \in R$. It is claimed that $\tilde{L} = L$, where $\tilde{L} = \{a \in R \mid aR \subseteq L\}$. For if $aR \subseteq L$, then $x_i aR \subseteq x_i L = 0$, or $x_i aR = 0$. If $x_i a \neq 0$, then $x_i aR = V$ because V is simple. So for all i, $x_i a = 0$ 'and hence $a \in L$. Thus $\tilde{L} = L$. Now by 6-2.3,

$$\mathrm{Hom}_R \left(\frac{R}{L}, \frac{R}{L} \right) \simeq \frac{N(L)}{\tilde{L}} = \frac{N(L)}{L}.$$

Note that $\cap_1^n x_i^{\perp} = \cap_1^n Dx_i^{\perp} = (Dx_1 + \cdots + Dx_n)^{\perp}$, or that $L = W^{\perp}$. Thus $W = W^{\perp \#} = L^{\#} = \{w \in V \mid wL = 0\}$.

Since $WN(W) \subseteq W$, $WN(W)L \subseteq WL = 0$. Thus $N(W)L \subseteq W^{\perp} = L$, or

$N(W) \subseteq N(L)$. For the converse, from $N(L)L \subseteq L$ follows $WN(L)L \subseteq WL = 0$. Thus $WN(L) \subseteq L^{\#} = W$; or $N(L) \subseteq N(W)$. Thus $N(W) = N(L)$.

It is claimed that $N(W)/L \cong D_n$. Map

$$\rho: N(W) \to \operatorname{Hom}_D(W, W) \cong D_n$$

$$a \to \rho a = \rho_a$$

where for any $w \in W$, $w\rho_a = wa$. For $a, b \in N(W)$, $\rho_{ab} = \rho_a \rho_b$. For any $y_1, \ldots, y_n \in W$, the density theorem gives an element $a \in R$ such that $x_i a = x_i \rho_a = y_i$ for all i. Thus the map ρ is onto. Its kernel is kernel $\rho = L$. Thus $N(W)/L \simeq \rho N(W) = \operatorname{Hom}_D(W, W)$. Finally, after combining our two ring isomorphisms,

$$\operatorname{Hom}_R \left(\frac{R}{L}, \frac{R}{L} \right) \cong \frac{N(L)}{L} = \frac{N(W)}{L} \cong D_n.$$

6-6.9   **Transitivity.** Consider an algebra A over a commutative field F (with $1 \in A$ or $1 \notin A$) and an algebra right A-module V. Set $Q = \operatorname{End}_A V$ and assume that $VA = V$ so that $F \subseteq \operatorname{center} Q$ by 6-5.9. Here "transitive" will mean with respect to F (i.e. $D = F$ in 6-5.6.) Then the following are all equivalent

   (i) A is 2-transitive on $_F V$.

  (ii) $\forall$ finite n, A is n-transitive on $_F V$, i.e. A is dense on V.

 (iii) V_A is simple and $Q = F$.

Proof. First of all, V is simple if and only if A is 1-transitive, and n-transitive implies $n - 1$ transitive. Thus (ii) $\Rightarrow$ (i).

(i) $\Rightarrow$ (iii). Let $0 \neq q \in Q$ and $0 \neq x \in V$ be arbitrary. Assume that $qx \notin xF = Fx$. Since x and qx are F-independent, by 2-transitivity, there exists an element $a \in A$ with $xa = x$ but $(qx)a = 0$. Since V is simple, Q is a division ring and $q^{-1} \in Q$. But then $0 = q^{-1}qxa = xa = x \neq 0$ is a contradiction. Hence there exists an $c(x) \in F$ with $qx = c(x)x$. If $k \in F$, then $q(kx) = q(xk) = (qx)k = c(x)xk = c(x)(kx)$.

Now let $x, y \in V$ be F-linearly independent, and write

$$qx = c(x)x, \quad qy = c(y)y, \quad q(x + y) = c(x + y)(x + y)$$

$$c(x), \quad c(y), \quad c(x + y) \in F.$$

Then $c(x + y)x + c(x + y)y = c(x)x + c(y)y$ implies that $c(x + y) = c(x) = c(y) \equiv c \in F$. Thus $q = c$ and $Q = F$.

(iii) $\Rightarrow$ (ii) By the density theorem V is Q-dense, and by hypothesis $Q = F$.

Since it is known that an algebraically closed field is infinite, the next proposition gives the same conclusion as 6-5.10(i) but under weaker hypotheses.

6-6.10   Proposition. *For an algebra A over an algebraically closed field F, suppose that V is a simple right algebra module. Assume that the dimension* $\dim_F V$ *of V over F satisfies* $\dim_F V \lneqq |F|$, *where $|F|$ denotes the cardinality of F. Then*

$$Q \equiv \operatorname{End}_A V = F,$$

i.e. V satisfies the previous three mutually equivalent conditions.

Proof. If I and J are index sets – I for an F-basis of Q, and; J for a Q-basis of V, then $I \times J$ indexes an F-basis of V. Hence $\dim_F V = |I \times J| \equiv |I||J| = \dim_F Q \dim_Q V$. Thus by hypothesis $\dim_F Q \dim_Q V \lneqq |F|$, and hence $\dim_F Q \lneqq |F|$.

The proof is by contradiction. So let $q \in Q$ but $q \notin F$. Then the set $\{(q - f)^{-1} \mid f \in F\} \subseteq Q$ must be linearly dependent over F, because otherwise if it is independent we would have that

$$|F| = |\{(q - f)^{-1} \mid f \in F\}| \leqslant \dim_F Q \lneqq |F|.$$

Thus for n distinct $f_1, \ldots, f_n \in F$ and some scalars $c_i \in F$, $\sum_1^n c_i (q - f_i)^{-1} = 0$. In the polynomial and rational function rings $F[x] \subset F(x)$ over F, define

$$p(x) = (x - f_1) \cdots (x - f_n) \in F[x]$$

$$r(x) = \frac{c_1}{x - f_1} + \cdots + \frac{c_n}{x - f_n} \in F(x)$$

$$h(x) = r(x)p(x) = \sum_{i=1}^n c_i \prod_{j \neq i} (x - f_j) \in F[x].$$

Then in the transcendental field extension $F \subset F(q)$, $r(q) = 0$, and hence $h(q) = r(q)p(q) = 0$. But this is a contradiction because q is not algebraic over F. Hence $Q = F$.

6-6.11   Example of a simple F-algebra module V with $F \neq \operatorname{End}_A V$. Let $F = \mathbb{R}$ be the reals, $V = A = \mathbb{C}$ the complexes with the natural operations: $v = v_1 + iv_2 \in V$, $z = x + iy \in A$, $vz = v_1 x - v_2 y + i(v_1 y + v_2 x)$ where v_1, v_2, x, $y \in \mathbb{R}$. Each $\lambda \in \mathbb{C}$ gives an element $\lambda \in \operatorname{End}_A V$ simply by $\lambda : V \to V$, $v \to \lambda v$. Thus $\mathbb{C} \subseteq \operatorname{End}_A V \neq F$.

In order to show that $\operatorname{End}_A V = \mathbb{C}$, identify $A = \mathbb{C}$ with the commutative 2×2 matrix ring, $z \to [x, y; -y, x]$, and $V = \mathbb{C}$ with $\mathbb{R}^2$, $v \to (v_1, v_2)$. With these identifications, $vz \to (v_1, v_2)[x, y; -y, x]$. For an arbitrary $\delta \in \mathrm{End}_A V$, define a, b, c, $d \in \mathbb{R}$ by $\delta(1, 0) = (a, b)$ and $\delta(0, 1) = (c, d)$. It follows from $\mathrm{End}_A V \subset \mathrm{End}_F V \cong \mathbb{R}_2$ that $\delta(v_1, v_2) = v_1(a, b) + v_2(c, d) = (v_1, v_2)[a, b; c, d]$. The fact that δ commutes with A means that $(\delta v)z = \delta(vz)$, or that $(v_1, v_2)[a, b; c, d][x, y; -y, x] = (v_1, v_2)[x, y; -y, x][a, b; c, d]$. Since the latter holds for all v_1, $v_2 \in \mathbb{R}$, $[a, b; c, d][x, y; -y, x] = [x, y; -y, x][a, b; c, d]$ for all x and y. When $x = 0$, $y = 1$, $[-b, a; -d, c] = [c, d; -a, -b]$, hence $\delta = [a, b; -b, a] \in \mathbb{C}$. Thus $\mathrm{End}_A V \cong \mathbb{C}$.

6-7 Examples

Now the emphasis is on constructing classes of primitive rings, although some general results about the structure of primitive rings are obtained first.

6-7.1 Definition. A ring R is *simple* if (i) $R^2 = 0$; and (ii) R has no ideals except the two trivial ones (0), and $R \lhd R$.

For nonprimitive simple rings, see 7-4.5.

6-7.2 Observation. A simple ring R which contains a regular proper right ideal is primitive.

Proof. By 6-1.5, let $L < R$ be a regular maximal right ideal with relative left identity e. Since $e \notin (R/L)^\perp \lhd R$, $(R/L)^\perp = 0$. Hence R/L is a simple faithful R-module.

The next lemma can be used to manufacture new primitive rings from known old ones.

6-7.3 Lemma. *If V is a simple faithful right R-module and $0 \neq I \lhd R$, then*

   (i)   *V is a simple faithful I-module, and*
   (ii) $\mathrm{Hom}_R(V, V) = \mathrm{Hom}_I(V, V)$.

Proof. (i) From $VI \neq 0$, and $VIR \subseteq VI$, it follows that $VI = V$. Suppose that $0 \neq W \subseteq V$ is an I-submodule of V. Since V is simple, $WR = V$. Thus $V = VI = WRI \subseteq WI$. Hence $W = V$.

   (ii) Always $\mathrm{Hom}_R(V, V) \subseteq \mathrm{Hom}_I(V, V)$. Let $\varphi \in \mathrm{Hom}_I(V, V)$, $v \in V$, and

$r \in R$ arbitrary. If v were of the form $v = xi$, $x \in V$, $i \in I$, then $\varphi(vr) = \varphi[x(ir)] = (\varphi x)ir$, since $ir \in I$. Thus $\varphi(vr) = [(\varphi x)i]r = [\varphi(xi)]r = (\varphi v)r$. Since by (i), $V = VI$, every $v \in V$ is a finite sum $v = \Sigma \, xi$, and the rest follows since φ is additive.

6-7.4   Definition. Let V be a simple right R-module with $D = \operatorname{End}_R V$. An element $r \in R$ is of *finite rank* if the dimension over D of the left D-vector space Vr is finite. The dimension of Vr is denoted by $\dim_D Vr$.

A right ideal $I \leqslant R$ is a *minimal right* ideal if (0) and I are the only right ideals of R contained inside I.

6-7.5   Proposition. *If V is a simple faithful right R-module with $D = \operatorname{End}_R V$, and $a \in R$ is of finite rank, then there exists a $b \in R$ such that*

(i)   *abR is a minimal right ideal.*

(ii)   *$\forall c \in R$, $\dim_D Vc = 1 \Leftrightarrow cR$ is a minimal right ideal.*

Proof. (i) and (ii) $\Rightarrow$: There are D-independent $x_1, \ldots, x_n \in V$ such that $Va = Dx_1 + \cdots + Dx_n$. By the Density Theorem for any $0 \neq y \in V$, there exists a $b \in R$ such hat $x_i b = y$ for all $i = 1, \ldots, n$. Thus $0 \neq Vab = Dy$.

Set $c = ab$. From now on $c \in R$ can be any element such that $Vc = Dy$ or $\dim_D Vc = 1$.

Suppose that $I < R$, $0 \neq I \subseteq cR$. The first step will be to show that there exists a $z \in V$ such that

$$zc = y \quad \text{and} \quad V = zI \neq 0.$$

From $0 \neq VI = V$, it follows that $V = wI$ for some $w \in V$. Since $I \subseteq cR$, $0 \neq wc \in Vc = Dy$. Thus $wc = ky$ for some $0 \neq k \in D$. Set $z = k^{-1}w$. Then $zc = y$ and $V = k^{-1}V = k^{-1}wI = zI$.

Take any $cs \in cR$, $s \in R$. Since $ys \in zI$, $ys = zi$ for some $i \in I$. But i is of the form $i = cq$ for some $q \in R$. Thus $ys = zcq$, and $zc(s - q) = 0$.

If $r \in R$ is arbitrary, then $(zr)c \in Vc = Dy$, and hence $zrc = fy = fzc$ for some $f \in D$. But then $zrc(s - q) = fzc(s - q) = 0$, or $zRc(s - q) = 0$. Since $zR = V$, $cs = cq$. Thus $I = cR$, and cR is minimal.

(ii) $\Leftarrow$: If $\dim_D Vc \geqslant 2$, then there exist D-independent elements $x_1 c$, $x_2 c \in Vc$ for some x_1, $x_2 \in V$. Take $t \in R$ such that $x_1 ct = x_1 c$ and $x_2 ct = x_1 c$. Then $(x_1 - x_2)ctR = 0$. Since $x_2 ctR \neq 0$, $ctR \neq 0$, and $ctR = cR$. Thus $(x_1 - x_2)cR = 0$. In particular, $x_1 c = x_2 c$, a contradiction. Hence $\dim_D Vc = 1$.

6-7.6   Definition. A ring R is *regular* (also called von Neumann regular) if for any $a \in R$ there exists a $b \in R$ such that $aba = a$.

In the next theorem it could be also shown that S is the sum of all the minimal left ideals, see [Koh 72].

6-7.7   Theorem. *Let R be any primitive ring, V any simple faithful right R-module, and $D = \mathrm{End}_R V$. Define S to be the set of all elements $a \in R$ of finite rank, i.e.*

$$S = \{a \in R \mid \dim_D Va < \infty\}.$$

Then

(i)   *S is the sum of all the minimal right ideals of R.*

(ii)   *S is a regular ring.*

(iii) *$S \lhd R$.*

(iv)   *S is a simple primitive ring.*

Proof. (i) and (ii). Let S_R be the sum of all the minimal right ideals of R. By 7.5(ii), $S_R \subseteq S$. Let $a \in S$ be arbitrary. Then Va is a left D-vector space direct sum $V = (\ker a) + Va$, and $Va = Dx_1 + \cdots + Dx_n$, where $x_1, \ldots, x_n \in V$ are D-independent.

Since $Va = Vaa = Dx_1 a + \cdots + Dx_n a$, it follows that $\{x_1 a, \ldots, x_n a\}$ is also a D-independent set. Hence for each i there exists a $b_i \in R$ such that $x_j ab_i = x_i \delta_{ij}$, where $\delta_{ij} = 1$ or 0 as $i = j$, or $i \neq j$. Next, $Vab_i = Dx_i$ is one dimensional, hence $ab_i R$ is a minimal right ideal of R by 7.4(ii). Define $b = b_1 + \cdots + b_n$ and $c = aba$. Then $c \in \Sigma ab_i R \subseteq S_R \subseteq S$. But $x_i(a - c) = x_i a - x_i ab_i a = 0$. Thus $V(a - c) = 0$, and hence $a = c$. Thus $a = aba \in S_R$. The above shows that $S = S_R$ and that S is a regular ring.

(iii) For any minimal right ideal $U < R$ it follows from $0 \neq VU = V$, that $0 \neq UR = U$ is a simple right R-module. Hence for any $t \in R$, either $tU = 0$, or $U \cong tU \subseteq S$. Thus $S \lhd R$.

(iv) Let $0 \neq K \lhd S$ be an ideal of the ring S. If $K \neq S$, then there exists a minimal right ideal $U \nsubseteq SKS$. Then $SKS \cap U = 0$. Since $S \lhd R$, also $SKS \lhd R$, and hence $USKS \subseteq U \cap SKS = 0$. By 7.4 V is a faithful S-module. Hence $V = VS$, $0 \neq VK = VSK$, and $V = VSKS$. Finally $0 \neq V = VU$, and $V = VUSKS$ is a contradiction. Hence S is a simple ring which is primitive by 7.3 (or even 7.2).

6-7.8   Remark. In the previous theorem condition (i) says that S is independent of the particular choice of V. In other words, the property of being a finite rank element does not depend upon which simple faithful module the rank is computed.

Later we will take the two ordinals ρ and σ below to be

$\rho = \sigma = \omega = \{0, 1, 2, \ldots\}$ the first infinite ordinal, and the reader may wish to do so throughout.

6-7.9   Notation. For each infinite cardinal number κ there exists a unique smallest limit ordinal ρ of cardinality $|\rho| = \kappa$. Such an ordinal ρ is then called an *initial ordinal*. The successor cardinal of κ is denoted by κ^+. Define ρ^+ to be the unique smallest ordinal such that $|\rho|^+ = |\rho^+|$. Throughout both ρ and σ are any initial ordinals satisfying $\omega \leqslant \rho \leqslant \sigma$. For any field F, let $V = \oplus \{F_i \mid i < \sigma\}$, all $F_i = F$ be the F-vector space with natural basis $\{e_i \mid 0 \leqslant i < \sigma\}$, where $e_i = (0, \_\, 0, 1, 0, \_\,)$ has all component zero except the i-th which is 1. Thus $e_0 = (1, 0, \_\,)$, $e_1 = (0, 1, 0, \_\,)$, etc. An infinite rectangular or square matrix is *row finite* if every row contains only a finite number of nonzero entries. Elements $T \in \mathrm{Hom}_F(V, V)$ act on the right of V so that V is a right $\mathrm{Hom}_F(V, V)$-module. Each T is realized in the usual way as a $\sigma \times \sigma$ row finite matrix.

For $0 \leqslant i, j < \sigma$, $E_{ij} \in \mathrm{Hom}_F(V, V)$ will be the linear transformation representing the usual matrix unit, $e_k E_{ij} = \delta_{ki} e_j$ for $0 \leqslant k < \sigma$. The matrix of E_{ij} has all zeroes except for a one in row i and column j.

6-7.10   Example. Abbreviate $H = \mathrm{Hom}_F(V, V)$; H is primitive with faithful simple right H-module V, and $\mathrm{End}_H V = F$. For $T \in H$, let $\dim_F VT$ denote the unique initial ordinal which can be used to index a vector space basis of the subspace $VT \subseteq V$. For $\omega \leqslant \rho \leqslant \sigma$ as before, by 6-7.3, $I_\rho = \{T \in H \mid \dim_F VT < \rho\} \lhd H$ defines a family of primitive rings.

In particular, the ideal I_ω of all $T \in H$ whose ranges VT are finite dimensional is a simple primitive ring without identity.

The ring I_ρ is isomorphic to the ring of all those $\rho \times \rho$ row finite matrices whose non-zero entries have cardinality strictly less than $|\rho|$.

The next two examples are stated without proof. However enough details are supplied that any reader wishing to do so could supply the proofs. The first example is one of the easiest examples of a primitive ring R with $S = 0$ as in 6-7.7.

6-7.11   Example. Take $\sigma = \omega$. If the range of any linear transformation T is infinite dimensional, then $PTQ = 1$ for some P and Q. From this it follows immediately that $R = \mathrm{Hom}_F(V, V)/I_\omega$ is a simple primitive ring which has some simple faithful right R-module W. Using that $1 \in R$, and using 6-7.5(ii) and 6-7.7(i) it now could be shown that the ideal $S \lhd R$ of finite rank elements on W is zero. The conclusion is that every nonzero element of R has infinite rank on W.

6-7.12 Example. Aside from (0) and the whole ring, every ideal of $\text{Hom}_F(V, V)$ is of the form I_ρ for some $\omega \leqslant \rho \leqslant \sigma$. The simple ring H/I_σ is primitive because it contains an identity. If $\omega \leqslant \rho < \rho^+ \leqslant \sigma$, then I_{ρ^+}/I_ρ is simple. That it is primitive can be shown from 6-7.2 because it is easy to construct regular right ideals in I_{ρ^+} and then pass to the quotient I_{ρ^+}/I_ρ.

From now on $\sigma = \omega$ and V is of countable dimension over F.

6-7.13 Example. Let $S, T \in \text{Hom}_F(V, V)$ be the back and forward shifts: $e_0 S = 0$, $e_i S = e_{i-1}$ for $i = 1, 2, \ldots$; $e_i T = e_{i+1}$, $i = 0, 1, 2, \ldots$ Then $TS = 1$ is the identity but $ST \neq 1$. Since $S^i T^j - S^{i+1} T^{j+1} = E_{ij}$, the algebra $F\langle S, T \rangle$ generated by S and T contains all matrices with a finite number of nonzero entries. The algebra $F\langle 1, S, T \rangle$ generated by 1, S, and T is also primitive.

The next two examples show among other things that neither the center nor a homomorphic image of a primitive ring need be primitive.

6-7.14 Example. For $\sigma = \omega$ let $\mathbb{1}$ denote the $\omega \times \omega$ identity matrix and $D \subset F$ be any subring (with $1 \in D$ or $1 \notin D$). Let R be the ring of all matrices of the form $M + d\mathbb{1}$, where $d \in D$, and M is any matrix having only a finite number of nonzero entries from F. Then R is primitive with center $R = D\mathbb{1}$. This example shows that the center of a primitive ring can be any commutative domain.

6-7.15 Example. For any integer n, take $A \subseteq F_n$ to be any subring of the ring F_n of all $n \times n$ matrices over F. For $N \in A$, let $\text{diag}(N, N, \ldots)$ be the matrix having an infinite number of copies of N along the diagonal. Let R be the primitive ring of all matrices of the form $M + \text{diag}(N, N, \ldots)$ where M is any matrix having only a finite number of nonzero F-entries and $N \in A$ is arbitrary. The set of all $M + \text{diag}(0, 0, \ldots) \in R$ is an ideal I, and $R/I \cong A$.

6-8 Exercises

1. In example 6-7.13, prove that the E_{ii} are pairwise orthogonal idempotents, and that $E_{ij} E_{pq} = \delta_{jp} E_{iq}$.

2. For a right vector space V over a division ring D, let $R = \text{End}_D V$ be the ring of all D-linear transformations of V written on the left. For $0 \neq v \in V$, let $V = vD \oplus W$ be a D-direct sum. Define $M = \{t \in R \mid tV \subseteq W\}$. (i) Prove that R/M is a simple right

R-module. (ii) Let $S = \{a \in R \,|\, a$ is of finite rank$\}$. Show that every maximal right ideal L of R is either of the form (i), or that $S \subseteq L$.

3. Let R be a ring with $R^2 = R$ such that every simple right R-module is injective. Prove that every maximal right ideal is a regular maximal right ideal. ([Koh 72; 2.23]).

4. Let V be a simple faithful right R-module and $D = \operatorname{Hom}_R(V, V)$. Let Γ be any left D-vector space basis of V. (i) Show that we may identify $\operatorname{Hom}_D(V, V) = \operatorname{End}_D V = \Pi_\Gamma V$. (ii) Endow $\Pi_\Gamma V$ with the product or Tychonoff topology with each V being discrete. (iii) Show that the product topology on $\operatorname{End}_D V = \Pi_\Gamma V$ is independent of the choice of basis Γ. (iv) Prove that $R \subseteq \operatorname{End}_D V$ is dense in this topology.

5. Let V be a left vector space over a division ring D. Define $V^* = \operatorname{Hom}_D({}_D V, {}_D D)$ where $f(dv) = df(v)$ for $f \in V^*$. For $a \in D$, define $a * f : V \to D$ by $(a * f)(v) = af(v)$. (i) Prove that V^* is a left D vector space. (ii) For any subsets $X \subset V$ and $Y \subset V^*$, define $X^\perp = \{f \in V^* \,|\, fX = 0\}$ and $Y^\# = \{v \in V \,|\, y(v) = 0 \,\forall y \in Y\} = \{v \,|\, Yv = 0\}$. Prove the following assertions:

(a) For $0 \neq f \in V^*$, $\{f\}^\# = f^\#$ is a hyperplane, i.e. $\dim_D V/f^\# = 1$.

(b) $\{F^\# \,|\, F \subset V^*,\ F$ is finite$\}$ is a neighborhood basis for an abelian group topology on V ($V \times V \to V$ is continuous) in which scalar multiplications (by single elements of D) are continuous.

(c) If $\dim_D V \nleq \infty$, then this topology is discrete.

(d) For $\varnothing \neq X \subseteq V$, $\varnothing \neq Y \subseteq V^*$, $X^\perp$ and $Y^\#$ are closed sets.

(e) For any subspace $W \subset V$, $W^{\perp\perp}$ is its closure.

(f) Any finite dimensional subspace $W \subset V$ is closed, i.e. $W^{\perp\perp} = W$.

(g) If $W \subseteq V$ is a D-subspace, then $V^*/W^\perp \cong W^*$.

(h) If $W \subseteq V$ is any D-subspace and $Y \subset V^*$ is a finite dimensional D-subspace, then $(W \cap Y^\#)^\perp = W^\perp + Y^{\#\perp}$.

(i) Give an example to show that there can exist nonclosed subspaces $W \subset V$, i.e. $W \neq W^{\perp\perp}$.

Remark. If V is a simple faithful right R-module and $D = \operatorname{End}_R V$, then we know that $R \subset \operatorname{Hom}_R(V, V)$. It can be shown that R is precisely the set of all continuous D-linear transformations on V with this topology on V. This remarkable fact (and more) is proved in [Ribenboim 67; pp. 115–129].

CHAPTER 7

The Jacobson radical

Introduction

The kind of theory developed here follows a pattern which is discernible in other such related theories. One starts with some given class Σ of modules, and then defines the radical corresponding to Σ to be the intersection of the annihilator ideals of all the modules in Σ.

Set theory is applicable mostly to sets, but possibly not always to the larger entities called classes, like the class of all (one sided) simple modules Σ used in this chapter. However, module isomorphism is an equivalence relation $\sim$ on Σ. The equivalence classes modulo this relation do form a set, the set $\Sigma/\sim$ of all isomorphism classes of simple modules. By the axiom of choice we can select one representative out of each equivalence class giving us a set Σ^* of simple modules. Every simple module is isomorphic to exactly one element of Σ^*. In all of what follows all the results could be rephrased in terms of Σ^*. However, for the sake of simplicity and directness, we will use Σ instead.

An ideal (such as a radical) could be described or characterized in the following three different ways: (i) by specifying what types of (one sided) ideals or what types of elements it necessarily must always contain; or (ii) as an intersection of certain types of (one sided) ideals. (iii) Necessary and sufficient conditions could be found in order for an arbitrary element of the ring to belong to the radical in question. The latter elementwise conditions seem to be particularly useful. However, for the Jacobson radical they require the concept of quasi-regularity which is introduced in section 1.

This chapter contains a lot of classical ring theory which today right now is needed just as much if not more than when first discovered. The whole theory of the Jacobson radical speaks for itself. It also ties in well with the much older theory of rings with the descending chain condition through proposition 7-1.36 which shows that the descending chain condition on right ideals in a ring forces its Jacobson's radical to be nilpotent. This would have been called a theorem if only its proof were not so easy. From the latter we derive the key Wedderburn (or Wedderburn-Artin) Theorem.

7-1   Characterizations

7-1.1   Quasi composition. As before $R^1 = R$ when $1 \in R$, and $R \equiv \{0\} \times R \lhd R^1 = \mathbb{Z} \times R$ if $1 \notin R$. Define a binary composition law called the quasi-product "$\circ$" by $R \times R \to R$, $(x, y) \to x \circ y = x + y - xy$. Here x, y, z, x_R^{-1}, x_L^{-1}, and x^0 will denote elements of R (but not elements of $R^1 \backslash R$). For any ring with an identity like R^1, $U(R^1)$ will denote the multiplicative group of units, $U(R^1) = \{\alpha \in R^1 \mid \exists \alpha^{-1} \in R^1, \; \alpha\alpha^{-1} = \alpha^{-1}\alpha = 1\}$. The element x is

  (i)   right quasi-regular: if $\exists x_R^{-1} \in R$, $x \circ x_R^{-1} = 0$;
  (ii) left quasi-regular: if $\exists x_L^{-1} \in R$, $x_L^{-1} \circ x = 0$;
  (iii) quasi-regular: if $\exists x^0 \in R$, $x^0 \circ x = x \circ x^0 = 0$.

The abbreviations (i) r.q.r., (ii) l.q.r., and (iii) q.r. will be used; x_R^{-1} is a right quasi-inverse of x; x_L^{-1} a left quasi-inverse, and x^0 is a quasi inverse of x..

7-1.2   Lemma. *In the previous notation for any* x, $y \in R$ *the following hold.*
  (i) $\langle R, 0 \rangle$ *is a semigroup with identity* 0.
  (ii) $\exists x_R^{-1}$ *and* $\exists x_L^{-1} \Rightarrow x_R^{-1} = x_L^{-1} = x^0$.
  (iii) $(x^0)^0 = x$; $xx^0 = x^0x$.
  (iv) $\{x \in R \mid \exists x^0 \in R\}$ *is a subgroup of* $\langle R, 0 \rangle$.
  (v) $x \circ y = 1 - (1-x)(1-y) \in R \lhd R^1$; *in particular*

$$x^0 \circ x = x \circ x^0 = 0 \Leftrightarrow (1 - x)^{-1} = 1 - x^0 \text{ in } R^1.$$

  (vi) $\{x \in R \mid \exists x^0 \in R\} \to U(R^1)$, $x \to 1 - x$ *is a monic group homomorphism.*
  (vii) $x^n = 0 \Rightarrow x$ *is quasi-regular and* $x^0 = -x - x^2 - \cdots - x^{n-1}$.

Proof. Verification is straightforward. Conclusion (vi) follows from (v). For (vii), one verifies that $(1 - x)^{-1} = 1 + x + \cdots + x^{n-1} \in R^1$ and then sets $x^0 = 1 - (1 - x)^{-1} \in R$.

7-1.3   Remarks. The alternate binary composition law $\square : R \times R \to R$, $(x, y) \to x \square y = x + y + xy$ could have been used equally well to produce results identical or similar to those of 7-1.1 and 7-1.2. However, nothing new would be gained, because

(1) $-(x \square y) = (-x) \circ (-y)$.

(2) $\langle R, \square \rangle \to \langle R, 0 \rangle$, $x \to -x$ is a semigroup isomorphism (which upon restriction and corestriction to the group of units is a group isomorphism).

(3) $\{1 - x \mid \exists x^0\} \lhd U(R^1)$ is a normal subgroup of index two; $U(R^1) = \{1 - x \mid \exists x^0\} \cup \{-1 + x \mid \exists x^0\}$.

7-1.4   Definitions and notation. Define $\Sigma(R)$ to be the class of all simple right R-modules; $\mathcal{M}(R)$ the set of all regular maximal right ideals of R; and $\mathcal{P}(R)$ the set of all primitive ideals of R. When R is either fixed or understood, abbreviate these as Σ, $\mathcal{M}$, and $\mathcal{P}$. Define the *Jacobson radical* of a ring R to be the ideal $J(R) = \cap \{V^\perp \mid V \in \Sigma\} \lhd R$.

The empty intersection by definition is always the whole space, e.g. if $\Sigma = \varnothing$, then $J(R) = R$.

7-1.5   Lemma. *For any $P \lhd R$,*

(i) *$P \lhd R$ is primitive* $\Leftrightarrow P = V^\perp$*, some $V \in \Sigma$.*

(ii) *$P \lhd R$ is primitive* $\Leftrightarrow P = R\!:\!M$*, some $M \in \mathcal{M}$.*

Proof. (i) $\Rightarrow$: By hypothesis, there exists a simple faithful $\bar{R} = R/P$-module $\bar{V}$. Define $V = \bar{V}$ as an abelian group, but $V \in \Sigma$ by $vr = v(r + P)$ for $v \in V$, $r \in R$. Then $V^\perp = P$. $\Leftarrow$: Given that $P = V^\perp$ for some $V \in \Sigma$, define $\bar{V}$ to be the simple $\bar{R} = R/P$-module, where as an additive group $\bar{V} = V$, but $v\bar{r} = vr$ for any $v \in V$, $\bar{r} \in \bar{R}$, where $r \in \bar{r}$ is arbitrary.

7-1.6   Theorem I. *For any ring R*

(i) $J(R) = \cap \{P \mid P \lhd R \text{ is primitive}\}$;

(ii) $J(R) = \cap \{L \mid L \in \mathcal{M}\}$.

Proof. (i) By the last lemma the following two sets of ideals $\{V^\perp \mid V \in \Sigma\} = \{P \mid P \in \mathcal{P}\}$ are equal, and hence so are also their intersections $J(R) = \cap \mathcal{P}$.

(ii) Recall that for any $L \in \mathcal{M}$, if $e \in R$ is its relative left identity satisfying $er - r \in L$ for all $r \in R$, then $x = e + L \in R/L \in \Sigma$, and $x^\perp = L$.

Thus $\{x^\perp \mid \exists V \in \Sigma, 0 \neq x \in V\} = \mathscr{M}$. Therefore

$$J(R) \equiv \bigcap_{V \in \Sigma} V^\perp = \bigcap_{V \in \Sigma} \left[\bigcap_{0 \neq x \in V} x^\perp \right] = \bigcap_{L \in \mathscr{M}} L$$

The following lemma also holds for groups with operators.

7-1.7   Lemma. *If $J < R$ is a fixed right ideal and $Y = \{L\}$ a given set of right ideals of R such that $J \subseteq L$ for all $L \in Y$, then*

$$\bigcap \left\{ \frac{L}{J} \,\middle|\, L \in Y \right\} = \frac{\cap \{L \mid L \in Y\}}{J}.$$

Proof. Evidently, $(\cap Y)/J \subset \cap\{L/J \mid L \in Y\}$. Conversely take any $\xi \in \cap \{L/J \mid L \in Y\}$, and then take any $z \in \xi$. For any $L \in Y$, $\xi = z + J \in L/J$. Hence $z \in L$, and $z \in \cap Y$. Thus $\xi = z + J \in (\cap Y)/J$, and equality follows.

7-1.8   Corollary 1 to Theorem I. *For every ring R, $J[R/J(R)] = 0$.*

Proof. Set $J = J(R)$. Then every regular maximal right ideal of R/J comes from one of R,

$$J\left(\frac{R}{J}\right) = \bigcap\left(\frac{L}{J}\right) = \frac{\cap\{L \mid J \subseteq L \in \mathscr{M}(R)\}}{J} = J \subseteq L \in \mathscr{M}(R)$$

$$= \frac{\cap \mathscr{M}(R)}{J} = \frac{J}{J} = \{J\} = 0.$$

Our next objective is to relate quasi-regularity with $J(R)$.

7-1.9   Definition. A right ideal K of R is *quasiregular* if every element of K is quasi-regular.

7-1.10   Remarks. (1) The concept of a regular right ideal has nothing to do with that of a quasi-regular one. The use of the term 'regular' for both is an unfortunate historical accident.

(2) The next lemma allows one to consider only right ideals and right quasi-regularity, the type of one-sided ideal being the same as the type of regularity.

(3) The next lemma shows that the property of K being a quasi-regular right ideal is independent of the ring R into which K is embedded as a right ideal.

7-1.11 Lemma. *If every element of a right ideal K of R is right quasi-regular, then*

    (i)   *K is quasi-regular, and*

    (ii) *K contains quasi-inverses of all of its members.*

Proof. (i) and (ii). For any $x \in K$, there exists an element $x_R^{-1} = y \in R$ such that $x \circ y = 0$, and hence $y = xy - x \in K < R$. Since now $y \in K$, by hypothesis there exists a $z \in R$ (actually $z \in K$) with $y \circ z = 0$. But then the element y has both a left quasi-inverse x, and a right one z, and hence the two are necessarily equal, $x = z$. Thus $x \circ y = y \circ x = 0$. I.e. for any $x \in K$, there exists $x^0 = y \in K$.

7-1.12 Lemma. *Let $K \leqslant R$ be any quasi-regular right ideal. Then for every regular maximal right ideal $L \leqslant R$, $K \subseteq L$.*

Proof. If not, there exists a $K \nsubseteq L \in \mathcal{M}$. Thus $L + K = R$. A relative left identity e for L is of the form $e = m + k$, $m \in L$, $k \in K$. Hene K contains an element $z \in K$ such that $k + z - kz = 0$. But

$$ez - z = (m + k)z - z = mz + k \in L,$$

and hence $k \in L$. But then $e = m + k \in L$, and $L = R$, a contradiction.

7-1.13 Remark. There are rings R with $\mathcal{M}(R) = \varnothing$. In this and similar cases, in general, a lemma of the above type is clearly vacuously true by default, because there would not possibly exist any objects which can falsify it.

7-1.14 Observation. For any element $y \in R$, form the regular right ideal $K = \{yr - r \mid r \in R\}$ with relative left identity y. Thus we know that $K = R$ if and only if $y \in K$. Then the following three conditions are all equivalent.

    (i) $y \in R$ is right quasi-regular.

    (ii) $K = R$.

    (iii) For any regular maximal right ideal L, there exists a $b \in R$ such that $y \circ b \in L$.

Proof. (ii) $\Rightarrow$ (i): If $K = R$, then $y \in R = K$, and so $y = yr - r$ for some $r \in R$. Thus $y \circ r = 0$. (i) $\Rightarrow$ (ii). If $y \circ r = 0$ for some $r \in R$, then $y = yr - r \in K$ by definition of K.

    (i) $\Rightarrow$ (iii). If $y \circ \lambda = 0$, then trivially $0 = y \circ \lambda \in L$ for all $L \leqslant R$.

(iii) $\Rightarrow$ (ii). If not, then $K \neq R$, and by Zorn's Lemma there exists a regular maximal right ideal $L < R$ with $K \subseteq L$. Thus y is also a relative left identity for L. By hypothesis, for some $b \in R$, $y \circ b = y - (yb - b) \in L$. Since $yb - b \in K \subseteq L$, also $y \in L$, and $L = R$, a contradiction. Thus $K = R$.

7-1.15   Theorem II. *For any ring R*

   (i)   *$J(R)$ is a quasi-regular ideal.*
  (ii) *$J(R)$ contains all quasi-regular right ideals of R. Consequently,*
 (iii) *$J(R)$ is the unique biggest quasi-regular right ideal of R.*

Proof. (i) If $y \in J(R)$ is not right quasi-regular, then $K = \{yr - r \,|\, r \in R\} \neq R$. By Zorn's Lemma there exists an $L \in \mathcal{M}$ with $K \subseteq L$. Now y is a relative identity for L also. But $y \in J(R) \subseteq L$. Hence $L = R$, a contradiction. Thus for every $y \in J(R)$, y is right quasi-regular.

(ii) Let $K \leqslant R$ be any quasi-regular right ideal. Then $K \subseteq L$ for all $L \in \mathcal{M}$. Hence $K \subseteq \cap \{L \,|\, L \in \mathcal{M}\} = J(R)$.

7-1.16   Corollary 1 to Theorem II. *For any ring R, $J[J(R)] = J(R)$.*

Proof. Regard $J(R) = S^1$ as a ring S^1. Since S^1 is a quasi-regular ideal of the ring S^1, $S^1 \subseteq J(S^1)$, i.e. $J(R) \subseteq J(J(R))$.

7-1.17   Definition. A subring $B \subseteq R$ is *nil* if for any $b \in B$ there exists an integer n which may depend on b such that $b^n = 0$.

An element $x \in R$ such that $x^m = 0$ for some m is called *nilpotent*; $e \in R$ is *idempotent* if $e^2 = e$.

A subring $B \subseteq R$ is *nilpotent* if there exists a positive integer N such that for any $b_1, \dots, b_N \in B$, $b_1 b_2 \dots b_N = 0$. The smallest such integer N with this property is called the index of nilpotency of B. A nilpotent subring B satisfies a property stronger than nil. Given any $b \in B$, simply take $b_1 = \cdots = b_N = b$. Then $b^N = 0$, where the same exponent N works for all elements $b \in B$.

7-1.18   Corollary 2 to Theorem II. *The radical $J(R)$ of any ring contains all nil right ideals of that ring R.*

7-1.19   Corollary 3 to Theorem II. *The radical $J(R)$ does not contain any nonzero idempotents.*

Proof. If $e = e^2 \in J(R)$, then there exists a $\lambda \in J(R)$ such that $e + \lambda - e\lambda = 0$. Hence $e = e^2 = e(e\lambda - \lambda) = 0$.

7-1.20   Corollary 4 to Theorem II (Nakayama). *Suppose that W is a finitely generated right R-module, i.e.* $W = \mathbb{Z}w_1 + \cdots \mathbb{Z}w_n + w_1 R + \cdots + w_n R$ *for some* $w_i \in W$, $\mathbb{Z} = 0, \pm 1, \dots$. *Then* $WJ(R) \subsetneqq W$.

Proof. Set $J = J(R)$. Select a minimal generating set $\langle w_1, \dots, w_n \rangle = W$ above so that n is as small as possible.

Now assume that $WJ = W$. Then it follows from this that $W = w_1 J + \cdots + w_n J$. In particular, $w_n \in W = WJ$ is of the form $w_n = w_1 z_1 + \cdots + w_n z_n$. There exists an element $\lambda \in J$ with $z_n \circ \lambda = 0$. Addition of the first two equations gives the third:

$$w_n - w_n z_n = \Sigma_1^{n-1} w_i z_i$$

$$-w_n \lambda + w_n z_n \lambda = -\Sigma_1^{n-1} w_i z_i \lambda$$

$$w_n = w_n - w_n(z_n \circ \lambda) = \Sigma_1^{n-1} w_i(z_i - z_i\lambda).$$

Consequently the generator w_n is superfluous in $W = \langle w_1, \dots, w_{n-1} \rangle$, thus contradicting the minimality of n. Therefore $WJ \neq W$.

7-1.21   Definition. For any modules $K < M$, the submodule K is *superfluous* in M if whenever $L \leqslant M$ is any submodule such that $K + L = M$, then already $L = M$.

For example, $(0) \leqslant M$ is superfluous. Another name sometimes used in the literature in place of superfluous is *small*.

7-1.22.   Nakayama's Lemma. *For this lemma only assume* $1 \in R$. *For any* $K < R$

(1)   (a) K *is q.r.* $\Leftrightarrow$ (b) $K < R$ *is superfluous.*
(2)   *Hence if K satisfies (1) (a), (b), then* $K \subseteq J(R)$.
(3)   *Furthermore, $J(R)$ is the unique largest superfluous right ideal of R.*

Proof. (1) (a) $\Rightarrow$ (b): If $K + L = R$ for some $L \leqslant R$, write $1 = k + m$ with $k \in K$, $m \in L$. Then there exists $(1 - k)^{-1} = 1 - k^0 \in R$ since $k \circ k^0 = 0$ for some $k^0 \in K$. Thus $1 = (1 - k)(1 - k^0) = m(1 - k^0) \in L$ and hence $L = R$.

(1) (b) $\Rightarrow$ (a): If $k \in K$, then $kR < R$ is superfluous too. But then from $kR + (1 - k)R = R$ it follows that $(1 - k)R = R$.   Thus

$(1 - k)(1 - \lambda) = 1$ for some element $1 - \lambda \in R$, which satisfies $k \circ \lambda = 0$. Thus K is quasi-regular by 7-1.11.

Conclusions (2) and (3) follow from 7-1.15.

All of our previous definitions and results could be formulated and have obvious analogues in terms of left modules and left ideals.

7-1.23   Corollary 4 to Theorem II. *Let* $N \subseteq R$ *be the left Jacobson radical of* R. *Then* $J(R) = N$.

Proof. We know that $N \lhd R$ and that N contains all quasi-regular left ideals of R. Since $J(R)$ is a quasi-regular left ideal, $J(R) \subseteq N$. On the other hand, N is a quasi-regular right ideal, and hence $N \subseteq J(R)$. Thus $N = J(R)$.

7-1.24   Theorem III. *Let* $n \in R$. *Then*

    (i) $n \in J(R) \Leftrightarrow \forall b \in R$, *nb is right quasi-regular.*
   (ii) $n \in J(R) \Leftrightarrow \forall a, b \in R$, *anb is right quasiregular.*

7-1.25   Corollary 1 to Theorem III. *For any* $n \in R$, *the following hold.*

    (i)   *If* $\forall b \in R$, *nb is r.q.r., then* $\Rightarrow \exists n^0 \in R$.
   (ii) *If* $\forall a, b \in R$, *anb is r.q.r., then* $\Rightarrow \exists n^0 \in R$.

Proof of 7-1.24 (i) $\Rightarrow$: If $n \in J(R)$, then $nb \in J(R)$, a quasi-regular ideal. Hence nb is right quasiregular.

    (i) $\Leftarrow$: It has to be shown that for any simple module V, $Vn = 0$. Suppose not. Then there exists a $w \in V$ with $wn \neq 0$. Therefore $wnR = V$. Thus there exists a $b \in R$ with $wnb = w$. By hypothesis there exists a $\lambda \in R$ such that $(nb) \circ \lambda = 0$. Then $\bar{0} = w0 = w(nb + \lambda - nb\lambda) = w + w\lambda - w\lambda$. Hence $w = 0$, a contradiction.

   (ii) $\Rightarrow$: If $n \in J(R) \lhd R$, then $anb \in J(R)$ is automatically quasi-regular.

  (iii) $\Leftarrow$: If $n \notin J(R)$, then $Vn \neq 0$ for some $V \in \Sigma$; $wn \neq 0$ for some $w \in V$. From $wR = V$, it follows that $wa = w$, for some $a \in R$. Thus $0 \neq wan = wn$. Similarly, there exists $b \in R$ such that $wanb = w$. Now by hypothesis, there exists a $\lambda \in R$ such that $(anb) \circ \lambda = 0$. Then $w0 = w(anb + \lambda - anb\lambda)$ reduces to $w + w\lambda - w\lambda = 0$, or consequently $w = 0$, a contradiction. Thus $Vn = 0$ and $n \in J(R)$.

In applying the last theorem, it is easier to verify that nb is merely right quasi-regular rather than both left and right quasi-regular.

7-1.26 Corollary 2 to Theorem III. *Any* $n \in R$ *satisfies the following*

  (i) $n \in J(R) \Leftrightarrow \forall\, b \in R,\ \exists\, (nb)^0 \in R.$

  (ii) $n \in J(R) \Leftrightarrow \forall\, a,\, b \in R,\ \exists\, (anb)^0 \in R.$

In view of the left right symmetry of the last condition, an alternate proof of a corollary of a previous theorem now follows immediately.

7-1.27 Corollary 3 to Theorem III. *The left and right Jacobson radicals of any ring coincide.*

The next corollary also gives an alternate proof of an already previous corollary to Theorem I. The reader should be aware that the same symbol "∘" serves double duty below for a binary composition law in two disjoint semigroups.

7-1.28 Corollary 4 to Theorem III. *For any ring R, $J[R/J(R)] = 0$.*

Proof. Set $J = J(R)$ and let $n + J \in J(R/J)$. For any $b + J \in R/J$, $(n + J)(b + J) = nb + J$ has a right quasi-inverse $c + J \in R/J$, i.e. $(nb + J) \circ (c + J) = (nb) \circ c + J = J$. Thus the element $z \equiv (nb) \circ c \in J$ has a quasi-inverse $z^0 \in R$ where $0 = [(nb) \circ c] \circ z^0 = (nb) \circ (c \circ z^0)$. Since for any $b \in R$, nb is right quasi-regular, we are allowed to conclude that $n \in J$.

Although the next lemma logically belongs immediately in front of Theorem II because it would give a stronger version of Theorem II, nevertheless it is presented only now much later because its proof is a little computational. When feasible we have tried to place the easier well motivated theorems early in each chapter.

7-1.29 Lemma. *Suppose that $A \subset R$ is a set of quasi-regular elements, and that $B \leqslant R$ is a quasi-regular right ideal. Then $A + B = \{a + b \mid a \in A, b \in B\}$ is a set of quasi-regular elements.*

Proof. If $a \in A$ and $b \in B$, then there exists $a^0 \in R$ and $b^0 \in B$. In general, for any $y \in R$, there exists $y^0 \in R$ if and only if there exists $(1 - y)^{-1} = 1 - y^0 \in R^1$. In particular there exists $z \equiv (a + b)^0 \in R$ if and only if $[1 + (a + b)]^{-1} = 1 - z \in R^1$. Let us factor $1 - a$ on the right in

$$1 - a - b = [(1 - a) - b(1 - a^0)(1 - a)]$$
$$= [1 - b(1 - a^0)](1 - a).$$

Define $x \equiv b - ba^0 \in B$. There exists $x^0 \in B$, and $(1 - x)^{-1} = 1 - x^0 \in R^1$. So now $1 - a - b = (1 - x)(1 - a)$. Hence

$$(1 - a - b)^{-1} = (1 - a^0)(1 - x^0) = 1 - a^0 \circ x^0 = 1 - (a + b)^0.$$

Thus $(a + b)^0 = a^0 \circ (b - ba^0)^0$.

7-1.30 Proposition. *For any ring R, the radical $J(R)$ is the sum of all the quasi-regular right ideals of R.*

Proof. Iteration of the previous lemma shows that the right ideal $Q = \Sigma\{K \mid K \leqslant R \text{ is q.r.}\} = \{a_1 + \cdots + a_n \mid n = 0, 1, 2, \ldots; \exists q.r. \ K_j \leqslant R, a_j \in K_j\}$ is quasi-regular. Since $J(R) \subseteq Q$, $Q = J(R)$.

Powers and products of sets occur frequently in ring theoretic arguments. To handle these, an adequate notation and symbolism is developed below and in the proof of the next proposition.

7-1.31 Let $X, Y, Z \subseteq R$ be nonempty subsets. Recall that the product XY is defined to be $XY = \{z \in R \mid \exists n, x_i \in X, y_i \in Y, z = \Sigma_1^n x_i y_i\}$. Multiplication of sets is associative, $(XY)Z = X(YZ)$. In particular, when $Y = X$, $X^2 = XX, \ldots, X^n = X^{n-1}X$. For $r \in R$, define $r^0 \equiv 1 \in R^1$; and if $X \neq \varnothing$, define $X^0 = \{1\} \subset R^1$.

7-1.32 Remark. There is a limited arithmetic of sorts on nonempty subsets of R. The sum is $X + Y = \{x + y \mid x \in X, y \in Y\}$. If $X = \{x_0\}$, then as usual $XY = x_0 Y$, $x_0 + Y = \{x_0\} + Y$. Set $-Y = \{-y \mid y \in Y\}$. The inclusion $X(Y + Z) \subseteq XY + XZ$ is an equality if $0 \in Y \cap Z$. If $0 \in Y$, then $Y \cup -Y \subseteq Y - Y$.

7-1.33 Proposition. *Suppose that $A \leqslant R$, $B \leqslant R$ are nilpotent, $A^p = 0$ and $B^q = 0$. Then $A + B$ is nil; moreover $(a + b)^n = 0$ for $n = p + q$ and $a \in A, b \in B$.*

Proof. For any $a \in A, b \in B, (a + b)^n = \Sigma \, a^{i(1)} b^{j(1)} \cdots a^{i(r)} b^{j(r)}$ where the sum is over all ordered sets $\{i(1), i(2), \ldots, i(r), j(1), \ldots, j(r)\}$ for varying r subject to the restrictions that $n = i(1) + \cdots + i(r) + j(1) + \cdots + j(r)$, where $0 \leqslant i(k), j(k) \leqslant n$, and the only ones of these which may possibly be zero are $i(1)$ and $j(r)$, the rest are larger or equal to one. One or both of the following two cases must occur.

Case 1. $j(1) + \cdots + j(r) \geqslant q$.

$$a^{i(1)}\underbrace{b^{j(1)}a^{i(2)}\cdots b^{j(r-1)}a^{i(r)}}_{B^{j(1)}}\underbrace{b^{j(r)}}_{B^{j(r)}} \in B^{j(1)+\cdots+j(r)} = (0).$$

If $j(r) = 0$, then $B^{j(r)} = \{1\}$.

Case 2. $i(1) + \cdots + i(r) \geqslant p$.

$$\underbrace{a^{i(1)}b^{j(1)}}_{A^{i(1)}}\underbrace{a^{i(2)}b^{j(2)}}_{A^{i(2)}}\cdots\underbrace{a^{i(r)}b^{j(r)}}_{A^{i(r)}} \in A^{i(1)+\cdots+i(r)} = (0).$$

Note that $A^{i(1)} = \{1\}$ if $i(1) = 0$.

7-1.34 Corollary. *For any ring R, define $N(R)$ to be the sum of all the nilpotent right ideals of R. Then*

  (i) $N(R) \lhd R$;
  (ii) $N(R)$ *is nil; and hence*
  (iii) $N(R) \subseteq J(R)$.

Proof. (i) If $A \leqslant R$ with $A^P = 0$, and $x \in R$, then $xA \leqslant R$, and $(xA)^P \subseteq xA^P = 0$. (ii) Every element of $N(R)$ is contained in a finite sum of nilpotent right ideals, which by an iteration of the last proposition is nil. (iii) Since $N(R)$ is quasi-regular, $N(R) \subseteq J(R)$.

7-1.35 Definition. A right R-module M satisfies the *descending chain condition* on right R-submodules (abbreviation D.C.C.) if one and hence all of the following equivalent conditions are satisfied.

  (i) For every integer indexed descending chain of submodules $M \geqslant M_1 \geqslant M_2 \geqslant \cdots \geqslant M_n \geqslant M_{n+1} \geqslant \cdots$ from some n on all the modules are equal, i.e. there exists an n such that $M_n = M_{n+i}$ for all $i = 1, 2, \ldots$
  (ii) Any nonempty set of submodules of M whatever always contains a minimal element.

Reversal of all inequalities in (i) and (ii) defines the *ascending chain condition* on submodules of M (abbreviation: A.C.C., or right A.C.C.)

If the set of all submodules of M satisfies the D.C.C., then M is said to be *Artinian* or *right Artinian*. Similarly if A.C.C., then M is *Noetherian*, or *right Noetherian*. The ring R itself is said to be right Artinian (or right Noetherian) if R_R regarded as a right R-module has these proper-

ties. Terms like right Artin ring, Artin module, or – less frequently –
Noether module are also used.

7-1.36   Proposition. *Suppose that the ring R satisfies the descending
chain condition on right ideals. Then $J(R) = N(R)$. Furthermore, $N(R)$ is
nilpotent.*

Proof. Set $J = J(R)$ and $N = N(R)$. By the descending chain condition,
there exists k such that $J \supseteq J^2 \supseteq \cdots \supseteq J^k = J^{k+1}$. Set $A = J^k$. We show
that $A = 0$. Suppose then that $0 \neq A = A^2$. Let $\mathscr{F}$ be the set of right
ideals $\mathscr{F} = \{L \mid L \leqslant R, \ LA \neq 0\}$. Since $A \in \mathscr{F}$, $\mathscr{F} \neq \varnothing$. By the D.C.C.,
there exists a minimal element $L \in \mathscr{F}$. Since $LA \neq 0$, there exists a $\lambda \in L$
with $\lambda A \neq 0$. Then $\lambda A = \lambda A^2 = (\lambda A)A$ shows that $\lambda A \in \mathscr{F}$. But
$0 \neq \lambda A \subseteq LA \subseteq L$, with L minimal in $\mathscr{F}$ requires that $\lambda A = L$. Thus there
exists an $a \in A$ with $\lambda a = \lambda \in L$. Since $a \in A = J^k$, a is quasi-regular, and
there is a $b \in J$ such that $a \circ b = 0$. Therefore $0 = \lambda 0 = \lambda(a + b - ab) =
\lambda + \lambda b - \lambda b = \lambda$, a contradiction. Hence $A = J^k = 0$. Thus $J \subseteq N$. In
every ring, $N \subseteq J$. Thus $J = N$.

We are finally in position to obtain a major result in the history of
ring theory – the Wedderburn theorems. The two ingredients which make
this remarkable theorem now accessible to us are the last proposition and
Jacobson's Density theorem. Here we want to derive the most substantial
core part of the Wedderburn theorem as expeditiously as possible, and
not get sidetracked in various corollaries, equivalent reformulations, and
different techniques of proof. Later on, we devote the whole of Chapter 10
to this important theorem, give all the refinements, and explore the side
issues. There the reader will find various characterizations and uniqueness questions, of say for example a certain division ring which here right
now will emerge from Jacobson's Density theorem.

Recall the definition of a simple ring (6-7.1).

7-1.37   Lemma. *For a finite number of simple rings S_i, let R be a subring
$R \subseteq S_1 \times \cdots \times S_m$ such that each projection $\pi_i \colon R \to S_i$ is onto. Then R is
ring isomorphic to a full direct product of some finite number of the S_i, i.e.
$R \cong S_{i(1)} \times \cdots \times S_{i(k)}$ for some indices $1 \leqslant i(1) \cdots < i(k) \leqslant m$, $1 \leqslant k \leqslant m$.*

Proof. By induction, assume that the above holds for $1, \ldots, m-1$. We
prove it for m by considering two cases. View $\pi_i R = S_i \equiv \{0\} \times \cdots \times S_i \times \cdots
\{0\} \subseteq \Pi_{i=1}^n S_i$.

Case 1. For any i, there exists an element $r \in R$, all of whose components are zero, except for the ith, i.e. $r = (0, \ldots, s, \ldots, 0) \in R$, $0 \neq \pi_i r = s \in S_i$, and $\pi_j r = 0$ for all $j \neq i$. Then for any $a, b \in S_i$, there exist $\alpha, \beta \in R$ such that $\pi_i \alpha = a$ and $\pi_i \beta = b$. Then $\pi_j(\alpha r \beta) = 0$ for $j \neq i$, while $\pi_i(\alpha r \beta) = (\pi_i \alpha)(\pi_i r)(\pi_i \beta) = asb \in S_i$. Since $0 \neq S_i s S_i \lhd S_i$, from the latter it follows that $\pi_i(RrR) = S_i s S_i = S_i \cong \{0\} \times \cdots \times S_i \times \cdots \{0\} \subseteq R$. Consequently, $R = \Pi_1^n S_i$.

Case 2 is when Case 1 fails for some index i which we may take without loss of generality to be $i = m$. Let $\rho: R \to S_1 \times \cdots \times S_{m-1}$ be the natural projection map onto the first $m - 1$ coordinates. Since $\ker \rho = \{r \in R \mid \pi_j r = 0, 1 \leqslant j \leqslant m - 1\} = 0$ in Case 2, ρ is a monomorphism. Thus $R \cong \rho R \subseteq S_1 \times \cdots \times S_{m-1}$, and the result follows by induction.

7-1.38 Let $R = \oplus_{i=1}^k M_{n(i)}(\Delta_i)$ where $M_{n(i)}(\Delta_i)$ is the full $n(i) \times n(i)$ matrix ring over a skew field Δ_i. Then

  (i) R satisfies the descending chain condition on right ideals; and
  (ii) $J(R) = 0$ and hence R has no nontrivial nilpotent ideals.

Proof. (i) Let $e_i = (0, \ldots, 1, \ldots, 0) \in R$ have all zero components except the i-th one which is the identity matrix. Then $1 = e_1 + \cdots + e_k$, $e_i e_j = 0$ for $i \neq j$, and $e_i^2 = e_i \in \text{center } R$ for all i. From this it follows that every right ideal $L \leqslant R$ is of the form $L = \oplus_{i=1}^k L_i = L_1 \oplus \cdots \oplus L_k$, where $Le_i = e_i L \cong L_i \leqslant M_{n(i)}(\Delta_i)$ is the right ideal of $M_{n(i)}(\Delta_i)$ obtained by projecting L into its ith component. Each right ideal of $M_{n(i)}(\Delta_i)$ is a vector space over $\Delta_i \cong \Delta_i \cdot 1$, where $1 \in M_{n(i)}(\Delta_i)$ is the identity matrix. Thus, first $M_{n(i)}(\Delta_i)$ satisfies the right D.C.C., and, secondly, so does also R.

(ii) In view of the latter fact, $J(R)$ is nilpotent and is the sum of all the nilpotent ideals of R. It suffices to show now that $J(R) = 0$. From 7-1.24 we conclude that $J(R) = \oplus_{i=1}^k J(M_{n(i)}(\Delta_i))$. Since $M_{n(i)}(\Delta_i)$ is a simple ring with identity (7-2.16), and since $1 \notin J(M_{n(i)}(\Delta_i)) \lhd M_{n(i)}(\Delta_i)$, we have $J(M_{n(i)}(\Delta_i)) = 0$. Thus $J(R) = 0$.

Some authors call the next theorem the Wedderburn–Artin theorem, while others including ourselves just the Wedderburn theorem.

7-1.39 Wedderburn Theorem I. *A simple ring R with the descending chain condition on right ideals is isomorphic $R \cong M_n(D)$ to the ring of all $n \times n$ matrices with entries in a division ring D, for some $1 \leqslant n$ and D. Conversely, any ring $M_n(D)$ is simple with the right D.C.C.*

Proof. Since $R^2 = R \neq 0$ and $J(R)^k = 0$ for some k (6-1.36), $R \neq J(R)$. Now by 6-7.2, R is primitive. By the Density theorem (6-5.7), there exists a simple faithful right R-module V. Set $D = \mathrm{End}_R V$. If $\{x_k\} \subset V$ are D-linearly independent, set $W_k = Dx_1 + \cdots + Dx_k$. Thus $x_{k+1} \notin W_k$ and $x_{k+1} W_k^\perp = V$. Since $W_{k+1} \supset W_k$, $W_{k+1}^\perp \subset W_k^\perp$, and the inclusion is proper because $x_{k+1} W_{k+1}^\perp = 0$ while $x_{k+1} W_k^\perp = V$. Thus $W_1^\perp \supset \cdots \supset W_k^\perp \supset W_{k+1}^\perp \supset \cdots$ is a properly descending chain of right ideals which by the D.C.C. shows that $\{x_i\}$ is finite. Set $\dim_D V = n < \infty$. Then by the Density theorem $R \cong M_n(D)$, the $n \times n$ matrix ring over D.

The above proof also established the following characterization of primitive rings with a one-sided descending chain condition.

7-1.40   Corollary. *Let R be a ring satisfying the right D.C.C. Then*

$$R \text{ is primitive} \Leftrightarrow R \text{ is simple} \Leftrightarrow R \cong M_n(D).$$

7-1.41   Wedderburn Theorem II. *Any ring R has no nonzero nilpotent ideals and satisfies the descending chain condition on right ideals if and only if R is isomorphic to a finite product of full matrix rings of finite ranks over skew fields.*

Proof. $\Leftarrow$. This was shown in 7-1.38.

$\Rightarrow$. Since $J(R) = \cap\{P \mid P \lhd R \text{ primitive}\}$, by the D.C.C. $J(R) = \bigcap_{i=1}^m P_i$ is an intersection of a finite number of primitive ideals $P_i \lhd R$ with m as small as possible. Define a ring homomorphism $\varphi: R \to \Pi_{i=1}^m R/P_i$ by $\varphi r = (r + P_1, \ldots, r + P_m) = (r + P_i)_i$. Set $\bar{R} = \varphi R$. Then $R \cong \bar{R}$ and each natural projection map $\bar{R} \to R/P_i$ is onto. The latter ring is simple by 7-1.40. Use of 7-1.37 and the minimality of m shows that $R \cong \bar{R} = R/P_1 \oplus \cdots \oplus R/P_m$ is a full direct product (or sum) of a finite number of simple rings. By Wedderburn's first theorem 7-1.39, each $R/P_i \cong M_{n(i)}(\Delta_i)$.

7-1.42   Corollary 1. *If R is a ring with Jacobson radical $J(R) = 0$ and with the D.C.C. on right ideals, then R contains an identity element.*

7-1.43   Corollary 2. *For a ring R with Jacobson radical $J(R) = 0$,*

$$\text{right D.C.C.} \Leftrightarrow \text{left D.C.C.}$$

7-2   Radicals of related rings

In this section, the radicals of various rings derived or constructed from R are related to $J(R)$. Some of the more specialized results of the
previous section, such as 7-1.24(ii) and 7-1.29 are needed here.

7-2.1   Lemma. *If* $I \lhd R$ *is an ideal, then*

$$\frac{J(R) + I}{I} \subseteq J\left(\frac{R}{I}\right).$$

Proof. Set $J = J(R)$. Then $(J + I)/I$ is a quasi-regular ideal of R/I.

Second Proof. For any $L/I \in \mathcal{M}(R/I) = \{L/I \mid I \subseteq L \in \mathcal{M}(R)\}$, $J(R) + I \subseteq L$.
Thus by use of 7-1.7, we have that

$$J\left(\frac{R}{I}\right) = \frac{\cap\{L \mid I \subseteq L \in \mathcal{M}(R)\}}{I} \supseteq \frac{J(R) + I}{I}.$$

7-2.2   Corollary. *For any* $I \lhd R$, *if* $J(R/I) = 0$, *then* $J(R) \subseteq I$.

7-2.3   Application. For any ring R (with $1 \in R$ or $1 \notin R$) form the ring
$\{0\} \times R \lhd \mathbb{Z} \times R$, $\mathbb{Z} = 0, \pm 1, \pm 2, \ldots$. Then $J[\mathbb{Z} \times R/\{0\} \times R] \cong J(\mathbb{Z}) = 0$.
Hence

   (i) $J(\mathbb{Z} \times R) \subseteq \{0\} \times R$; in particular,
   (ii) $J(R) = J(R^1)$.

7-2.4   Definition. A ring R is *semiprimitive* if the intersection of all the
primitive ideals of R is zero.

7-2.5   Remarks. (1) A ring is *semiprimitive* if and only if $J(R) = 0$. (2)
Some authors and treatises have used the term semisimple for our
semiprimitive.

7-2.6   Proposition. *For any right ideal* $L \leqslant R$,

$$J(L) = \{z \in L \mid zL \subseteq J(R)\}.$$

Proof. Set $N = \{z \in L \mid zL \subseteq J(R)\}$. (i) Since $J(L)L \subseteq J(L) \lhd L$ the elements of $J(L)L$ are quasi-regular, and since $J(L)LR \subseteq J(L)L$, $J(L)L$ is a quasi-regular right ideal of R. Hence $J(L)L \subseteq J(R)$, and thus $J(L) \subseteq N$.

Conversely, for any $n \in N$ and $b \in L$, the element $nb \in J(R)$ is quasi-regular in R, i.e. for some $\lambda \in R$, $nb + \lambda - nb\lambda = 0$. Since $n \in L$, also $\lambda \in L$. So nb is actually quasi-regular in the ring L for all $b \in L$. By 7-1.24, $n \in J(L)$. Thus $N \subseteq J(L)$, and $N = J(L)$.

Although (i) and (ii) below are immediate consequences of the formula for $J(L)$, nevertheless they follow easily from more basic principles, without invoking the last proposition.

7-2.7   Corollary. *For $L \leqslant R$, denote by $^{\perp}L$ the left ideal $^{\perp}L = \{r \in R \mid rL = 0\}$. Then*

(i) $J(R) \cap L \subseteq J(L)$.
(ii) $^{\perp}L \cap L \subseteq J(L)$.
(iii) *If R is semiprimitive, then $J(L) = {}^{\perp}L \cap L$.*

Proof. (i) Since $J(R) \cap L$ is a quasi-regular right ideal of R, the quasi-inverse of every element of $J(R) \cap L$ also belongs to $J(R) \cap L$. But then $J(R) \cap L \lhd L$ is a quasi-regular ideal in the ring L, and as such $J(R) \cap L \subseteq J(L)$.

(ii) From $R(^{\perp}L) \subseteq {}^{\perp}L$ and $^{\perp}LL = \{0\} \subseteq L$ it follows that $^{\perp}L \cap L \lhd L$ is a nilpotent ideal of L; hence surely $^{\perp}L \cap L \subseteq J(L)$.

7-2.8   Proposition. *For any $K \lhd R$, $J(K) = J(R) \cap K$.*

Proof. By previous results $J(R) \cap K \subseteq J(K) = \{z \in K \mid zK \subseteq J(R)\}$. Note that as a consequence of the fact that K is a left ideal, the above set $J(K)$ is an ideal $J(K) \lhd R$ in R. In particular then, $J(K)$ is a quasi-regular right ideal of R, and hence $J(K) \subseteq J(R)$. Thus $J(K) = J(R) \cap K$.

7-2.9   Corollary. *Every ideal of a semiprimitive ring is also semiprimitive.*

Our next objective is to see how the regular maximal right and primitive ideal spaes $\mathcal{M}(K)$ and $\mathcal{P}(K)$ of a given fixed ideal $K \lhd R$ are obtained from those of R.

7-2.10   Lemma. *For $K \lhd R$ and $M \in \mathcal{M}(R)$ with $K \nsubseteq M$, R/m is a simple right K-module.*

Proof. Since $K \lhd R$, the set $Q = \{r \in R \mid rK \subseteq M\}$ is a right ideal of R

with $M \subseteq Q$. If $Q = R$, then $RK \subseteq M$ and $K \subseteq R : M \subseteq M$, a contradiction. So by the maximality of M, $Q = M$.

Hence for any $x \in R \backslash M$, $x \notin Q$, or $xK \nsubseteq M$. Since $xK + M$ is a right ideal of R, $xK + M = R$. Thus for any $0 \neq x + M \in R/M$, $(x + M)K = (xK + M)/M = R/M$, and R/M is a simple K-module.

7-2.11   Theorem IV.   *For $K \lhd R$.*

(i) $\mathcal{M}(K) = \{M \cap K \mid K \nsubseteq M \in \mathcal{M}(R)\}$;

(ii) $\mathcal{P}(K) = \{P \cap K \mid K \nsubseteq P \in \mathcal{P}(R)\}$.

Proof. (i) For M as above in (i), $K/(M \cap K) \cong (M + K)/M = R/M$ is an isomorphism of right R-modules, and hence automatically one of right K-modules. Since R/M is a simple right K-module, so is $K/(M \cap K)$. This requires that $M \cap K \subset K$ be a maximal right K-ideal of the ring K. The relative left identity e of M can be expressed as $e = m + u \in M + K = R$ with $m \in M$, $u \in K$. The element $u \in K$ is also a relative left identity for M in R. Then $M \cap K$ is regular in K with $uk - k \in M \cap K$ for all $k \in K$. Thus $M \cap K \in \mathcal{M}(K)$.

Conversely, let $A \subset K$ be any regular maximal right K-ideal of the ring K. It is claimed that $AR \subseteq A$. For if not, and if $ar \in K \backslash A$ for some $a \in A$ and $r \in R$, then $arK + A = K$ because K/A is a simple K-module. But $rK \subseteq K$ and $arK \subseteq aK \subseteq A$. Thus $A = K$, a contradiction. So A is a right R-ideal.

There exists a $\mu \in K$ with $\mu k - k \in A$ for all $k \in K$. Set $(\mu - 1)R = \{\mu r - r \mid r \in R\}$. Next it is shown that the right R-ideal $A + (\mu - 1)R$ is proper. For if not, then $\mu = a + \mu b - b \in A + (\mu - 1)R = R$, for some $a \in A$, $b \in R$. For any $k \in K$, $\mu k = ak + \mu bk - bk \in A$ because $ubk - bk \in A$. Hence every $k = k - \mu k + \mu k \in A$, or $K = A$, a contradiction. Thus by Zorn's lemma there exists an $M \in \mathcal{M}(R)$ such that $A + (\mu - 1)R \subseteq M$, where the element $\mu \in K \cap (R \backslash M)$ is also a relative left identity for M in R. Thus $K \nsubseteq M$. Now the latter fact by the very first part of this proof implies that $M \cap K \in \mathcal{M}(K)$. Since $A \subseteq M \cap K$, the maximality of A as a right K-ideal of K requires that $A = K \cap M$ as was to be shown. (Alternatively, it is easily seen from the modular law that $M \cap K = A + [(\mu - 1)R] \cap K$, and then directly, that $(\mu - 1)R \cap K = (\mu - 1)K \subseteq A$.)

(ii) For $P \in \mathcal{P}(R)$ with $K \nsubseteq P$, write $P = R : M$ for some $M \in \mathcal{M}(R)$. If $K \subseteq M$, then $RK \subseteq K \subseteq M$ implies that $K \subseteq R : M = P$. So $K \nsubseteq M$. Hence $M \cap K \in \mathcal{M}(K)$. Since $R(P \cap K) \subseteq M \cap K$, also $K(P \cap K) \subseteq M \cap K$. Thus $P \cap K \subseteq K : (K \cap M) \equiv \{k \in K \mid Kk \subseteq K \cap M\}$. Conversely, $R = M + K$, and $R[K : (K \cap M)] = (M + K)[K : (K \cap M)] \subseteq K \cap M$. So

$K:(K \cap M) \subseteq P \cap K$. Thus $P \cap K$ is of the form $P \cap K = K:(K \cap M)$ which is a primitive K-ideal of the ring K.

Next, take $B \in \mathscr{P}(K)$. By the first part of the proof $B = K:(K \cap M)$ for some $M \cap K \in \mathscr{M}(K)$ with $K \nsubseteq M \in \mathscr{M}(R)$. This allows us to obtain $P = R:M \in \mathscr{P}(R)$. Then $(M + K)[K:(K \cap M)] \subseteq K \cap M \subseteq M$ shows that $K:(K \cap M) \subseteq P$, and $K:K \cap M \subseteq P \cap K$. Conversely, $R(P \cap K) \subseteq M \cap K$, so $P \cap K \subseteq K:K \cap M$. Thus $B = K:(K \cap M) = P \cap K$.

7-2.12 Corollary 1 to Theorem IV. *Let* $K \lhd R$. *For any* R-*module* V, *let* $V = V_K$ *be* V *regarded as a* K-*module. Then the class* $\Sigma(K)$ *of all simple* K-*modules is* $\Sigma(K) = \{V_K \mid V \in \Sigma(R),\ VK \neq 0\}$.

Proof. For any V in the latter set, $V \cong R/M$ as an R-module for some $M \in \mathscr{M}(R)$. Since $RK \nsubseteq M$, certainly $K \nsubseteq M$, and by a previous lemma R/M is a simple right K-module. Hence $V \in \Sigma(K)$.

Conversely, any module $W \in \Sigma(K)$ is K-isomorphic to one of the form $W \cong K/(M \cap K)$ for some $K \nsubseteq M \in \mathscr{M}(R)$. But as was seen in the last proof. $K/(M \cap K) \cong (M + K)/M = R/M$ are isomorphic as R-modules and hence automatically as K-modules. Thus $W \cong (R/M)_K$.

7-2.13 Corollary 2 to Theorem IV. *For any* $K \lhd R$, $J(K) = K \cap J(R)$ (see 7-2.8).

Second proof. First, the middle term may be omitted in

$$K \cap J(R) = K \cap \left[\bigcap_{K \subseteq M \in \mathscr{M}(R)} M \right] \cap \left[\bigcap_{K \nsubseteq M \in \mathscr{M}(R)} \right]$$

to yield

$$K \cap J(R) = \bigcap_{K \nsubseteq M \in \mathscr{M}(R)} (K \cap M) = J(K)$$

In this proof $\mathscr{M}(R)$ could equally well be replaced by $\mathscr{P}(R)$.

7-2.14 Matrix rings. For any integer $n \geqslant 1$, and any ring R whatever, R_n will denote the ring of $n \times n$ matrices over R. If $1 \in R$, then e_{ij} denotes the usual matrix unit with 1 in row i column j and zeros everywhere else. If $1 \notin R$, then $e_{ij} \in (R^1)_n$. Note that there are natural inclusions $R_n \lhd (R_n)^1 \subseteq (R^1)_n$.

Irrespective of whether $1 \in R$ or $1 \notin R$, write $W = \|w_{ij}\| = \Sigma w_{\alpha\beta}e_{\alpha\beta} \in R_n$, where one sums over all twice repeated indices and $w_{ij} \in R$ is the entry in row i column j of the matrix W. For some of the subsequent computations the use of the Kronecker symbol in matrix products $e_{ij}e_{pq} = e_{iq}\delta_{jp}$ is helpful. An element $r \in R$ will be written on either side in $re_{ij} = e_{ij}r$ depending upon which is more useful. For $a_i \in R$, $\text{diag}(a_1, \ldots, a_n) = a_1e_{11} + a_2e_{22} + \cdots + a_ne_{nn}$ denotes a diagonal matrix.

7-2.15 Proposition. *For any ring R and any integer n, $J(R_n) = (J(R))_n$.*

Proof. Set $J = J(R)$. Form the right ideals $U_i = e_{i1}J + e_{i2}J + \cdots + e_{in}J$ of R_n. In order to show that each $J_n \subseteq J(R_n)$, it suffices to show that each U_i is a quasi-regular right ideal, because a sum $J_n = U_1 + \cdots + U_n$ of quasi-regular right ideals is quasi-regular. Take $Z = e_{i1}z_{i1} + e_{i2}z_{i2} + \cdots + e_{in}z_{in} \in U_i$ with all $z_{ij} \in J$, $j = 1, \ldots, n$. Since z_{ii} is right quasi-regular, $(z_{ii} - 1)R = R$. Thus each z_{ij} is of the form $z_{ij} = (z_{ii} - 1)\lambda_{ij}$, $j = 1, \ldots, n$. Set $\Lambda = e_{i1}\lambda_{i1} + e_{i2}\lambda_{i2} + \cdots + e_{in}\lambda_{in}$. Then $Z \circ \Lambda = Z + \Lambda - Z\Lambda = \Sigma_j e_{ij}(z_{ij} + \lambda_{ij} - z_{ii}\lambda_{ij}) = 0$, where there is no sum on i. Thus $J_n \subseteq J(R_n)$.

In order to show that $J(R_n) \subseteq J_n$ take any $W = \|w_{\alpha\beta}\| \in J(R_n)$. It suffices to show that for each fixed p, q for all a, $b \in R$, the element $aw_{pq}b$ is right quasi-regular. Form $P = \Sigma_\alpha ae_{\alpha p}We_{q\alpha}b = (e_{11} + e_{22} + \cdots + e_{nn})aw_{pq}b = \text{diag}(aw_{pq}b, \ldots, aq_{pq}b)$. Since $P \in J(R_n)$, there exists $P^0 = Q = \|q_{\alpha\beta}\| \in J(R_n)$ with $P \circ Q = 0$. For any i, the (i, i)-entry of $P + Q - PQ = 0$ is $0 = aw_{pq}b + q_{ii} - aw_{pq}bq_{ii} = (aw_{pq}b) \circ q_{ii}$. Thus $w_{pq} \in J(R)$ and $W \in J_n$. Hence $J(R_n) = J_n$.

No introduction to the ring R_n could be complete or even adequate, if it failed to mention the next basic fact below.

7-2.16 Lemma. *Assume that the ring R has the property that for any $w \in R$, $w \in RwR$. Then every ideal $I \lhd R_n$ is of the form $I = B_n$ for some ideal $B \lhd R$.*

Proof. Define a set $B \subset R$ by $B = \{w \in R \mid \exists W = \|w_{\alpha\beta}\| \in I$ such that $w = w_{pq}$ for some p, $q\}$. For any a, $b \in R$ and any i, $j \leq n$, $ae_{ip}(\Sigma w_{\alpha\beta}e_{\alpha\beta})e_{qj}b = aw_{pq}be_{ij}$. The latter together with the hypothesis on R can now be used to show that B is an ideal by successively verifying the following in the order listed, for any w, x, $y \in B$, and any a, c, $d \in R$: $-w \in B$; $x + y \in B$; $cw \in B$, $wd \in B$.

Next, for any i, j, $Be_{ij} \subseteq I$, and also $B_n \subseteq I$. Always, $I \subseteq B_n$.

7-2.17 The ring eRe. In any ring R for any idempotent $e = e^2 \in R$, $J(eRe) = eJ(R)e$.

Proof. Abbreviate $J = J(R)$. If V is a simple R-module such that $Ve \neq 0$, then for any $0 \neq ve \in Ve$, $veR = V$. Hence $0 \neq ve(eRe) = Ve$, and Ve is a simple right eRe-module. But in that case $VJ(eRe) = VeJ(eRe) = 0$, and thus $J(eRe) \subseteq \{V^\perp \mid V \in \Sigma(R)\} = J$. Hence $J(eRe) \subseteq J \cap eRe = eJe$.

If $a \in eJe \subseteq J$, then there exists a $\lambda \in J$ with $a \circ \lambda = 0$. Thus also $e(a \circ \lambda)e = (eae) \circ (e\lambda e) = 0$. Since every element of the ideal $eJe \triangleleft eRe$ has a quasi-inverse which also belongs to the ring eRe, it follows that $eJe \subseteq J(eRe)$. Hence $J(eRe) = eJe$.

7-2.18   For any ring R, $R[x]$ denotes the ring of polynomials in x with coefficients in R. For $b \in R$, write $bx^i = x^i b$, where $bx^0 = x^0 b = b \in R[x]$. Thus $R \subset R[x]$. Note that $x^i \in R[x]$ if and only if $1 \in R$.

7-2.19   Lemma. *Suppose that $I \triangleleft R[x]$ is an ideal and that $\alpha = a_0 + \cdots + a_\mu x^\mu \in I$, $a_\mu \neq 0$, is a nonzero polynomial of least degree μ which is contained in I. If $b \in R$, if $\beta \in R[x]$ and $1 \leqslant k$ an integer, then*

  (i) $a_\mu^k b = 0 \Rightarrow \alpha a_\mu^{k-1} b = 0$, $a_\mu^{k-1} \alpha b = 0$;
  (ii) $a_\mu^k \beta = 0 \Rightarrow \alpha a_\mu^{k-1} \beta = 0$, $a_\mu^{k-1} \alpha \beta = 0$.

Proof. (i) Since $\alpha a^{k-1} b \in I$ and $a^{k-1} \alpha b \in I$, then $\deg(\alpha a_\mu^{k-1} b) \leqslant \mu - 1$ and $\deg(a_\mu^{k-1} \alpha b) \leqslant \mu - 1$, would contradict the minimality of μ.

(ii) The condition $a_\mu^k \beta = 0$ translates into the condition $a_\mu^k b_j = 0$ on the coefficients b_j of $\beta = b_0 + \cdots b_j x^j + \cdots$. Thus $a_\mu^{k-1} \alpha \beta = \Sigma a_\mu^{k-1} \alpha b_j x^j = 0$ by (i), and similarly for $\alpha a_\mu^{k-1} \beta = 0$.

The next theorem is due to S. Amitsur.

7-2.20   Theorem. *Assume that R has no nil ideals. Then $R[x]$ is semiprimitive.*

Proof. Suppose that $0 \neq \alpha = a_0 + \cdots + a_\mu x^\mu \in J(R[x])$ with $\deg \alpha$ minimal among the nonzero polynomials of $J(R[x])$. Let $\mathcal{S} = \{p \in R[x] \mid 0 \neq p \in J(R[x]), \deg p = \mu\}$ be the set of nonzero polynomials of minimal degree in $J(R[x])$. Now let I be the leading coefficients of all $p \in \mathcal{S}$ together with zero; i.e. $I = \{0\} \cup \{c_\mu \in R \mid \exists 0 \neq c_0 + \cdots + c_\mu x^\mu \in J(R[x])\}$. Then it is easy to verify that $I \triangleleft R$ is an ideal. It suffices to show that I is nil.

Since $\alpha x \delta$ is right quasi-regular for every $\delta \in R[x]$, also $\alpha x \in J(R[x])$ by 7-1.24(i). Thus there exist a $\gamma \in J(R[x])$ such that $(\alpha x) \circ \gamma = \gamma \circ (\alpha x) = 0$.

From this it follows that γ is of the form $\gamma = \beta x$ for some $\beta \in R[x]$. Cancellation of x gives

$$\alpha + \beta - \alpha\beta x = 0$$

$$\alpha + \beta - \beta\alpha x = 0.$$

It will be shown that $a_\mu^K \beta = 0$ for all sufficiently large K. Assume not. Set $m = \text{minimum}\{\deg a_\mu^K \beta \mid K = 1, 2, \ldots\}$, where the latter is a monotone decreasing sequence of nonnegative integers, because $\deg a_\mu^{K+1} \beta \leqslant \deg a_\mu^K \beta$. Thus there exists an integer n such that for $K_0 \leqslant K$, $\deg a_\mu^K \beta = m$. From now on K denotes an arbitrary integer $K \geqslant K_0$. Write $\beta = \beta_1 + x^{m+1}\beta_2$, where $\beta_1 = b_0 + \cdots + b_m x^m \in R[x]$ and $\beta_2 \in R[x]$. From $\deg a_\mu^K \beta = m$ it follows that $a_\mu^K \beta_2 = 0$, $a_\mu^K b_m \neq 0$, and that $a_\mu^K \beta = a_\mu^K \beta_1$. By the previous lemma, $a^{K-1}\alpha\beta_2 = 0$. Therefore,

$$0 = a_\mu^K(\alpha + \beta - \alpha\beta x) = a_\mu^K \alpha + a_\mu^K \beta_1 - a_\mu^K \alpha\beta_1 x.$$

The coefficient of the highest power of x, that is $x^{\mu+m+1}$, is $a_\mu^{K+1} b_m$ because the degrees of the first two terms are μ and m. Thus $a_\mu^{K-1} b_m = 0$ is a contradiction. Thus $a_\mu^K \beta = 0$. But then

$$0 = a_\mu^K(\alpha + \beta - \beta\alpha x) = a_\mu^K \alpha.$$

Once more the coefficient of the highest power $x^{\mu+1}$ of x is $a_\mu^{K+1} = 0$. Thus $I \neq 0$ is a nil ideal, a contradiction. Hence $J(R[x]) = 0$.

7-2.21 Observations. (1) For any ring R and any $I \lhd R$, set $\bar{R} = R/I$. Then $R[x]/I[x] \cong \bar{R}[x]$.

Proof. For $r \in R$, let $\bar{r}$ denote the coset $\bar{r} = r + I$. The kernel of the map $R[x] \to \bar{R}[x]$, $r_0 + \cdots + r_n x^n \to \bar{r}_0 + \cdots + \bar{r}_n x^n$ is $I[x]$. Hence $R[x]/I[x] \cong \bar{R}[x]$.

(2) In any commutative ring C the sum of two nil ideals is nil. Hence the sum $N = \Sigma\{I \mid I \lhd C, I \text{ is nil}\} = \{a \in C \mid \exists n = n(a), a^n = 0\}$ of all the nil ideals of C is the unique biggest maximal nil ideals of C. Furthermore, C/N has no nil ideals.

7-2.22 Corollary. *Suppose that C is any commutative ring, that $N \lhd C$ is its maximal nil ideal, and that $\bar{C} = C/N$. Then*

(i) $N[x]$ *is the unique maximal nil ideal of $C[x]$.*
(ii) $J(C[x]) = N[x]$.
(iii) $C[x]/J(C[x]) \cong \bar{C}[x]$.

Proof. (iii) Conclusion (iii) follows from (i) and (ii). (i) Let $M \lhd C[x]$ be the set of nilpotent elements of $C[x]$. An induction on the degree of a polynomial in $N[x]$ shows that $N[x] \subseteq M$. A direct computation shows that $\bar{C}[x]$ contains no nilpotent elements. Since $M/N[x]$ is isomorphic to a nil ideal of $\bar{C}[x]$, $M = N[x]$.

(ii) The last proposition shows that $J(\bar{C}[x]) = 0$. Now by 7-2.2 it follows that $J(C[x]) \subseteq N[x]$. But by (i), $N[x] \subseteq J(C[x])$. Thus $J(C[x]) = N[x]$.

7-2.23   For any ring R, let $R[[x]]$ denote the ring of all power series over R. Then $J(R[[x]]) = J(R) + xR[[x]]$.

Proof. First, assume $1 \in R$. The units of $R[[x]]$ are all those power series whose constant term is invertible in R. For if $\alpha = a + a_1 x + a_2 x^2 + \cdots$; $a^{-1}, a, a_1, a_2, \ldots \in R$; then $\alpha = a(1 - \lambda)$, where $\lambda = -a^{-1}a_1 x - a^{-1}a_2 x^2 - \cdots \in xR[[x]]$. Then $(1 - \lambda)^{-1} = 1 + \lambda + \lambda^2 + \cdots \in R[[x]]$. Hence $\alpha^{-1} = (1 - \lambda)^{-1}a^{-1} \in R[[x]]$. For any $\gamma = c + c_1 x + c_2 x^2 + \cdots \in R[[x]]$, $1 - \gamma\beta$ is a unit in $R[[x]]$ for all $\beta = b + b_1 x + b_2 x^2 + \cdots \in R[[x]]$ if and only if $(1 - cb)^{-1} \in R$ for $b \in R$, i.e. if and only if $c \in J(R)$. Thus by 7-1.26(i), $J(R[[x]]) = J(R) + xR[[x]]$.

Next, when $1 \notin R$, then note that $R[[x]]^1 \subset R^1[[x]]$ is a proper containment, and that $R[[x]] \cap xR^1[[x]] = xR[[x]]$. By 7-2.3, $J(R^1) = J(R)$. Since $R[[x]] \lhd R^1[[x]]$ is an ideal, its radical is given by $J(R[[x]]) = R[[x]] \cap J(R^1[[x]]) = R[[x]] \cap \{J(R) + xR^1[[x]]\} = J(R) + xR[[x]]$.

7-3   Local rings

A frequently occurring class of noncommutative rings called local rings are introduced and studied by focusing on their Jacobson radicals. Here in section 3 only, the ring R will be assumed to have an identity.

7-3.1   Nonunits. For a ring R with $1 \in R$, $U = U(R) = \{u \in R \mid \exists w \in R, uw = wu = 1\}$ denotes the multiplicative group of units. The sets N_R and N_L of right and left noninvertible elements are $N_R = \{a \in R \mid \nexists b \in R, ab = 1\}$ and $N_L = \{d \quad R \mid \nexists c \in R, cd = 1\}$. The set of nonunits N of R is $N = R \backslash U = N_R \cup N_L$. For every proper ideal such as $J(R)$, always $J(R) \subseteq N_R \cap N_L$. Also, $N_R = \{a \in R \mid aR \neq R\}$ and $N_L = \{d \mid Rd \neq R\}$.

7-3.2   Definition. A ring R is *local* if $1 \in R$ and N is an additive group.

A cyclic proof of the equivalence of eight conditions is like a long closed chain – if one link is missing, nothing is left. For this reason first a proof of the three pivotal properties (a), (b), and (c) is given.

7-3.3   Proposition. *For a ring R and N_R, N, and U as in 7-3.1, the following are all equivalent.*

   (a) $R/J(R)$ *is a division ring.*
   (b) $J(R) = N$.
   (c) $\forall x \in R$, *either* $x \in U$, *or* $1 - x \in U$.
   (d) R *is local.*
   (e) $\exists!$ *maximal right ideal.*
   (f) $J(R)$ *is a maximal right ideal.*
   (g) N_R *is an additive group.*
   (h) $J(R) = N_R$.

Proof. Set $J = J(R)$. (c) $\Rightarrow$ (a): For any $n \in N$, if $n \in N_R$, then also $nb \in N_R$ for all $b \in R$. By (c), $1 - nb \in U$, and hence $n \in J$, by 7-1.24(i). Thus $N \subseteq J$, and hence $N = J$.

(a) $\Rightarrow$ (b): For any $x \in R \backslash J$, $(x + J)R = R/J$, and $xR + J = R$. Since J is superfluous, $xR = R$. Similarly, $Rx = R$, and hence $x \in U$. Thus $N \subseteq J$ and $J = N$. Alternate proof of (a) $\Rightarrow$ (b): For $x \in R \backslash J$, $xb + n = 1$ for some $b \in R$, $n \in J$. If $n \circ n^0 = n^0 \circ n = 0$, then $xb(1 - n^0) = (1 - n)(1 - n^0) = 1$. Hence $xR = R$ and similarly $Rx = R$. Thus $x \notin N$ and $N = J$.

(b) $\Rightarrow$ (c): Since N is an additive group, for any $x \in R$, $1 = x + (1 - x)$ shows that either $x \in U$, or $1 - x \in U$.

The conclusions (b) $\Rightarrow$ (d) $\Rightarrow$ (c) are easy. Hence all that remains is to show the equivalence of (d)–(h).

(d) $\Rightarrow$ (e): Suppose that $L_1 \neq L_2$ are two maximal right ideals. Always $L_i \subseteq N$. By (d), $R = L_1 + L_2 \subseteq N$, a contradiction.

(e) $\Leftrightarrow$ (f): Since J is the intersection of the maximal right ideals, conditions (e) and (f) are equivalent.

(f) $\Rightarrow$ (g): If $x, y \in N_R$, then $xR \subseteq N_R$ and $yR \subseteq N_R$. Hence there exist maximal right ideals L_1, L_2 such that $xR \subseteq L_1$ and $yR \subseteq L_2$. But $L_1 = L_2 = J$, and hence $x - y \in J$. Consequently $R \neq (x - y)R \subseteq J$ shows that $x - y \in N_R$.

(g) $\Rightarrow$ (h): For any $n \in N_R$ and any $b \in R$, also $nb \in N_R$. Since $1 = nb + (1 - nb) \notin N_R$ and N_R is an additive group, necessarily $1 - nb \notin N_R$ for all b. Thus $n \in J$ and $N_R \subseteq J$. Hence $N_R = J$.

(h) $\Rightarrow$ (d): For any $x \in R \backslash J$, $x \notin N_R$, and hence $xR = R$. Since every element of R/J is right invertible, R/J is a division ring.

7-3.5   Corollary. *If R is a local ring, then $N = N_R = N_L$.*

A special case of the construction below will produce commutative local rings. Throughout the process below, the reader should keep in mind how the rationals are constructed from the integers.

7-3.6   Ring of fractions. Start with a commutative ring C and a multiplicative semigroup $S \subset C$. Define an equivalence relation $\sim$ on the set $C \times S$ by $(a, x) \sim (b, y)$ if $(ay - bx)s = 0$ for some $s \in S$. Denote the equivalence class of (a, x) by $a/x = \{(b, y) \,|\, (a, x) \sim (b, y)\}$. The set $Q = C \times S/\sim$ is called the *ring of quotients* of C with respect to S, where

(i) $a/x + b/y = (ay + bx)/xy \qquad a, b \in C$;
(ii) $(a/x)(b/y) = ab/xy \qquad x, y \in S$;
      $x/x = y/y = 1 \in Q, \quad 0/x = 0/y = 0 \in Q$;
      $a/x = 0 \Leftrightarrow as = 0, \quad$ some $s \in S$.

These definitions are independent of the particular choice of equivalence class representatives.

There is a natural ring homomorphism $\phi: C \to Q$, $\phi a = at/t$, $a \in C$, where $t \in S$ is any element. (If $1 \in S$, we may take $t = 1$.) Thus kernel $\phi = \{a \in C \,|\, as = 0$ for some $s \in S\} \triangleleft C$. For $x \in S$, $(\phi x)^{-1} = t/xt \in Q$. Any ideal $I \triangleleft C$ generates a corresponding ideal $\phi(I)Q \triangleleft Q$ in Q. By putting an element of $\phi(I)Q$ over a common denominator in S as in equation (i) above, since $I \triangleleft C$, the numerator will be in I. Therefore $\phi(I)Q = \{a/x \,|\, a \in I, x \in S\}$.

If $0 \in S$, then $Q = 0$. The map ϕ is monic if and only if the elements of S are not zero divisors $(as = 0, a \in C, s \in S \Rightarrow a = 0)$. In this case then identify each $a \in C$ with $a = a = at/t \in Q$, and regard $C \subset Q$.

7-3.7   Applications. Below are three special cases of $S \subset C$ for this construction.

(1)   The set S of all nonzero divisors is a semigroup if $S \neq \emptyset$.
(2)   In a commutative ring only, an ideal $P \triangleleft C$ such that its complement $R \backslash P$ is a multiplicative semigroup is called a *prime ideal*. With $S = R \backslash P$, the resulting Q is denoted by $Q = C_P = \{a/x \,|\, a, x \in C; x \notin P\}$. In this case $P \subset C \subset C_P$ are subrings of C_P. Hence PC_P is the ideal generated by P in C_P.

(3)   When C is a domain, since $(0) \lhd C$ is a prime ideal, $S = C \setminus \{0\}$ can be viewed as a special case of either (1) or (2). The resulting Q in this case is just the quotient field of C.

7-3.8   Suppose that C is a commutative ring with $P \lhd C$ any prime ideal, and $(C/P)_0$ the quotient field of the domain C/P. Then

(i) C_P is a local ring with unique maximal ideal $PC_P \lhd C_P$; $J(C_P) = PC_P$.

(ii)   The natural map

$$\frac{C}{P} \to \frac{C + PC_P}{PC_P}, \quad a + P \to a + PC_P, \quad a \in C,$$

is monic.

(iii) $C_P / PC_P \cong (C/P)_0$; moreover, every element of C_P / PC_P is of the form $\alpha \beta^{-1}$ for $\alpha, \beta \in (C + PC_P) \setminus PC_P$.

Proof. (i) Each element of $C_P \setminus PC_P$ is of the form a/x with both a, $x \in C \setminus P$. But then also $x/a \in C_P$, and $(a/x)(x/a) = 1$.

(ii) From $PC_P = \{b/y \,|\, b \in P, \, y \in C \setminus P\}$ and $C = \{at/t \,|\, a \in C, \, t \in S\}$, it follows that if $at/t = b/y \in C \cap PC_P$, then $aty = bt$. Since $ty \in C \setminus P$ but $aty \in P$, by the primeness of P, $a \in P$. Thus $C \cap PC_P = P$. Hence

$$\frac{C}{P} = \frac{C}{C \cap PC_P} \cong \frac{C + PC_P}{PC_P}.$$

(iii) Every element $a/x + PC_P \in C_P / PC_P$, $a \notin P$, is of the form $(a + PC_P)(x + PC_P)^{-1}$. If $p_1, p_2 \in P$ and $t \in C \setminus P$, then $(xt + p_1)/(yt + p_2)$ $- x/y \in PC_P$. This shows that $(C/P)_0 \to C_P / PC_P$, $(x + P)/(y + P) \to x/y + PC_P$; $x, y \in C \setminus P$; is a well defined ring homomorphism, independent of the choices of the two coset and one equivalence class representatives. Its restriction to $C/P \subset (C/P)_0$ is the natural inclusion map of C/P into C_P / PC_P in (ii). Since a nonzero homomorphism of a field is monic, $(C/P)_0 \cong C_P / PC_P$.

7-4   Examples

7-4.1   For any ring S, the following conditions are all equivalent.

(i) $\Sigma(S) = \varnothing$;
(ii) $\mathcal{M}(S) = \varnothing$;
(iii) $\mathcal{P}(S) = \varnothing$;
(iv) $J(S) = S$.
(v) $\forall n \in S$, n is right quasi-regular.

A ring satisfying (i)–(iv) is said to be a radical ring.

The next example gives a local ring without zero divisors whose radical is its own square. This local ring contains a radical subring S for which it is possible to see explicitly and directly that S has no regular right ideals whatsoever.

7-4.2   Example. For any skew field F, let $\mathbb{Q}^+$ be the additive semi-group of nonnegative rational numbers. Let $F[[\mathbb{Q}^+]]$ be the ring of all power series α of the form

$$\alpha = ax^0 + \alpha(1)x^{q(1)} + \cdots + \alpha(i)x^{q(i)} + \cdots$$

$$= x^0 a + x^{q(1)}\alpha(1) + \cdots$$

$$x^0 \equiv 1 \qquad a \equiv \alpha(0) \in F;$$

$$0 \neq \alpha(i) \in F\backslash\{0\} \quad \text{for all } i = 1, 2, \ldots;$$

where the exponents $\{0 < q(1) < \cdots < q(n) < \cdots\} \subset \mathbb{Q}^+$ form a proper ascending finite or infinite sequence without any cluster points in $\mathbb{Q}^+$. The latter exponent restriction guarantees that the formal product of two such series is (i) well defined, and (ii) satisfies the above descending chain condition on exponents. Define the lower degree $l.\deg \alpha \in \mathbb{Q}^+$ to be

$$l.\deg \alpha = \begin{cases} 0 & \text{if } a \neq 0 \\ q(1) & \text{if } a = 0. \end{cases}$$

As in 7-2.23, the element α is a unit in $\alpha \in U(F[[\mathbb{Q}^+]])$ if and only if $\alpha(0) = a \neq 0$ in which case

$$\alpha = \alpha(0)(1 - \lambda), \quad \lambda = -\alpha(0)^{-1}[\alpha - \alpha(0)],$$

$$\alpha^{-1} = (1 + \lambda + \lambda^2 + \cdots)\alpha(0)^{-1} \in F[[\mathbb{Q}^+]],$$

$$l.\deg \lambda^n = nq(1).$$

Consequently, this is a local ring with $J(F[[\mathbb{Q}^+]]) = \{\alpha \mid 0 \nleq l.\deg \alpha$, i.e. $\alpha(0) = 0\} \cup \{0\}$ as its ideal of nonunits.

For any $0 \neq \alpha$, $0 \neq \beta \in F[[\mathbb{Q}^+]]$ with $l.\deg \alpha = p \leq l.\deg \beta = q$, express $\alpha = x^p \alpha_1 = \alpha_1 x^p$, $\beta = x^q \beta_1$ where $\alpha_1, \beta_1 \in U(F[[\mathbb{Q}^+]])$ are invertible. Thus α divides β on either side $\beta = \alpha(\alpha_1^{-1}\beta_1 x^{q-p}) = (x^{q-p}\beta_1\alpha_1^{-1})\alpha$.

For any right ideal $A < F[[\mathbb{Q}^+]]$ define $q_A = \text{infimum}\{l.\deg \alpha \mid 0 \neq \alpha \in A\}$. Then either

Case 1. $\exists \alpha \in A$, $l.\deg \alpha = q_A$, and $A = \alpha F[[\mathbb{Q}^+]] = F[[\mathbb{Q}^+]]\alpha \lhd A$; or

Case 2. $A = \{\alpha \in A \mid q_A \nleq l.\deg \alpha\} \cup \{0\} \lhd A$.

Thus every one-sided ideal is automatically two sided, and the set of ideals of this ring forms a chain. Furthermore $J(F[[\mathbb{Q}^+]])^2 = J(F[[\mathbb{Q}^+]])$.

Next, let S be the ring $S = J(F[[\mathbb{Q}^+]])$ of all power series without a constant term. The same arguments as above now prove the weaker conclusion that if $0 \neq \alpha$, $0 \neq \beta \in S$ with $l.\deg\alpha \lneq l.\deg\beta$, that then $\beta = \alpha\gamma = \delta\alpha$, with γ, $\delta \in S$ also. From this it follows that the ring S does not have any maximal right ideals at all. Hence $\mathcal{M}(S) = \emptyset$ and $J(S) = S$. These conclusions are of course consistent with our previous theory which tells us that S must be a radical ring because $J(J(R)) = J(R)$ for any ring R, and here S was defined to be $S = J(F[[\mathbb{Q}^+]])$.

7-4.3   Example. Let S be that subring of the rationals consisting of all those rational numbers with even numerators but odd denominators, i.e. $S = \{2n/(2m + 1) \mid n, \ m \in \mathbb{Z} = 0, \ \pm1, \ \pm2, \ldots\}$. To discover the quasi inverse z^0 of any element $z = 2n/(2m + 1) \in S$, we work in S^1, where $(1 - z)^{-1} = 1 - z^0$, and $z^0 = 1 - (1 - z)^{-1} = (-2n)/[2(m - n) + 1]$. Thus $J(S) = S$.

This ring has two additional interesting properties. A subring of a radical ring need not be a radical ring. In $2\mathbb{Z} \subset S$, for any odd prime p, $2p\mathbb{Z} \triangleleft 2\mathbb{Z}$ is a regular maximal ideal with relative identity $p + 1$. Thus $J(2\mathbb{Z}) = 0$.

For any odd prime p, the two ideals $(2/p)S \subset (2/p)S^1 \triangleleft S$ are both prime. Hence the intersection of the prime ideals of S is zero.

7-4.4   Recall that a ring R is *simple* if $R^2 \neq 0$ and R has only the two trivial ideals (0) and R.

7-4.5   Observations. Any simple ring R has the following properties:

(1) $\forall 0 \neq x \in R$, $RxR = R$; in particular,
(2) $R^2 = R$;
(3) R is either (i) primitive, or a radical ring, (ii) $J(R) = R$.

Proof. (1) The left $l(R) = \{z \in R \mid zR = 0\} \triangleleft R$ and right $r(R) = \{z \in R \mid Rz = 0\} \triangleleft R$ annihilator ideals of R satisfy $l(R)^2 \subseteq l(R)R = 0$ and $r(R)^2 \subseteq Rr(R) = 0$. Since $R^2 \neq 0$, $l(R) \neq R$ and $r(R) \neq R$. Thus they are both zero, $l(R) = 0$ and $r(R) = 0$. Hence for any $0 \neq x \in R$, $0 \neq xR \nsubseteq r(R)$. Thus $0 \neq RxR \triangleleft R$ implies that $RxR = R$. (2) For any $0 \neq x \in R$, $R = RxR \subseteq R^2$.

(3) If $J(R) \neq R$, then $J(R) = 0$, in which case there exists a regular

138   The Jacobson radical

maximal right ideal $L < R$. But then R/L is a simple faithful right R-module and R is primitive.

There are examples of rings primitive on the right but not the left ([Bergman 64]). There also exist simple radical rings ([Sasiada and Cohn 67]).

7-5   Exercises

Prove the following:

1. For any a, $b \in R$, $a \circ ((-a) \circ b) = (a \circ (-a)) \circ b = a^2 \circ b = (-a) \circ (a \circ b) = ((-a) \circ a) \circ b$.
2. If ax is r.q.r. for every $x \in R$, then a is r.q.r.
3. If $a \in R$ is both r.q.r. and also left quasi regular, and $a + a^0 - aa^0 = 0$, then $aa^0 = a^0 a$.
4. Give a direct computational proof that the set J defined as $J = \{a \in R \mid \forall x \in R,\ ax$ is r.q.r.$\}$ is an ideal by following these steps:
   (i) $\forall a \in J$, a is r.q.r.; $\forall r \in R$, $\forall a \in J$, $ar \in J$.
   (ii) $\forall a \in J$, $-a \in J$.
   (iii) Show that $\forall a, b, x, u, w \in R$

   $$[(a-b)x] \circ (u \circ w) = [a(x-xu)] \circ w + (-(bx)) \circ u - [(-bx) \circ u]w.$$

   (iv) Let a, $b \in J$; $x \in R$ arbitrary. Select u, $w \in R$ such that $(-bx) \circ u = 0$, and $[a(x - xu) \circ w] = 0$.
   (v)   Show that $a - b \in J$ and J is a right ideal.
   (vi) $\forall a \in J$, $\lambda \in R$, if $a + \lambda - a\lambda = 0$, then $\lambda \in J$ and $a\lambda = \lambda a$.
   (vii) $\forall r \in R$, $\forall a \in J$, $ra \in J$ and $J \lhd R$.
5. If every element of R, except for one element e, is r.q.r. then R is a division ring. (Hint: Show $e^2 \circ a \neq 0$ for all $a \in R$. If $e \circ a \neq e$, show that then $0 = (e \circ a) \circ b = e \circ (a \circ b)$ for some $b \in R$. Conclude that $e \circ a = e$ for every $a \in R$. Show that $(e - a)(e - b) = e - a \circ b$.)
6. Give an example of a ring R such that $J(R) = R$ and a subring $S \subset R$ such that $J(S) = 0$.
7. ([Andrunakievitch 58]). For any ring R, let $J \lhd I \lhd R$; let J' be the ideal of R generated by J. Then $(J')^3 \subseteq J$.
8. Let R be an algebra over a commutative field F. (i) if $L < R$ is a regular maximal right ideal of R, then $FL \subseteq L$. (ii) Set $V = R/L$. Show that V is an F vector space.
9. For a ring R that is an algebra over a field, define the concept: the Jacobson radical of R as an F-algebra. Then prove that the latter concept is the same as the purely ring concept $J(R)$.

The following sequence of exercises introduces the fundamentals of group algebras. The solutions to many problems can be found in [Passman 77; p. 5–6].

Let G be a group, H a subgroup of G, and F a field. Let $F[G]$ be the group algebra (see Chapter 1, exercise 25), where $\alpha \in F[G]$ is either viewed as $\alpha: G \to F$, or $\alpha = \Sigma\{g\alpha(g) \mid g \in G, \alpha(g) \in F\}$. A *left transversal* $X \subset G$ is a set such that $G = \cup\{xH \mid x \in X\}$ and $x_1 H \neq x_2 H$ whenever $x_1 \neq x_2$.

An *involution* "$*$" on any ring R is a multiplication reversing isomorphism of order two: $(a-b)^* = a^* - b^*$, $(ab)^* = b^*a^*$, $a^{**} = a$; $a, b \in R$. Let $\rho: F[G] \to F[H]$ by $\rho\alpha = \alpha \mid H \dots$ the restriction of α to H.

10. Show that $\rho(\Sigma\{g\alpha(g) \mid g \in G\}) = \Sigma\{h\alpha(h) \mid h \in H\}$. Show that $F[G]$ is a left right $F[H]$-bimodule, and that ρ is a bimodule homomorphism, and in particular that $\rho(a\alpha b) = a(\rho a)b$ for $a, b \in F[H]$.

11. Prove that $\alpha = \Sigma x k_x$, $k_x \in F[H]$ uniquely, $x \in X$. Show that $F[G]$ is a free right or left $F[H]$-module.

12. For the complex numbers $\mathbf{C}$, let $\bar{e}$ be the complex conjugate of $c \in \mathbf{C}$. For $\alpha = \Sigma g\alpha(g) \in \mathbf{C}[G]$, define $\alpha^* = \Sigma g^{-1}\alpha(g)$ and $\alpha^\# = \Sigma g^{-1}\overline{\alpha(g)}$. Prove that "$\#$" and "$*$" are involutions.

13. If $H \lhd G$, let $\pi: G \to G/H$. Prove that $\bar{\pi}: F[G] \to F[G/H]$, $(\bar{\pi}\alpha)(gH) = \alpha(g)$ is a well-defined ring homomorphism with kernel $F[H]$. Conclude that $F[H] \lhd F[G]$ and that $F[G]/F[H] \cong F[G/H]$.

14. Prove that $F[G] \to F$, $\alpha \to \Sigma\{\alpha(g) \mid g \in G\}$, is a homomorphism with kernel $K = \{\alpha \mid \Sigma \alpha(g) = 0\} \lhd F[G]$. Show that $K = \Sigma\{(g-1)F \mid g \in G\}$.

15. If $H \lhd G$, prove that $\rho(g^{-1}\alpha g) = g^{-1}(\rho\alpha)g$ for $g \in G$.

16. Assume that for any $g \in G$, $\tilde{g} = \{x^{-1}gx \mid x \in G\}$ is finite. Define $\hat{g} \in F[G]$ as the sum of the finite number of distinct conjugates of g. Prove that the center of $F[G]$ consists of all elements of the form $\Sigma_g \hat{g}k_g$, where $k_g \in F$.

17*. Prove the Maschke theorem. Suppose that G is a finite group and F a field whose characteristic does not divide the order $|G|$ of G. Then the Jacobson radical of $F[G]$ is zero, i.e. $J(F[G]) = 0$.

CHAPTER 8

Subdirect product decompositions

Introduction

A subdirect product of modules is a module $\bar{V}$ that is a submodule $\bar{V} \subset \Pi\{V_i \mid i \in I\}$ of a direct product of modules such that each projection $\pi_i : \bar{V} \to V_i$ is onto. Given a module V belonging to some specific class of modules, the immediate objective is to represent V as being isomorphic to a subdirect product module $\bar{V}$, $V \cong \bar{V} \subset \Pi V_i$ where all the modules V_i are of some simpler and better understood class than V. The second objective is to find conditions under which a subdirect product $\bar{V} \subset \Pi V_i$ is in some sense close to being the whole product, i.e. a so called dense subdirect product.

One such choice of the V_i is subdirectly irreducible modules. A module V is subdirectly irreducible if V is *not* isomorphic to any subdirect product $V \cong \bar{V} \subset \Pi\{V_i \mid i \in I\}$ whatever such that every projection $\pi_i : \bar{V} \to V_i$ has a nonzero kernel in $\bar{V}$.

Although this chapter is phrased for rings and modules, much of the theory of subdirect products and subdirect irreducibility could be done in a more general abstract setting called universal algebras which include rings, modules, ordinary algebras, groups with operators, and semigroups as special cases. For example every appropriately defined abstract algebra belonging to a certain class of abstract algebras is a subdirect product of subdirectly irreducible algebras in the class, provided the class is closed under epimorphic images, subalgebras, and products. ([Pierce 68; p. 52, Proposition 4.1]).

Thus every definition, lemma or proposition can be stated for modules and rings. To avoid unilluminating repetition some definitions and results are stated for rings and others for modules. It is left to the reader to supply the missing counterpart. If we had simply done everything in

140

terms of either modules or rings the reader would not really be convinced
that the results hold also for the other.

Everything in this chapter is easy to prove. One of the most widely
quoted results is Birkhoff's Theorem (8-1.9) which says that every module
or ring is a subdirect product of subdirectly irreducible modules or rings.
It also holds for a wide variety of other algebraic objects including groups
and semigroups.

8-1   Subdirect products

The previous notation is continued for a ring R with or without
identity (so that $R \lhd R^1$), for the Jacobson radical $J(R)$, and the set $\mathscr{P}(R)$
of primitive ideals of R.

8-1.1   Notation. For an indexed family $\{R_i \mid i \in I\}$ of rings, elements
$\alpha \in \Pi\{R_i \mid i \in I\} = \Pi R_i$ will sometimes be regarded as functions $\alpha: I \to \cup R_i$,
$i \to \alpha(i) \in R_i$ of the index set I into the disjoint union $\cup R_i$ of the rings. The
other description $\alpha = \{\alpha(i)\}_{i \in I} = (\alpha(i))$ will also be used. As usual, the
projection map $\pi_i \colon \Pi R_i \to R_i$ is the evaluation at i, $\pi_i \alpha = \alpha(i)$.

8-1.2   Definition. $\bar{R}$ is a *subdirect product* of $\{R_i \mid i \in I\}$ if $\bar{R} \subseteq \Pi R_i$, and
if the restrictions of all projection maps to $\bar{R}$, also denoted by π_i,
$\pi_i \colon \bar{R} \to R_i$, are all onto. Note that $\oplus R_i \subseteq \Pi R_i$ is always a subdirect
product.

Although taking 'is' to be synonymous with 'is isomorphic to' is an
abuse of terminology, R will be said to be a subdirect product of the R_i,
if R is isomorphic $R \cong \bar{R}$ to some actual subdirect product in $\bar{R} \subset \Pi R_i$.

8-1.3   Proposition. *A ring R is a subdirect product of rings $\{R_i\} \Leftrightarrow$ there
exist ideals $\mathcal{O}_i \lhd R$ such that for all i.*

(i) $R/\mathcal{O}_i \cong \mathscr{R}_i$, and
(ii) $\bigcap \mathcal{O}_i = 0$.

Proof. $\Rightarrow$: When $R \subseteq \Pi R_i$ is subdirect, if $\mathcal{O}_i \lhd R$ is the kernel of $R \to R_i$,
$\alpha \to \alpha(i)$, then (i) $R/\mathcal{O}_i \cong R_i$ because π_i is onto. For $\alpha \in R$,

$$\alpha = 0 \leftrightarrow \forall i, \quad \alpha(i) = 0 \in R_i \leftrightarrow \forall i, \quad \alpha \in \mathcal{O}_i \leftrightarrow \alpha \in \cap \mathcal{O}_i.$$

Thus (ii) $\cap \mathcal{O}_i = 0$.

$\Leftarrow$: Set $R_i = R/\mathcal{O}_i$ and form ΠR_i. Map $R \to \bar{R} \subseteq \Pi R_i$ by $r \to \bar{r}$, where
$\bar{r}(i) = r + \mathcal{O}_i \in R_i$. Alternatively, $r \to (\underline{\quad}, r + \mathcal{O}_i, \underline{\quad}) \in \Pi R/\mathcal{O}_i$. Since

by hypothesis (ii) $\cap \mathcal{O}_i = 0$, $\bar{r} = 0$ if and only if $r = 0$. Thus $R \cong \bar{R} \subseteq \Pi R_i$ is subdirect, since the composition of $R \to \bar{R}$ with π_i is the natural quotients map of $R \to R/\mathcal{O}_i = R_i$.

8-1.4   Proposition. *A ring R is semiprimitive if and only if R is a subdirect product of primitive rings.*

Proof. $\Rightarrow$: Take $\mathcal{P}(R) = \{\mathcal{O}_i \mid i \in I\}$. Then each $R_i = R/\mathcal{O}_i$ is a primitive ring, and $J(R) = \cap \mathcal{O}_i = 0$ by hypothesis. Hence $R \to \bar{R} \subseteq \Pi R_i$, $r \to \bar{r} = (\underline{\quad}, r + \mathcal{O}_i, \underline{\quad})$, is a ring isomorphism of R onto a subring $\bar{R}$ which is subdirect by the previous proposition.

$\Leftarrow$: It is given that $R \subseteq \Pi R_i$ is subdirect, and that each R_i is primitive. Set $\mathcal{O}_i = \{\alpha \in R \mid \alpha(i) = 0\} \lhd R$. Then $\pi_i \colon R \to R_i$, $\pi \alpha = \alpha(i)$, is onto by hypothesis, with kernel $\pi_i = \mathcal{O}_i$. Thus $R/\mathcal{O}_i \cong \pi_i R = R_i$, and since $R/\mathcal{O}_i$ is a primitive ring, $\mathcal{O}_i \lhd R$ is a primitive ideal. Since $J(R) \subseteq \cap \mathcal{O}_i = 0$, R is semiprimitive.

8-1.5   Corollary. *A commutative ring C is primitive if and only if C is a subdirect product of fields.*

8-1.6   Definition. A module V is *subdirectly irreducible* if and only if whenever, for any indexed family of modules $\{V_i \mid i \in I\}$, V is a subdirect product $V \cong \bar{V} \subset \Pi V_i$ of the V_i, then necessarily there exists a $j \in I$ such that the projection map $\pi_j \colon V \to V_j$ is an isomorphism.

The module V is *subdirectly reducible* if it is not subdirectly irreducible.

8-1.7   Example. (i) The ring of integers $\mathbb{Z} = 0, \pm 1, \ldots$, is a subdirect product $\mathbb{Z} \to \bar{\mathbb{Z}} \subset \Pi\{\mathbb{Z}/p\mathbb{Z} \mid p \text{ is a prime}\}$, $n \to (\underline{\quad}, n + p\mathbb{Z}, \underline{\quad})$. Since $\mathbb{Z} \not\cong \mathbb{Z}/p\mathbb{Z}$ for every single p, $\mathbb{Z}$ is subdirectly reducible.

(ii) For any nonempty index set I, the diagonal map $\mathbb{Z} \to \Pi\{\mathbb{Z} \mid I\}$ is the monic map that sends n to the constant vector $\hat{n} = (\underline{\quad}, n, n, n, \underline{\quad})$ all of whose components are equal to n. Thus $\mathbb{Z} \cong \hat{\mathbb{Z}} = \{\hat{n} \mid n \in \mathbb{Z}\} \subset \Pi\{\mathbb{Z} \mid I\}$ is a subdirect product.

8-1.8   Lemma. *For any ring R, set $M = \cap\{K \mid 0 \neq K \lhd R\}$. Then R is subdirectly irreducible if and only if $M \neq 0$.*

Proof. $\Rightarrow$: If $M = 0$, by 8-1.3, R is the subdirect product

$$R \to \bar{R} \subset \Pi\{R/K \mid 0 \neq K \lhd R\}$$

$$r \to \bar{r} = (\underline{\quad}, r + K, \underline{\quad}).$$

In addition, for any $0 \neq K \lhd R$, the projection may $\bar{R} \to R/K$ is not an isomorphism. This particular decomposition of R shows that R is subdirectly reducible, a contradiction. Thus $M \neq 0$.

$\Leftarrow$: It is given that $M \neq 0$. Suppose that R is not subdirectly irreducible. Then there exist rings $\{R_i \mid i \in I\}$ so that R is isomorphic to a subdirect product $\bar{R}$, where $R \cong \bar{R} \subset \Pi\{R_i \mid i \in I\}$. Furthermore, for any index $i \in I$, $\pi_i : R \to R_i$ is not an isomorphism and hence each $0 \neq \pi_i^{-1}(0) = \mathcal{O}_i \lhd R$. Without loss of generality, $R_i = R/\mathcal{O}_i$ and $R \cong \bar{R} \subset \Pi R/\mathcal{O}_i$ with $\cap \mathcal{O}_i = 0$ by 8-1.3. But then $0 \neq M \subset \cap \mathcal{O}_i$ is a contradiction. Hence R is subdirectly irreducible.

8-1.9 Birkhoff theorem. *Every ring is a subdirect product of subdirectly irreducible rings.*

Proof. For any $0 \neq a \in R$, the set $\mathcal{S} = \{\mathcal{O} \mid \mathcal{O} \lhd R, \ a \notin \mathcal{O}\} \neq \varnothing$ is not empty because $(0) \in \mathcal{S}$. An application of Zorn's lemma to $\mathcal{S}$ shows that there exists an ideal $\mathcal{O}_a \lhd R$ in $\mathcal{S}$ maximal with respect to $a \notin \mathcal{O}_a$. Then $R/\mathcal{O}_a$ has a (nonzero) minimal ideal $M = (R^1 a R^1 + \mathcal{O}_a)/\mathcal{O}_a$ generated by $a + \mathcal{O}_a$. By the previous lemma, the ring $R/\mathcal{O}_a$ is subdirectly irreducible. Since for any $0 \neq b \in R$, $b \notin \mathcal{O}_b$, also $b \notin \cap\{\mathcal{O}_a \mid 0 \neq a \in R\}$. Thus $\cap\{\mathcal{O}_a \mid 0 \neq a \in R\} = 0$, and $R \cong \bar{R}$, where $\bar{R}$ is the subdirect product

$$R \to \bar{R} \subset \prod_{0 \neq a \in R} R/\mathcal{O}_a$$

$$r \to (\underline{\quad}, r + \mathcal{O}_a, \underline{\quad})$$

of subdirectly irreducible rings.

The next three results say that a finite subdirect product of 'simple objects' is actually a direct product of some of these.

8-1.10 Lemma. *Suppose that a module V is a subdirect product $V \subseteq \Pi_1^n V_i = \oplus_1^n V_i$ of a finite number of simple modules $V_1, \ldots, V_n$. Then there exists an integer $1 \leq k \leq n$, and indices $1 \leq i(1) < i(2) < \cdots < i(k) \leq n$ such that the natural projection π onto these components is an isomorphism $V \cong \pi V = V_{i(1)} \times \cdots \times V_{i(k)}$.*

Proof. If $n = 1$, then $V = V_1$. By induction assume that for this particular V, the lemma holds with n replaced by $1, \ldots$, or $n - 1$.

Consider case 1 when for every index i, there exists an element $(0 \underline{\quad} 0, v_i, 0 \underline{\quad} 0) \in V$ having all components zero except the i-th $0 \neq v_i \in V_i$. Due to the simplicity of the V_i,

$$\oplus_1^n (0 \underline{\quad} 0, v_i, 0 \underline{\quad} 0) R = \oplus_1^n V_i.$$

So it remains to consider when for at least one index – call it n without loss of generality – there does *not* exist an element of the form $(0 \underline{\quad} 0, v_n) \in V$ with $0 \neq v_n \in V_n$. Equivalently, the projection map $\rho_0: V \to V_1 \times \cdots \times V_{n-1}$, $\rho_0(v_1, \ldots, v_{n-1}, v_n) = (v_1, \ldots, v_{n-1})$ onto the first $n - 1$ components is monic. Thus $V \cong \rho_0 V \subseteq V_1 \times \cdots \times V_{n-1}$ is still a subdirect product. By induction, there exists a $k \leqslant n - 1$ and $1 \leqslant i(1) < \cdots < i(k) \leqslant n - 1$ such that the projection map ρ_1 of $\rho_0 V$ onto these components is an isomorphism $\rho_0 V \cong \rho_1 \rho_0 V = V_{i(1)} \times \cdots \times V_{i(k)}$. Now $\pi = \rho_1 \rho_0$ gives the required isomorphism $V \cong \rho_0 V \cong \pi V = V_{i(1)} \times \cdots \times V_{i(k)}$,

8-1.11   Lemma. *If a ring R is a subdirect product $R \subseteq S_1 \times \cdots \times S_n$ of a finite number of simple rings S_i (7-4.4), then R is naturally isomorphic to a full direct product of a finite number of them $R \cong \pi R = S_{i(1)} \times \cdots \times S_{i(k)}$ under the natural projection π.*

Proof. In case 1, for every i, there exists $0 \neq (0 \underline{\quad} 0, s_i, 0 \underline{\quad} 0) \in R \cap [(0) \times \cdots \times S_i \times \cdots \times (0)] \subseteq \Pi_1^n S_i$. By 7-4.5(1), $S_i s_i S_i = S_i$.

For any $a, b \in S_i$, since $\pi_i: R \to S_i$ is onto, there exist $\alpha, \beta \in R$ such that $\pi_i \alpha = a$, and $\pi_i \beta = b$. Thus the components of α and β other than i do not influence $\pi_i [\alpha(0 \underline{\quad} 0, s_i, 0 \underline{\quad} 0) \beta] = a s_i b \in S_i$. Hence $R(0 \underline{\quad} 0, s_i, 0 \underline{\quad} 0) R = (0) \times \cdots \times S_i \times \cdots \times (0) \subset R$, and thus $\oplus_1^n R(0 \underline{\quad} 0, s_i, 0 \underline{\quad} 0) = \oplus_1^n S_i$, and the corollary holds in case 1.

Case 2 can be proved by induction by exactly the same argument used in the previous lemma for the same case 2 there.

8-1.12   Corollary. *Any subdirect product $A \subseteq D_1 \times \cdots \times D_n$ of a finite number of division rings D_i is isomorphic to $A \cong D_{i(1)} \times \cdots \times D_{i(k)}$ a direct product of some of these under the natural projection.*

The next lemma reduces to the previous lemma 8-1.10 if below all the $R_i = R$. The proof is almost the same.

8-1.13   Lemma. *Let V_i be a simple right module over a ring R_i for $i = 1, \ldots, n$; and suppose that $\bar{R} \subseteq \Pi_1^n R_i$ and $\bar{V} \subseteq \Pi_1^n V_i$ are sub-*

direct products as in 8-2.8 *with* $VR \subseteq V$. *Here* $\Pi_1^n V_i$ *is a right* $\Pi_1^n R_i$-*module under the componentwise action. Then there exist indices* $1 \leqslant i(1) < \cdots < i(k) \leqslant n$ *such that*

$$\bar{V} \cong \pi\bar{V} = V_{i(1)} \times \cdots \times V_{i(k)}$$

are isomorphic as $\bar{R}$-*modules under the natural projection* $\pi : \Pi_1^n V_i \to V_{i(1)} \times \cdots \times V_{i(k)}$, *where* $(\pi\bar{v})\bar{r} \equiv \pi(\bar{v}\bar{r})$ *for* $\bar{v} \in \bar{V}, \bar{r} \in \bar{R}$.

8-2   Dense subdirect products

Any product ΠV_i carries a product topology with all the V_i discrete, and an R-submodule $\bar{V} \subseteq \Pi V_i$ is dense in this topology if and only if it is a dense subdirect product in the sense of the next definition.

8-2.1   Definition. A ring R is a *dense* subdirect product of an indexed family $\{R_i \mid i \in I\}$ of rings, if

> $\forall n$ and $\forall n$ distinct $i(1), \ldots, i(n) \in I$;   and
>
> $\forall a(i(k)) \in R_{i(k)}$ $k = 1, \ldots, n;$ $\to$
>
> $\exists \alpha \in R$ such that
>
> $\alpha(i(k)) = a(i(k)), \quad k = 1, \ldots, n.$

8-2.2   Lemma. *Suppose that the ring* R *is a subdirect product* $R \subset \Pi\{R_i \mid i \in I\}$ *of rings* R_i, *and that* $\mathcal{O}_i \lhd R$ *is the kernel in* R *of the projection* $\pi_i : R \to R_i$, $\pi_i(\alpha) = \alpha(i)$. *Then* R *is a dense subdirect product* $\Leftrightarrow$

> $\forall n$, $\forall$ *ordered subset of distinct* $\{i(1), \ldots, i(n)\} \subset I$,
>
> $$\mathcal{O}_{i(1)} + \bigcap_{k=2}^{n} \mathcal{O}_{i(k)} = R.$$

Proof. $\Rightarrow$: Given $\beta \in R$, define $a(i(1)) = \beta(i(1))$, and set the rest equal to zero, $a(i(2)) = \cdots = a(i(n)) = 0$. There exists an $\alpha \in R$ such that

$$\alpha(i(1)) = \beta(i(1)) = a(i(1))$$
$$\alpha(i(2)) = \quad 0 \quad = a(i(2))$$
$$\vdots$$
$$\alpha(i(n)) = \quad 0 \quad = a(i(n)).$$

Then $(\beta - \alpha)(i(1)) = 0$, while $(\beta - \alpha)(i(k)) = \beta(i(k))$ for $k = 2, \ldots, n$. Thus $\beta = (\beta - \alpha) + \alpha$ with $(\beta - \alpha) \in \mathcal{O}_{i(1)}$, and $\alpha \in \cap_2^n \mathcal{O}_{i(k)}$.

$\Leftarrow$: It suffices to show that for any n, any $i(1), \ldots, i(n) \in I$; and $a(i(k)) \in R_{i(k)}$, that there exists a $\gamma \in R$ such that $\gamma(i(1)) = a_{i(1)}$, while $\gamma(i(2)) = \cdots = \gamma(i(n)) = 0$. For the general case follows by adding n such γ, one for each $i(k)$. Take any $\alpha \in R$ with $\alpha(i(1)) = a(i(1))$. By hypothesis $\alpha = \beta + \gamma$ with $\beta \in \mathcal{O}_{i(1)}$, and $\gamma \in \cap_2^n \mathcal{O}_{i(k)}$. Then $\gamma = \alpha - \beta$ works $\gamma(i(2)) = \cdots = \gamma(i(n)) = 0$, and $\gamma(i(1)) = \alpha(i(1)) - \beta(i(1)) = \alpha(i(1)) = a_{i(1)}$.

8-2.3   Proposition. *Suppose that R is a ring containing ideals $R_i \lhd R$ for $i = 1, \ldots, n$ such that*

(a) $(R/R_1)^2 = R/R_1$; *and*
(b) $R_1 + R_k = R$ *for any* $k = 2, \ldots, n$.

Then

$$R = R_1 + R_2 R_3 \cdots R_n = R_1 + \bigcap_{k=2}^{n} R_k.$$

Proof. First note that since always $R_2 R_3 \cdots R_n \subseteq \cap_2^n R_k$, if $R = R_1 + R_2 R_3 \cdots R_n$ then necessarily also $R = R_1 + \cap_2^n R_k$.

The proposition is true for $n = 2$ and assume by induction that it holds when n is replaced by $2, \ldots, n - 1$. Given $R_1, \ldots, R_{n-1}, R_n \lhd R$ as above, the first $n - 1$ ideals $R_1, \ldots, R_{n-1} \lhd R$ alone do satisfy the above hypotheses (a) and (b). Hence by induction $R = R_1 + R_2 R_3 \cdots R_{n-1}$. However, in addition also $R = R_1 + R_n$. Condition (a) is equivalent to $R = R^2 + R_1$. Thus

$$R = R_1 + R^2 = R_1 + (R_1 + R_2 R_3 \cdots R_{n-1})(R_1 + R_n)$$
$$= R_1 + R_2 R_3 \cdots R_n.$$

8-2.4   Corollary 1. *Suppose that $M_i \lhd R$, $i \in I$, is a collection of distinct maximal ideals such that*

$$\bigcap_{i \in I} M_i = 0, \quad \text{and} \quad (R/M_i)^2 \neq 0 \quad \text{all} \quad i \in I.$$

Then R is isomorphic to a dense subdirect product $R \cong \bar{R} \subseteq \Pi R/M_i$.

Proof. By hypothesis, $R \cong \bar{R} \subset \Pi R/M_i$ under the map $r \to \bar{r}$, where $\pi_i \bar{r} = r + M_i \in R/M_i$. Let n; $\{i(1), \ldots, i(n)\} \subset I$; and n cosets $a_k + M_{i(k)} \in R/M_{i(k)}$ be given as in 2.1. For simplicity set $k = i(k)$, and let $1 = i(1)$ be any arbitrary index from these n indices. It has to be shown that there is a single $r \in R$ such that $a_i + M_i = r + M_i$ for all $i = 1, \ldots, n$. It suffices to show this for $a_1 \in R$ arbitrary, and $a_2 = \cdots = a_n = 0$. The sum of n such elements r, one for each index k, will work in the general case.

Since R/M_1 is a simple ring, the hypothesis (a) $(R/M_1)^2 = R/M_1$ holds in 8-2.3. By their maximality and distinctness, (b) $M_1 + M_k = R$ for $k \neq 1$. Now the last proposition implies that $R = M_1 + M_2 M_3 \cdots M_n$. Thus $a_1 = m_1 + r$ with $m_1 \in M_1$ and $r \in M_2 M_3 \cdots M_n$. For $j \geqslant 2$, $r + M_j = M_j$, and $r + M_1 = a_1 + M_1$ as required.

8-2.5   Corollary 2 (*Chinese remainder theorem.*) *Suppose that* $m_1, \ldots, m_n$ *is any finite set of pairwise relatively prime nonzero integers. Given any integers* $a_1, \ldots, a_n$, *the n simultaneous congruences* $x \equiv a_i \, \mathrm{mod}(m_i)$ *have a common solution* x.

Proof. It suffices to prove that there exists an x such that $x \equiv a_1 \, \mathrm{mod}(m_1)$ and $x \equiv 0 \, \mathrm{mod}(m_i)$ for $i \geqslant 2$. Set $R = \mathbb{Z} = 0, \pm 1, \pm 2, \ldots$ and $R_i = m_i \mathbb{Z} = (m_i) \lhd \mathbb{Z}$. By the last proposition $\mathbb{Z} = (m_1) + (m_2 m_3 \cdots m_n)$. Hence $1 = m_1 s + (m_2 m_3 \cdots m_n)t$ for some s, $t \in \mathbb{Z}$, and $a_1 = m_1 s a_1 + (m_2 m_3 \cdots m_n)t a_1$. Set $x = (m_2 m_3 \cdots m_n)t a_1$. Then $x \equiv a_1 \, \mathrm{mod}(m_1)$ while $x \equiv 0 \, \mathrm{mod}(m_i)$ for $i \geqslant 2$.

8-3   Exercises

1. Show that the ring of integers is isomorphic to the following subdirect products of rings.
   (a) $\bar{\mathbb{Z}} = \{(n, n, \ldots, n, \ldots) \mid n \in \mathbb{Z}\} \subset \Pi_1^\infty \mathbb{Z}$.
   (b) $\bar{\mathbb{Z}} = \{(\ldots, n + (p), \ldots) \mid n \in \mathbb{Z}\} \subset \Pi\{\mathbb{Z}/(p) \mid p = 2, 3, 5, 7, \ldots\}$.
   (c) $\bar{\mathbb{Z}} = \{(\ldots, n + (p^k), \ldots) \mid n \in \mathbb{Z}\} \subset \Pi\{\mathbb{Z}/(p^k) \mid p = 2, 3, 5, \ldots;$
         $k = 1, 2, \ldots\}$.

   Determine which of these are dense subdirect products.
2. Let $n = p_1^{\alpha(1)} \cdots p_n^{\alpha(n)}$ where p_i are distinct primes and $1 \leqslant \alpha(i)$. Prove that $\mathbb{Z}/n\mathbb{Z} \cong \mathbb{Z}/(p^{\alpha(1)}) \times \cdots \times \mathbb{Z}/(\rho^{\alpha(n)})$. Generalize this to commutative principal ideal domains.
3. Let D_i, $i \in I$ be a family of division rings and let $\mathscr{F}$ be an ultrafilter on I containing all singletons. (See exercises for Chapter 0.) Define $\Pi^{\mathscr{F}}\{D_i \mid i \in I\} = \{(d_i) \in \Pi\{D_i \mid \{i \mid d_i = 0\} \in \mathscr{F}\}\}$. Prove that $\Pi\{D_i \mid i \in I\}/\Pi^{\mathscr{F}}\{D_i \mid i \in I\}$ is a division ring.
4. Let $Z_4 = \mathbb{Z}/4\mathbb{Z}$, $e = (1, 1, \ldots, 1, \ldots) \in \Pi_1^\infty Z_4$, and define $R = \oplus_1^\infty 2Z_4 + Z_4 e$. (a) Show that $R \subset \Pi_1^\infty Z_4 = T$ is a (i) subdirect product, but (ii) not a dense subdirect product. (iii) Show that $R < T$ is not an essential extension of R-modules.
5. For a fixed prime $p \in \mathbb{Z}$, show that $R \subset \Pi_1^\infty \mathbb{Z}$ is a nondense subdirect product, where $\Pi_1^\infty (p) \subset R = \{(r_i) \mid \forall i, j, r_i - r_j \in (p)\}$. Generalize this to any commutative principal ideal domain.

CHAPTER 9

Primes and Semiprimes

Introduction

Suppose that C is some property that an ideal of a ring may or may not have, and that $\mathscr{C}$ is the class of all ideals satisfying C, for all possible rings. Thus C and $\mathscr{C}$ completely determine each other. Then the C-radial C-rad R of R is defined to be $C\text{-rad } R = \cap\{P \mid P \triangleleft R, P \text{ has } C\}$. Examples of this are $C = $ primitive, or $C = $ prime; the corresponding radicals are the Jacobson radical Rad R, and the prime radical rad R. The method used on primitive ideals could be applied to other classes of ideals $\mathscr{C}$ with corresponding property C, and is summarized below

Property C	$\mathscr{C}$	$C\text{-rad } R = \cap\{P \triangleleft R \mid P \in \mathscr{C}\}$	$C\text{-rad } R = 0$; R is semi-C
primitive	primitive ideals	$J(R) = \text{Rad } R$ Jacobson radical	semiprimitive
prime	prime ideals	rad R prime radical	semiprime

The type of questions asked and answered in both cases is similar: (i) For an arbitrary $P \triangleleft R$, to find necesssary and sufficient conditions in order for P to possess property C. (ii) Various descriptions and characterizations of C-rad R are found. If available, particularly useful are elementwise descriptions giving necessary and sufficient conditions on an arbitrary element $z \in R$ in order that $z \in C\text{-rad } R$. (iii) In both cases, $C\text{-rad}[C\text{-rad } R] = C - \text{rad } R$ and $C\text{-rad}[R/(C\text{-rad } R)] = 0$. Also, useful and illuminating are theorems that assert that certain types of one or two

148

sided ideals N always necessarily contained inside the C-radical, i.e. $N \subseteq C\text{-rad } R$.

In general, a ring R is said to have a certain property applicable to ideals, if and only if the zero ideal in that ring has this property, e.g. primitive, semiprimitive, prime, and semiprime. After having studied property C and supposedly understood the class $\mathscr{C}$, we now factor out $C\text{-rad } R$. Then $R/(C\text{-rad } R)$ is a subdirect product of C-rings (primitive or prime rings). Again, having studied the formation of subdirect products from the individual rings forming it, all that now remains is to investigate a C-ring (a primitive or prime ring).

In a commutative ring the prime radical is just the set of nilpotent elements. In a noncommutative ring the latter need not even form an ideal (e.g. a matrix ring), but the strongly nilpotent ones do, and in fact they are equal to the prime radical in general (theorem 9-2.8). This theorem and theorem 9-2.16 which gives three characterizations of a prime ring R are two of the more fundamental and important aspects of this chapter. The fact that primeness and semiprimeness can be extended to noncommutative rings in an interesting way which does not amount to just mimicking the commutative case seems striking and unexpected. This gave an impetus to non-commutative ring theory to develop analogues of other commutative concepts.

At the very end of the chapter we quote and explain the Hilbert Nullstellensatz. Although it is a commutative ring theory theorem, a subject not covered in this book, we use this theorem as an example to illustrate the concepts of this chapter.

9-1 Prime ideals

Most of this section (9-1.4–9-1.8) develops various methods whereby one can prove that a given ideal is or is not prime.

9-1.1 Notation. As before R is a completely arbitrary ring with $1 \in R$ or $1 \notin R$; $R = R^1$ if $1 \in R$, and $R \equiv \{0\} \times R \lhd R^1 = \mathbb{Z} \times R$ if $1 \notin R$, see 1-3.1 to 1-3.5.

9-1.2 Definition. An ideal $P \lhd R$ is *prime* if $P \neq R$ and if

$$\forall A \lhd R, B \lhd R, AB \subseteq P \Rightarrow A \subseteq P, \quad \text{or} \quad B \subseteq P.$$

The ring R is a *prime ring* if $(0) \lhd R$ is a prime ideal.

From this it follows that $P \lhd R$ is a prime ideal if and only if R/P is a prime ring.

An equivalent definition of $P \lhd R$ being prime is that for any ideals A and B of R

if $A \nsubseteq P$ and if $B \nsubseteq P$, then $\Rightarrow$ also $AB \nsubseteq P$.

9-1.3   Definition. A nonempty subset $S \subset R$ is an *m-system* if

$a, b \in S \Rightarrow \exists x \in R$ such that $axb \in S$.

9-1.4   Lemma. For $P \lhd R$, the following are equivalent

(i) $P \lhd R$ is prime.
(ii) $\forall\, a, b \in R$, if $aRb \subset P$, then $\Rightarrow a \in P$ or $b \in P$.
(iii) $R \backslash P$ is an *m*-system.

Proof. The property (ii) is equivalent to

$\forall\, a, b \in R \backslash P, \qquad aRb \cap (R \backslash P) \neq \varnothing.$

Thus (ii) $\Leftrightarrow$ (iii).

(i) $\Rightarrow$ (ii): If $aRb \subseteq P$, then $(RaR)(RbR) \subseteq P$, where $RaR \lhd R$. So $RaR \subseteq P$ or $RbR \subseteq P$. Set $B = R^1 bR^1 = Rb + bR + RbR + \mathbb{Z}b$, and verify that $B^3 \subseteq RbR$. If $RbR \subseteq P$, then $B^3 = B^2 B \subseteq P$; thus $B^2 \subseteq P$, and so $B \subseteq P$. Hence $b \in B \subseteq P$.

(ii) $\Rightarrow$ (i). Given ideals A, B or R with $A \subseteq P$, and $B \subseteq P$, take $a \in A \backslash P$, $b \in B \backslash P$. Then $AB \supseteq aRb \nsubseteq P$.

9-1.5   Observations

(1)   An ideal $P \lhd R$ is prime if and only if $\forall\, A \leqslant R, B \leqslant R$, if $AB \subseteq P$, then $\Rightarrow A \subseteq P$ or $B \subseteq P$.

   Proof. Since $R^1 AR^1 \lhd R$, it follows from $(R^1 AR^1)(R^1 BR^1) = AB + RAB \subseteq P$, that either $A \subseteq R^1 AR^1 \subseteq P$, or that $B \subseteq P$.

(2)   For any two right ideals $A \leqslant R, B \leqslant R$ and any ideal $P \lhd R$,

$AB \subseteq P \Leftrightarrow (A + P)(B + P) \subseteq P;$

$AB \nsubseteq P \Leftrightarrow P \subsetneqq AB + P.$

(3)   Hence any $P \lhd R$ is prime if and only if $\forall\, P \subseteq A' \leqslant R, \forall\, P \subseteq B' \leqslant R$, if $A'B' \subseteq P$, then $\Rightarrow A' = P$ or $B' = P$.

9-1.6   Lemma. *An ideal $P \lhd R$ is prime if and only if for any ideals C, $D \lhd R$ such that $P \subsetneqq C$ and $P \subsetneqq D$, also $P \subsetneqq CD + P$.*

Proof. ⇒: For any $a \in C \backslash P$ and $b \in D \backslash P$, also $aRb \nsubseteq P$ by 9-1.4(ii), but $aRb \subseteq CD$. Thus $P \neq DC + P$.

⇐: Given A, $B \lhd R$ with $A \nsubseteq P$ and $B \nsubseteq P$, set $C = A + P \supsetneq P$ and $D = B + P \supsetneq P$. By hypothesis $P \subsetneq CD + P = AB + AP + PB + P^2$. Since $P \lhd R$, also $AP + PB + P^2 \subseteq P$, and thus $AB \nsubseteq P$. Now, P is prime by the equivalent form of 10-1.2.

9-1.7   Corollary. *An ideal $P \lhd R$ is prime ⇔ for any ideals $P \subsetneq C \lhd R$, $P \subsetneq D \lhd R$, it follows that $CD \nsubseteq P$.*

9-1.8   Corollary. *The previous Lemma 9-1.6 and Corollary 9-1.7 also hold if $C, D \lhd R$ are replaced by merely right ideals.*

9-1.9   Proposition. *Let $S \subset R$ be an m-system with $0 \notin S$. By Zorn's lemma let $P \lhd R$ be an ideal of R maximal with respect to $P \cap S = \varnothing$. Then $P \lhd R$ is prime.*

Proof. Let $P \subsetneq C \lhd R$ and $P \subsetneq D \lhd R$. By the maximality of P, take $x \in C \cap S \neq \varnothing$, and $y \in D \cap S \neq \varnothing$. Since S is an m-system, there exists an $r \in R$ with $xry \in xRy \cap S \neq \varnothing$. So $xry \notin P$ but $xry \in CD$. Therefore $CD \nsubseteq P$ (i.e. $P \subsetneq CD + P$). So P is a prime ideal by the last lemma.

9-1.10   Lemma. *For any ring R and any $M \lhd R$ a proper ideal, the following hold:*

   (i)   *M is primitive ⇒ M is prime.*
   (ii) *M is maximal as an ideal, and M is regular as a right ideal ⇒ M primitive.*

Proof. (i) By hypothesis $M = V^{\perp}$ for some simple right R-module V. Suppose that $A, B \lhd R$ with $AB \subseteq M$. If $VA \neq 0$, then $VA \leqslant V$ is a submodule and hence $V = VA$. But then $VB = VAB = 0$, $B \subseteq M$, and $M \lhd R$ is prime.

   (ii) By Zorn's lemma, there exists a regular maximal right ideal L with $M \subseteq L$. Thus $RM \subseteq M \subset L$, and consequently $M \subseteq (R/L)^{\perp}$, where $(R/L)^{\perp} \lhd R$ is a primitive ideal. But by the maximality of M, $M = (R/L)^{\perp}$ is primitive.

9-2   Semiprime ideals and the prime radical

Among other things, this section gives various necessary and sufficient conditions for an ideal $I \lhd R$ to be semiprime. However, the reader has to translate the results from the case considered here, $I = 0$.

9-2.1   Definition. The *prime radical* $\operatorname{rad} R$ of R is defined to be the intersection $\operatorname{rad} R = \cap \{ P \mid P \lhd R$ is prime$\}$ of all the prime ideals of R. The empty intersection by definition and convention is all of R.

The ring R is *semiprime* if $\operatorname{rad} R = 0$; $I \lhd R$ is a *semiprime ideal* if R/I is a semiprime ring.

9-2.2   Lemma. *Let* $K \lhd R$ *and* $P \lhd R$ *with* $K \subset P$. *Then*

(i) $P \lhd R$ *is prime* $\Leftrightarrow P/K \lhd R/K$ *is prime.*

(ii)   *For any* $K \lhd R$

$$\operatorname{rad} \frac{R}{K} = \frac{\cap \{ P \mid K \subseteq P,\ P \lhd R \text{ is prime} \}}{K}$$

(iii)   *prime* $\Rightarrow$ *semiprime.*

Proof. Conclusions (i) and (iii) are clear. (ii) By (i), $\operatorname{rad}(R/K) = \cap \{ P/ K \mid K \subseteq P,\ P \lhd R \text{ prime} \} = [\cap \{ P \mid K \subseteq P,\ P \lhd R \text{ prime} \}]/K$, where the last equality follows from a set theoretic argument showing inclusion both ways.

9-2.3   Corollary. *For any ring, let* $\mathscr{S} = \{ K \lhd R \mid R/K \text{ is semiprime} \}$ *be the set of semiprime ideals of* R. *Then*

(i) $\operatorname{rad}(R/\operatorname{rad} R) = 0$, *i.e.* $\operatorname{rad} R \in \mathscr{S}$;

(ii)   *moreover,* $\operatorname{rad} R$ *is the unique smallest element of* $\mathscr{S}$.

Proof. Conclusion (i) follows by taking $K = \operatorname{rad} R$ above in 9-2.2(ii). For any $K \in \mathscr{S}$, by 9-2.2(ii), $\operatorname{rad} R/K = 0$ means that $K = \cap \{ P \mid K \subseteq P,\ P \lhd R \text{ prime} \} \supseteq \operatorname{rad} R$.

9-2.4   Definition. A sequence $b_0, b_1, b_2, \ldots$ in R indexed by the non-negative integers is an *m-sequence* if there exist $c_0, c_1, c_2, \ldots \in R$ such that $b_i = b_{i-1} c_{i-1} b_{i-1}$ for all $i = 1, 2, \ldots$.

9-2.5   Consequences. Let $b_0, b_1, b_2, \ldots$ be an m-sequence in R.

(1)   Then for any $i < j$, $b_j \in b_i R b_i$. Each element of an m-sequence is simultaneously divisible on both sides by all its predecessors in

the sequence. In particular, if one element of an m-sequence is zero, then all the subsequent terms are identically zero.

(2) An m-sequence is an m-system.

Proof. It has to be shown that $b_j R b_i \cap \{b_n\} \neq \emptyset$ for all i and j. Suppose that $i \leqslant j = i + k$. Then $b_{j+1} = b_j c_j b_j = b_j c_j b_{j-1} c_{j-1} b_{j-1} = b_j c_j b_{j-1} c_{j-1} b_{j-2} c_{j-2} b_{j-2} = \cdots = b_j z b_i$ for some $z \in R$.

(3) For any m-system S, and for any $x \in S$ whatever, if $a_0 = x \in S$, then $a_1 = a_0 c_0 a_0 \in S$, $a_2 = a_1 c_1 a_1 \in S$, $a_3 = a_2 c_2 a_2 \in S$, etc., for some c_0, c_1, $c_2, \ldots \in R$; and thus there exists an m-sequence $\{a_n\} \subseteq S$ which passes through $x = a_0$.

9-2.6   Definition. An element $x \in R$ is *strongly nilpotent* if for any m-system S in R containing $x \in S$, necessarily also $0 \in S$.

Thus completely equivalently, $x \in R$ is strongly nilpotent if and only if for any m-sequence $\{a_n\}$ such that $a_0 = x$, eventually $0 \in \{a_n\}$.

9-2.7   Consequences. (1) Let $T \subset R$ be a subring of R. (i) An m-system or sequence in T is automatically one in R. (ii) An element $t \in T$ which is strongly nilpotent in the subring T need not be strongly nilpotent when viewed as an element $t \in R$ of the bigger ring R.

(2) Any multiplicative semigroup $S \subset R$ is an m-system.

(3) Hence in particular for any $a \in R$, $S = \{a, a^2, a^3, \ldots\}$ is an m-system.

(4) Any strongly nilpotent element $a \in R$ is nilpotent. For $\{a, a^2, \ldots\}$ is an m-system containing a, and hence $0 = a^N \in \{a, a^2, \ldots\}$ for some N. Or, alternatively in terms of m-sequences, $a_0 = a$, $a_1 = a^3$, $a_2 = a^7, \ldots$, $a_{n+1} = a_n^2 a, \ldots$ is an m-sequence (which in general *is* not a semigroup) which eventually is zero if a is strongly nilpotent.

(5) For a commutative ring D; an element $a \in D$ is strongly nilpotent if and only if a is nilpotent.

Proof. $\Leftarrow$:   Let $a^N = 0$.   Given   any m-sequence $\{a_n\}$, $a_0 = a$, $a_1 \in a_0 D a_0 = a^2 D$, $a_2 \in a^4 D, \ldots, a_n \in a^{2^n} D = 0$ if $N \leqslant 2^n$.

9-2.8   Theorem. *For an element $x \in R$, the following are all equivalent.*

(1) $x \in \mathrm{rad}\, R$.

(2) $\forall$ m-system S, if $x \in S$, then $\Rightarrow 0 \in S$.

(3) $\forall$ m-sequence $\{a_n\}$ such that $a_0 = x$, there exists an N such that $0 = a_N = a_{N+1} = \cdots$.

Proof. The above shows that $(2) \Leftrightarrow (3)$. $(1) \Rightarrow (2)$: If not, let S be an m-system with $x \in S$ but $0 \notin S$. Take $P \lhd R$ maximal with respect to $P \cap S = \varnothing$. Then $P \lhd R$ is prime, but $x \notin P \supseteq \operatorname{rad} R$, a contradiction. So $0 \in S$. $(2) \Rightarrow (1)$: For any prime $P \lhd R$, $0 \notin S = R \backslash P$, which is an m-system. Hence $x \notin R \backslash P$, or $x \in P$ for every prime ideal P. Thus $x \in \operatorname{rad} R$.

The next corollary merely rephrases the last theorem.

9-2.9   Corollary. *The prime radical* $\operatorname{rad} R$ *of any ring* R *is exactly the set of all strongly nilpotent elements of* R.

9-2.10   Proposition. *A ring* R *is semiprime if and only if for every* $0 \neq x \in R$, *also* $xRx \neq 0$.

Proof. $\Rightarrow$: If $x \neq 0$, there exists a prime ideal $P \lhd R$ with $x \notin P$. Since $x \in R \backslash P$, and $R \backslash P$ is a m-system, $xRx \cap (R \backslash P) \neq \varnothing$.
$\Leftarrow$: If $x \neq 0$, set $a_0 = x$; select $0 \neq a_1 \in xRx$, $0 \neq a_2 \in a_1 R a_1, \ldots$ and get an m-sequence with $0 \notin \{a_n\}$. Thus $x \notin \operatorname{rad} R$ by 9-2.8(3).

9-2.11   Corollary. *The ring* R *is semiprime* $\Leftrightarrow$ *for any right ideal* $0 \neq A \leqslant R$, $A^2 \neq 0$.

Proof. $\Rightarrow$: For any $0 \neq a \in A$, $aRa \neq 0$, $(aRa)R(aRa) \neq 0$, and hence $0 \neq aR(aR) \subset A \neq 0$.
$\Leftarrow$: If $xRx = 0$ for $x \in R$, then $xRxR = 0$. By hypothesis $xR = 0$. Hence $x^2 = 0$. But then $xR^1xR^1 = (\mathbb{Z}x + xR)(\mathbb{Z}x + xR) = \mathbb{Z}x^2 + x^2R + xRx + xRxR = 0$. Thus $x \in xR^1 = (0)$ again by hypothesis.

9-2.12   Remark. Let R be a semiprime ring. Then

  (i) $A, B \leqslant R$, $AB = 0 \Rightarrow BA = 0$.
  (ii) $C \leqslant R$, $x \in R$, $Cx = 0 \Rightarrow xC = 0$.

Proof. (i) For $x \in BA$, $xRx \subseteq BARBA = 0$. (ii) Since $xC \leqslant R$, $(xC)^2 = 0$.

In the commutative case, a straightforward argument shows that the ring is prime if and only if it is a domain, while finally 9-2.7(5) and 9-2.8(3) imply that it is semiprime if and only if it has no nilpotent elements.

9-2.13 **Proposition.** *Let D be any commutative ring. Then*

(i) $P \triangleleft D$ *is prime* $\Leftrightarrow \begin{cases} \forall\ a,\ b \in D \ \textit{if}\ ab \in P, \\ \textit{then}\ a \in P\ \textit{or}\ b \in P. \end{cases}$

(ii) $\operatorname{rad} D = \bigcap_{P \triangleleft D \text{ prime}} P = \{x \in D \mid \exists\, n = n(x),\ x^n = 0\}$

Proof. (i) $\Rightarrow$: Suppose that $ab \in P$. As a consequence of commutativity, then $aDb \subseteq P$, and either $a \in P$ or $b \in P$ by primeness of P. (i) $\Leftarrow$: This converse does not require D to be commutative. If $A, B \triangleleft D$ with $AB \subseteq P$ but $A \nsubseteq P$, then take $a \in A \backslash P$. Thus $aB \subseteq P$. By hypothesis now $B \subseteq P$, and P is prime.

(ii) For any $x \in D$ such that $x^n \neq 0$ for all n, let $P \triangleleft D$ be maximal with respect to $P \cap \{x^n \mid n = 1, 2, \dots\} = \varnothing$. Then $x \notin P \supset \operatorname{rad} D$. Conversely, let $x^n = 0$ for $n \geqslant 2$. Then for any prime $P \triangleleft D$, $0 = x^{n-1}x \in P$, and $x \in P$, or $x^{n-1} \in P$ by (i) above. Thus $x \in \operatorname{rad} D$ and (ii) has been proved.

Alternatively, but less directly, this follows immediately from 9-2.9 and 9-2.7(5).

9-2.14 **Corollary.** *Suppose that $\mathscr{U}$ and $\mathscr{L}$ are upward and downward directed sets of prime ideals in a commutative ring D in the sense that*

$$\forall\ A_1, A_2 \in \mathscr{U} \to \exists\ A_3 \in \mathscr{U}, A_1 + A_2 \subseteq A_3, \quad \textit{and}$$

$$\forall\ A_1, A_2 \in \mathscr{L} \to \exists\ A_3 \in \mathscr{L}, A_3 \subseteq A_1 \cap A_2.$$

Then $\cup \mathscr{U} \equiv \cup \{P \mid P \in \mathscr{U}\} \triangleleft D$ *and* $\cap \mathscr{L} \equiv \cap \{P \mid P \in \mathscr{L}\} \triangleleft D$ *are also prime ideals.*

Now we return again to a completely arbitrary ring R.

9-2.15 **Remark.** For any ring R, define ideals $R_l, R_r, R_m \triangleleft R$ as $R_l = \{x \in R \mid xR = 0\}$, $R_r = \{y \mid Ry = 0\}$, and $R_m = \{z \mid RzR = 0\}$. Then

(i) $R_l^2 = 0$, $R_r^2 = 0$, $RR_m \subseteq R_l$, and $R_m R \subseteq R_r$.

(ii) In particular, if R has no nilpotent ideals, then $R_m \subseteq R_r \cap R_l = 0$; and for any $0 \neq a \in R$, also $RaR \neq 0$.

9-2.16 **Theorem.** *For any ring R, the following are all equivalent.*

(1) (0) *is the only nilpotent ideal of R.*

(2) $\cap \{P \mid P \triangleleft R \text{ is prime}\} = (0)$.

(3) $\forall\ A, B \triangleleft R$, *if $AB = 0$, then $\Rightarrow A \cap B = 0$.*

Proof. $(1) \Rightarrow (2)$: For any $0 \neq a \in R$, $RaR \neq 0$, and hence $0 \neq (RaR)(RaR) \subseteq RaRaR$ shows that $aRa \neq 0$. Set $a_0 = a$, and select an m-sequence $S = \{a_n\}$ with $0 \neq a_1 \in a_0 R a_0$, $0 \neq a_2 \in a_1 R a_1, \dots$. Then $a \in S$, $0 \notin S$, and there exists a prime ideal $P \triangleleft R$; $P \cap S = \varnothing$, and $a \notin P \supseteq \operatorname{rad} R$. Thus $\operatorname{rad} R = 0$.

$(2) \Rightarrow (3)$: For any prime $P \triangleleft R$, $0 = AB \subseteq P$. Since $A \subseteq P$ or $B \subseteq P$, certainly $A \cap B \subseteq P$. Thus $A \cap B \subseteq \operatorname{rad} R = 0$.

$(3) \Rightarrow (1)$: Let $A \triangleleft R$ with $A^n = 0$. Successive applications of hypothesis (3) gives the following implications: $AA^{n-1} = 0 \rightarrow A^{n-1} = A \cap A^{n-1} = 0$; $AA^{n-2} = 0 \rightarrow A^{n-2} = A \cap A^{n-2} = 0, \dots$ till $A = 0$.

9-2.17 Corollary. For any ring R,

(i) $\operatorname{rad} R = \cap \{K \mid K \triangleleft R, R/K \text{ has no nilpotent ideals}\}$; hence

(ii) $\operatorname{rad} R$ is the unique smallest ideal of R modulo which the quotient ring has no nilpotent ideals.

Proof. (i) and (ii). The first and second equals signs below follow from 9-2.3 and 9-2.16(1) respectively.

$$\operatorname{rad} R = \cap \{K \mid K \triangleleft R, R/K \text{ is semiprime}\}$$
$$= \cap \{K \mid K \triangleleft R, R/K \text{ has no nilpotent ideals}\}.$$

9-3   Nil radicals

Corollary 9-2.16 and the fact that $\operatorname{rad} R$ is always a nil ideal of R suggests that nil and nilpotent ideals should be studied at the same time as prime and semiprime ones in order that these four concepts can be interrelated where they overlap.

9-3.1   (1) If R is a nil (nilpotent) ring so is every subring and every homomorphic image of R.

(2) If $B \triangleleft R$ such that B and R/B are nil (nilpotent), then R is nil (nilpotent).

(3) The sum of two nil (nilpotent) ideals of a ring is a nil (nilpotent) ideal.

(4) The sum $A + B$ of two nilpotent right ideals $A, B \leqslant R$ is a nilpotent right ideal $A + B \leqslant R$.

Proof. (2) Let $B \triangleleft R$ and R/B be nilpotent. There exist p and q such that $b_1 b_2 \dots b_q = 0$ for any $b_i \in B$, and $x_1 x_2 \dots x_p \in B$ for any $x_i \in R$. For $1 \leqslant i \leqslant n = pq$, let $y_i \in R$ be arbitrary. Set $b_1 = y_1 y_2 \dots y_p$, $b_2 =$

$y_{p+1} y_{p+2} \cdots y_{2p}, \ldots b_q = y_{(q-1)p+1} \cdots y_{qp} \in B$. Then $y_1 y_2 \ldots y_n = b_1 b_2 \ldots b_q = 0$. To prove the nil case, take in the above argument $b_1 = b_2 = \cdots = b_q = b \in B$, $x_1 = \cdots = x_p = x \in R$, and $y_1 = \cdots = y_n = y \in R$.

(3) Let $A, B \lhd R$ be nil (nilpotent). Then $(A + B)/A \cong B/(A \cap B)$. The latter is a nil (nilpotent) ring. Since both A and $(A + B)/A$ are nil (nilpotent) rings, by (2) above, so is also $A + B$.

(4) Let $A^p = 0$ and $B^q = 0$. For any $x \in R$, $(xA)^p \subseteq xA^p = 0$, and also $(RA)^p = 0$. Consequently $(RA + A)^p = (RA)^p + (RA)^{p-1}A + \cdots + (RA)A^{p-1} + A^p \subseteq RA^p + A^p = 0$. In view of $A + RA \lhd R$, now by (3) above $A + B + RA + RB \lhd R$ is a nilpotent ideal of R. Hence $A + B$ is nilpotent.

Remark. For $A, B \leqslant R$ assume that A and B are nil, and that $A \cap B \lhd B$. Since both A and $(A + B)/A$ are nil rings, so is $A + B$.

9-3.2 For any ring R whatever as before, define $N(R) = \Sigma\{A \mid A \leqslant R, A \text{ is nilpotent}\}$ to be the sum of all the nilpotent right ideals of R. For the fixed ring R, the ideal $U \lhd R$ is defined as $U = \Sigma\{I \mid I \lhd R, I \text{ is nil}\}$. Then

(i) $N(R) = \Sigma\{I \mid I \lhd R \text{ is nilpotent}\}$;
(ii) $N(R) \subseteq U$;
(iii) U is nil.

Proof. (i) and (ii). As previously noted, for any $A \leqslant R$ with $A^P = 0$, $A + RA \lhd R$, and $(A + RA)^P \subseteq RA^P + A^P = 0$. (iii) Any element of U belongs to a finite sum of nil ideals, which by 9-3.1(3) is a nil ideal.

9-3.3 **Definition.** An ideal $B \lhd R$ is a *nil radical* if (i) B is nil and (ii) R/B contains no ($\neq 0$) nilpotent ideals.

9-3.4 **Consequence.** By 9-2.3, rad R is the unique smallest nil radical of R which is contained in every nil radical of R.

9-3.5 **Notation.** For an arbitrary but fixed ring R, define $N(0) = 0$, $N(1) = N(R)$, and $N(2) \lhd R$ as $N[R/N(1)] = N(2)/N(1)$. Assume that there is an ordinal β such that for every $\alpha < \beta$, $N(\alpha) \lhd R$ has already been defined, and that $N(\gamma) \subseteq N(\alpha)$ if $\gamma < \alpha$. If β is not a limit ordinal, then $\beta = \alpha + 1$ and define $N(\beta)$ by

$$N[R/N(\alpha)] = N(\alpha + 1)/N(\alpha).$$

Otherwise, when β is a limit ordinal, simply set $N(\beta) = \cup\{N(\alpha) \mid \alpha < \beta\} \lhd R$. There exists a unique smallest ordinal τ such that $N(\alpha) \subsetneqq N(\tau)$ for $\alpha < \tau$, but $N(\tau) = N(\tau + 1) = \cdots$.

9-3.6   Proposition. *With the above notation,* rad $R = N(\tau)$.

Proof. Since $N[R/N(\tau)] = 0$, the ring $R/N(\tau)$ has no nilpotent ideals. By 9-2.3, rad $R \subseteq N(\tau)$. By 9-2.17, it suffices to show that for any ideal $K \lhd R$ such that R/K has no ($\neq 0$) nilpotent ideals, $N(\tau) \subseteq K$. First, $N(0) \subseteq K$. It is easy to see by using 9-3.1 (3) that $N(1) \subseteq K$. Assume by induction that for some ordinal β it has already been shown that $N(\alpha) \subseteq K$ for all $\alpha < \beta$. If β is a limit ordinal, then $N(\beta) = \cup\{N(\alpha) \mid \alpha < \beta\} \subseteq K$ trivially. If $\beta = \alpha + 1$, then $N(\alpha + 1) = \Sigma\{I \mid N(\alpha) \subseteq I \lhd R, \ I/N(\alpha)$ is nilpotent$\}$. Thus $N(\alpha + 1) = \Sigma\{I \mid I \lhd R, \ \exists p, \ I^p \subseteq N(\alpha)\}$. In particular for every I appearing in the sum for $N(\alpha + 1)$, $I^p \subseteq N(\alpha) \subseteq K$.

Now, if $N(\alpha + 1) \nsubseteq K$, then there is an I as above, $I \nsubseteq K$, and hence the ideal $(I + K)/K \lhd R/K$ is nilpotent, because $(I + K)^p \subseteq I^p + K \subseteq K$, a contradiction. Thus also $N(\alpha + 1) \subseteq K$, and by induction $N(\tau) \subseteq K$. To repeat, this means that $N(\tau) \subseteq \cap\{K \mid K \lhd R, \ R/K$ has no nilpotent ideals$\} = $ rad R. Thus $N(\tau) = $ rad R.

9-4   Primes and semiprimes in derived rings

9-4.1   Lemma. *Suppose that $K \lhd R$ is a fixed ideal, and that $P, Q \lhd R$ denote two other ideals. Then*

(i) $P \lhd R$ *prime* $\Rightarrow P \cap K \lhd K$ *is prime.*
(ii) $Q \lhd R$ *semiprime* $\Rightarrow Q \cap K \lhd K$ *is semiprime.*

Proof. (i) Let $xKy \subseteq P \cap K$ for $x, y \in K$. Suppose that $y \notin P$. Then since P is prime in R, it follows from $xKy \supseteq xKRy = (xK)Ry \subseteq P$ that $xK \subseteq P$. But then $Rx \subseteq K$, $xRx \subseteq xK \subseteq P$, and $x \in P$. Thus $x \in P \cap K$, and the latter is prime in K.

(ii) If not, then $xKx \subseteq Q \cap K$ for some $x \in K$, but $x \notin Q$. Then since $xKRx \subseteq Q$, also $(xK)R(xK) \subseteq Q$, and hence $xK \subseteq Q$ by the semiprimeness of Q in R. But then in view of $Rx \subseteq K$, $xRx \subseteq xK \subseteq Q$ contradicts the semiprimeness of Q, in R, since $x \notin Q$. Thus $Q \cap K \lhd K$ is semiprime.

In the last lemma we saw how properties of the big ring R induced similar properties in the subring K. The next lemma goes in the opposite direction.

9-4.2   Lemma. *Suppose that* $K \lhd R$ *is any fixed ideal, then any semi-prime ideal* $L \lhd K$ *of the ring* K *is an ideal* $L \lhd R$ *in the larger ring* R.

Proof. For any $r \in R$ and $\lambda \in L$, at least $\lambda r \in LR \subseteq K$, and it remains only to show that $\lambda r \in L$. (The proof that $r\lambda \in L$, being symmetric, will be omitted.) Thus $(\lambda r K)(\lambda r K) = \lambda(rK)\lambda(rK) \subseteq \lambda K \lambda K \subseteq L$, where λrK is a right K-ideal of the ring K. The semiprimeness of L in K implies that $\lambda r K \subseteq L$. But then also $\lambda r K \lambda r \subseteq \lambda r K \subseteq L$ with $\lambda r \in K$. A second application of the semiprimeness of L, gives that $\lambda r \in L$.

9-4.3   Lemma. *If* $T \subset R$ *is any subring of* R, *then* $T \cap \operatorname{rad} R \subseteq \operatorname{rad} T$.

Proof. For an $x \in T \cap \operatorname{rad} R$, any m-system in R containing x also contains zero. But an m-system in T is automatically an m-system in R. Hence any m-system in T containing x also necessarily contains zero. But any element $x \in T$ having this property is a strongly nilpotent element of T, and hence belongs to $x \in \operatorname{rad} T$ the prime radical of T.

9-4.4   Corollary. *For any ring* R, $\operatorname{rad}(\operatorname{rad} R) = \operatorname{rad} R$.

Proof. For $T = \operatorname{rad} R$, the last lemma becomes $\operatorname{rad} R \subseteq \operatorname{rad}(\operatorname{rad} R)$.

9-4.5   Prime radical of an ideal. *If* $K \lhd R$ *is an ideal in any ring* R, *then the prime radical* $\operatorname{rad} K$ *of* K *viewed as a ring in its own right is* $\operatorname{rad} K = K \cap \operatorname{rad} R$.

Proof. By 9-4.1(i), $\operatorname{rad} K \subseteq K \cap \operatorname{rad} R$ while 9-4.3 gives the opposite inclusion.

9-4.6   Lemma. *Let* R *be any ring and* R_n *the ring of* $n \times n$ *matrices over* R *for some integer* n. *Then* R_n *is a prime ring if and only if* R *is a prime ring.*

Proof. The usual matrix conventions of summing over all repeated indices, of using $e_{ij} \in (R^1)_n$ to denote the usual matrix units, as well as $\|a_{ij}\| = \Sigma e_{\alpha\beta}a_{\alpha\beta} \in R_n$, $a_{\alpha\beta} \in R$, will be used.

R_n prime $\Rightarrow R$ prime: If $aRb = 0$ for a, $b \in R$, then for any $\|r_{ij}\| \in R_n$, if $\operatorname{diag}(a, \ldots, a) \in R_n$ is the diagonal matrix with a on the diagonal, then we have $\operatorname{diag}(a, \ldots, a)\|r_{ij}\|\operatorname{diag}(b, \ldots, b) = \|ar_{ij}b\|$. Thus $\operatorname{diag}(a, \ldots, a)R_n$ $\operatorname{diag}(b, \ldots, b) = 0$, and $a = 0$ or $b = 0$.

R prime $\Rightarrow R_n$ prime: Let $\|a_{i\alpha}\|(R_n)\|b_{\beta j}\| = 0$. Then for any $p, q \leqslant n$ and any $r \in R$, $\|a_{i\alpha}\|(re_{pq})\|b_{\beta j}\| = \Sigma e_{ij}a_{ip}rb_{qj} = 0$. Thus $a_{ip}rb_{qj} = 0$ for any i, j. Now, if $a_{ip} \neq 0$ for some fixed i and p, then $a_{ip}Rb_{qj} = 0$ for all choices of q and j. Thus $\|b_{qj}\| = 0$.

9-4.7   Lemma. *If* $B \triangleleft R$, *then there is a natural isomorphism* $(R/B)_n \rightarrow R_n/B_n$.

Proof. Let E_{ij} and e_{ij} denote the matrix units in the rings $(R/B)_n$ and R_n respectively. (I.e. $E_{ij} \in ((R/B)^1)_n$ while $e_{ij} \in (R^1)_n$). The required isomorphism is $\psi(\Sigma E_{ij}(r_{ij} + B)) = (\Sigma e_{ij}r_{ij}) + B_n$.

9-4.8   Prime radical of R_n. For any integer n, let R_n denote the ring of all $n \times n$ matrices with entries in R, for any ring R whatever. Suppose that the ring R satisfies the restrictive hypothesis that for any $w \in R$, also $w \in RwR$. Then

  (i)   the prime ideals of R_n are precisely of the form $P_n \triangleleft R_n$ where $P \triangleleft R$ is prime; and
  (ii) $\text{rad}(R_n) = (\text{rad } R)_n$.

Proof. (i) As seen before in 7-2.16, the additional hypothesis guarantees that every ideal of R_n is of the form $B_n \triangleleft R_n$ for some $B \triangleleft R$. It follows from the last two lemmas that the ring $R_n/B_n \cong (R/B)_n$ is prime if and only if $B \triangleleft R$ is prime.

  (ii) If $\mathscr{S}$ is any set of ideals of R whatsoever, then $\cap \{B_n \mid B \in \mathscr{S}\} = [\cap \{B \mid B \in \mathscr{S}\}]_n$.

Sometimes a question about semiprime rings can be reduced to a more easily answerable question regarding prime rings by the following application of 8-1.4.

9-4.9   Observation. Any semiprime ring R is a subdirect product of prime rings as follows:

$$R \xrightarrow{\ \cong\ } \bar{R} \subseteq \Pi\{R/P \mid P \triangleleft R \text{ is prime}\},$$

$$r \rightarrow \bar{r} = (\ldots, r + P, \ldots), \ r \in R.$$

We will quote without proof from commutative ring theory one of the more famous and historically influential early theorems which still today is frequently needed in algebraic geometry. It will not be used in any subsequent chapter in this book, but is used merely as an illustration of the ideas of this chapter.

9-4.10   Hilbert Nullstellensatz. For a commutative algebraically closed field F, let $F[x_1, \ldots, x_n]$ be the polynomial ring in n-indeterminates, and $I \lhd F[x_1, \ldots, x_n]$ any ideal whatsoever. Define $V(I) \subseteq F^n = F \times F \times \cdots \times F$ by $V(I) = \{(c_1,\ c_2, \ldots, c_n) \in F^n \mid \forall f \in I,\ f(c_1, \ldots, c_n) = 0\}$. Suppose that $g \in F[x_1, \ldots, x_n]$ is any polynomial which vanishes $g(c_1, \ldots, c_n) = 0$ for all $(c_1, \ldots, c_n) \in V(I)$, i.e. $g \mid V(I) \equiv 0$. Then for some integer m, $g^m \in I$.

Above, if $I = F[x_1, \ldots, x_n]$, then $V(I) = \varnothing$. Conversely, suppose that $I \lhd F][x_1, \ldots, x_n]$ is such that $V(I) = \varnothing$. Then $g = 1$ vacuously satisfies the requirement $g \mid V(I) = 0$ above. Hence $g = 1 \in I = F[x_1, \ldots, x_n]$. Let us summarize the conclusion of the foregoing argument in a useful corollary.

9-4.11   Corollary 1. *For F algebraically closed and $I \lhd F[x_1, \ldots, x_n]$*

$$I \neq F[x_1, \ldots, x_n] \Leftrightarrow V(I) \neq \varnothing.$$

9-4.12   Definition. For any commutative ring D (with or without identity) and any ideal $I \lhd D$, define $\sqrt{I} = \{g \in D \mid \exists m = m(g),\ g^m \in I\}$. The ideal $\sqrt{I} \lhd D$ is sometimes called the *radical* of I.

We now translate the Nullstellensatz into a statement about the prime radical of the quotient ring $F[x_1, \ldots, x_n]/I$.

9-4.13   Example. *For any commutative ring D and any $I \lhd D$*, by 9-2.13 the prime radical of D/I is $\mathrm{rad}(D/I) = \sqrt{I}/I$.

From now on take $D = F[x_1, \ldots, x_n]$ and $I \lhd D$. The previous theorem then is the statement

$$\{g \in D \mid g \mid V(I) = 0\} = \sqrt{I} = \{g \in D \mid g + I \in \mathrm{rad}(D/I)\}.$$

9-4.14   Remarks. (1) According to the Hilbert basis theorem 3-5.4, $D = F[x_1, \ldots, x_n]$ is Noetherian, and hence any ideal $I \lhd D$ is finitely generated $I = f_1 D + \cdots + f_k D$ by some $f_k \in D$.

(2) Therefore 9-4.10 says that $g^m = f_1 g_1 + \cdots + f_k g_k$ for some $g_k \in D$.

(3) Define $M = (x_1 - \lambda_1)D + \cdots + (x_n - \lambda_n)D \lhd D$ for any $\lambda_i \in F$. Then $D/M \cong F$ and M is a maximal ideal.

(4) Suppose now that $I \lhd D$, $I \neq D$ is proper. The first corollary above shows that there exists at least one point $(\lambda_1, \ldots, \lambda_n) \in V(I) \neq \varnothing$. If $I = f_1 D + \cdots + f_k D$, then $f_i \mid V(I) = 0$, and hence $f_i(\lambda_1, \ldots, \lambda_n) = 0$. From this, by use of some elementary facts about polynomials, we can conclude

that $f_i \in M = (x_1 - \lambda_1)D + \cdots + (x_n - \lambda_n)D$, $1 \leqslant i \leqslant k$, and hence that $I \subseteq M$.

The above says that any proper ideal I is contained in a maximal ideal M of a certain particular kind as above.

What are the maximal ideals of $F[x_1, \ldots, x_n]$? For $n = 1$, since the polynomial ring $F[x]$ is a principal ideal domain, the maximal ideals are all $(x - \lambda)F[x]$, $\lambda \in F$, because F is algebraically closed. It is now easy and instructive to deduce from the Nullstellensatz what are the maximal ideals of $F[x_1, \ldots, x_n]$, which is no longer a principal ideal domain when $n \geqslant 2$. If $I \lhd D$ is maximal then by the general remark 9-4.14(4), $I \subseteq M$. Since both are maximal, $I = M$. Let us recapitulate this as a second corollary.

9-4.15   Corollary 2. For an algebraically closed field F, there is a bijective correspondence between the maximal ideals $M \lhd D = F[x_1, \ldots, x_n]$ and the points of F^n given by

$$M = (x_1 - \lambda_1)D + \cdots + (x_n - \lambda_n)D \to (\lambda_1, \ldots, \lambda_n) \in F^n; \; \lambda_i \in F.$$

9-5   Exercises

Below, $\mathbf{Z}$, $\mathbf{Q}$, and J_p are the rings of integers, rationals, and p-adic numbers. Find the prime radicals of the following rings in exercises 1-3.

1. $\mathbf{Z}/(m)$, $m = p_1^{\alpha(1)} \cdots p_k^{\alpha(k)} \in \mathbf{Z}$, p_i distinct primes, $1 \leqslant \alpha(i)$.

2. (a) $\begin{vmatrix} \mathbf{Z} & \mathbf{Z} \\ 0 & \mathbf{Z} \end{vmatrix}$ (b) $\begin{vmatrix} \mathbf{Z} & 0 \\ \mathbf{Q} & \mathbf{Z} \end{vmatrix}$ (c) $\begin{vmatrix} J_p & Z(p^\infty) \\ 0 & J_p \end{vmatrix}$.

3. In the ring $R = \mathbf{Z}e_{11} + \mathbf{Z}e_{12} + p\mathbf{Z}e_{21} + \mathbf{Z}e_{22}$ show that both $J \subset I$ are prime ideals where $J = p\mathbf{Z}e_{11} + \mathbf{Z}e_{12} + p^2\mathbf{Z}e_{21} + p\mathbf{Z}e_{22}$ and $I = \mathbf{Z}e_{11} + \mathbf{Z}e_{12} + p\mathbf{Z}e_{21} + p\mathbf{Z}e_{22}$. (See [Dauns 78; p. 180, 6.11].)

4. Find $\text{rad}(\Pi_1^\infty \mathbf{Z}/(\oplus_1^\infty \mathbf{Z}))$.

5. For any rings $\{R_i \mid i \in I\}$, find (i) $\text{rad} \, \Pi_{i \in I} \, R_i$, and (ii) $\text{rad}(\oplus_{i \in I} R_i)$ in terms of $\text{rad} \, R_i$.

CHAPTER 10

Projective modules and more on Wedderburn theorems

Introduction

At this point the reader might wish to refamiliarize her or himself with the basic Wedderburn theory, covered in 7-1.35 through 7-1.43. Since we first use projective modules in this chapter (in Theorem 10-4.4), they are discussed at the beginning of this chapter. The reader who is mainly interested in the refinements and additions to Wedderburn theory of this chapter, need only read 10-1.1 through 10-1.13. Since projective modules are important in their own right, and since we need them in many subsequent chapters, the rest of the elementary theory of projectives is also developed in the first section of this chapter.

Projective modules are generalizations of free modules, because a free module is projective. For this reason logically, in successively studying more and more general classes of modules, they could very well be studied right after the free modules and before the injectives. In our treatment of free, projective, and injective modules we are assuming the ring R has an identity $1 \neq 0$ and all modules are unital. However, from then on in sections 10-2 throughout 10-7, we most decidedly do not assume that the ring has an identity. This chapter can be looked upon as being an improvement and extension of previous more basic material – first, the projective modules generalize the free ones, and then we develop in more detail the fundamental Wedderburn theory already covered in 7-1.35 through 7-1.43.

Proposition 10-1.5 proves three equivalent characterizations of projective modules, one of which is that a module is projective if and only if it is a direct summand of some free module. This shows that projective modules are very close relatives of free modules.

163

In various module theoretic proofs it is desirable to produce appropriate direct summands of a given module. And what makes the projective and injective modules very basic and fundamental classes of modules is that both classes give rise to direct summands. Another characterization of a projective module P is that if there is any surjective module map $M \to P$, then P is a direct summand of M (10-1.5(2)).

Since free modules have bases, and projectives generalize the free ones, it is logical to ask whether there is a concept of a basis for a projective module P. The answer is given by Proposition 10-1.14 which sometimes is called the Dual Basis Lemma, which is a slight misnomer because an arbitrary projective module P can not have a basis, for if it did it would have to be free. The one defining property of a basis that is preserved is that every element of P is a finite R-linear combination of the basis elements. This latter property alone is of course not enough, for the whole set P has it. In the special case when $R = F$ is a field and P a vector space over F, the Dual Basis Lemma gives a vector space basis of P, and the dual basis for the dual vector space $\text{Hom}_F(P, F)$.

Schanuel's Lemma is used to show that the projective dimension of a module is well defined (10-2.5). It is with regret that the $\text{Ext}_R^n(\cdot, \cdot)$ functors are not covered. As a substitute, a rather explicit and detailed sequence of exercises together with some hints guide the reader clear through projective resolutions of complexes, homology groups, $\text{Ext}_R^n(\cdot, \cdot)$ functors, and the long exact sequence associated with them.

This chapter focuses on semiprimitive rings with a descending chain condition on one-sided ideals. The Wedderburn Theorems say that a ring R belongs to this class if and only if it is isomorphic to a finite direct sum of complete finite matrix rings over division rings. These theorems were derived quickly as a consequence of the Jacobson Density Theorem already in Chapter 7. Here some alternate approaches to the Wedderburn Theorems are given. Thus it is assumed that a finite intersection $J(R) = \cap_1^m P_i = 0$ of primitive ideals P_i is zero, and from this a completely independent direct proof given that then R is a finite direct sum of simple rings.

In this chapter one does not want to assume an identity element for the ring R. Indeed, it will be a consequence of the two hypotheses on the ring that R must have an identity.

A brief summary of chain conditions is given, which ideally are formulated and proved in the context of groups with operators.

Ring theoretically one of the more important parts of this chapter, is section three which treats rings which contain a minimal right ideal, and for this the descending chain condition is not assumed.

In reading the literature, care must be exercised with the term "semi-simple" – different authors define it differently. Here this term is not used.

Finite dimensional algebras over a field appear often in other areas of mathematics, e.g. in the theory of irreducible representations of finite groups, and even in physics. The Wedderburn structure theory presented here is an indispensable tool for understanding these semiprimitive finite dimensional algebras. For example, let G be any finite group, and F any field whose characteristic does not divide the order of G. In particular, F may have characteristic zero. Then the group algebra $F[G]$ (see Exercise 1-25(d)) is known to be semiprimitive and is of finite dimension $|G|$ over F. It then follows from the first Wedderburn Theorem (6-1.39) that $F[G] = S_1 \oplus \cdots \oplus S_m$, with each $S_i \cong M_{n(i)}(D_i)$, where the latter is the $n(i) \times n(i)$ matrix ring over a division ring D_i with $F \subseteq$ center D_i. Then it follows from this chapter that every simple $F[G]$-module is isomorphic to some minimal right ideal of some $M_{n(i)}(D_i)$. The latter may be taken to consist of all matrices in $M_{n(i)}(D_i)$ whose first row is arbitrary, but the remaining $n(i) - 1$ rows, zero. For more details, see [Ribenboim 69, p. 143, Theorem 20] and [Curtis and Reiner 62; p. 186, Theorem 27.22].

10-1 Projective modules

Free modules are a special case of the larger class of projective modules. Unlike for free modules, in the next two definitions the map $\bar{\mu}$ producing the commutative diagram need not be unique.

10-1.1 Definition. An R-module P is *projective* if for any module map $\mu: P \to B$, and for any epimorphism $\alpha: A \to B$, there exist a homomorphism $\bar{\mu}: P \to A$ such that $\alpha\bar{\mu} = \mu$. In other words, given any data as in the first diagram, it can be completed with a dashed arrow so that the second one commutes.

If $\mu = \alpha\bar{\mu}$, then μ is said to factor through α.

10-1.2 Recall that R-module I is *injective* if for every module homomorphism $\alpha: B \to I$ and every monomorphism $\alpha: B \to A$, there exists an R-map $\bar{\mu}: A \to I$ such that $\bar{\mu}\alpha = \mu$.

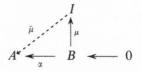

10-1.3   Duality. Replacement of "P" by "I" and reversal of all arrows in the diagrams in 10-1.1 defining a projective results in the diagram 10-1.2 which defines an injective. Similarly, if in the categorical definition 1-2.2(ii) of a direct sum, all arrows are reversed and the direct sum "$\oplus$" is replaced by the direct product "Π", one obtains the categorical definition of the direct product.

If a theorem and its proof about projectives is expressed categorically without reference to elements, and if it involves only dualizable objects and concepts, then one obtains a theorem about injectives for free by this process of dualization. Dualizing twice brings one back where one started from. The concept of a free module does not seem to have a natural or useful dual concept.

In order to dualize some of the proofs in this and the next section only the four correspondences above the dotted line are required.

Duality

direct sum $\oplus$	⋯⋯	direct product Π
epic	⋯⋯	monic
quotient module	⋯⋯	submodule
projective	⋯⋯	injective
⋯⋯⋯⋯⋯⋯⋯⋯⋯⋯⋯⋯⋯⋯⋯⋯⋯⋯⋯⋯⋯⋯⋯⋯		
direct limit	⋯⋯	inverse limit
cokernel	⋯⋯	kernel

10-1.4   Lemma. *A free R-module F is projective.*

Proof. Let $X \subset F$ be a basis of F, and let any R-map $\mu: F \to B$ and an epimorphism $\alpha: A \to B$ be given. Our objective will be to produce the dotted arrow "$\bar{\mu}$" so that the diagram below commutes

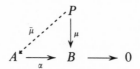

For each $x \in X$, select an element $a \in A$ such that $\alpha a = \mu x$. This is possible because α is onto, i.e. $\mu x \in B = \alpha A$. First, define $\bar{\mu} x = a$. Each $f \in F$ is uniquely of the form $f = x_1 r_1 + \cdots + x_n r_n$, for n distinct $x_i \in X$, and $0 \neq r_i \in R$. Elements $a_i \in A$ have already been selected such that $\bar{\mu}(x_i) = a_i$. If $\bar{\mu}$ is to be an R-homomorphism, it must be defined as $\bar{\mu} f = (\bar{\mu} x_1) r_1 + \cdots + (\bar{\mu} x_n) r_n = a_1 r_1 + \cdots + a_n r_n$. Now it can be readily verified that $\bar{\mu}$ is an R-module homomorphism.

10-1.5   Proposition. For an R-module P, the following are all equivalent

(1) P is projective.
(2)   Every exact sequence $M \xrightarrow{\rho} P \to 0$ splits.
(3) P is a direct summand of a free R-module F.

Proof. $(1) \Rightarrow (2)$. If $i = 1_P$ is the identity map on P, then the projectivity of P guarantees that i factors through ρ by a map $\mu: P \to M$ with $i = \rho \mu$. In other words, there exists a "μ" such that the diagram below commutes

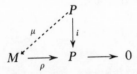

$(2) \Rightarrow (3)$. As seen previously (2-1.9) for any generating set $T \subset P$ (with $P = \langle T \rangle$), if $F = \oplus \{R \mid T\}$ is the free R-module on T, then there is an epimorphism $\pi: F \to P \to 0$. With $M = F$, by (2), the latter sequence splits. Hence $F \cong P \oplus \ker \pi$, and (3) holds.

$(3) \Rightarrow (1)$. Let a map $\mu: P \to B$ and an epic $\alpha: A \to B$ be given. By hypothesis (3), there exists a free module F such that $F = P \oplus Q$ where $Q \subset F$ is some (not necessarily unique) submodule of F. Let $j: P \to F$ be the inclusion map $jp = (p, 0) \in P \oplus Q = F$ and $\pi: P \oplus Q \to P$ the projection $\pi(p, q) = p$ (along Q). Since F being free is projective, for the previous α and for $\mu\pi: F \to B$, there exists a $\gamma: F \to A$ such that $\alpha\gamma = \mu\pi$. Define $\bar{\mu} = \gamma j: P \to A$.

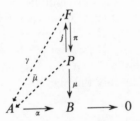

Then $\alpha\bar{\mu} = \alpha\gamma j = \mu\pi j = \mu 1_P = \mu$, since $\pi j = 1_P$ is the identity map on P.

The previous characterization (2) of 10-1.5 of projectives is so frequently used that for emphasis it is reformulated below as a corollary.

10-1.6   Corollary 1. *If $M \to P$ is epic with P projective, then $M \cong P \oplus N$ for some submodule $N \subseteq M$.*

10-1.7   Corollary 2. *A direct summand of a projective module is projective.*

Proof. Let P be projective and $P = Q \oplus L$ By (3) of 10-1.5 without loss of generality we may assume that $F = G \oplus P$ where F is free. Hence $F = G \oplus Q \oplus L$, and hence again by (3) of 10-1.5, both Q and L are projective.

Conclusion (3) of 10-1.5 can be used not only to prove that a given module P is projective by showing that it is a direct summand of a free or projective module, but it can be used also in reverse. Given a projective module P, sometimes it is useful to embed it as a direct summand of a free module F, in which case we might want to take F as small as possible. This is underscored by the next corollary.

10.1.8   Corollary 3. *If P is a projective module with a set T of generators, $P = \langle T \rangle$, then P is a direct summand of the free module $F = \oplus \{R \mid T\}$ on T.*

10-1.9   Corollary 4. *For any idempotent $e = e^2 \in R$, eR is a projective right R-module.*

Proof. Since eR is a direct summand of the free module $R = eR \oplus (1 - e)R$, it is projective.

10-1.10   Notation. Let $f: \oplus \{P_i \mid i \in I\} \to B$ be any module homomorphism, and let $j_i: P_i \to \oplus P_i$ be the inclusion map. For any such f, define $f_i: P_i \to B$ to be the restriction $f_i = fj_i = f \mid P_i$ of f to P_i. For any $x = (x_i) \in \oplus P_i$, $x = \Sigma j_i x_i$. Thus $fx = \Sigma fj_i x_i = \Sigma f_i x_i$. For $(b_i)_{i \in I} \in \oplus \{b \mid I\}$, define $s: \oplus \{B \mid I\} \to B$ by $s[(b_i)_{i \in I}] = \Sigma b_i \in B$. Then f is the composite $f = so(\oplus f_i)$ as in 1-2.1. Conversely, given only the map $f_i: P_i \to B$, they define a unique map $f = so(\oplus f_i): \oplus P_i \to B$.

10-1.11   Lemma. *Let $\{P_i \mid i \in I\}$ be any family of R-modules for an arbi-*

trary index set I. Then

$$\bigoplus_{i\in I} P_i \text{ is projective} \Leftrightarrow \text{each } P_i \text{ is projective.}$$

Proof. ⇒: Given that $\oplus P_i$ is projective, it will be shown that for any k, P_k is projective. Let $\pi_k: \oplus P_i \to P_k$ be the projection, and suppose that a map $f_k: P_k \to B$ and an epic $\alpha: A \to B \to 0$ are given. Since $\oplus P_i$ is projective, $f_k \pi_k$ factors through a map $\gamma: \oplus P_i \to A$ such that $\alpha \gamma = f_k \pi_k$. If $j_k: P_k \to \oplus P_i$ is the inclusion map, then $\pi_k j_k$ is the identity on P_k and hence $f_k = f_k \pi_k j_k = \alpha \gamma j_k$ where $\gamma j_k: P_k \to A$. Thus P_k is projective.

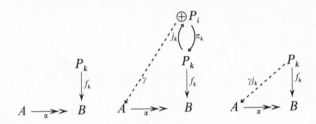

⇐: Assume all the P_i are projective and let $f: \oplus P_i \to B$ and an epimorphism $\alpha: A \to B$ be given. For each i, let $f_i = f j_i: P_i \to B$ (see 10-1.10). Then by the projectivity of P_i there is a map $\bar{f}_i: P_i \to A$ such that $f_i = \alpha \bar{f}_i$. As in 10-1.10, simply define $\bar{f} = \text{so}(\oplus\{\bar{f}_i | i \in I\})$, i.e. for any $x = (x_i)_{i \in I} \in \oplus P_i$, $\bar{f}x = \Sigma \bar{f}_i x_i$. Then $\alpha(\bar{f}x) = \Sigma \alpha \bar{f}_i x_i$. It now follows from $\alpha \bar{f}_i = f_i$ and from $f = \text{so}(\oplus\{f j_i | i \in I\}) = \text{so}(\oplus f_i)$ that $\alpha(\bar{f}x) = \Sigma f_i x_i = fx$. Thus $\alpha \bar{f} = f$, and hence $\oplus P_i$ is projective:

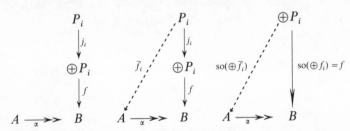

10-1.12 Example. If $\mathbb{Z} = 0, \pm 1, \pm 2, \ldots$ and $R = \mathbb{Z}/6\mathbb{Z} = \{\bar{0}, \bar{1}, \bar{2}, \bar{3}, \bar{4}, \bar{5}\}$, then $R = \{\bar{0}, \bar{3}\} \oplus \{\bar{0}, \bar{2}, \bar{4}\} \cong \mathbb{Z}/2\mathbb{Z} \oplus \mathbb{Z}/3\mathbb{Z}$. The module $\mathbb{Z}/2\mathbb{Z}$ is projective because it is a direct summand of the free module R. However it is not free because every free R-module being a direct sum of R's, must have at least six elements.

10-1.13 Example. Let K be any skew field or division ring and $R = K_2$, the ring of 2×2 matrices over K. The right ideal

$$M = \left\{ \begin{vmatrix} a & b \\ 0 & 0 \end{vmatrix} \middle| a, b \in K \right\} = \begin{vmatrix} K & K \\ 0 & 0 \end{vmatrix} = \begin{vmatrix} 1 & 0 \\ 0 & 0 \end{vmatrix} R = eR$$

is a projective right R-module because $e^2 = e$. If M were free, then $M \cong \oplus R$ and hence the dimension of M over K would be four or larger.

For a right R-module M, the left R-module $\mathrm{Hom}_R(M, R)$, where for $\varphi: M \to R$, $a \in R$, $m \in M$, $(a * \varphi)(m) = \varphi ma$, is called the *dual* of M. Two indexed sets of elements $\{y_\alpha\} \subset M$ and $\{\varphi_\alpha\} \subset \mathrm{Hom}_R(M, R)$ are called a *dual basis* of M if they satisfy the two conditions (a) and (b) below. A dual basis need not be a basis.

10-1.14 Dual basis lemma. *A module P is projective if and only if there exists an index set Γ and a subset $\{y_\alpha \mid \alpha \in \Gamma\} \subset P$, and right R-module homomorphisms $\varphi_\alpha: P \to R$, $\alpha \in \Gamma$, such that for any $q \in P$*

(a) $|\{\alpha \in \Gamma \mid \varphi_\alpha(q) \neq 0\}| \lneqq \infty$ *is finite, and*
(b) $q = \Sigma \, y_\alpha \varphi_\alpha(q)$.

Proof. $\Rightarrow$: There is a module $F = \oplus x_\alpha R$ free on $\{x_\alpha \mid \alpha \in \Gamma\}$ such that $F = P \oplus C$ for some submodule $C \subset F$. Furthermore as seen in 10-1.8, we may choose Γ of the same size or cardinality as any generating set of P. Write $x_\alpha = y_\alpha + z_\alpha$ with $y_\alpha \in P$, and $z_\alpha \in C$.

Each $q \in P \subset F$ is uniquely of the form $q = \Sigma \, x_\alpha r_\alpha$, $r_\alpha \in R$. Since $\Sigma \, z_\alpha r_\alpha = q - \Sigma y_\alpha r_\alpha \in P \cap C = 0$, also $q = \Sigma \, y_\alpha r_\alpha$. Sine for any α, the projection $F \to x_\alpha R$ followed by the natural isomorphism $x_\alpha R \to R$ is an R-homomorphism, its restriction $\varphi_\alpha: P \to R$ to P is likewise. Thus $\varphi_\alpha(q) = r_\alpha$. Hence (a) and (b) $q = \Sigma \, y_\alpha \varphi_\alpha(q)$ hold.

$\Leftarrow$: Given Γ and $\{y_\alpha, \varphi_\alpha \mid \alpha \in \Gamma\}$ satisfying (a) and (b), form the free module $F = \oplus x_\alpha R$ on a set of free generators $X = \{x_\alpha \mid \alpha \in \Gamma\}$ indexed by Γ, where $x_\alpha R \cong R$. It suffices to show that P is isomorphic to a direct summand of F.

Given an element $q \in P$, since the $\varphi_\alpha(q)$ are uniquely determined, there is a unique expression $q = \Sigma \, y_\alpha \varphi_\alpha(q)$ by (a) and (b), and hence the formula $jq = \Sigma \, x_\alpha \varphi_\alpha(q)$ gives a well defined function $j: P \to F$. (Note that possibly there is also another finite sum $q = \Sigma \, y_\alpha r_\alpha$ with one or several $r_\alpha \neq \varphi_\alpha(q)$.) It follows from the fact that the φ_α are R-homomorphisms that so also is j. It still remains to show that $P \cong jP$ and that jP is a summand of F.

Let $g: X \to P$ be the function $gx_\alpha = y_\alpha$, and $\bar{g}: F \to P$ its (unique)

extension to a module homomorphism with $\bar{g} | X = g$, i.e. $\bar{g}x_\alpha = gx_\alpha = y_\alpha$.
For any $q \in P$, $\bar{g}(jq) = \Sigma(\bar{g}x_\alpha)(\varphi_\alpha q) = \Sigma y_\alpha \varphi_\alpha(q) = q$. Thus $\bar{g}j = 1_P$ is the
identity on P, and the exact sequence $F \to P \to 0$ splits. Hence P is
projective.

10-1.15 Corollary. *If P is projective and T is any set of generators of P,
then the index set Γ above (in 10-1.14) may be chosen to be of the same
cardinality $|\Gamma| = |T|$ as T.*

If $ab = 0$ in a ring with $a \neq 0$ and $b \neq 0$, then the elements a and b are
said to be the proper *zero divisors*. A ring (commutative, noncommutative) is called a *domain* if it has no proper zero divisors.

The dual basis lemma is frequently used to characterize finitely
generated projective modules as will be done in the next proposition.

10-1.16 Application. Let D be a commutative domain with $1 \in D$ and K
its field of quotients. Suppose that $I \subset K$ is a D-submodule, i.e. $ID \subseteq I$.
Define $I^{-1} = \{a \in K \mid Ia \subseteq D\}$. Then I^{-1} is also a D-submodule of K, but
$I^{-1} \neq \{1/y \mid y \in I\}$ in general.

10-1.17 Definition. A D-submodule $I \subset K$ is *invertible* if $II^{-1} = D$.

10-1.18 Proposition. *For $I \subset K$ and D as above, I is invertible if and only
if I is a projective D-module.*

Proof. $\Rightarrow$: If $II^{-1} = D$, then there exist $y_i \in I$, $z_i \in I^{-1}$ such that
$1 = \Sigma_1^n y_i z_i \in II^{-1} = D$. Define $\varphi_i : I \to D$ by $\varphi_i q = z_i q$ for every $q \in I$. Since
$q = \Sigma_1^n y_i z_i q = \Sigma_1^n y_i(\varphi_i q)$, I is projective.
$\Leftarrow$: If I is projective, there exist $\{y_\alpha\} \subset I$, $\varphi_\alpha : I \to D$ where α ranges over
some index set, such that $x = \Sigma y_\alpha(\varphi_\alpha x)$ for every $0 \neq x \in I$. Thus

$$1 = \Sigma y_\alpha(\varphi_\alpha x)x^{-1} = \Sigma_1^n y_i(\varphi_i x)x^{-1}.$$

Here the commutativity of D is required to show that for any $0 \neq a \in I$,
$\varphi_i(xa) = (\varphi_i x)a = \varphi_i(ax) = (\varphi_i a)x$, and hence that the ratio $(\varphi_i x)/
x = (\varphi_i a)/a$ is independent of the choice of $0 \neq a$ or $0 \neq x \in I$: Thus
$[(\varphi_i x)/x]a = \varphi_i a \in D$ for every $a \in I$. Consequently $(\varphi_i x)/x \in I^{-1}$. Lastly,
for any $d \in D$, $y_i d \in I$, and $d = \Sigma_1^n (y_i d)[(\varphi_i x)/x] \in II^{-1}$. Hence $II^{-1} = D$.

10-1.19 Corollary. *For D and $I \subset K$ as above*

$$I \text{ is projective} \Leftrightarrow \exists\, y_i \in I,\, z_i \in K \text{ such that}$$
$$Iz_i \subseteq D, \text{ and } \Sigma_1^n\, y_i z_i = 1.$$

10-1.20 Remark. A nonprincipal proper ideal I in any commutative ring D is not free as a D-module. For suppose that $I = \oplus\{xD \mid x \in X\}$ is free on $X \subset I$. Since I is not principal, $|X| \geqslant 2$, let $x_1 \neq x_2 \in X$. Then the basis property of X together with $x_1 x_2 + x_2(-x_1) = 0$ is a contradiction.

10-1.21 Example. Let $\mathbb{Z} \subset Q$ be the integers and rationals, let $D = \mathbb{Z}[\sqrt{-5}]$ and $I = \{3x + (2 + \sqrt{-5})y \mid x, y \in D\}$ the ideal $I = (3, 2 + \sqrt{-5})$ generated by 3 and $2 + \sqrt{-5}$. It can be shown directly (or seen in [Jones 75; p. 234]) that I is a proper nonprincipal ideal. Then $D \subset I^{-1} \subset K = Q[\sqrt{-5}]$ and $z = (1/3)(2 - \sqrt{-5}) \in I^{-1}$. Thus $3z + (2 + \sqrt{-5})1 + (-2 - \sqrt{-5})z = 1$ shows that $II^{-1} = D$. Hence I is a projective D-module that is not free.

10-2   Projective dimension

The connotation of the word "dimension" as size or cardinality may be misleading. The projective dimension of the ring of integers modulo four $\mathbb{Z}/4\mathbb{Z}$ viewed as a module over itself is infinite, while that of any finite or infinite field is zero.

10-2.1 Every unital right R-module M is of the form $M \cong F_0/K_0$ where

$$0 \longrightarrow K_0 \xrightarrow{\beta_1} F_0 \xrightarrow{\alpha_0} M \longrightarrow 0$$

is an exact sequence with F_0 free. Similarly let

$$0 \longrightarrow K_1 \xrightarrow{\beta_2} F_1 \xrightarrow{\gamma_1} K_0 \longrightarrow 0$$

be exact with F_1 free. Then

$$0 \longrightarrow K_1 \xrightarrow{\beta_2} F_1 \xrightarrow{\alpha_1 = \beta_1 \gamma_1} F_0 \longrightarrow M \longrightarrow 0$$

is an exact sequence. This process can be continued to yield an exact sequence

$$0 \longrightarrow K_{n-1} \xrightarrow[\beta_n]{} F_{n-1} \xrightarrow[\alpha_{n-1}]{} F_{n-2} \longrightarrow \cdots$$

$$\cdots \longrightarrow F_1 \xrightarrow[\alpha_1]{} F_0 \xrightarrow[\alpha_0]{} M \longrightarrow 0$$

with the F_i free. If also K_{n-1} happens to be free, then the exact sequence is called a free resolution of finite length of M. If F is a free module, then $0 \to F \to F \to 0$ and $0 \to 0 \to 0 \to F \to F \to 0$ are finite free resolutions of F. Thus if we have any finite free resolution of a module, we can always prolong it by tacking on some more zeros on the left. In the special case when M has a free resolution of finite length, logically we could define the "free dimension of M" to be n, if n is the smallest possible integer n for which there exists a free resolution of M (with K_{n-1} free by definition, and of course $K_{n-1} \neq 0$).

If in the above process we replace the free modules F_i now more generally with projective modules P_i, and if K_{n-1} happens to be projective, then we would have a finite projective resolution of M of length n. If M does not have a finite projective resolution, then the projective dimension of M is said to be infinite, projective dimension $M = \infty$. Otherwise, the projective dimension of M is the smallest integer proj. dim. $M = n$ such that M has some *projective resolution* of length n; i.e. there exists some exact sequence with all P_i projecture and $P_n \neq 0$,

$$0 \to P_n \to P_{n-1} \to \cdots P_0 \to M \to 0$$

Why do we discuss the projective dimension and not the free dimension? For at least two reasons. The class of modules having a finite projective resolution is larger than those possessing finite free resolutions. The other more surprising reason is found in 10-2.6.

The reason for not considering "dimensions of M" for other classes of modules is that the above process of forming a free or a projective resolution of M can at least be started, if not finished for every module M, because every module is of the form $M \cong P_0/K_0$ where P_0 is projective. In fact, we may take P_0 to be even free.

The module D constructed below is called the pullback of A and B and is useful in other situations.

10-2.2 Construction. Suppose that given are modules A, B, and C and module homomorphisms $\alpha: A \to C$, $\beta: B \to C$, and $\lambda: A \to B$ such that

$\alpha = \beta\lambda$. Define $D = \{(a, b) \in A \times B \,|\, \alpha a = \beta b\}$. Then $D \cong A \oplus \ker \beta$.

Proof. Define $\tilde{A} = \{(a, \lambda a) \,|\, a \in A\}$. Then $\tilde{A} \subseteq D$, $\tilde{A} \cong A$, and $[(0) \times \ker \beta] \cap \tilde{A} = (0)$. For any $(a, b) \in D$, $(a, b) = (a, \lambda a) + (0, b - \lambda a)$, where $\beta(b - \lambda a) = \beta b - \alpha a = 0$. Thus $D = \tilde{A} + (0) \times \ker \beta = \tilde{A} \oplus [(0) \times \ker \beta]$.

10-2.3   Schanuel's lemma. *If $M \cong \tilde{M}$ under γ, and the sequences*

$$0 \longrightarrow K \longrightarrow P \xrightarrow{\;\alpha\;} M \longrightarrow 0, \quad and$$
$$\downarrow \gamma$$
$$0 \longrightarrow L \longrightarrow Q \xrightarrow{\;\beta\;} \tilde{M} \longrightarrow 0$$

are exact with P and Q projective, then $P \oplus L \cong Q \oplus K$.

Proof. Without loss of generality, take $M = \tilde{M}$. Since β is epic, the projectivity of P gives a map $\lambda: P \to Q$ such that $\alpha = \beta\lambda$, and similarly there is a map $\mu: Q \to P$ such that $\beta = \alpha\mu$.

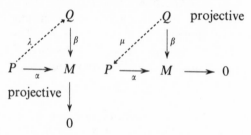

Thus the so-called pullback $D = \{(a, b) \in P \times Q \,|\, \alpha a = \beta b\}$ can be expressed in two ways as $D \cong P \oplus \ker \beta = P \oplus L$, and also as $D \cong Q \oplus \ker \alpha = Q \oplus K$. Hence $P \oplus L \cong Q \oplus K$.

Below the Q's and the P's alternate. If n is odd, $K \oplus Q_n \oplus \ldots$ ends with P_0, if even $\ldots$ with Q_0.

10-2.4   Corollary. *Suppose that*

$$0 \to K_n \to P_n \to \cdots \to P_1 \to P_0 \to M \to 0,$$
$$0 \to L_n \to Q_n \to \cdots \to Q_1 \to Q_0 \to M \to 0$$

are exact sequences with the P_i and Q_j projective. Then

$$K_n \oplus Q_n \oplus P_{n-1} \oplus \cdots \cong L_n \oplus P_n \oplus Q_{n-1} \oplus \cdots$$

Proof. The case $n = 1$ is Schanuel's lemma. By induction assume the corollary holds for $n - 1$, we prove it for n. Form the sequences

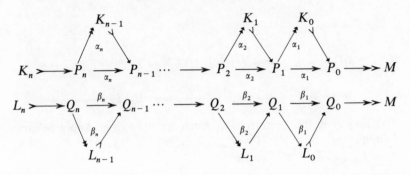

Define K_0 as the kernel of $P_0 \to M \to 0$. Thus $\alpha_1 P_1 = K_0$. The corestriction $P_1 \to K_0$ of α_1 to its image is also written simply as α_1: next $K_1 = \ker \alpha_1$ and $\alpha_2 \colon P_2 \to K_1$, etc.

Form the table where the modules on the same line will be isomorphic, where for sake of concreteness n is even.

$K_0 \oplus Q_0$	$L_0 \oplus P_0$	0
$K_1 \oplus Q_1 \oplus P_0$	$L_1 \oplus P_1 \oplus Q_0$	1
$K_2 \oplus Q_2 \oplus P_1 \oplus Q_0$	$L_2 \oplus P_2 \oplus Q_1 \oplus P_0$	2
$K_{n-1} \oplus \overbrace{Q_{n-1} \oplus P_{n-2} \oplus \cdots \oplus P_0}^{S_{n-1}}$	$L_{n-1} \oplus \overbrace{P_{n-1} \oplus Q_{n-2} \oplus \cdots \oplus Q_0}^{T_{n-1}}$	$n-1$ odd
$K_n \oplus Q_n \oplus P_{n-1} \oplus \cdots \oplus Q_0$	$L_n \oplus P_n \oplus Q_{n-1} \oplus \cdots \oplus P_0$	n even
$K_n \oplus S_n$	$L_n \oplus T_n$	

The modules $S_0 = Q_0$, $S_1 = Q_1 \oplus P_0$, $S_2 = Q_2 \oplus P_1 \oplus Q_0$, S_n are defined by the table, and similarly for the T's. However, note that

$$K_n \oplus S_n = K_n \oplus Q_n \oplus T_{n-1}, \quad \text{and}$$
$$L_n \oplus T_n = L_n \oplus P_n \oplus S_{n-1}.$$

Below, it follows from the exactness of the top and bottom sequences that the middle sequences are also exact, where $\alpha_n \oplus 1$ acts as the identity on the second component of $P_n \oplus S_{n-1}$, and where $\ker(\alpha_n \oplus 1) = K_n \subset P_n \oplus S_{n-1}$. The dual applies to β, L, and T.

$$0 \longrightarrow K_n \longrightarrow P_n \xrightarrow{\ \alpha_n\ } K_{n-1} \longrightarrow 0$$

$$0 \longrightarrow K_n \longrightarrow P_n \oplus S_{n-1} \xrightarrow{\alpha_n \oplus 1} K_{n-1} \oplus S_{n-1} \longrightarrow 0$$

$$\cong \Big\downarrow \gamma$$

$$0 \longrightarrow L_n \longrightarrow Q_n \oplus T_{n-1} \xrightarrow{\beta_n \oplus 1} L_{n-1} \oplus T_{n-1} \longrightarrow 0$$

$$0 \longrightarrow L_n \longrightarrow Q_n \xrightarrow{\ \beta_n\ } L_{n-1} \longrightarrow 0.$$

The map γ is the isomorphism given by induction. Apply Schanuel's Lemma with $K = K_n$, $P = P_n \oplus S_{n-1}$, $\alpha = \alpha_n \oplus 1$, $M = K_{n-1} \oplus S_{n-1}$, $\tilde{M} = L_{n-1} \oplus T_{n-1}$, etc., to yield

$$K_n \oplus Q_n \oplus T_{n-1} \cong L_n \oplus P_n \oplus S_{n-1},$$

$$S_n = Q_n \oplus T_{n-1} \quad T_n = P_n \oplus S_{n-1}, \qquad \text{and}$$

$$K_n \oplus S_n \cong L_n \oplus T_n.$$

10-2.5   Corollary. *Suppose that*

$$0 \longrightarrow K_{n-1} \longrightarrow P_{n-1} \longrightarrow \cdots \longrightarrow P_0 \longrightarrow M \longrightarrow 0$$

$$0 \longrightarrow L_{n-1} \longrightarrow Q_{n-1} \longrightarrow \cdots \longrightarrow Q_0 \longrightarrow M \longrightarrow 0$$

are exact sequences with the P_i and Q_j projective. If one of K_{n-1} or L_{n-1} is projective, then so is the other.

Proof. Let K_{n-1} be projective. Define $P_n = K_{n-1}$. By 10-2.4, the module $L_{n-1} \oplus P_{n-1} \oplus Q_{n-1} \oplus \cdots$ is projective because it is isomorphic to a direct sum of projectives, $L_{n-1} \oplus P_{n-1} \oplus \cdots \cong P_n \oplus Q_{n-1} \oplus \cdots$. Thus L_{n-1}, being a direct summand of a projective is itself projective.

The last corollary guarantees that the following is well defined.

10-2.6   Projective dimension. Let M be any module. Define the *projective dimension* of M to be *proj. dim. $M = n$* if there exists some exact sequence of projective modules P_i with $P_n \neq 0$

$$0 \longrightarrow P_n \longrightarrow P_{n-1} \longrightarrow \cdots \longrightarrow P_1 \longrightarrow P_0 \longrightarrow M \longrightarrow 0,$$

and proj. dim $M = \infty$ if there does not exist such a finite sequence.

10-2.7 **Example.** For a prime $p \in \mathbb{Z} = 0, \pm 1, \pm 2, \ldots$ let $R = \mathbb{Z}/p^2\mathbb{Z}$ and $M = p\mathbb{Z}/p^2\mathbb{Z}$. Then proj. dim. $M = \infty$ as shown by the following infinite projective resolution:

$$\longrightarrow R \longrightarrow R \longrightarrow R \longrightarrow p\mathbb{Z}/p^2\mathbb{Z} \longrightarrow 0.$$
$$1 \longrightarrow p + p^2\mathbb{Z}$$

The maps $R \to R$ are multiplication by p.

For any module M, $0 \to M \to M \to 0$ is a projective resolution of M if and only if M is projective, if and only if proj. dim. $M = 0$.

10-2.8 **Example.** Let R be the ring of all $n \times n$ upper triangular matrices over a field F, and $J \lhd R$ be the ideal of matrices all of whose diagonal entries are zero. It will be shown that the projective dimensions of the right R-modules J and R/J are proj. dim. $J = 1$ and proj. dim. $R/J = 2$. If $e_{ij} = \|\delta_{ij}\|$ denotes the usual matrix unit with all entries zero except for an entry 1 in row i and column j, then $J = fR$, where $f = \Sigma_{i=1}^{n} e_{i,i+1} = e_{12} + e_{23} + \cdots + e_{n-1,n} \in R$ is the matrix with all zero entries except for ones right next to the main diagonal. Define a surjective R-map $\alpha: R \to J$ by $\alpha r = fr$, $r \in R$. Since $R = e_{11}R \oplus (1 - e_{11})R$, $K = \ker \alpha = e_{11}R$ is projective, and $0 \to K \xrightarrow{\beta} R \xrightarrow{\alpha} J \to 0$ is a projective resolution of J, where β is the inclusion map. All of the above is verified by the following computations for $n = 4$, but the same would hold for any n:

$$r = \begin{pmatrix} a_1 & a_2 & a_3 & a_4 \\ 0 & b_2 & b_3 & b_4 \\ 0 & 0 & c_3 & c_4 \\ 0 & 0 & 0 & d_4 \end{pmatrix} \in R, \quad f = \begin{pmatrix} 0 & 1 & 0 & 0 \\ 0 & 0 & 1 & 0 \\ 0 & 0 & 0 & 1 \\ 0 & 0 & 0 & 0 \end{pmatrix} \quad fr = \begin{pmatrix} 0 & b_2 & b_3 & b_4 \\ 0 & 0 & c_3 & c_4 \\ 0 & 0 & 0 & d_4 \\ 0 & 0 & 0 & 0 \end{pmatrix}$$

Thus proj. dim. $J = 1$.

If π is the natural projection, and α, β are as before then $0 \to K \xrightarrow{\beta} R \xrightarrow{\alpha} R \xrightarrow{\pi} R/J \to 0$ is exact, because $\alpha R = J = \ker \pi$, and $\beta K = K = \ker \alpha$. Hence proj. dim. $R/J = 2$.

10-3 Minimal right ideals

At first, without invoking any restrictive hypotheses on the ring, this section develops some basic relationships between minimal one-sided ideals and idempotents.

10-3.1 **Definition.** A right ideal $U \leq R$ is a *minimal* right ideal if (i)

$U \neq 0$ and (ii) U contains only the two trivial R-submodules, (0) and all of U. Consequently, for a minimal U, either $UR = 0$, or $UR = U$; furthermore if $UR \neq 0$, then U is a simple right R-module.

Thus for any right ideal $W \leqslant R$ such that $WR \neq 0$,

$$W \text{ is a minimal right ideal} \iff W \text{ is a simple right } R\text{-module.}$$

10-3.2   Lemma. *Let $e = e^2$ and $f = f^2 \in R$. For $y \in R$, L_y denotes left multiplication by y. Then there is an additive abelian group isomorphism*

$$\mathrm{Hom}_R(eR, fR) = \{L_y \mid y \in fRe\} \cong fRe.$$

If $f = e$, then this is a ring isomorphism.

Proof. Clearly $\{L_y \mid y \in fRe\} \subseteq \mathrm{Hom}_R(eR, fR)$. Conversely, for any R-homomorphism $\phi: eR \to fR$, define y by $y = \phi e \in fR$. Then $fy = y$; and $ye = (\phi e)e = \phi(e^2) = y$ because ϕ is a right R-homomorphism. So $y = fye \in fRe$. For any $x \in eR$, $\phi x = \phi(ex) = (\phi e)x = L_y x$. Hence $\phi = L_y$. For $y \in fRe$, $L_y e = y$; hence $L_y = 0$ if and only if $y = 0$.

10-3.3   Lemma. *If $0 \neq U \subset R$ is a minimal right ideal, then either $U^2 = 0$, or there exists a $0 \neq e = e^2 \in U$ such that*

   (i)   *$U = eR$, and eRe is a division ring.*
   (ii) *Furthermore, $\mathrm{End}_R eR \cong eRe$ as rings.*

Proof. (i) There exists a $0 \neq a \in U$ such that $0 \neq aU \subseteq U^2 \neq 0$, and hence $aU = U$. Therefore $a^\perp \cap U \subsetneqq U$, and again $a^\perp \cap U = 0$. There exists $e \in U$ such that $ae = a$. Thus $a(e^2 - e) = 0$, $e^2 - e \in a^\perp \cap U = 0$, and $e^2 = e$. Again $0 \neq eR \subseteq U$; and $U = eR$.

(ii) For any $0 \neq eae \in eRe$, $0 \neq (eae)e \in eaeR \subseteq eR$ implies that $eaeR = eR$. So there exist a $b \in R$ such that $eaeb = e$, and hence $eae(ebe) = e$.

By the last lemma $\mathrm{End}_R eR \cong eRe$. Since eR is simple right R-module, Schur's Lemma gives us an alternative but indirect proof that eRe is a division ring.

10-3.4   Corollary. *In a semiprime ring R, any idempotent $e = e^2 \in R$ satisfies the following:*

   (i)   *eR is a minimal right ideal if and only if eRe is a division ring.*
   (ii) *eR is a minimal right ideal if and only if Re is a minimal left ideal.*

Proof. (i) ⇐: If $0 \neq exR \subseteq eR$, for some $x \in R$, then by the semiprimeness of R, $0 \neq (exR)^2 \subseteq exRe$. Hence $exye \neq 0$ for some $y \in R$. Thus $0 \neq exye(eze) = e$ since eRe is a division ring. Thus $exR = eR$ is minimal.

(ii) The hypothesis that eRe is a division ring is right left symmetric, and hence also equivalent to Re being a minimal left ideal.

10-3.5 Observation. A prime ring with a minimal right ideal is primitive.

Proof. Let $U \leqslant R$ be minimal. In a prime ring the product of two nonzero right ideals is not zero. Thus first, $U^2 \neq 0$, and secondly, $UU^{\perp} = 0$ implies that $U^{\perp} = 0$. Thus U is a simple faithful right R-module.

10-3.6 Theorem. *Let R be a primitive ring containing a minimal right ideal $U \subseteq R$. Let V be any simple faithful right R-module whatever with skew field $D = \text{End}_R V$. Then there exists a $0 \neq e = e^2 \in R$ such that*

(i) *$U = eR$ and $\text{End}_R U \cong eRe$;*
(ii) *$V \cong U$;*
(iii) *$D \cong eRe$.*

Proof. (i) If $U^2 = 0$, then $U \subseteq J(R) = 0$. Thus $U = eR$ and $\text{End}_R U \cong eRe$ acts on U by left multiplication for some idempotent e.

(ii) Since $V^{\perp} = 0$, there exists an $x \in V$ with $0 \neq xU \subseteq V$ a right R-submodule of V. Therefore $V = xU$. Thus left multiplication by x, $\phi = L_x : U \to V$, $\phi u = xu$ is a nonzero homomorphism of simple modules U and V, and hence an isomorphism. Thus $U \cong xU = V$.

(iii) In general, any isomorphism $\phi : U \to V$ of any modules whatever, induces a canonical ring isomorphism $\phi^* : \text{End}_R U \to \text{End}_R V$ such that $\phi(\delta u) = (\phi^*\delta)\phi u$ for any $\delta \in \text{End}_R U$ and $u \in U$. Simply set $\phi^*\delta = \phi\delta\phi^{-1} : V \to V$. In particular, $D \cong eRe$.

The above two isomorphisms show that there is a bijective correspondence between arbitrary left D-linear combinations from V and left eRe combinations of elements of U.

10-3.7 Corollary 1. *For R as above* $\dim_D V = \dim_{eRe} U$.

10-3.8 Corollary 2. *For division rings D and Δ, and integers n and m*

suppose that $\psi: D_n \to \Delta_m$ *is a ring isomorphism of the* $n \times n$ *matrix ring over* D *onto the* $m \times m$ *one over* Δ. *Then* $n = m$ *and* $D \cong \Delta$.

Proof. Set $R = D_n$. For $E_{ij} \in R$ and $e_{ij} \in \Delta_n$ the usual matrix units, let $U = E_{11}D + \cdots + E_{1n}D < R$ and $V = e_{11}\Delta + \cdots + e_{1m}\Delta$ be all matrices with arbitrary entries in row one, and zeroes everywhere else. They are minimal right ideals of R and Δ_m. By 10-3.3 for $E_{11} = E_{11}^2$ we have $U = E_{11}R$ while $\mathrm{End}_R\, U \cong E_{11}RE_{11} = E_{11}D$ is just left multiplication by D.

For $r \in R$ and $v \in V$, $v \cdot r = v(\psi r)$ defines $V = V_R$ as a simple faithful R-module. Since $\psi R = \Delta_m$ is onto, every element of $\mathrm{End}_R\, V$ commutes with right multiplication of V by Δ_m. Hence

$$\mathrm{End}_R\, V \subseteq \mathrm{Hom}_{\Delta_m}(V, V) = e_{11}\Delta \cong \Delta,$$

and the reverse inclusion always holds. Thus $\mathrm{End}_R\, V \cong \Delta$. By the last theorem $D \cong \mathrm{End}_R\, U \cong \mathrm{End}_R\, V \cong \Delta$, and $n = \dim_D U = \dim_\Delta V = m$.

10-4   Main theorems

10-4.1   Composition series. For a module V, a chain of submodules $V = V_0 \supset V_1 \supset \cdots \supset V_s \supset V_{s+1} = (0)$ is a *composition series* if all the quotients V_i/V_{i+1}, $i = 0, \ldots, s$; are simple. The total number s of such quotients which are called composition factors of V is defined to be the *length* of the composition series.

Any module V has a composition series (for some finite $s < \infty$) $\Leftrightarrow V$ satisfies both the descending chain condition (D.C.C.) as well as the ascending chain condition (A.C.C.) on R-submodules.

The Jordan-Hölder Theorem states that any two composition series of a module V are of the same length s, and furthermore, that there is a one-to-one correspondence between the two sets of composition factors such that the corresponding composition factors are isomorphic. For a proof, see [P. B. Bhattacharya, S. K. Jain, and S. R. Nagpaul 86; p. 116, Theorem 1.3].

10-4.2   Definition. A set $\mathscr{S} = \{e, f, \ldots\} \subset R$ of idempotents $e^2 = e$ is an *orthogonal* set of idempotents if for any $e \neq f \in \mathscr{S}$, $ef = fe = 0$. In this case any finite sum $e + f + \cdots + g$ of distinct elements of $\mathscr{S}$ is again an idempotent which commutes with all elements of $\mathscr{S}$.

Any idempotent $0 \neq g = g^2 \in R$ is *primitive* if g cannot be written further as a sum of two nonzero orthogonal idempotents of R.

10-4.3 Definition. A module C is *indecomposable* if whenever $C = A \oplus B$ for A, $B \leqslant C$, then either $C = A$ or $C = B$.

10-4.4 Lemma. *Let* $0 \neq g = g^2 \in R$. *Then*

$$g \text{ is primitive} \Leftrightarrow gR \text{ is indecomposable}.$$

Proof. $\Rightarrow$ If $gR = A \oplus B$ for A, $B \leqslant R$ with $g = a + b$, $a \in A$, $b \in B$, then $a = ga = a^2 + ba$, and $a - a^2 = ba \in A \cap B = 0$. Thus $a^2 = a$ and $ba = 0$. Similarly also $b^2 = b$ and $ab = 0$. If $A \neq 0$ and $B \neq 0$, then also $a \neq 0$ and $b \neq 0$. The converse is clear.

If $\mathscr{S} \subset R$ is a set of orthogonal idempotents, then $\Sigma\{eR \mid e \in \mathscr{S}\} = \oplus\{eR \mid e \in \mathscr{S}\}$ is a direct sum of right ideals. The next lemma is a partial converse.

10-4.5 Lemma. *Suppose that R is a ring with $1 \in R$ and that R is a direct sum $R = U_1 \oplus \cdots \oplus U_q$ of a finite number of right ideals $U_i < R$. If $1 = e_1 + \cdots + e_q$, with $e_i \in U_i$, then for all i and j*

(i) $e_i^2 = e_i$, $e_j e_i = 0$, $i \neq j$.

(ii) $U_i = e_i R$.

(iii) *UNIQUENESS: If $f_1, \ldots, f_q$ is a set of orthogonal idempotents such that $U_i = f_i R$ then $e_i = f_i$, $i = 1, \ldots, q$.*

(iv) *If in addition all $U_i \lhd R$ are ideals; $i = 1, \ldots, q$; then all $e_i \in$ center R, and e_i is the identity element of the ring U_i.*

Proof. (i) By the directness, from $0 = 1 \cdot e_i - e_i = e_1 e_i + \cdots + e_{i-1} e_i + (e_i^2 - e_i) + e_{i+1} e_i + \cdots + e_q e_i$ it follows that $e_i^2 - e_i = 0$ and $e_j e_i = 0$ if $i \neq j$.

(ii) Since $e_i \in U_i$, $e_i R \subseteq U_i$. Conversely for any $x \in U_i$, again from $0 = 1 \cdot x - x = e_1 x + \cdots + (e_i x - x) + \cdots + e_q x$ it follows that $e_j x = 0$ for $j \neq i$, and $e_i x - x = 0$.

(iii) For any $x \in R$, $x = f_1 x_1 + \cdots + f_q x_q$ for some $x_j \in R$. But $f_i x = f_i x_i$ and we may let $x_i = f_i x$, or $x = f_1 x + \cdots + f_q x$. For $x = 1$, this becomes $1 = f_1 + \cdots + f_q = e_1 + \cdots e_q$. Since $f_i - e_i \in U_i$ and their sum is direct, $f_i = e_i$.

(iv) For any $r \in R$, $0 = 1 \cdot r - r \cdot 1 = \Sigma(e_i r - r e_i)$ together with $e_i r - r e_i \in U_i$ implies that $r e_i = e_i r$.

10-4.6 Lemma. *Assume that $1 \in R$ and assume that $R = U_1 \oplus \cdots \oplus U_q$ is* *a direct sum of minimal right ideals. Then as a right R-module*

(i)   *R has a composition series and hence the D.C.C. on right ideals.*
(ii) $J(R) = 0$.

Proof. (i) The following is a composition series:

$$R = U_1 + \cdots + U_q \supset U_2 + \cdots + U_q \supset \cdots \supset U_q \supset (0).$$

(ii) Since $1 \in R$, U_i is a simple right R-module. Hence $U_i J(R) = 0$. Thus $J(R) = J(R) \cdot 1 \subseteq R J(R) = (U_1 + \cdots + U_q) J(R) = 0$.

Alternate proof of (ii). Let $M_i = U_1 + \cdots + U_{i-1} + U_{i+1} + \cdots + U_q$. Then M_i is a maximal right ideal because $R/M_i \cong U_i$ is a simple right R-module. Hence $J(R) \subseteq \bigcap_1^q M_i = 0$.

10-4.7   Applications. (1) in the finite $n \times n$ matrix ring $M_n(\Delta)$ over a division ring Δ, the set $U_i = \Sigma_{j=1}^n e_{ij} \Delta$ is a minimal right ideal, where e_{ij} is the matrix $e_{ij} = \|\delta_{ij}\|$, and U_i consists of all matrices having arbitrary entries in row i and zeroes everywhere else. Thus $M_n(\Delta) = U_1 \oplus \cdots \oplus U_n$ has the D.C.C. on right ideals and is semiprimitive, $J(M_n(\Delta)) = 0$. Note that U_i has the structure as in 10-3.3; $U_i = e_{ii} M_n(\Delta)$ with $e_{ii}^2 = e_{ii}$, and $e_{ii} M_n(\Delta) e_{ii} = e_{ii} \Delta \cong \Delta$ is a division ring. Moreover in addition, $e_{11} + \cdots e_{nn} = 1$ with $e_{ii} e_{jj} = 0$ for $i \neq j$. Lastly, $M_n(\Delta)$ is a simple ring (7-4.4).

(2) For any product $R = \Pi(R_i | i \in I)$ of rings R_i over any finite or infinite index set I, $J(R) = \Pi J(R_i)$ (7-1.24). Let $\Delta_1, \ldots, \Delta_k$ be division rings and $1 \leqslant n(1), \ldots, n(k)$ an indexed sequence of integers. Form the full $n(i) \times n(i)$ matrix rings $M_{n(i)}(\Delta_i)$ over Δ_i. Then also $\oplus_{i=1}^k M_{n(i)}(\Delta_i)$ is a semiprimitive ring with the D.C.C. on right ideals.

A ring satisfying the conditions of the next proposition is automatically semiprimitive.

10-4.8   Proposition. *For a ring R with the D.C.C. on right ideals the following are all equivalent.*

(i)   *R is prime.*
(ii) *R is primitive.*
(iii) *R is simple.*

Proof. (ii) $\Rightarrow$ (i). Every primitive ring is prime (9-1.10). (i) $\Rightarrow$ (ii). The D.C.C. implies that R has a minimal right ideal, which together with (i) forces R to be primitive as in 10-3.5.

(ii) ⇔ (iii). This was established in the proof of Wedderburn's theorem I (see 7-1.40).

The reader at this point should recall 7-1.38 through to 7-1.42.

10-4.9 Example. If $\mathbb{Q} \subset \mathbb{R}$ are the rationals and the reals, then the ring R has a composition series

$$R \equiv \begin{bmatrix} \mathbb{Q} & \mathbb{R} \\ 0 & \mathbb{R} \end{bmatrix} \supset \mathbb{Q}e_{11} + \mathbb{R}e_{12} \supset \mathbb{R}e_{12}.$$

Hence R satisfies both the right D.C.C. and right A.C.C. Since every $\mathbb{Q}$ vector subspace of $J(R) = \mathbb{R}e_{12}$ is a left ideal, R satisfies neither the left A.C.C. or D.C.C. The sum of the minimal right ideals $\mathbb{R}e_{12} + \mathbb{R}e_{22}$ as a ring has the structure of the next example.

10-4.10 Example. For D a division ring and $2 \leqslant n$ an integer, the ring $R = D \times \cdots \times D$ has componentwise addition and the associative multiplication $(a_1, \ldots, a_n)(c_1, b_2, \ldots, b_n) = (a_1 c_1, a_2 c_1, \ldots, a_n c_1)$. Let $e_i = (0 \underline{\quad} 0, 1, 0 \underline{\quad} 0)$ where the i-th component is one, the rest zero. Since $R = \oplus_1^n e_i D$ is a direct sum of minimal right ideals $e_i D$, R has both the right D.C.C. and right A.C.C. Every additive subgroup of $J(R) = e_2 D + \cdots + e_n D$ is a left ideal. If D contains the rationals, then R has neither of the two left chain conditions.

In view of their importance, we remind the reader of the Wedderburn theorems (7-1.39 and 7-1.41) by reformulating the second one in different words, and adding the new necessary and sufficient condition (3) below. The second one includes the first as a special case.

10-4.11 Wedderburn Theorem II. *For a ring R, the following three conditions are all equivalent.*

(1) *R has the right D.C.C. and $J(R) = 0$.*
(2) *$R \cong \oplus_{i=1}^m M_{k(i)}(\Delta_i)$; $M_{k(i)}(\Delta_i) \lhd R$ is the $k(i) \times k(i)$ matrix ring over a division ring Δ_i, $i = 1, \ldots, m$.*
(3) *$\exists \, 1 \in R$; $R = U_1 \oplus \cdots \oplus U_q$, $U_k < R$ are minimal right ideals, $k = 1, \ldots, q$.*

10-4.12 Corollary to Theorem II. *If R satisfies (3) above, then each $U_i = e_i R$ for $e_i^2 = e_i \in R$, where $1 = e_1 + \cdots + e_q$ is a sum of orthogonal primitive idempotents.*

Proof. (2) $\Rightarrow$ (3). This is 10-4.7(2). (3) $\Rightarrow$ (1). This follows from Lemma 10-4.6. (1) $\Rightarrow$ (3). Since $J(R) = \cap\{P \mid P \triangleleft R \text{ primitive}\}$, by the D.C.C., $J(R) = \cap_{k=1}^{m} P_k$ is an intersection of a finite number of primitive ideals $P_i \triangleleft R$ with m as small as possible. Thus $R \cong \bar{R} \subseteq \Pi_1^m R/P_k$ where $\bar{R}$ is a subdirect product. But each R/P_k is a simple ring by 10-4.4, and a finite subdirect product of simple rings is isomorphic to a full direct product of some of them (7-1.11), which by the minimality of m in this case is all of them. Thus $R \cong \bar{R} = R/P_1 \oplus \cdots \oplus R/P_m$. By 10-4.8, 7-1.39 and 7-1.40 each $R/P_i \cong M_{k(i)}(\Delta_i)$.

10-5   Direct proofs

An alternate proof of part of 10-4.11(2) is given. It is based on useful techniques involving intersections of ideals and as a result expresses the simple matrix rings occurring in 10-4.11(2) as certain intersections of maximal ideals. Then an independent self-contained proof is given of a comprehensive theorem (10-5.4) which contains more than the two previous Wedderburn theorems.

10-5.1   Theorem. *Suppose that R is a ring such that a finite intersection of maximal ideals is zero. Then*

(i) $R = S_1 \oplus \cdots \oplus S_m$ *is a direct sum where each $S_i \triangleleft R$ is a simple ring.*

10-5.2   Corollary. *Suppose that $P_i \triangleleft R$ are maximal ideals such that the intersection $\cap_1^m P_i = 0$ is irredundant, i.e. $P_j \nsubseteq \cap\{P_i \mid j \neq i \leqslant m\}$ for all j. The notation* " $\hat{}$ " *means that the term under it is missing. Then in* (i), *the S_i are*

(ii) $S_i = P_1 \cap \cdots \cap \hat{P}_i \cap \cdots \cap P_m$ $i = 1, \ldots, m$;
(iii) $R = S_1 \oplus \cdots \oplus S_n \oplus [P_1 \cap \cdots \cap P_n]$ $1 \leqslant n \leqslant m$.

Proof. To begin, define $S_i \triangleleft R$ by the above formula (ii), and define $Q_n = P_1 \cap \cdots \cap P_n$. Thus $S_i \cap P_i = Q_m = 0$. Note also that (i) now is the special case $n = m$ of (iii); (i) is also simply the special case $n = m - 1$ of (iii), when $Q_{m-1} = S_m$. Since (ii) is a definition, it suffices to prove only (iii).

Although logically this is not needed for the proof, let us first see that the theorem holds for $m = 1$ and 2. If $m = 1$, $P_1 = 0$, $Q_1 = 0$, and the empty intersection is $S_1 = R$. For $m = 2$, $P_1 \cap P_2 = 0$, $S_1 = P_2$, $S_2 = P_1 = Q_1$, and $R = P_2 \oplus P_1 = S_1 \oplus Q_1$.

Let m be arbitrary but nonvariable throughout the proof. By induction assume that for the ring R (iii) holds for $1, \ldots, n$. Then we have to prove it for $n + 1$.

When $n = 1$, $Q_1 = P_1$ and $P_1 \cap S_1 = Q_m = 0$. Since P_1 is maximal, $R = S_1 \oplus Q_1$, which is (iii) for $n = 1$.

Next, let us dispose of the cases when $n \geqslant m - 1$. Always, $n \leqslant m$. In the case when $n = m$, the formula (iii) is the same as for the case when $n = m - 1$. If $n = m - 1$, then by induction $R = S_1 \oplus \cdots \oplus S_{m-1} \oplus Q_{m-1}$. But $Q_{m-1} = S_m$ by definition. The induction hypothesis now becomes the formula (iii) for $m + 1$, i.e. $R = S_1 \oplus \cdots \oplus S_m = S_1 \oplus \cdots \oplus S_{n+1} \oplus Q_{n+1}$, where $Q_{n+1} = Q_m = 0$. So we may take $2 \leqslant n \leqslant m - 2$.

For $1 \leqslant j$, the maximal R-ideal $P_{n+j} \nsubseteq Q_n = P_1 \cap \cdots \cap P_n$ from the irredundancy, and hence

$$\frac{R}{P_{n+j}} = \frac{Q_n + P_{n+j}}{P_{n+j}} \cong \frac{Q_n}{Q_n \cap P_{n+j}}$$

is a simple ring. This implies that $Q_n \cap P_{n+j} \lhd Q_n$ is also a maximal ideal.

It is asserted that

(1) $\quad Q_n \cap P_{n+1} \nsubseteq \bigcap\limits_{j=2}^{m-n} [Q_n \cap P_{n+j}]$.

For, assume otherwise, then

$$Q_{n+1} = Q_n \cap P_{n+1} \subseteq Q_n \cap \bigcap\limits_{j=2}^{m-n} P_{n+j} = P_1 \cap \cdots \cap \hat{P}_{n+1} \cap \cdots \cap P_m$$

$$= S_{n+1}.$$

Next, intersect both sides with P_{n+1}, where $Q_{n+1} \subseteq P_{n+1}$, and get

$$Q_{n+1} \cap P_{n+1} = Q_{n+1} \subseteq S_{n+1} \cap P_{n+1} = 0.$$

Since $n + 1 \leqslant m - 1$, $Q_{n+1} = 0$ contradicts the irredundancy of the intersection of the P_i. Thus the assertion (1) has been established.

Since $Q_{n+1} = Q_n \cap P_{n+1} \subset Q_n$ is a maximal ideal of Q_n, equation (1) implies that

(2) $\quad Q_n = Q_{n+1} + \left[\bigcap\limits_{j=2}^{m-n} Q_n \cap P_{n+j} \right]$.

This sum is actually direct, because

$$Q_{n+1} \cap \left[\bigcap\limits_{2}^{m-n} Q_n \cap P_{n+j} \right] = (P_1 \cap \cdots \cap P_{n+1}) \cap [P_{n+2} \cap \cdots \cap P_m]$$

$$= 0,$$

where Q_n was dropped from the intersection since $Q_{n+1} \subset Q_n$, and only $P_{n+2}, \ldots, P_{n+(m-n)}$ remained. Furthermore

$$\bigcap_{j=2}^{m-n} Q_n \cap P_{n+j} = Q_n \cap \left[\bigcap_2^{m-n} P_{n+j} \right] = S_{n+1}$$

Thus (2) becomes $Q_n = Q_{n+1} \oplus S_{n+1}$. This and the induction statement $R = S_1 \oplus \cdots \oplus S_n \oplus Q_n$ for n concludes the proof of $R = S_1 \oplus \cdots \oplus S_{n+1} \oplus Q_{n+1}$.

10-5.3 Definition. For any right R-module M, the *socle* of M is the sum $\operatorname{soc} M = \Sigma\{V \mid V \leqslant M, V \text{ is simple}\}$ of all the simple submodules of M.

A module M is *completely reducible* (abbreviation: c.r.) if M can be written in some way as a direct sum $M = \oplus\{V_i \mid i \in I\}$ of simple submodules for some family $V_i \leqslant M$, $i \in I$.

The direct proof below does not use the density theorem.

10-5.4 Theorem. *For a ring R with an identity element the following six conditions are all equivalent*

(1) *Every R-module is projective.*
(2) *Every short exact sequence splits.*
(3) *Every R-module is injective.*
(4) *Every nonzero R-module is completely reducible.*
(5) $R = U_1 \oplus \cdots \oplus U_q$ *for minimal right ideals U_i.*
(6) $R \cong \oplus_{i=1}^m M_{k(i)}(\Delta_i)$ *is a direct sum of finite matrix rings.*

Proof. The equivalence of (1), (2), and (3) is shown first. $(2) \Rightarrow (1)$. Given any module P in order to show that P is projective, take any short exact sequence $A \xrightarrow{\alpha} B \to 0$ and any map $f: P \to A$. By (2), there exists a homomorphism $\beta: B \to A$ with $\alpha\beta = 1_B$ the identity. Then $\beta f: P \to A$ is the required map because $\alpha\beta f = 1_B f = f$.

$(1) \Rightarrow (2)$. It suffices to show that any short exact sequence of the form $A \xrightarrow{\alpha} B \to 0$ splits. For the projective $P = B$ and the identity map $f = 1_B: P \to B$, there exists a map $\beta: P \to A$ such that $f = \alpha\beta$. Thus $\beta: B \to A$ with $\alpha\beta = 1_B$.

$(2) \Rightarrow (3)$. Given any module Q, any short exact sequence $0 \to B \xrightarrow{\beta} A$, and any map $g: B \to Q$, g has to be extended to A. By (2), there exists a map $\alpha: A \to B$ such that $\alpha\beta = 1_B$ is the identity on B. Then $g\alpha: A \to Q$ is the required extension because $g\alpha\beta = g1_B = g$.

$$Q \qquad\qquad \alpha\beta = 1_B \qquad\qquad Q$$

(3) $\Rightarrow$ (2), it has to be shown that any short exact sequence $0 \to B \xrightarrow{\beta} A$ splits. For the injective $Q = B$ and the identity map $g = 1_B: B \to Q$, there exists an $\alpha: A \to Q$ such that $\alpha\beta = g$. Thus $\alpha: A \to B$ with $\alpha\beta = 1_B$ as required.

The plan now is to prove (2) $\Rightarrow$ (4) $\Rightarrow$ (5) $\Rightarrow$ (6) $\Rightarrow$ (5) $\Rightarrow$ (2).

(2) $\Rightarrow$ (4). First it will be shown that any nonzero module contains a simple submodule. For any $0 \neq x \in M$, let $\mathcal{S} = \{N \mid x \notin N < M\}$ be the set of submodules of M excluding x. Since $(0) \in \mathcal{S} \neq \varnothing$, Zorn's lemma produces a maximal element $N \in \mathcal{S}$. By (2), $0 \to N \to M$ splits as $M = N \oplus F$ for some $F < M$ with $F \cong M/N$. If F is not simple it contains a proper nonzero submodule $0 \neq G \lneq F$, and again by (2), $F = G \oplus H$ for some $0 \neq H < F$. By the maximality of $N \in \mathcal{S}$, $x \in N \oplus G \notin \mathcal{S}$ as well as $x \in N \oplus H \notin \mathcal{S}$. Thus we get the contradiction that

$$x \in (N \oplus G) \cap (N \oplus H) = N \oplus [(N \oplus G) \cap H] = N,$$

where the modular law was used. Hence F is simple.

Let $\mathcal{Y} = \{\mathcal{F}\}$ be the set whose elements $\mathcal{F} = \{U\}$ are families of simple submodules $U < M$ whose sum is direct $\Sigma\{U \mid U \in \mathcal{F}\} = \oplus\{U \mid U \in \mathcal{F}\}$. Note that $\varnothing \in \mathcal{Y} \neq \varnothing$. Or, less artificially, for our previous simple $F < M$, $\{F\} \in \mathcal{Y}$. Under set inclusion $\mathcal{Y}$ becomes a partially ordered set, i.e. for $\mathcal{F}, \mathcal{F}' \in \mathcal{Y}$, $\mathcal{F}' < \mathcal{F}$, iff $\mathcal{F}' \subset \mathcal{F}$, iff $\mathcal{F} = \mathcal{F}' \cup \mathcal{F}''$ for some $\mathcal{F}'' \in \mathcal{Y}$. If $\{\mathcal{F}_\alpha\} \subset \mathcal{Y}$ is a chain in $\mathcal{Y}$ indexed by α, define $\mathcal{C}$ to be the following set of simple submodules, $\mathcal{C} = \cup_\alpha \mathcal{F}_\alpha = \{U < M \mid \exists \alpha$ such that $U \in \mathcal{F}_\alpha\}$. If $U_0 \cap \Sigma\{U \mid U_0 \neq U \in \mathcal{C}\} \neq 0$, there exists a finite number $U_0 \neq U_i \in \mathcal{C}$, $i = 1, \dots, n$, with $U_0 \cap (U_1 + \cdots + U_n) \neq 0$. Since $\mathcal{C} = \cup \mathcal{F}_\alpha$, each $U_i \in \mathcal{F}_{\alpha(i)}$ for some $\alpha(i)$ for $i = 0, 1, \dots, n$. Since $\{\mathcal{F}_\alpha\}$ is totally ordered, let $\alpha = \text{maximum } \{\alpha(0), \alpha(1), \dots, \alpha(n)\}$. Then $U_i \in \mathcal{F}_{\alpha(i)} \subseteq \mathcal{F}_\alpha$ for all $i = 0, 1, \dots, n$; i.e. all $U_0, U_1, \dots, U_n \in \mathcal{F}_\alpha$. However, now $U_0 \cap (U_1 + \cdots + U_n) \neq 0$ contradicts the definition of $\mathcal{F}_\alpha \in \mathcal{Y}$. Hence $\mathcal{C} \in \mathcal{Y}$.

Zorn's lemma gives a maximal element $\mathcal{F} \in \mathcal{Y}$. Set $N = \oplus\{U \mid U \in \mathcal{F}\}$. The third time by (2), $0 \to N \to M$ splits as $M = N \oplus P$ for some $P < M$. If $P \neq 0$, by the first part, P contains a simple submodule $V \leqslant P$. But then $\mathcal{F} \subsetneqq \mathcal{F} \cup \{V\} \in \mathcal{Y}$ contradicts the maximality of $\mathcal{F}$ in $\mathcal{Y}$. Thus $M = N = \oplus\{U \mid U \in \mathcal{F}\}$.

(4) $\Rightarrow$ (5). As a right R-module $R = \oplus\{U_\alpha \mid \alpha \in I\}$ is a direct sum of simple submodules $U_\alpha < R$ indexed by $\alpha \in I$. Since $1 \in R$, $1 = e_1 + \cdots + e_q$ for $0 \neq e_i \in U_{\alpha(i)}$, where $\alpha(1), \ldots, \alpha(q) \in I$ are distinct. For any $x \in R$, $x = 1 \cdot x = e_1 x + \cdots + e_q x \in U_{\alpha(1)} \oplus \cdots \oplus U_{\alpha(q)} = R$. The fact that for any $\beta \in I$, $0 \neq U_\beta = U_\beta \cap R = U_\beta \cap (U_{\alpha(1)} \oplus \cdots \oplus U_{\alpha(q)})$ implies that $I = \{\alpha(1), \ldots, \alpha(q)\}$. Rename $\alpha(1) = 1, \ldots, \alpha(q) = q$. The rest is clear from Lemmas 10-4.6 and 10-4.4.

(5) $\Rightarrow$ (6). Given that $R = U_1 \oplus \cdots \oplus U_q$, define $R_1 = \Sigma\{U_i \mid U_i \cong U_1\} < R$. If $k(1)$ is the smallest integer such that $U_{k(1)} \not\subseteq R_1$, then define $R_2 = \Sigma\{U_i \mid U_i \cong U_{k(1)}\}$. If $R_1 + R_2 \neq R$, then there exists a smallest integer $k(2)$ with $U_{k(2)} \not\subseteq R_1 + R_2$. After a finite number m of such steps, $R = R_1 + R_2 + \cdots + R_m$.

Remark. At this point we know that for any $U_i \cong U_{k(1)}$, $U_i \not\subseteq R_1$, and hence that $U_i \cap R_1 = 0$. However, we do not yet know that $R_1 \cap R_2 = 0$.

It is claimed that $R_i R_j = 0$ if $i \neq j$. Without loss of generality we prove that $R_1 R_2 = 0$. The case of $R_2 R_1$ or $R_i R_j$ will be the same. It suffices to show that $UW = 0$ where $U \subset R_1$, $W \subset R_2$ are of the special form U, $W \in \{U_1, \ldots, U_q\}$. If $UW \neq 0$, then there exists an $x \in U$, and a $y \in W$ such that $xy \neq 0$. Define $\phi = L_x : W \to U$ to be left multiplication by x, $\phi w = xw \in U$ for all $w \in W$. Since $\phi y = xy \neq 0$, $\phi \neq 0$. Hence ϕ is an isomorphism which is a contradiction. Thus $RR_i = R_i^2 \subseteq R_i$, and $R_i \lhd R$ for all i.

By 10-4.6, R has no nonzero nilpotent right ideals. Thus

$$[R_i \cap (R_1 + \cdots + \hat{R}_i + \cdots + R_m)]^2 \subseteq R_i(R_1 + \cdots + \hat{R}_i + \cdots + R_m) = 0$$

shows that $R = R_1 \oplus \cdots \oplus R_m$.

Now it suffices to assume that $R = U_1 \oplus \cdots \oplus U_n$ for isomorphic minimal right ideals $U_1 \cong \cdots \cong U_n$, and show that $R \cong M_n(\Delta)$, where $\Delta \cong \mathrm{End}_R U_1$. By 10-4.6, $1 = e_1 + \cdots + e_n$, where $e_i e_j = e_i \delta_{ij}$ are orthogonal idempotents, and $U_i = e_i R$. For $x \in R$, $L(x): R \to R$ will denote in this proof the map $L(x)r = xr$, that is left multiplication by x.

Recall that $\mathrm{Hom}_R(U_i, U_j) = \{L(e_j c e_i) \mid c \in R\} \cong e_j R e_i$ from 10-3.2, where $c \in R$ need not be unique, but $e_j c e_i$ is. Thus there exist elements $g_j \in e_j R e_1$ and $\tilde{g}_j \in e_1 R e_j$ such that

$$U_1 \xrightarrow{L(g_j)} U_j \xrightarrow{L(\tilde{g}_j)} U_1$$

$L(\tilde{g}_j g_j) =$ identity on U_1; $L(g_j \tilde{g}_j) =$ identity on U_j, where $L(\tilde{g}_j) = L(g_j)^{-1}$. Thus not only $L(\tilde{g}_j g_j) = L(e_1)$ and $L(g_j \tilde{g}_j) = L(e_j)$, but even $\tilde{g}_j g_j = e_1$ and $g_j \tilde{g}_j = e_j$. Define $\Delta = e_1 R e_1 \subset R$.

Let $r, a, b \in R$ be arbitrary. The composite endomorphism

$$U_1 \xrightarrow{L(g_j)} U_j \xrightarrow{L(e_i r e_j)} U_i \xrightarrow{L(\tilde{g}_i)} U_1$$

defines a unique element $\tilde{g}_i e_i r e_j g_j \in e_1 R e_1$ of the division ring Δ. Map $R \to M_n(\Delta)$ by $r \to M(r) = \|\tilde{g}_i e_i r e_j g_j\|$ where the latter element of Δ is the (i, j) entry $M(r)_{ij}$ of the $n \times n$ matrix $M(r)$. Trivially $M(a - b) = M(a) - M(b)$. Define $m_{ij} \in \Delta$ by $M(a)M(b) = \|m_{ij}\|$. Thus in view of $g_k \tilde{g}_k = e_k$,

$$m_{ij} = \sum_{k=1}^{n} M(a)_{ik} M(b)_{kj} = \sum_{k=1}^{n} (\tilde{g}_i e_i a e_k g_k)(\tilde{g}_k e_k b e_j g_j)$$

$$= \tilde{g}_i e_i a \left(\sum_{k=1}^{n} e_k e_k e_k \right) b e_j g_j = \tilde{g}_i e_i a \cdot 1 \cdot b e_j g_j = M(ab)_{ij}.$$

Thus $R \to M_n(\Delta)$, $r \to M(r)$ is a ring homomorphism.

For $r \in R$, if all $\tilde{g}_i e_i r e_j g_j = 0$, then since $g_i \tilde{g}_i = e_i$ and $g_j \tilde{g}_j = e_j$, it follows that $e_i r e_j = 0$. Thus

$$0 = \sum_{i=1}^{n} \sum_{j=1}^{n} e_i r e_j = \left(\sum_{i=1}^{n} e_i \right) r \left(\sum_{j=1}^{n} e_j \right) = 1 \cdot r \cdot 1 = r.$$

Hence the map $r \to M(r)$ is monic.

To see that it is onto, take any $\delta \in \Delta = e_1 R e_1$. It suffices to show that there exists an $r \in R$ such that $M(r) = \delta E_{ij}$, where $E_{ij} = \|\delta_{ij}\|$. The composite endomorphism

$$U_j \xrightarrow{L(\tilde{g}_j)} U_1 \xrightarrow{L(\delta)} U_1 \xrightarrow{L(g_i)} U_i$$

is $L(g_j \delta \tilde{g}_j)$. The required element is $r = g_i \delta \tilde{g}_j$. Since $g_i \in e_i R$ while $\tilde{g}_j \in R e_j$, it follows that

$$e_p g_i = \delta_{pi} g_i \qquad \text{and} \qquad \tilde{g}_j e_k = \delta_{jk} \tilde{g}_j$$

for any indices p and k. Consequently,

$$M(r)_{pk} = \tilde{g}_p e_p r e_k g_k = \tilde{g}_p e_p g_i \delta \tilde{g}_j e_k g_k$$

$$= \tilde{g}_p \delta_{pi} g_i \delta \delta_{jk} \tilde{g}_j g_k = \begin{cases} \tilde{g}_i g_i \delta \tilde{g}_j g_j & (p, k) = (i, j) \\ 0 & (p, k) \neq (i, j) \end{cases}$$

But $\tilde{g}_i g_i = e_1 = \tilde{g}_j g_j$, and thus $M(r)_{ij} = e_1 \delta e_1 = \delta$. Thus $R \cong M_n(\Delta)$.

(6) $\Rightarrow$ (5). Without loss of generality $R \cong M_n(\Delta)$, and the rest follows from 10-4.7(1).

(5) $\Rightarrow$ (2). Let $R = U_1 \oplus \cdots \oplus U_q$. It has to be shown that for any R-module M and any submodule $N < M$, that N is a summand of M.

Let $P \leqslant M$ be maximal with respect to $P \cap N = 0$. Thus $N \oplus P \leqslant M$ is an essential extension. Suppose that $x \in M$ but $x \notin N \oplus P$. Since $1 \in R$, $xR \nsubseteq N \oplus P$. There exists i such that $xU_i \nsubseteq N \oplus P$. But $L_x: U_i \to xU_i$ being a nonzero homomorphism of a simple module onto xU_i implies that it is an isomorphism and that $xU_i \cap (N \oplus P) = 0$. Thus $N \oplus (P \oplus xU_i)$ contradicts the maximality of P. Hence $M = N \oplus P$.

10-6   Uniqueness

The following lemma and corollary show that in the Wedderburn theorems, the representations $R = S_1 \oplus \cdots \oplus S_m$, $S_i = P_1 \cap \cdots \cap \hat{P}_i \cap \cdots \cap P_m$ are unique (10-4.11(2), 10-5.1(i), 10-5.2(ii)). We already know that in $S_i \cong M_{k(i)}(\Delta_i)$ the $k(i)$ is unique and Δ_i up to isomorphism (10-3.8).

10-6.1   Lemma. *Suppose that a ring* $R = \oplus\{S_i | i \in I\} = \oplus\{R_j | j \in J\}$ *is a direct sum of ideals* S_i, $R_j \lhd R$ *all of which are simple rings. Then*

(i)   *there is a bijection* $\pi: I \to J$ *such that* $S_i = R_{\pi i}$.
(ii)   *Every ideal* $K \lhd R$ *can be written uniquely* $K = \oplus\{S_i | i \in I, S_i \subseteq K\}$ *as the direct sum of those* S_i *which* K *contains.*

Proof. (i) For any i, j, $S_i R_j \lhd S_i$ and $S_i R_j \lhd R_j$. Hence either $S_i R_j = 0$, or $S_i = S_i R_j = R_j$. Since $S_i = S_i^2 = S_i R = \oplus\{S_i R_j | j \in J\}$, for each i there exists a unique j such that $S_i = S_i R_j = R_j$. Set $\pi i = j$.

(ii) Let $\pi_i: R \to S_i$ be the natural projection. In a simple ring $(\pi_i K)S_i = 0$ if and only if $\pi_i K = 0$. But $\pi_i K \lhd S_i$ and $(\pi_i K)S_i = KS_i \lhd S_i$. So if $\pi_i K \neq 0$, since $K \lhd R$, we have

$$S_i = \pi_i K = (\pi_i K)S_i = KS_i \subseteq K.$$

Always $K \subseteq \oplus\{\pi_i K | i \in I\}$, and hence in this case $K = \oplus\{S_i | i \in I, \pi_i K \neq 0\} = \oplus\{S_i | i \in I, S_i \subseteq K\}$.

10-6.2   Corollary. *Every primitive ideal of the above ring R is uniquely of the form* $P_j = \oplus\{S_i | j \neq i \in I\}$ *for some unique* $j \in I$. *Each* $S_i = \cap\{P_j | i \neq j \in I\}$ *is an irredundant intersection of a unique set of primitive ideals* $P_j \lhd R$, $j \in I, j \neq i$.

10-6.3   Lemma. *For any ring R and a right R-module M suppose that*

$M = \oplus_j^n V_i \cong \oplus_1^n W_j$ where V_i and W_j are simple modules, and n and m are integers. Then

    (i) $n = m$ and there is a permutation π of $\{1, \ldots, n\}$ such that
   (ii) $V_i \cong W_{\pi i}$ $i = 1, \ldots, n$.

Proof. If $n = 1$, then $V_1 = W_1 \oplus \cdots \oplus W_m$. Since each W_j is a nonzero submodule of the simple module V_1, $m = 1$ and $V_1 = W_1$.

For a fixed m and R, assuming the above result for $1, \ldots, n - 1$ we prove it by induction for n. There is a projection $\pi_k : M = \oplus_1^m W_i \to W_k$ such that $\pi_k V_n \neq 0$. By renumbering the W_j if necessary, without loss of generality we may assume that $k = m$ and that $\pi_m(V_n) \neq 0$. Set $B = W_1 \oplus \cdots \oplus W_{m-1}$. Then $\ker \pi_m = B_m$. The restriction $(\pi_m \mid V_n) : V_n \to W_m$ is an isomorphism. Hence $\ker(\pi_m \mid V_n) = B \cap V_n = 0$. Thus $M = W_m \oplus B \supseteq V_n \oplus B$. This latter inclusion is an equality; because $0 \neq (V_n \oplus B)/B \subseteq (W_m \oplus B)/B \cong W_m$ is simple, it follows that $V_n \oplus B = W_m \oplus B$. Thus $V_n \cong W_m$ and let $\pi n = m$.

Set $\bar{M} = V_1 \oplus \cdots \oplus V_{n-1}$. Then $M/V_n \cong \bar{M}$ and also $M/V_n = (V_n \oplus B_k)/V_n \cong B$. The induction hypothesis applied to

$$M = V_1 \oplus \cdots \oplus V_{n-1} \cong W_1 \oplus \cdots \oplus W_{m-1}$$

yields that $n - 1 = m - 1$, and that $V_i \cong W_{\pi i}$ for some permutation π of $\{1, \ldots, n - 1\}$ for all $i \leqslant n - 1$.

10-7   Rings with D.C.C. and idempotents

All the Wedderburn theory, and more, can be derived independently by idempotent methods that are distinct from previous sections. In order to make this section self-sufficient none of the more involved parts of our previous theory of the Jacobson radical will be used.

Throughout this section the restrictive hypotheses on the ring R are linearly ordered. At first, there are no restricted assumptions on the ring R in 10-7.1–10-7.4, and these are frequently used tools in many other ring theoretic contexts. Then in the middle some results are established which do not assume $J(R) = 0$. Only in the end portion is it assumed that R is a semiprime ring with a chain condition.

10-7.1   Lemma. Let $I \lhd R$ that is of the form $I = eR = Re$ for some $e = e^2 \in R$. Then

    (i) $e \in \text{center } R$;

(ii) *if also $R = fR$ (or $R = Rf$) for some $f = f^2 \in R$, then $f = e$.*
(iii) *Any right, left, or two-sided I-ideal of the ring I is, respectively, also one of R.*

Proof. (i) For any $r \in R$, the elements $x = er \in I$ as well as $y = re \in I$. Thus $x \in Re$ and $y \in eR$ imply that $x = xe = ey = y$, i.e. that $er = re$, or that necessarily $e \in$ center R.

(ii) If $R = Rf$, then $e \in Rf$ implies that $e = ef$. But $f \in eR$ implies that $f = ef$. Thus $e = f$. The case $R = fR$ is similarly proved.

(iii) Let $A \subseteq Re$ be a right I-ideal of the ring I. Then $A = Ae$, $eR = I$, and hence $AR = AeR = AI \subseteq A$. The two remaining cases are similar.

10-7.2 Idempotent lifting lemma. *Suppose that $I \lhd R$ is a nil ideal, and $0 \neq a + I = (a + I)^2 \in R/I$ is an idempotent element, where $a \in R$. Let $\mathbb{Z}[x]$ be the polynomial ring with integer coefficients. Then there exists an idempotent $e = e^2 \in a\mathbb{Z}[a] \subseteq R$ upstairs such that $a + I = e + I$.*

Proof. If n is the index of nilpotency of $a - a^2 \in I$, then $0 = (a - a^2)^n = a^n - a^{n+1}f(a)$ for some $f(x) \in \mathbb{Z}[x]$, or $a^n = [a^n]af(a)$. Now we use a repetitive argument that is like lifting yourself up by your own bootstraps; substitute the first a^n for the second one in the brackets to get $a^n = a^{n+2}f(a)^2$. Repetition shows that $a^n = a^{n+k}f(a)^k$ for all k. In particular, for $k = n$, $a^n = a^{2n}f(a)^n$. Set $e = a^n f(a)^n$. Then

$$\left. \begin{array}{l} e^2 = a^{2n}f(a)^n f(a)^n \\ a^{2n}f(a)^n = a^n \end{array} \right\} \Rightarrow e^2 = e.$$

Use of $a^n = a^{n+1}f(a)$ and $a^{n+1} + I = a + I$ yields

$$a + I = a^n + I = a^{n+1}f(a) + I = (a^{n+1} + I)(f(a) + I)$$
$$= (a + I)(f(a) + I) = af(a) + I.$$

Finally

$$\left. \begin{array}{l} a + I = af(a) + I \\ a + I = (a + I)^n \end{array} \right\} \Rightarrow a + I = a^n f(a)^n + I = e + I.$$

10-7.3 Consequence. The above is not merely an abstract existence proof, the idempotent e can be constructed from the following concrete formulae:

$$(a - a^2)^n = 0, \quad e = a^n f(a)^n,$$

$$f(a) = \sum_{i=1}^{n} (-1)^{i-1} \binom{n}{i} a^{i-1} \in \mathbb{Z}[a].$$

The proof of the previous lemma is also valid if $a + I = I$, in which case, however, the idempotent e is zero.

10-7.4 Corollary. *In any ring R, assume that $a \in R$ is a nonnilpotent element such that $(a - a^2)^n = 0$ for some n. Then $0 \neq e = a^n f(a)^n \in R$ is a nonzero idempotent for some polynomial $f(x) \in \mathbb{Z}[x]$.*

Proof. In the commutative ring $\mathbb{Z}[a]$, $I = (a - a^2)\mathbb{Z}[a]$ is a nilpotent principal ideal. The ideal I has been specifically so defined that $a + I = a^2 + I \in \mathbb{Z}[a]/I$ is an idempotent in the quotient ring. Now lift it to $e = e^2 = a^n f(a)^n \in a\mathbb{Z}[a] \subseteq R$ an idempotent in R.

Suppose that $e = 0$; then $a + I = e + I$ implies that $a \in I$, a nilpotent ideal. Hence a would be nilpotent, a contradiction. So $e = e^2 \neq 0$.

The next theorem is a key result from which an amazing variety of consequences will follow easily.

10-7.5 Brauer theorem. *Suppose that R is a ring satisfying the right D.C.C. and $L < R$ is a nonnilpotent right ideal. Then L contains a nonzero idempotent $0 \neq e = e^2 \in L$.*

Proof. Form the set $\mathscr{S}$ consisting of all right ideals $A \leqslant R$, with $A \subseteq L$, but A not nilpotent. Since $L \in \mathscr{S}$, $\mathscr{S} \neq \varnothing$, and hence by the chain condition, $\mathscr{S}$ contains a minimal number A. Then A^2 is not nilpotent because A is not. From $A \supseteq A^2 \in \mathscr{S}$ and the minimality of A it follows that $A = A^2$.

Next form the set $\mathscr{F} = \{B \leqslant R \mid B \subseteq A,\ BA \neq 0\}$. Since $A \in \mathscr{F} \neq \varnothing$, again let $B \in \mathscr{F}$ be a minimal element. Thus $BA \neq 0$ and there exists a $b \in B$ with $bA \neq 0$. In view of $B \supseteq bA = (bA)A \neq 0$, $bA \in \mathscr{F}$, and the minimality of $B \in \mathscr{F}$ requires that $B = bA$. Hence $b = ba$ for some $a \in A$. Moreover, $0 \neq b = ba = ba^2 = ba^n$ for all n. Thus a is not nilpotent (and hence A is not nil).

Form the right ideal $b^{\perp} \cap B \leqslant R$. From $b^{\perp} \cap A \subseteq A$, yet with $0 = b(b^{\perp} \cap A) \subsetneqq bA = B$, it follows that $b^{\perp} \cap A \subsetneqq A$. Now the minimality of $A \in \mathscr{S}$ tells us that $b^{\perp} \cap A$ must be nilpotent. In particular, $b(a - a^2) = 0$, and the element $a - a^2 \in b^{\perp} \cap A$ is nilpotent while a is not. An appeal to the last corollary now produces the required nonzero $0 \neq e = e^2 \in L$ idempotent.

10-7.6 Corollary 1. *In a right D.C.C. ring R, for any right ideal $L < R$,*

L is nilpotent $\Leftrightarrow$ L is nil.

Proof. $\Leftarrow$: If L is not nilpotent, then by the last theorem there exists a $0 \neq e = e^2 \in L$. Hence L is not nil, a contradiction.

10-7.7   Corollary 2. *For a ring R with the right D.C.C., let N be the sum of all the nilpotent right ideals. Then*

(i) $N \lhd R$ *is a nilpotent ideal;*
(ii) N *is the unique largest nil right ideal of R.*
(iii) N *is the sum of all the nilpotent left ideals of R.*

Proof. First, recall that in any ring, the sum of a finite number of nilpotent right ideals is a nil right ideal, and hence that the sum N of all the nilpotent right ideals is nil. Also $N \lhd R$. If $I \subset R$ is a nilpotent left ideal, then $I + IR \lhd R$ is a nilpotent ideal of R. Hence (iii) holds in general.

By the last corollary, the concepts nil right ideal and nilpotent right ideal coincide in the present special ring. (i) Hence N is nilpotent. (ii) By its definition, N contains every nil right ideal, and hence is the largest such.

10-7.8   Corollary 3. *For a ring R with the D.C.C. on right ideals, the following are all equivalent; R contains no nonzero*

(i)   *nil right ideals;*
(ii) *nilpotent right ideals;*
(iii) *nil ideals;*
(iv) *nilpotent ideals;*
(v)   *nilpotent left ideals.*

10-7.9 Proposition. *For a nonnilpotent right ideal $L \leqslant R$ in a right D.C.C. ring R, form the set of right R-ideals $\mathscr{S} = \{e^{\perp} \cap L \mid 0 \neq e = e^2 \in L\}$. If $0 \neq e \in L$ is such an idempotent that $e^{\perp} \cap L \in \mathscr{S}$ is a minimal member of $\mathscr{S}$, then*

(i) $e^{\perp} \cap L$ *is nilpotent;*
(ii) $L = eR \oplus (e^{\perp} \cap L)$ *is a direct sum of right ideals.*

Proof. (i) Since L contains nonzero idempotents, $\mathscr{S} \neq \varnothing$, and by the D.C.C. an e as above exists. If $e^{\perp} \cap L$ were not nilpotent, then there exists a $0 \neq f = f^2 \in e^\perp \cap L$. Set $g = f + e - fe \in L$. Use $ef = 0$ to compute that not only $g = g^2$, but also that

$$eg = e \to g^\perp \subseteq e^\perp;$$
$$gf = f \to f \notin g^\perp, f \in e^\perp.$$

Since $f \in L$, the inclusion $g^\perp \cap L \subsetneqq e^\perp \cap L$ is proper. Since $g \in L$, also $g^\perp \cap L \in \mathscr{S}$, which contradicts the minimality of $e^\perp \cap L \in \mathscr{S}$. Thus $e^\perp \cap L$ is nilpotent.

(ii) For any $x \in L$, write $x = ex + (x - ex)$. Note that $x - ex \in e^\perp \cap L$, and that every element of $e^\perp \cap L$ is of this form. Since $eR \cap e^\perp = 0$, $L = eL \oplus (e^\perp \cap L)$, As always $eR = e(eR) \subseteq eL$, and $eR = eL$.

10-7.10 Remark. The previous proposition when applied to any ring R satisfying the right D.C.C. shows that $R = eR \oplus K$, where K is a nilpotent right ideal. If R is nilpotent, $R = K$. If R has an identity $1 \in R$, then $e = 1$ and $K = 0$.

For simplicity from now on it will be assumed that R contains no nonzero nilpotents ideals (10-7.8) – i.e. that R is semiprime.

10-7.11 Corollary 1. *Any ideal $I \lhd R$ in a semiprime right D.C.C. ring is of the form $I = eR = Re$ for some unique central idempotent $e = e^2 \in \text{center } R$.*

Proof. By the last proposition, $I = eR \oplus (e^\perp \cap I) = eR$, where $e = e^2 \in R$, and $e^\perp \cap I = 0$. The set $B = \{y - ye \mid y \in I\}$ is a left ideal of R since $y = ey$ for $y \in I = eR$, $B \subseteq e^\perp \cap I = 0$. (Alternatively, $BB \subseteq BI = 0$, and $B = 0$ by 10-7.8(v).) Thus $I = Ie$. Since $e \in I \lhd R$, also $Re \subseteq I$, and $I = Re$. Thus $I = eR = Re$. In general, the latter requires that $e \in \text{center } R$ and that such an e is unique (10-7.1).

The last corollary says that any ideal I has an identity, and in the next corollary we take I to be $I = R$ all of R.

10-7.12 Corollary 2. *A ring without nonzero nilpotent ideals and the right D.C.C. contains an identity element.*

The fact that $1 \in R$ together with Lemma 10-7.1 gives us the next corollary.

10-7.13 Corollary 3. *Suppose that R is a semiprime right D.C.C. ring.*

Then any ideal $I \lhd R$ *is a direct summand of* R,

(i) $R = I \oplus J, J \lhd R$.
(ii) *Any ideal* $I \lhd R$ *(and* J*) of the ring* R *is also a ring of the exact same type as* R – *semiprime with the right D.C.C.*
(iii) *Any minimal ideal of* R *is a simple ring with identity.*

Proof. Write $I = eR$ and $J = (1 - e)R$ where e and $(1 - e)$ are central idempotents of R.

10-7.14 Theorem (Wedderburn). *Assume that* R *is a ring with no nilpotent nonzero ideals and with the descending chain condition on right ideals. Then* R *is the finite direct sum of all of its minimal ideals, i.e.*

(i) $R = S_1 \oplus \cdots \oplus S_m, 0 \neq S_i \lhd R$ is a minimal ideal, $i = 1, \ldots, m$.
(ii) *Every minimal ideal of* R *is one of the* S_i.
(iii) *Each* S_i *is a simple semiprime right D.C.C. ring.*

Proof. (i) and (iii). Take $S_1 \in \{I \mid 0 \neq I \lhd R\}$ to be any minimal member. Write $R = S_1 \oplus R_1$, with $R_1 \lhd R$ and $S_1 R_1 = R_1 S_1 = 0$. By our preceeding corollaries, not only does R_1 inherit the hypotheses of R, but for additive subgroups of R_1, the concepts (minimal) R_1-ideal and (minimal) R-ideal coincide. If $R_1 \neq 0$, repeat with R replaced by R_1, to get $R_1 = S_2 \oplus R_2$, and $R = S_1 \oplus S_2 \oplus R_2$. If necessary repeat the above argument, till $R = S_1 \oplus S_2 \oplus \cdots \oplus S_m \oplus R_m$ where all the $S_i \lhd R$ are minimal R-ideals and $R_m \lhd R$. By the chain condition, for some m, $R_m = 0$.

(ii) If $I \lhd R$ is any minimal ideal, then since $1 \in R$, $I = RI = S_1 I \oplus \cdots \oplus S_m I$. By the minimality of $0 \neq I \lhd R$, there exists one and only one unique i for which $S_i I \neq 0$. But $I = S_i I \lhd S_i$, and the minimality of S_i requires that $I = S_i$.

An attempt has been made in this section to present techniques of proof that are different from the previous one. The rest of the Wedderburn structure theory could also be developed by starting with this approach, but having made this point, we will stop here.

10-8   Exercises

1. Show that every cyclic projective module P is of the form $P = eR, e^2 = e \in R$.

2. If P is a projective module show that there exists a free module G such that $P \oplus G$ is free. (Hint: There is a free module $F = P \oplus Q$, consider $G = \oplus_1^\infty F$.)

3. Prove that the ideal I in 10-1.16 is proper and nonprincipal.

4. For $R = \mathbf{Z} = 0, \pm 1, \pm 2, \ldots$, prove that every projective module is free. ([Fuchs; Vol. I, p. 75]).

5. For $P = I$ as in 10-1.16 find the subset $\{y_\alpha \mid \alpha \in \Gamma\} \subset I$ and the maps $\{\varphi_\alpha \mid \alpha \in \Gamma\}$ guaranteed by the Dual Basis Lemma 10-1.14.

6. Given are exact sequences of R-modules $0 \to A \to B \to C \to 0$, $0 \to K_1 \to P_1 \to A \to 0$, and $0 \to K_3 \to P_3 \to C \to 0$, with P_1 and P_3 projective. Prove that there exists a commutative diagram with exact rows and exact columns, and with P_2 projective:

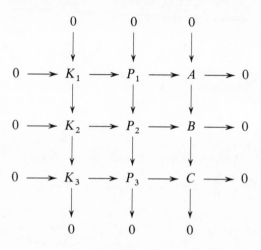

(Hint: Put $P_2 = P_1 \oplus P_3$, use the projectivity of P_3 to get a map $P_3 \to B$.)

The next definitions and sequence of exercises informally introduce some basic concepts of homological algebra.

Consider sequences of modules (A) and (B) and module homomorphisms with all $d_{n-1}d_n = 0$ and $\partial_{n-1}\partial_n = 0$, and homomorphisms f_n giving a commutative diagram

(A) $\cdots \longrightarrow A_n \xrightarrow{d_n} A_{n-1} \longrightarrow \cdots$

$\qquad\qquad\quad \downarrow f_n \qquad\quad \downarrow f_{n-1} \qquad\qquad f_{n-1}d_n = \partial_n f_n$

(B) $\cdots \longrightarrow B_n \xrightarrow{\partial_n} B_{n-1} \longrightarrow \cdots$

Such a sequence (A) is called a *complex*, and $f = (f_n):(A) \to (B)$ a *map of complexes*. The n-th *homology group* of (A) is $H_n(A) =$ kernel d_n/image d_{n+1}.

7*. Prove that f induces an R-homomorphism $f_*: H_n(A) \to H_n(B)$.

Consider a homomorphism $f: A \to B$ of R-modules and (exact) projective resolutions of these two modules:

$$(1) \cdots \longrightarrow P_n \longrightarrow P_{n-1} \longrightarrow \cdots \longrightarrow P_1 \longrightarrow P_0 \longrightarrow A \longrightarrow 0$$
$$\downarrow \qquad \downarrow \qquad\qquad \downarrow \qquad \downarrow \qquad \downarrow f$$
$$(2) \cdots \longrightarrow Q_n \longrightarrow Q_{n-1} \longrightarrow \cdots \longrightarrow Q_1 \longrightarrow Q_0 \longrightarrow B \longrightarrow 0$$

From some point on all the P_n or Q_n could be possibly all zero. For a module C, form the complex

$$(3) \quad 0 \to \mathrm{Hom}_R(P_0, C) \xrightarrow{d_1^*} \mathrm{Hom}_R(P_1, C) \xrightarrow{d_2^*} \cdots \xrightarrow{d_n^*} \mathrm{Hom}_R(P_n, C) \xrightarrow{d_{n+1}^*} \cdots$$

Denote the n-th homology group by $\mathrm{Ext}_R^n(A, C) = \ker d_{n+1}^* / \mathrm{im}\, d_n^*$.

8*. Prove that f above can be extended to a map $(f_n):(1) \to (2)$ of complexes, i.e. the vertical arrows can be filled in to give a commutative diagram above.

9*. Prove that there exists an induced map of complexes

$$(4) \quad 0 \longrightarrow \mathrm{Hom}_R(Q_0, C) \longrightarrow \mathrm{Hom}_R(Q_1, C) \longrightarrow$$
$$f_0^* \downarrow \qquad\qquad\qquad f_1^* \downarrow$$
$$(3) \quad 0 \longrightarrow \mathrm{Hom}_R(P_0, C) \longrightarrow \mathrm{Hom}_R(P_1, C) \longrightarrow$$

10*. Prove that there is an induced abelian group homomorphism $f^n: \mathrm{Ext}_R^n(B, C) \to \mathrm{Ext}_R^n(A, C)$.

> **Remark.** It can be shown that the definition of $\mathrm{Ext}_R^n(A, C)$ is independent of the particular projective resolution of A used, and so are the maps f_n.

> Given are three complexes (A), (B), (C) of R-modules and maps of complexes $f = (f_n)(A) \to (B)$, $g = (g_n):(B) \to (C)$ such that $0 \to A_n \xrightarrow{f_n} B_n \xrightarrow{g_n} C_n \to 0$ is exact for all n. Then $(0) \to (A) \xrightarrow{f} (B) \xrightarrow{g} (C) \to (0)$ is called an *exact sequence of complexes* where $(0) \to (A)$ and $(C) \to (0)$ are complex maps for the zero complex.

11*. Show that for each n there is an R-homomorphism $\psi: H_n(C) \to H_{n-1}(A)$. Hint: Use the diagram below for "diagram chasing".

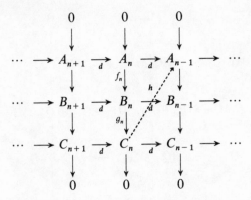

The same "d" is used in all three sequences. For $z \in H_n(C)$, choose $c \in z$ so that $dc = 0$. Find $b \in B_n$ such that $g_n b = c$, etc.

Remark. It can be shown that ψ is independent of choice of coset representatives and that then we get an exact sequence of R-modules

$$H_{n+1}(C) \xrightarrow{\psi} H_n(A) \xrightarrow{f_*} H_n(B) \xrightarrow{g_*} H_n(C) \xrightarrow{\psi} H_{n-1}(A) \xrightarrow{f_k} \cdots$$

$$\longrightarrow H_{n-1}(R) \xrightarrow{g_*} H_{n-1}(C) \xrightarrow{\psi} \cdots$$

12*. Given an exact sequence of R-modules $0 \to A \xrightarrow{\alpha} B \xrightarrow{\beta} C \to 0$, and projective resolutions of A and C, use exercise 6 to prove that there exists a projective resolution of B so that the diagram below is commutative with split exact columns and exact rows:

$$
\begin{array}{ccccccccc}
 & 0 & & 0 & & 0 & & 0 & \\
 & \downarrow & & \downarrow & & \downarrow & & \downarrow & \\
\cdots \to & P_2 & \xrightarrow{d_2} & P_1 & \xrightarrow{d_1} & P_0 & \xrightarrow{\varepsilon_A} & A & \to 0 \\
 & \downarrow & & \downarrow & & \downarrow & & \downarrow{\scriptstyle \alpha} & \\
\cdots \to & P_2 \oplus Q_2 & \to & P_1 \oplus Q_1 & \to & P_0 \oplus Q_0 & \to & B & \to 0 \\
 & \downarrow & & \downarrow & & \downarrow & & \downarrow{\scriptstyle \beta} & \\
\cdots \to & Q_2 & \xrightarrow{\partial_2} & Q_1 & \xrightarrow{\partial_1} & Q_0 & \xrightarrow{\varepsilon_c} & C & \to 0 \\
 & \downarrow & & \downarrow & & \downarrow & & \downarrow & \\
 & 0 & & 0 & & 0 & & 0 &
\end{array}
$$

13*. For any right R-module M, apply $\operatorname{Hom}_R(\cdot, M)$ (abbreviation $\operatorname{Hom}_r(X, M) \equiv (X, M)$) to the diagram in the last exercise to obtain the following exact sequence of complexes (see exercise 6) with split exact columns:

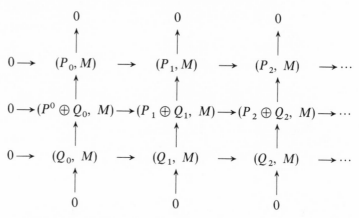

14*. Using the last two exercises show that any short exact sequence of modules $0 \to A \xrightarrow{\alpha} B \xrightarrow{\beta} C \to 0$ induces an exact sequence of abelian groups:

$$\cdots \to \operatorname{Ext}_R^{n-1}(A, M) \xrightarrow{\psi} \operatorname{Ext}_R^n(C, M) \xrightarrow{\beta^*} \operatorname{Ext}_R^n(B, M)$$

$$\xrightarrow{\alpha^*} \operatorname{Ext}_R^n(A, M) \xrightarrow{\psi} \operatorname{Ext}_R^{n+1}(C, M) \to \cdots$$

Here the map ψ (see exercise 11) is called the *connecting homomorphism*, and the above sequence the *long exact sequence in the first variable of Ext*.

15*. For any right R-module M let $\cdots \to P_n \to \cdots \to P_1 \xrightarrow{d} P_0 \xrightarrow{\varepsilon} M \to 0$ be a projective resolution of M. For a short exact sequence $0 \to A \xrightarrow{\alpha} B \xrightarrow{\beta} C \to 0$ of modules, prove that there exists a commutative diagram where the rows are complexes (see exercise 6) and the columns are short exact sequences.

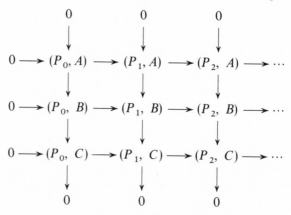

16*. Use the last exercise to prove that for any short exact sequence of modules $0 \to A \xrightarrow{\alpha} B \xrightarrow{\beta} C \to 0$, there is an induced exact sequence called the *long exact sequence* in the second variable of *Ext*:

$$\cdots \to \mathrm{Ext}_R^{n-1}(M, C) \xrightarrow{\psi} \mathrm{Ext}_R^n(M, A) \xrightarrow{\alpha^*} \mathrm{Ext}_R^n(M, B)$$

$$\xrightarrow{\beta^*} \mathrm{Ext}_R^n(M, C) \to \mathrm{Ext}_R^{n+1}(M, A) \to \cdots$$

17*. If the previous short exact sequence of modules $0 \to A \xrightarrow{\alpha} B \xrightarrow{\beta} C \to 0$ splits, then prove that the following are short exact split sequences for all $n = 0, 1, 2, \ldots$ and any module M:

(a) $0 \longrightarrow \mathrm{Ext}_R^n(C, M) \xrightarrow{\beta^*} \mathrm{Ext}_R^n(B, M) \xrightarrow{\alpha^*} \mathrm{Ext}_R^n(A, M) \longrightarrow 0$

(b) $0 \longrightarrow \mathrm{Ext}_R^n(M, A) \xrightarrow{\alpha^*} \mathrm{Ext}_R^n(M, B) \xrightarrow{\beta^*} \mathrm{Ext}_R^n(M, C) \longrightarrow 0$.

18*. Let A and M be R-modules. Prove:

(a) $\mathrm{Ext}_R^1(A, M) = 0$ for all $M \Leftrightarrow A$ is projective.
(b) $\mathrm{Ext}_R^n(A, M) = 0$ for all M and all $n \geqslant 1 \Leftrightarrow A$ is projective.
(c) $\mathrm{Ext}_R^{n+1}(A, M) = 0$ for all $M \Leftrightarrow$ projective dimension of $A = n$.
(d) $\mathrm{Ext}_R^{n+i}(A, M) = 0$ for all M and all $i \geqslant 1 \Leftrightarrow$ projective dimension of $A = n$.
(e) $\mathrm{Ext}_R^1(M, A) = 0$ for all $M \Leftrightarrow A$ is injective.
(f) $\mathrm{Ext}_R^n(M, A) = 0$ for all M and all $n \geqslant 1 \Leftrightarrow A$ is injective.

Remark. The *right global dimension of a ring* R (abbreviation: r.gl.dim R) is defined as the supremum

r.gl.dim $R = \sup\{\text{proj.dim } M \mid M = M_R\}$

of the projective dimensions proj.dim.M of all right R-modules. The *left global dimension* l.gl.dim R is defined similarly.

19. For $R = \mathbf{Z}$, prove that r.gl.dim $\mathbf{Z} = \infty$.

The next sequence of exercises pertain to sections 10-3 through 10-7.

In exercises below R is a ring with $1 \in R$ and with the D.C.C. on right ideals and whose largest nilpotent ideal (see 9-3.2) is N. Prove the following assertions.

1. If R has no zero divisors, R is a division ring.
2. $N = \mathrm{rad}\, R = J(R)$.
3. If $ab = 1$ for $a, b \in R$, then $ba = 1$. Give examples to show that this fails if R does not have the D.C.C.

4. An idempotent $0 \neq e = e^2 \in R$ is primitive if and only if eR is a minimal right ideal.

5. If $G \subset N$ is an additive subgroup, then $G^2 \neq G$.

6. R/N is a division ring if and only if 1 is the only nonzero idempotent in R.

7. Every right ideal of R can be written as a finite direct sum of indecomposable right ideals.

8. There exist indecomposable right ideals $U_i \leqslant R$ such that $R = U_1 \oplus \cdots \oplus U_n$.

9. There exist pairwise orthogonal idempotents $e_i e_j = \delta_{ij} e_i$ such that $1 = e_1 + \cdots + e_n$.

10. Every indecomposable direct summand U of R_R has a greatest proper submodule $U \cap N$.

11. Suppose that U is an indecomposable right ideal of R with $U \nsubseteq N$. (i) Then $R = A \oplus U$ for some $A \leqslant R$. (ii) For any decomposition $R = U_1 \oplus \cdots \oplus U_n$ where $U_i \leqslant R$, U_i are indecomposable, $U \cong U_i$ for some i.

12. Let U and W be two indecomposable direct summands of R_R. Then $U \cong W$ if and only if the two simple modules $U/U \cap N$ and $W/W \cap N$ are isomorphic.

13. Prove that exercises 1, 2, 4, and 8 do not require the hypothesis that $1 \in R$.

14. Classify up to isomorphism all two dimensional algebras A over a field F into eight classes according to the following outline.
    (a) $A^2 = 0$.
    (b) $\dim_F A^2 = 1$. $A = Fa + Fa^2$, $0 \neq a^2 \in A$, $a^3 = 0$.
    (c) $J(A) = 0$. (i) Either $A = F[a]$ is a quadratic extension field of F; $a \neq F$, $a^2 = k \in F$, $k \neq c^2$ for any $c \in F$. (ii) Or $A = Fe_1 \oplus Fe_2$, $e_1^2 = e_1$, $e_2^2 = e_2$, $e_1 e_2 = e_2 e_1 = 0$.
    (d) $\dim_F J(A) = 1$. Show that there is a subalgebra $S \subset A$ and a vector space direct sum $A = S + J(A)$. Let $1_S = e \in S$ and $0 \neq n \in J(A)$, $A = Fe + Fn$, $n^2 = 0$, $e^2 = e$.

    Case 1. $en = ne = 0$.

    Case 2. $en \neq 0$, $ne = 0$. $A \cong \begin{vmatrix} F & F \\ 0 & 0 \end{vmatrix}$.

    Case 3. $ne \neq 0$, $en = 0$. $A \cong \begin{vmatrix} F & 0 \\ F & 0 \end{vmatrix}$.

    Case 4. $en \neq 0$, $ne \neq 0$. $A \cong F \times F$ additively, $(k, c)(\bar{k}, \bar{c}) = (k\bar{k}, k\bar{c} + c\bar{k})$; $k, c, \bar{k}, \bar{c} \in F$.

15. A ring R is *locally nilpotent* if any finite subset of R generates a nilpotent subring. Prove that there does not exist a simple locally nilpotent ring.

CHAPTER 11

Direct sum decompositions

Introduction

This chapter contains more major results of module theory than any previous chapter. It discusses chain condition in modules and direct sum decompositions of modules. The two are closely connected. Chain conditions in modules and rings have significant influence on the structure of a module or ring. For example, the D.C.C. together with semiprimitivity leads to the Wedderburn Theorems. And a ring R is right Artin semiprimitive if and only if every module is a direct sum of simple modules. In another direction, we saw that a ring R was right Noetherian if and only if every direct sum of injective modules remains injective. Note that in both examples above direct sum decompositions of modules appear together with chain conditions.

So far we understand semiprimitive right Artinian rings R completely through the Wedderburn Theorems. At this point quite naturally, two questions arise. What happens if we drop the semiprimitivity restriction? And what can we say about the structure of a right Noetherian ring, with some appropriate auxiliary restrictions in addition to the A.C.C.? Both questions are answered by the Hopkins–Levitzki Theorem (11-3.8). Hopkins proved that a right Artinian ring is right Noetherian. However, the latter fact alone does not give us a picture of what an arbitrary right Artinian ring looks like. Here a theorem of Levitzki supplies the missing pieces in this jigsaw puzzle. Levitzki Theorem (11-3.6) states that in a right Noetherian ring, every nil one-sided ideal is nilpotent. In algebra in general, results which allow one to go from 'nilpotent' to 'nil' seem to be hard to prove and also useful. In this case it gives us a necessary and sufficient condition for a ring R to be right Artinian namely, (i) R is right

Noetherian; (ii) the Jacobson radical $J = J(R)$ of R is nil, and (iii) R/J is semiprimitive right Artinian. In this case we can then deduce that J is actually nilpotent, while all the modules J^n/J^{n+1}, $n = 0, 1, \ldots$ are finite direct sums of simple modules (11-3.6). The above description gives us a good idea of what a right Artinian ring looks like.

In order to understand the structure of a module M, sometimes one expresses M as a direct sum $M = \oplus\{M_\alpha \mid \alpha \in \theta\}$ of modules M_α of a simpler, better understood type. In connection with direct sum decompositions, the endomorphism rings of modules play a role in three ways. (i) In module theoretic proofs, it is necessary to construct endomorphisms of a module having certain prescribed properties. Sometimes such constructions can best be carried out by decomposing the module M as a direct sum of other modules, and carry out the required construction separately on each direct summand. (ii) Properties of endomorphisms reflect chain conditions. A monic endomorphism of an Artinian module is epic. An epic endomorphism of a Noetherian module is monic. Fitting's Lemma says that an endomorphism of a module satisfying both chain conditions is either invertible or nilpotent. (iii) The so-called Krull–Remark–Schmidt Theorem says that the decomposition of a module as a direct sum of indecomposable modules is unique up to isomorphism, provided that the endomorphism ring of each sum summand is local.

Section one starts out with the completely reducible modules, that is modules which can be expresed as direct sums of simple modules. This class of modules includes the socle of any module. This chapter includes several topics which at first glance may appear to have no connection with direct sum decompositions, but after a more thorough look are a logical consequence of such direct sums. The dual of the socle is the radical of a module, which generalizes the Jacobson radical to modules. It is natural to treat this topic immediately after the socle, although directly the radical is not related to direct sums. The radical of a unital module however, is the sum of all the superfluous submodules.

Next, direct sums of indecomposable modules are considered. A module M which is either Artinian or Noetherian can be written as a direct sum of indecomposables. Therefore, modules with chain conditions are investigated, as well as their endomorphisms. A slight generalization of the Krull–Schmidt Theorem is proved, with particular emphasis on the fact that the permutation in that theorem is independent of the choice of submodules (11-4.8(iii) and 11-4.10(iii)). Then the Krull–Remak–Schmidt Theorem is proved. If one had to single out one or two major results of this chapter, it would be the later theorem and the Hopkins–Levitzki Theorem. However the reader will find here several topics which relate to the readers previous knowledge and are so to speak built on the previous chapters. Here follows a brief description of only one of several such topics.

Suppose that $f: M \to M$ is an R-module endomorphism of a module M. It is shown that if M is an Artinian and Noetherian module, that then f is monic if and only if f is epic (11-3.6). This immediately leads to a direct sum decomposition of M as a direct sum of two submodules $M = K \oplus I$ which are invariant under f, i.e. $fK \subseteq K$ and $fI \subseteq I$. Moreover on I, $f \mid I : I \to I$ acts as an isomorphism, while f is nilpotent $f^n \mid K = 0$ on K for some integer n (11-3.7). These abstract module theoretic theorems are immediately applicable to the special case when f is a linear operator on a finite dimensional vector space M over a field F by use of 1-1.14, where $R = F[x]$. In this case the resulting f-invariant subspaces K and I are called the stable kernel and the stable range of f.

11-1   Completely reducible modules

11-1.1   Notation. Here the ring may or may not contain an identity element. The socle of any right R-module M was defined (10-5.3) as the submodule $\operatorname{soc} M = \Sigma\{V \mid V \leqslant M, V \text{ is a simple module}\}$.

11-1.2 Recall (10-3.3) that a module M is completely reducible if there exist some family of simple submodules $M_\alpha \leqslant M$, $\alpha \in \theta$, such that $M = \oplus\{M_\alpha \mid \alpha \in \theta\}$.

11-1.3   Lemma. *For some given modules $N < M$ suppose that there is an indexed set of simple submodules $M_\alpha < M$, $\alpha \in \theta$ having the property that $M = N + \Sigma\{M_\alpha \mid \alpha \in \theta\}$. Then there exists a subset $\mathscr{L} \subseteq \theta$ such that $M = N \oplus (\oplus\{M_\beta \mid \in \mathscr{L}\})$.*

Proof. By Zorn's Lemma there exists a subset $\mathscr{L} \subseteq \theta$ of θ maximal with respect to the two properties that their corresponding sum is a direct sum (i) $\Sigma\{M_\beta \mid \beta \in \mathscr{L}\} = \oplus\{M_\beta \mid \beta \in \mathscr{L}\}$ which at the same time meets N trivially. (ii) $N \cap \Sigma\{M_\beta \mid \beta \in \mathscr{L}\} = 0$. For every $\alpha \in \theta$, if $M_\alpha \cap [N \oplus (\oplus M_\beta)] = 0$, then $\mathscr{L} \subsetneqq \mathscr{L} \cup \{\alpha\}$, where the latter satisfies (i) and (ii), and thus contradicts the maximality of $\mathscr{L}$. Therefore $M = N \oplus (\oplus M_\beta)$.

A special case of the above lemma with $N = 0$ gives the next basic corollary.

11-1.4   Corollary. *A module M which is a sum $M = \Sigma\{M_\alpha \,|\, \alpha \in \theta\}$ of an indexed subset of simple submodules $M_\alpha < M$, $\alpha \in \theta$, is a direct sum of some of these $M = \oplus\{M_\beta \,|\, \beta \in \mathscr{L}\}$ for some subset $\mathscr{L} \subseteq \theta$.*

11-1.5   Corollary. *The socle soc M of any module M is a direct sum of simple submodules of M.*

11-1.6   Corollary. *Suppose that $M = \oplus\{M_\alpha \,|\, \alpha \in \theta\}$ is a completely reducible module, with all M_α simple and that $N < M$ is an arbitrary submodule. Then there exists a subset $\mathscr{L} \subset \theta$ such that*

  (i) $N \cong \oplus\{M_\alpha \,|\, \alpha \in \theta \backslash \mathscr{L}\}$ *and*
  (ii) $M/N \cong \oplus\{M_\beta \,|\, \beta \in \mathscr{L}\}$.

11-1.7   Corollary. *Let V, $M_\alpha \in \Sigma$ for all $\alpha \in \theta$. If $V < M = \oplus M_\alpha$, then $V \cong M_\delta$ for some $\delta \in \theta$.*

Proof. If $N = V$ in the last corollary then necessarily $\theta \backslash \mathscr{L} = \{\delta\}$ is a singleton, and $V \cong M_\delta$.

Property (3) below reflects the ordinary connotation of the expression 'completely reducible' as also does (2), which is its definition.

11-1.8   Proposition. *For an R-module M the following are all equivalent*

  (1) $M = \operatorname{soc} M$.
  (2)   *M is a direct sum of simple submodules.*
  (3)   *Every submodule of M is a direct summand.*
  (4)   *Every short exact sequence of modules $0 \to N \to M \to Q \to 0$ splits.*

Proof. The implication $(3) \Leftrightarrow (4)$ is immediate, and $(1) \Leftrightarrow (2) \Rightarrow (3)$ follow from 11-1.5 and 11-1.6.

$(3) \Rightarrow (2)$: By hypothesis $M = (\operatorname{soc} M) \oplus D$ for some $D \leqslant M$. It suffices to show that every nonzero submodule of M contains a simple submodule. Since every submodule of M inherits the property (3), it suffices to show that M contains a simple submodule.

Take any $0 \neq m \in M$, and by Zorn let $N < M$ be maximal with respect to $m \notin N$. By (3), $M = N \oplus C$ for some $0 \neq C < M$. If C is not simple then it contains a proper nonzero submodule $0 \neq A < C$, and hence again by (3), $C = A \oplus B$ with also $B \neq 0$. From

$N + A + B = N \oplus A \oplus B$, and the maximality of N, we obtain the contradiction that $m \in (N \oplus A) \cap (N \oplus B) = N$. Therefore, C is simple.

11-1.9   Homogeneous components. Define an equivalence relation '$\sim$' on the class Σ by $V \sim W$ if $V \cong W$. Let Γ be the resulting set of equivalence classes of isomorphic simple modules.

Let M be a completely reducible module. For $\gamma \in \Gamma$, the γth *homogeneous component* M_γ of M is defined as the submodule

$$M_\gamma = \Sigma \{V \mid V \leqslant M, V \in \gamma\}.$$

A module is called *homogeneous* if it is equal to a single homogeneous component. Then the following hold.

   (1)   If $M = \oplus \{V_\alpha \mid \alpha \in \theta\}$ with all $V_\alpha \in \Sigma$, then $M_\gamma = \oplus \{V_\alpha \mid V_\alpha \in \gamma\}$.
   (2)   For any simple submodule $W < M$, $W \cong V_\alpha$ for some $V_\alpha \in \gamma$. In particular, $W \in \gamma$.
   (3)   For any module P and any module homomorphism $\phi : M \to P$, $\phi M_\gamma \subseteq (\operatorname{soc} P)_\gamma$.
   (4) $M = \oplus \{M_\gamma \mid \gamma \in \Gamma\}$.

Proof. (1) Set $Q = \oplus \{V_\alpha \mid V_\alpha \in \gamma\}$. Write $M = Q \oplus (\oplus \{V_\beta \mid \beta \in \mathcal{L}\})$ where $\mathcal{L} \subset \theta$ is exactly the set of all those $\beta \in \theta$ such that $V_\beta \notin \gamma$. Clearly, $Q \subseteq M_\gamma$. Conversely, suppose that $V < M$ with $V \in \gamma$. If $V \cap Q = 0$, then $V \cong (V \oplus Q)/Q < M/Q \cong \oplus V_\beta$. Hence by 11-1.7, $V \cong V_\beta$ for some $\beta \in \mathcal{L}$. Since $V_\beta \notin \gamma$, this is a contradiction. Hence $M_\gamma = Q$.

   (2) It now follows from (1) and 11-1.7 that $W \cong V_\alpha$ for some $V_\alpha \in \gamma$.
   (3) For any $V_\alpha \in \gamma$, $\phi V_\alpha = 0$ or $\phi V_\alpha \cong V_\alpha$. By definition of $(\operatorname{soc} P)_\gamma$, $\phi V_\alpha \subseteq (\operatorname{soc} P)_\gamma$.
   (4) This is immediate from (1).

11-1.10 Properties of the socle. Let M be any module, $\operatorname{soc} M \leqslant M$ its socle, and $\operatorname{soc} M = \oplus \{(\operatorname{soc} M)_\gamma \mid \gamma \in \Gamma\}$ its decomposition into its homogeneous components. Then the following hold.

   (1) $\operatorname{soc} M = \cap \{N \mid N \leqslant M \text{ is essential}\}$.
   (2)   For any module homomorphism $\phi : M \to P$ into any module P, $\phi \operatorname{soc} M \subseteq \operatorname{soc} P$. In particular, $\operatorname{soc} R \lhd R$.
   (3)   For any submodule $N < M$, $\operatorname{soc} N = N \cap \operatorname{soc} M$. In particular, $\operatorname{soc}(\operatorname{soc} M) = \operatorname{soc} M$.
   (4)   For any module map $\phi : M \to P$ and any equivalence class $\gamma \in \Gamma$ of isomorphic simple modules, $\phi(\operatorname{soc} M)_\gamma \subseteq (\operatorname{soc} P)_\gamma$. In particular $(\operatorname{soc} RF)_\gamma \lhd R$.

Proof. (1) Let Q be the intersection in (1). If $S \leqslant M$ is any simple submodule, then for any large submodule $N \leqslant M$, $N \cap S \neq 0$, and hence $S \leqslant N$. Thus $\operatorname{soc} M \subseteq Q$.

It suffices to show that every submodule $A < Q$ is a direct summand of Q. Take any $B \leqslant M$ such that $A \oplus B \leqslant M$ is large. Then $A < Q \leqslant A \oplus B$. Use of the modular law then gives $Q = Q \cap (A \oplus B) = A \oplus (Q \cap B)$.

(2) If $V < M$ is any simple submodule, then either $\phi V = 0$, or, $V \cong \phi V$ is an isomorphism. Thus $\phi \operatorname{soc} M \subseteq \operatorname{soc} P$.

(3) Since a simple submodule of N is also a simple submodule of M, $\operatorname{soc} N \subseteq \operatorname{soc} M$ by the definition of 'socle'. Thus $\operatorname{soc} N \subseteq N \cap \operatorname{soc} M$.

Conversely, since $N \cap \operatorname{soc} M < \operatorname{soc} M$ is a submodule of the completely reducible module $\operatorname{soc} M$, by 10-1.6, also $N \cap \operatorname{soc} M$ is completely reducible. Therefore $N \cap \operatorname{soc} M \subseteq \operatorname{soc} N$.

(4) This follows from 11-1.9 (3).

11-2 Radical of a module

For any module M, the radical of M will be defined as a submodule $J(M) < M$. If $1 \in R$ and $M = R$ is viewed as a right module over itself, then $J(R)$ in this new module sense turns out to be the already familiar Jacobson radical of R.

Take a mathematical statement about the socle $\operatorname{soc} M$ of a certain limited type, which here will not be precisely specified. Then systematically replace all appropriate operations, properties, or concepts by their duals. The result will be a statement about the radical $J(M)$. For example, reversing all arrows and replacing 'image' with 'kernel' and 'sum' with 'intersection' in the definition of the socle yields the definition of the radical given below.

Concept	Dual
Image	Kernel
Sum	Intersection
Essential or large submodule	Superfluous or small submodule (see 11-2.2)

The careful reader will note that although the dual of the fact that "the socle is the intersection of all the large submodules (11-1.8 (1))" is that "the radical is the sum of all the small submodules (11-2.8)", nevertheless the latter radical statement requires more restrictive hypotheses than the former socle statement (11-2.11).

11-2.1   Definition. Let θ be any class or set of modules whatsoever, and M any R-module. The *trace* and *reject* of θ in M are defined to be the following submodules of M:

(i) $\text{Tr}_\theta M = \Sigma\{fV \mid V \in \theta, f \in \text{Hom}_R(V, M)\}$;
(ii) $\text{Rej}_\theta M = \cap\{\text{Kernel } f \mid V \in \theta, f \in \text{Hom}_R(M, V)\}$.

In other words, the trace of θ in M is the sum of all the homomorphic images of modules of θ in M. The reject of θ in M is the intersection of all the kernels of all possible homomorphisms of M into modules of θ.

Now specialize θ to be $\theta = \Sigma$ the class of all simple right R-modules. Obviously, $\text{Tr}_\Sigma M = \text{soc } M$. The *radical* of M, denoted by $J(M) = JM$, is defined as the submodule

$$J(M) = \text{Re } j_\Sigma M = \cap\{N \mid N < M, M/N \in \Sigma\}.$$

The empty intersection is always understood to be the whole module. By 7-1.6(ii), if $1 \in R$ and $M = R_R$, then $J(R)$ is our previous old Jacobson radical of the ring R.

Recall that $R \lhd \mathbb{Z} \times R$ if $1 \notin R$. Recall also that a subset $X \subset M$ generates the module M if $M = \Sigma\{xR^1 \mid x \in X\}$ (see 1.24). If $1 \in R$, then $R = R^1$, and of course this definition agrees with the usage of 'generated' in Chapter 2. In particular in all cases ($1 \in R$ or $1 \notin R$), a module M is finitely generated if $M = x_1 R + \cdots + x_n R + \mathbb{Z}x_1 + \cdots + \mathbb{Z}x_n$ for some finite subset $\{x_1, \ldots, x_n\} \subset M$.

11-2.2   Remark. Recall the definitions of small submodule and superfluous submodule (Definition 7-1.21). Note also that the concept 'superfluous' does not apply to a single module, but to pairs of modules ordered by inclusion. The zero submodule of course is always superfluous.

The following lemma and its corollaries are generally useful in other module proofs in other contexts.

11-2.3   Lemma. *Let M be any module and $A < M$ any proper submodule such that the quotient module M/A is finitely generated. Then there exists a maximal submodule $B < M$ of M containing $A \subset B$.*

Proof. If $\{y_1 + A, \ldots, y_n + A\}$ with $y_i \in M \backslash A$ is any minimal generating set of M/A, with n as small as possible, then the submodule $C = A + y_1 R^1 + \cdots + y_{n-1} R^1 < M$ is proper. For otherwise the smaller set $\{y_1 + A, \ldots, y_{n-1} + A\}$ would already generate M/A. By Zorn's Lemma there exists a submodule $B < M$ maximal with respect to the two properties that (1) $C \subseteq B$, but (2) $y_n \notin B$.

First, since $M = A + y_1 R^1 + \cdots + y_n R^1$ and since

$$\{y_1, \ldots, y_n\} \subseteq C + y_n R^1,$$

it follows that $M = C + y_n R^1$. Secondly, suppose that $D \leqslant M$ is any submodule containing $B \subset D$ properly. By the maximality of B with respect to (2), $y_n \in D$. But then $M = C + y_n R^1 \subseteq D$, and $M = D$. Hence B is a maximal submodule of M.

11-2.4   Corollary 1. *If above $MR = M$, then M/B is a simple module.*

A slight modification of the proof of the last lemma ($C = A, y = y_n$) immediately gives the next corollary.

11-2.5   Corollary 2. *For any module M suppose that $A < M$ is any proper submodule and that $y \in M \backslash A$ is any element of M not in A. Then let $B < M$ be any submodule given by Zorn's Lemma as being maximal with respect to (1) $A \subseteq B$ but (2) $y \notin B$. Then*

  (i) $B < B + yR^1$ is a maximal submodule of the module $B + yR^1$.
  (ii)   In particular, if $M = B + yR^1$, then B is a maximal submodule $B < M$ in M.

11-2.6   Corollary 3. *With the notation and hypotheses of the last corollary*

  (i)   *if $y \in yR$, then $\Rightarrow (B + yR^1)/B \in \Sigma$;*
  (ii)   *if $MR = M = B + yR^1$, then $\Rightarrow M/B \in \Sigma$.*

The next corollary follows by taking $A = 0$ in the last lemma.

11-2.7   Corollary 4. *Any finitely generated module M contains a maximal submodule. If in addition $MR = M$, then $J(M) \neq M$.*

11-2.8   Proposition. *If M is any module with $MR = M$, then*

$$J(M) = \Sigma \{S \mid S < M \text{ is superfluous}\}.$$

Proof. Define Y to be the latter sum. It will be shown that for any submodule $N < M$ with a simple quotient $M/N \in \Sigma$, the submodule N contains all superfluous submodules $S < M$, i.e. $S \subseteq N$. For if $S \neq N$, then $S + N = M$. But then the superfluousness of S gives the contradiction that $N = M$. Thus $Y \subseteq J(M)$.

It is asserted that for any $y \in J(M)$, $yR^1 < M$ is superfluous. For if not,

then for some proper submodule $A < M$, $A + yR^1 = M$. Now as before
Zorn's Lemma produces a proper submodule $B < M$ maximal in M with
respect to $A \subseteq B$ and $y \notin B$. Also $B + yR^1 = M$. By Corollary 11-2.6(ii),
M/B is simple. But the $y \in J(M) \subseteq B$ is a contradiction. Thus $yR^1 < M$
is superfluous, and $J(M) \subseteq Y$. Hence $J(M) = Y$.

11-2.9   Definition. A module M is *uniserial* if the set of all of its
submodules is linearly ordered by set inclusion.

Note that any proper submodule $S < M$ of any uniserial module M is
superfluous in M. This fact together with the last proposition proves the
assertions below.

11-2.10   Let M be any uniserial module with $MR = M$. Then

  (i)   *either* $J(M) < M$ is the unique maximal submodule;
  (ii)   *or* M contains no maximal submodules and $J(M) = M$.

11-2.11   Counterexample. Let the ring R be $R = \mathbb{Z} = 0, \pm 1, \ldots$ additively,
but with all products zero, $R^2 = 0$. Over such a ring, any module M is a
radical module $M = J(M)$. Let M be the uniserial module $M = \mathbb{Z}/4\mathbb{Z}$.
Then $2\mathbb{Z}/4\mathbb{Z} < M$ is the only nonzero superfluous submodule of M.
Hence 11-2.8 is false if the hypothesis that $MR = M$ is omitted.

11-2.12   Terminology. Historically, the terms 'irreducible module' and
'semisimple module' have been used so extensively that we had better
explain how they were used and how they are now being used.

  Irreducible module is the same as simple module, and *semisimple
module* is a synonym for completely reducible module.

  The former is not to be confused with $C < M$ being a meet irreducible
submodule in M which means that if A, $B \leqslant M$ with $C = A \cap B$, then
either $C = A$, or $C = B$.

  To avoid confusion upon reading other literature the reader should
note the following. If the ring R is right semisimple and $1 \in R$, then R is
semiprimitive with the right D.C.C. However, if $1 \notin R$, then R may be
right semisimple, but not satisfy the right D.C.C. (Historically, a small
minority of authors have required semisimple modules to satisfy the
D.C.C.) Some authors (e.g. Lambek) do not use 'semisimple'.

  We avoid the term 'semisimple module' because it conflicts with the
following uniform notational scheme which seems to be gaining some
favor. Given a property P of (right) modules, a module M is semi P if

the intersection of all submodules of M possessing property P is zero. Next, a submodule $K \leqslant M$ is P in M iff M/K has property P. Consequently, M has P iff $(0) < M$ is P. We use this scheme for primitive–semiprimitive, prime versus semiprime (but *not* for a hereditary and semihereditary). The remark preceding 11-1.8 gives another reason why here the term 'semisimple module' is not used.

11-3 Artinian and Noetherian modules

A module is Noetherian if it satisfies the A.C.C. on submodules (3-5.1) and Artinian—if the D.C.C. (7-1.35). There are alternate completely equivalent ways of defining these two properties (Noetherian: 3-5.2(i), (ii) and (iii); Artinian: 7-1.35(i) and (ii).

To put this chapter into perspective from a broader point of view the reader should recall some of the following peripheral properties that describe submodules or modules:

proper, maximal and minimal submodule (6-1.4),
indecomposable module (10-2.3),
composition series (10-2.1),
composition length (10-2.1).

Although this section does not depend on Chapter 10, we shall consider here conditions on modules like 10-3.4(2) and (4) – but for arbitrary rings, i.e. with $1 \in R$ or $1 \notin R$, and without any chain conditions on R.

11-3.1 Observation. For any module M, suppose that $A, B, N < M$ are three submodules, and that $\pi : M \to M/N$ is the natural projection. Then

$$\left. \begin{array}{l} B \subseteq A \\ N \cap B = N \cap A \\ \pi B = \pi A \end{array} \right\} \Rightarrow A = B$$

Proof. For any element $a \in A$, there is a $b \in B$ and $n \in N$ such that $b = a + n$. Since $n = b - a \in A \cap N = B \cap N$, it follows that $a = b - n \in B$. Thus $A = B$.

11-3.2 Proposition. *For any modules $N < M$, (i) M is Artinian $\Leftrightarrow$ both N and M/N are Artinian; (ii) M is Noetherian $\Leftrightarrow$ both N and M/N are Noetherian.*

Proof. Let $\pi: M \to M/N$ be the natural projection. Only (ii) will be proved because the proof of (i) is the same. $\Leftarrow$: Any strictly ascending sequence of submodules $0 < A_1 < A_2 < \cdots \leqslant M$ of M gives rise to two sequences each of which by hypothesis become stationary after p and q steps:

(1) $A_1 \cap N \leqslant \cdots < A_p \cap N = A_{p+1} \cap N = \cdots$;
(2) $\pi A_1 \leqslant \cdots < \pi A_q = \pi A_{q+1} = \cdots$.

If $n = \text{maximum}\ (p, q)$, then by 11-3.1 $A_n = A_{n+1} = \cdots$.

(ii) $\Rightarrow$: Trivially, N is Noetherian. Since there exists a one to one correspondence between submodules of M/N and those submodules of M containing N, also M/N must be Noetherian.

11-3.3 Corollary. *Suppose that the module M is a sum $M = N_1 + \cdots + N_k$ of a finite number of Artinian (Noetherian) modules N_i. Then also M is likewise Artinian (Noetherian).*

Proof. It suffices to consider the Artinian case for $k = 2$ and $M = N_1 + N_2$. Then $M/N_1 = (N_1 + N_2)/N_1 \cong N_2/N_1 \cap N_2$ is Artinian because it is an epimorphic image of the Artinian module N_2. A second application of 11-3.2 shows that M is Artinian.

11-3.4 Corollary. *If the ring R is right Noetherian (Artinian), and M is a finitely generated right R-module, then M likewise is Noetherian (Artinian).*

Proof. By hypothesis, $M \cong F/N$ where $F = x_1 R \oplus \cdots \oplus x_n R$ is free on the finite set $\{x_1, \ldots, x_n\}$. Since $x_i R \cong R$, both F and N are Noetherian (Artinian), and hence so is also M.

So far we have considered the concepts right Artinian and right Noetherian separately. Our next step is to intertwine them. As can be seen below 'Artinian' and 'Noetherian' are not interchangeable. The ring $R = \mathbf{Z}$ and module $M = \mathbf{Z}$ are both Noetherian, but not Artinian.

11-3.5. Proposition. *Let R be a right Artinian ring. Then a right R-module M with $MR = M$ is right Noetherian if and only if M is right Artinian.*

Proof. Set $J = J(R)$. Then $J^k = 0$ for some k by 7-1.36. Form the chain of submodules $M \supseteq MJ \supseteq MJ^2 \supseteq \cdots \supseteq MJ^{k-1} \supseteq 0$. Since each MJ^i/MJ^{i+1} is an R/J-module in a natural way, it is a direct sum of simple R/J-modules which are also simple R-modules by 10-3.4(4). But then MJ^i/MJ^{i+1} is a finite direct sum of simple modules if and only if it satisfies both of the chain conditions D.C.C. and A.C.C. The latter fact follows from either 11-1.3 or from 10-2.1. An inductive argument based on 11-3.2 now proves the proposition.

The next theorem is due to Levitzki.

11-3.6   Theorem. *For a right Noetherian ring, let $N(R)$ be the sum of all nilpotent right ideals. Then*

(i)   *every nil right or left ideal is nilpotent. In particular*
(ii) *$N(R)$ is nilpotent, and contains all nil (nilpotent) one-sided ideals.*
(iii) *$R/N(R)$ is a right Noetherian ring which contains no nonzero nil one-sided ideals.*

Proof. (ii) As noted in 7-3.3, 3.4, $N(R) \lhd R$, and also $N(R)$ is the sum of all nilpotent left ideals. By 3-5.2(iii), $N(R)$ is equal to some finite sum of nilpotent right ideals. However, a finite sum of nilpotent right ideals is nilpotent by 9-3.1 (4).

(i) Let B be a nil right or nil left ideal of R. Since $(B + N(R))/N(R) \subseteq R/N(R)$ is also nil, and $N(R/N(R)) = 0$, without loss of generality assume that $N(R) = 0$, i.e. that R is a right Noetherian ring with no nonzero nilpotent one-sided ideals. Let $r \in R$. When $B \leqslant R$ is a nil right ideal, then $rB \leqslant R$ is also a nil right ideal. If B is a nil left ideal, then $rB \subseteq B$, and rB is a nil abelian subgroup.

By 3-5.2(iii) the set of right annihilator ideals of nonzero elements of B has a maximal element $b^{\perp}$ where $0 \neq b \in B$. In order to show that $B = 0$, it suffices to show now that only $b = 0$. It is asserted that $bRb = 0$. If $Rb = 0$, then surely also $bRb = 0$. Otherwise, take any $r \in R$ whatsoever for which $rb \neq 0$. Since $rb \in rB$ and rB is nil, there exists an integer $n \geqslant 1$ (depending on rb) such that $(rb)^n \neq 0$ but $(rb)^{n+1} = 0$. But $b^{\perp} \subseteq [(rb)^n]^{\perp}$ for all n. By the maximality of $b^{\perp}$, $b^{\perp} = [(rb)^n]^{\perp}$. But $rb \in [(rb)^n]^{\perp}$ and hence $rb \in b^{\perp}$, or $brb = 0$. Consequently $bRb = 0$. If $C = \mathbf{Z}b + Rb + bR + RbR \lhd R$ is the smallest ideal of R containing b, then $C^3 \subseteq bRb = 0$. Since R has no nilpotent ideals, $C = 0$, and hence $b = 0$. Thus $B \subseteq N(R)$ and (iii) now follows.

11-3.7   Corollary. *For R a right Noetherian ring and its maximal nilpotent ideal $N(R) \lhd R$, let $J = J(R)$, rad $R \lhd R$ be the Jacobson and the prime radicals of R. Then*

(i)   rad $R = N(R) \subseteq J$.
(ii) $N(R) = J \Leftrightarrow J^k = 0$ *for some k. In this case $J(R/J^n) = J/J^n$, $n \leqslant k$.*

Proof. (i) Always, $N(R) \subseteq J$ by 7-1.34(iii). A strongly nilpotent element is nilpotent (9-2.7 (4)), and rad R is precisely the set of all strongly nilpotent elements of R (9-2.8 (1)). Consequently rad R is nil. Thus rad $R \subseteq N(R)$ by the last theorem. The reverse inclusion follows from the fact that rad R contains all nilpotent ideals of R ((9-2.17 (i)). Hence rad $R = N(R)$.

(ii) Again, since $N(R)$ is nilpotent by the last theorem, first $N(R) \subseteq J$; and secondly, $N(R) = J$ if and only if $J^k = 0$ for some k.

Now let $N(R) = J$ with $J^k = 0$. Then R/J^n is still a right Noetherian ring, and its unique maximal nilpotent ideal is

$$N\left(\frac{R}{J^n}\right) = \frac{N(R)}{J^n} = \frac{J}{J^n} \subseteq J\left(\frac{R}{J^n}\right);$$

and

$$J\left(\frac{R/J^n}{J/J^n}\right) \cong J\left(\frac{R}{J}\right) = 0.$$

Use of 7-2.2 (with $I = J/J^n$) now shows that $J(R/J^n) \subseteq J/J^n$ and consequently the two are equal, thus proving (ii).

We now use the above theorem (but surprisingly not the last corollary) to obtain the theorem of Hopkins–Levitzki, which gives us a satisfactory picture of right Artinian rings (with or without an identity element). It also tells us that any right Artinian ring is automatically right Noetherian.

11-3.8   Theorem. *For any ring R and its Jacobson radical $J = J(R)$ the following two conditions* (A) *and* (N) *are equivalent.*

(1)   (A) *R is right Artinian.*
       (N) (i) *R is right Noetherian;*
            (ii) *J is nil; and*
            (iii) *R/J is a finite direct sum of full finite matrix rings over division rings.*

(2)   *If* (A) *and hence* (N) *holds, then $J^k = 0$ for some k, while each J^n/J^{n+1}, $0 \leqslant n \leqslant k$, is a finite direct sum of simple R-modules. Moreover, each right ideal of R is finitely generated.*

11-3.9   Remark. Condition (N)(iii) is equivalent to condition (N)(iii)′ R/J satisfies the right D.C.C. as a ring (or as a right R-module).

Proof. (1)(A) $\Rightarrow$ (N). By 11-3.5, (i) R_R has the A.C.C. By the D.C.C., (ii)

$J^k = 0$ for some k. Since R_R is Artinian any quotient module of it such as R/J also satisfies the D.C.C. on right R-submodules, which coincide with right R/J – ideals. Thus (N)(iii)$'$ and N(iii) hold.

(1) (N) $\Rightarrow$ (A). By condition (N)(ii) and 11-3.6, $J^k = 0$ for some k. From the short exact sequence of right R-modules $0 \to J^{k-1} \to R \to R/J^{k-1} \to 0$ we see that R is right Artinian if and only if J^{k-1} and R/J^{k+1} are (11-3.2(ii)). Both of the latter are modules over the ring R/J which is a semiprimitive ring with the right D.C.C. by the condition (N)(iii). Thus both J^{k-1}, R/J^{k-1} are direct sums of simple R/J and hence also simple R-modules. At this point we need the condition N(i) to conclude that both J^{k-1} and R/J^{k-1} are right Noetherian R-modules (11-3.2(ii)). But then each one of these latter two modules is a finite direct sum of simple modules, which in turn forces J^{k-1} and R/J^{k-1} to be right Artinian R-modules. By a third application of 11-3.2(ii) we finally conclude that R is right Artinian.

Recall that a module C is directly indecomposable, or direct sum indecomposable, and here just indecomposable, if C cannot be decomposed further as a direct sum of two nonzero submodules.

11-3.10 Lemma. *Any module M which is either Artinian or Noetherian can be expressed as a finite direct sum of indecomposable modules M_i.*

Proof. First it will be shown that any such module M has an indecomposable summand. (1) If M is indecomposable, we are done. (2) Otherwise $M = M_1 \oplus M_2$ with both $0 \neq M_i \neq M$. If M_1 is indecomposable, we are done. (3) If not, $M_1 = M_{11} \oplus M_{12}$ nontrivially, and $M = M_{11} \oplus M_{12} \oplus M_2$. If M_{11} is indecomposable we may stop. (4) Otherwise $M_{11} = M_{111} \oplus M_{112}$, and at the fourth step $M = M_{111} \oplus M_{112} \oplus M_{12} \oplus M_2$. At the nth step, M will be a direct sum of n nonzero submodules. If either chain condition holds, this process terminates in an indecomposable direct summand of M.

So let $M = M_1 \oplus M_1'$ where M_1 is indecomposable. If $M_1' = 0$, or if M_1' is indecomposable, $s = 2$. If not, $M_1' = M_2 \oplus M_2'$, with M_2 indecomposable, and $M = M_1 \oplus M_2 \oplus M_2'$. Thus at the sth step, $M = M_1 \oplus \cdots \oplus M_s \oplus M_s'$, and hence either one of the two chain conditions guarantees that for some s, $M_s' = 0$.

11-3.11 Proposition. *Let $f : M \to M$ be an endomorphism of a module M. For any integer $1 \leqslant n$, define $I_n = $ image f^n and $K_n = $ kernel f^n. Then*

(a1) $I_n = I_{n+1} \Rightarrow M = I_n + K_n$.

(b2) $K_n = K_{n+1} \Rightarrow I_n \cap K_n = 0$.
(a3) M Artinian $\Rightarrow M = I_n + K_n$ for all sufficiently large n.
(b4) M Noetherian $\Rightarrow I_n \cap K_n = 0$ for all large n.
(a5) M Artinian and f monic $\Rightarrow f$ is epic.
(b6) M Noetherian and f epic $\Rightarrow f$ is monic.

Proof. (a1) Since $I_n = f(I_n) = f(I_{n+1})$, $I_n = I_{2n}$. Let $x \in M$ be arbitrary. Then $f^n x \in f^{2n} M$, and hence $f^n x = f^{2n} y$ for some $y \in M$. Thus $x = f^n y + (x - f^n y)$, where $f^n(x - f^n y) = 0$. Thus $x \in I_n + K_n$, and $M = I_n + K_n$.

(b2) Let x be any element in $x \in I_n \cap K_n$. Then $x = f^n y$ for some $y \in I_n$. Since $x \in K_n$, $0 = f^n x = f^{2n} y$. But $K_n = K_{n+1}$ implies that kernel f^{2n} = kernel f^n, and hence that already $0 = f^n y = x$. Thus $I_n \cap K_n = 0$.

(a3) Since $I_{n+1} \subseteq I_n$, for all sufficiently large n, $I_{n+1} = I_n$. By (a1), $M = I_n + K_n$.

(b4) Since $K_n \subseteq K_{n+1}$, for all sufficiently large n, $K_n = K_{n+1}$. Now by (b2), $I_n \cap K_n = 0$.

11-3.12 Corollary. *Let $f : M \to M$ be any endomorphism of a module M that is both Artinian and Noetherian. Then for some integer*

(i) $M = K_n \oplus I_n$, where $K_n = K_{n+1}$, $I_n = I_{n+1}$, and
(ii) $fK_n \subseteq K_n$, $fI_n = I_n$.
(iii) *The restriction and corestriction of f to $f : I_n \to I_n$ is an isomorphism.*
(iv) *There exists an endomorphism $g : M \to M$ such that $fgf = f$; in particular*
(v) $\text{End}_R M$ *is a regular ring.*

Proof. (i) and (ii). Let n be the unique smallest integer for which both (a3) and (b4) hold. Actually even $fK_n \subseteq K_{n-1} \subseteq K_n$. (iii) The restriction and corestriction $(f | I_n) : I_n \to I_n$ is onto because $fI_n = I_{n+1} = I_n$. Since $\ker(f | I_n) = K_1 \cap I_n \subseteq K_n \cap I_n$, the map $(f | I_n)$ is monic. (iv) and (v). Let g be defined by $gK_n \subseteq K_n$, $gI_n \subseteq I_n$, where on I_n, g is the inverse of f, and on K_n, g can be arbitrary, say zero.

Recall that a module is of finite composition length if and only if it is both Noetherian and Artinian.

11-3.13 Fitting lemma. *Let M be an indecomposable module with a finite composition series. Then any endomorphism $f : M \to M$*

(i)   *is either invertible, or nilpotent.*

(ii)   *The sum of any two nilpotent endomorphisms is also nilpotent. In particular,*

(iii) $\text{End}_R M$ *is a local ring whose unique maximal ideal* $J(\text{End}_R M)$ *is nil.*

Proof. (i) Let f be noninvertible. By (a5) and (b6), f is neither monic nor epic. Let n and $M = I_n \oplus K_n$ be as above in 12-3.12(i). First suppose that $I_n \neq 0$. Since M is indecomposable, $K_n = 0$. Since $K_1 \subseteq K_n$, $\ker f = 0$. So now f is monic, while by (a5) it is also epic, and hence an isomorphism. This is a contradiction and hence $I_n = 0$, or equivalently, $M = K_n$ and $f^n = 0$.

(ii) and (iii). Let $f, g \in \text{End}_R M$ be nilpotent and $\phi \in \text{End}_R M$ arbitrary. Then ϕf or $f\phi$ being noninvertible by (i) above are automatically nilpotent. Suppose that $h = f + g$ is not nilpotent. Again by (i), h has an inverse h^{-1}. Then $\alpha = fh^{-1}$ and $\beta = gh^{-1}$ are noninvertible, and a third time by (i) the maps α and β are nilpotent, yet with $\alpha + \beta = 1_M$ the identity. Take any n such that $\alpha^n = \beta^n = 0$, say the smallest such n. Since $\beta = 1_M - \alpha$, $\alpha\beta = \beta\alpha$ commute, and thus

$$1_M = (\alpha + \beta)^{2n} = \sum_{i=0}^{2n} \binom{2n}{i} \alpha^i \beta^{2n-i} = 0$$

is a contradiction. Hence $h = f + g$ is nilpotent.

The following simple but general observation in conjunction with 11-3.2 will allow us to conclude that free modules over Artinian or Noetherian rings have unique rank, and at the same time also a subsequent fact about injective modules.

11-3.14 Observation. Suppose that A, B, and C are right R-modules and $\phi : A \to A$ is a monomorphism satisfying the following two hypotheses:

(a) $A = C \oplus B$ and   (b) $\phi A \subseteq B$.

Now define $D = C + \phi C + \phi^2 C + \cdots + = \Sigma \{\phi^i C \mid i = 0, 1, \ldots\}$. Then

(i) $D = C \oplus \phi C \oplus \phi^2 C \oplus \cdots \oplus$; in particular

(ii) $D = C \oplus \phi D$; if $C \neq 0$ then

(iii) A does not satisfy either the A.C.C. or D.C.C. on submodules.

Proof. From $\phi(C \oplus B) = \phi C \oplus \phi B \subseteq B$ and $B \cap C = 0$ we get $C \oplus \phi C \oplus \phi B \leqslant A$. Application of ϕ to the latter yields $\phi C \oplus \phi^2 C \oplus \phi^2 B \subseteq \phi A \subset B$. And since $B \cap C = 0$, we have a direct sum of submodules $C \oplus \phi C \oplus \phi^2 C \oplus \phi^2 B \leqslant A$. Repetition of this argument proves (i), from which (ii) follows.

(iii) Consequently,

$$\left\{ \bigoplus_{i=0}^{k} \phi^i C \right\}_{k=0,1,\dots} \quad \text{and} \quad \left\{ \bigoplus_{j=k}^{\infty} \phi^j C \right\}_{k=0,1,\dots}$$

are properly infinite ascending and descending chains of submodules of A.

11-3.15 Consequence 1. Over a nonzero right Artinian or right Noetherian ring, free right R-modules have unique rank.

Proof. If not, then $R^m \cong R^n$ for $n < m$. Repeated application of 11-3.2 shows that whenever R is right Artin or right Noether so is also R^m, for any finite m. In the last observation, take $A = R^m = C \oplus B$ where $R^n \cong B = \{0\} \oplus \cdots \oplus \{0\} \oplus R^n$, and $C = R^{m-n} \oplus \{0\} \oplus \cdots \oplus \{0\}$. There is a monomorphism $\phi: A \to A$ with $\phi A = B$ because $A \cong R^m \cong R^n \cong B$. Then 11-3.14(iii) gives the required contradiction.

The common thread in the two consequences is that we have a module which is isomorphic to a proper direct summand of itself.

11-3.16 Consequence 2. Suppose that A and B are injective modules with A isomorphic to a submodule of B, and B isomorphic to a submodule of A. Then $A \cong B$.

Proof. Without loss of generality, take $A = C \oplus B$. There exists a monomorphism $\phi: A \to A$ with $\phi A \subseteq B$. Thus $\phi D \leqslant B$ as in 11-3.14. Since B is injective, there exist $B_1, B_2 \leqslant B = B_1 \oplus B_2$ where B_1 is an injective hull of ϕD. Then $C \oplus \phi D \leqslant C \oplus B_1$ is an essential extension, and hence $C \oplus B_1$ is an injective hull of $C \oplus \phi D$.

But isomorphic modules have isomorphic injective hulls, and hence

$$C \oplus \phi D = D \cong \phi D \Rightarrow C \oplus B_1 \cong B_1.$$

The rest of the argument now is easy:

$$A = C \oplus B = (C \oplus B_1) \oplus B_2 \cong B_1 \oplus B_2 = B \quad \text{or} \quad A \cong B.$$

11-3.17. Remarks. The last consequence is due to Bumby [65]. An easy example in Goodearl [76, p. 25, Ex. 7] shows that it is false for noninjective modules.

11-4   Direct sums of indecomposables

At this point it would be natural to formulate and prove a uniqueness result for the decomposition of a module M into a finite sum of indecomposables analogous to 10-4.3 or X-4.4 for simple modules. Such a proof requires that endomorphism rings of indecomposable direct summands of M be local rings. This can be guaranteed if M is assumed to satisfy both chain conditions. The point is that in view of 11-3.5 just one chain condition alone already tells us that M can be expressed as some direct sum of indecomposables, yet nevertheless both chain conditions are required for the uniqueness.

Thus in this section we will dispense with both chain conditions, and instead eventually assume that the module M is expressible as some (finite or infinite) direct sum of indecomposable modules each one having a local endomorphism ring. Various properties of such modules including uniqueness of such decompositions are formulated and proved.

11-4.1 Properties of direct sum decompositions. Let M be a module and $M = \oplus\{M_i \mid i \in I\}$ and $M = \oplus\{N_j \mid j \in J\}$ be two direct sum decompositions of M.

(1) A direct summand K of $M = N \oplus K$ is a *maximal direct summand* if and only if some (or any) complementary summand N is indecomposable.

(2) A decomposition $M = \oplus M_i$ will be said to *complement* (maximal) direct summands of M if for every (maximal) direct summand K of M, there exists a subset $\Gamma \subset I$ such that $M = K \oplus (\oplus\{M_i \mid i \in \Gamma\})$.

(3) The above two direct sum decompositions of M are *equivalent* if their exists a bijection $\rho : I \to J$ such that $M_i \cong N_{\rho i}$ under some isomorphism $f_i : M_i \to N_{\rho i}$. In this case the induced sum map $\oplus\{f_i \mid i \in I\} : M \to M$ is an automorphism of M.

11-4.2 Lemma. *For maps of modules* $P \xrightarrow{f} M \xrightarrow{g} Q$, *suppose that their composite gf is an isomorphism. Then* $M = \text{image}(f) \oplus \text{kernel}(g)$.

Proof. The fact that gf is monic alone is enough to guarantee that $fP \cap g^{-1}0 = 0$. For any $m \in M$, $m = f(gf)^{-1}gm + [m - f(gf)^{-1}gm]$ shows that $M = fP \oplus g^{-1}0$.

11-4.3 Lemma. *Suppose that* $M = C \oplus B$ *and that* $\pi : M \to C$ *is the* *corresponding projection. If $A \leqslant M$ is any submodule and $(\pi \mid A): A \to C$ is the restriction of π to A, then*

   (i) $\pi A = C \cap (A + B)$,
   (ii) $M = A \oplus B \Leftrightarrow \pi \mid A$ *is an isomorphism of A onto C.*

Proof. (i) Since $\ker \pi = B$, $\pi A = \pi[A + B] = \pi[(A \oplus B) \cap (C \oplus B)]$. By the modular laws, $\pi A = \pi[((A + B) \cap C) \oplus B] = (A + B) \cap C$.

   (ii) The map $\pi \mid A$ is monic if and only if $A \cap B = 0$, and $A + B = A \oplus B \leqslant M$. Now by (i), $\pi \mid A$ is onto if and only if $\pi A = C \cap (A + B) = C$. Thus

$$\pi A = C \Leftrightarrow C \subseteq A + B \Leftrightarrow M = A + B.$$

Thus $\pi \mid A$ is both monic and onto, if and only if $M = A \oplus B$.

11-4.4 **Lemma.** *Suppose that $M = \oplus\{A_\alpha \mid \alpha \in \theta\}$ is a direct sum decomposition of a module M with each $\mathrm{End}_R A_\alpha$ a local ring. Suppose that $F = \{\alpha(1), \dots, \alpha(s)\} \subset \theta$ and $E = \{f, g, \dots, h\} \subset \mathrm{End}_R M$ are finite sets, where $\Sigma\{f \mid f \in E\} = f + g + \cdots + h = 1_M : M \to M$ is the identity map on M, and all the $\alpha(i)$ are distinct. Then for every index $\alpha(i) \in F$, there exists an $f_i \in E$ (with repetitions allowed) such that*

$$M = f_1 A_{\alpha(1)} \oplus \cdots \oplus f_s A_{\alpha(s)} \oplus \left[\bigoplus_{\alpha \in \theta \backslash F} A_\alpha \right].$$

Furthermore, each f_i induces an automorphism $f_i : A_{\alpha(i)} \to f_i A_{\alpha(i)}$ by restriction and corestriction.

Proof. First, take $s = 1$, $\alpha(1) = \alpha$, and in this special case we will let $f_1 = f$ later. Let $\pi_\alpha^A : M \to A_\alpha$ and $\phi_\alpha^A : A_\alpha \to M$ be the projection and inclusion maps defined by the direct sum decomposition $M = \oplus A_\alpha$. Let $1_\alpha^A : A_\alpha \to A_\alpha$ and $1_M : M \to M$ be the identity maps of A_α and of M. Then

$$1_\alpha^A = \pi_\alpha^A 1_M \phi_\alpha^A = \sum_{f \in E} \pi_\alpha^A f \phi_\alpha^A.$$

Since $\mathrm{End}_R A_\alpha$ is a local ring, there exists at least one $\pi_\alpha^A f \phi_\alpha^A \in \mathrm{End}_R A_\alpha$ which is an automorphism.

   Lemma 11-4.2 gives $M = f \phi_\alpha^A A_\alpha \oplus \ker \pi_\alpha^A = f_\alpha^A \oplus (\oplus\{A_\gamma \mid \gamma \in \theta \backslash \{\alpha\}\})$. Since $\pi_\alpha^A f \phi_\alpha^A$ is an automorphism, $f \phi_\alpha^A$ is monic, and the induced map $f : A_\alpha \to f A_\alpha$ is an automorphism

$$A_\alpha \xrightarrow{f \phi_\alpha^A} M \xrightarrow{\pi_\alpha^A} A_\alpha.$$

Since $\mathrm{End}_R fA_\alpha \cong \mathrm{End}_R A_\alpha$ is local, the above process may be continued for the new decomposition

$$M = f_1 A_{\alpha(1)} \oplus (\oplus\{A_\alpha \mid \alpha \in \theta\backslash\{\alpha(1)\}\})$$

and the new set $F\backslash\{\alpha(1)\} \subset \theta$, but still the same old $E \subset \mathrm{End}_R M$. In this construction the old ϕ_α^A, π_α^A must be replaced by the new ones; say ϕ_α^B, π_α^B defined by this new decomposition of M. The result will be the selection of an index $\alpha(2) \in F\backslash\{\alpha(1)\}$ and an $f_2 \in E$ such that $f_1 A_{\alpha(1)} \subseteq \ker \pi_{\alpha(2)}^B$, and $M = f_1 A_{\alpha(1)} \oplus f_2 A_{\alpha(2)} \oplus (\oplus\{A_\alpha \mid \alpha \in \theta\backslash\{\alpha(1), \alpha(2)\}\})$, where again $f_2 : A_{\alpha(2)} \to f_2 A_{\alpha(2)}$ induces an automorphism. The rest now is clear.

11-4.5 Lemma. *Let ϕ_α^A and π_α^A be the inclusion and projection maps corresponding to $M = \oplus\{A_\alpha \mid \alpha \in \theta\}$ where all $\mathrm{End}_R A_\alpha$ are local. Suppose also that $M = C \oplus B$ where C is indecomposable, and let ϕ_C and π_C be the corresponding inclusion and projection maps for C. Then there exists an $\alpha \in \theta$ and an A_α such that*

(i) $M = A_\alpha \oplus B$;
(ii) $\pi_C \phi_\alpha^A$ *is an isomorphism of A_α onto C:*

$$A_\alpha \xrightarrow{\phi_\alpha^A} M \xrightarrow{\pi_C} C.$$

(iii) *The set of all α such that the map $\pi_\alpha^A \phi_C$ is an isomorphism*

$$C \xrightarrow{\phi_C} M \xrightarrow{\pi_\alpha^A} A_\alpha$$

is finite.

Proof. Take any $0 \neq x \in C$ and express it as $x = x_{\alpha(1)} + \cdots + x_{\alpha(s)}$ with $0 \neq x_{\alpha(i)} \in A_{\alpha(i)}$, $1 \leq s$.

(i) and (ii) Set $f = \phi_C \pi_C : M \to M$, and write $1_M = f + (1_M - f)$. Now use the last lemma (with $F = \{\alpha(1), \ldots, \alpha(s)\}$ and $E = \{f, 1_M - f\}$) to express

$$M = f_1 A_{\alpha(1)} \oplus \cdots \oplus f_s A_{\alpha(s)} \oplus D, \quad D = \oplus\{A_\alpha \mid \alpha \in \theta\backslash\{\alpha(1), \ldots, (s)\}\}$$

where each f_i is either $f_i = f$, or $f_i = 1_M - f$.

Define $g : M \to M$ to be the map

$$g(y_{\alpha(1)} + \cdots + y_{\alpha(s)} + d) = f_1 y_{\alpha(1)} + \cdots + f_s y_{\alpha(s)} + d,$$

where

$$y_{\alpha(i)} \in A_{\alpha(i)} \text{ and } d \in D.$$

Intuitively, g is the automorphism $g = f_1 \oplus \cdots \oplus f_s \oplus 1_D : M \to M$ which is the identity on D, while each f_i acts only in the $\alpha(i)$th summand $A_{\alpha(i)}$. Alternatively, in rigorous notation

$$g = \sum_{i=1}^{s} \phi_{\alpha(i)}^A f_i \pi_{\alpha(i)}^A + \left(1_M - \sum_{i=1}^{s} \phi_{\alpha(i)}^A \pi_{\alpha(i)}^A\right).$$

Thus the nonzero element x is mapped by this automorphism g onto th nonzero element

$$g(x) = \sum_{i=1}^{s} f_i x_{\alpha(i)} \neq 0.$$

Moreover, even each $f_i x_{\alpha(i)} \neq 0$.

Suppose that for all i, $f_i = 1_M - f$, which is the projection onto B. Then $fx = \phi_C \pi_C x = x$, and consequently

$$0 = (1_M - f)x = (1_M - f)\sum_{i=1}^{s} x_{\alpha(i)} = \sum_{i=1}^{s} f_i x_{\alpha(i)} = gx$$

is a contradiction.

Hence there exists an i for which $f_i = f = \phi_C \pi_C$, and such that for $\alpha = \alpha(i)$, the induced map is an isomorphism $f : A_\alpha \to fA_\alpha = \pi_C A_\alpha$. Hence $\pi_C A_\alpha \subseteq C$ and both are direct summands of M; write $M = \pi_C A_\alpha \oplus G$ where actually

$$G = \oplus\{f_j A_{\alpha(j)} \mid 1 \leqslant j \leqslant s, j \neq i\} \oplus D.$$

Therefore

$$C = C \cap M = C \cap (\pi_C A_\alpha \oplus G) = \pi_C A_\alpha \oplus (C \cap G).$$

Since C is indecomposable and $\pi_C A_\alpha \neq 0$, $C = \pi_C A_\alpha$. Thus $\pi_C \phi_\alpha^A$ is an isomorphism

$$A_\alpha \xrightarrow{\phi_\alpha^A} M \xrightarrow{\pi_C} C$$

mapping A_α one to one onto C. By 11-4.3, $M = A_\alpha \oplus B$.

(iii) If α is any index such that $\pi_\alpha^A \phi_C$ is an isomorphism, then $\pi_\alpha^A \phi_C x \neq 0$. But $x \in C$, and $\pi_\alpha^A \phi_C x = \pi_\alpha^A x = \pi_\alpha^A(x_{\alpha(1)} + \cdots + x_{\alpha(s)})$ and hence $\alpha = \alpha(i)$ for some i.

Repeated application of the last lemma gives the next corollary.

11-4.6 Corollary. *Suppose that $M = \oplus\{A_\alpha \mid \alpha \in \theta\}$ with $\mathrm{End}_R A_\alpha$ local for all α and assume in addition that $M = C_1 \oplus \cdots \oplus C_s \oplus D$ where each C_i is* *indecomposable. Then there exist distinct indices* $\alpha(1), \ldots, \alpha(s) \in \theta$ *such that*

(i) $M = A_{\alpha(1)} \oplus \cdots \oplus A_{\alpha(s)} \oplus D$, *and*

(ii) $C_i \cong A_{\alpha(i)}$ *for each i.*

11-4.7 **Remark.** In the last corollary let

$$\eta_i^A : M = A_{\alpha(1)} \oplus \cdots \oplus A_{\alpha(s)} \oplus D \to A_{\alpha(i)}$$

and

$$\eta_i^C : M = C_1 \oplus \cdots \oplus C_s \oplus D \to C_i$$

be the corresponding projections. Then in view of 11-4.3, the isomorphism $C_i \cong A_{\alpha(i)}$ is given by $\eta_i^A : C_i \to A_{\alpha(i)}$; and $\eta_i^C : A_{\alpha(i)} \to C_i$ is also an isomorphism.

11-4.8 **Theorem.** *Assume that M is a module which can be decomposed as a direct sum* $M = \oplus \{M_\alpha \,|\, \alpha \in \theta\}$ *of modules* M_α *of which all the endomorphism rings* $\text{End}_R M_\alpha$ *are local rings for all* α. *Assume also that* $M = \oplus \{N_\beta \,|\, \beta \in \mathcal{L}\}$ *with each* N_β *an indecomposable module. Then it follows that all* $\text{End}_R N_\beta$ *are also local.*

Suppose that $D < M$ *is any submodule such that*

(a) $M = M_1 \oplus \cdots \oplus M_s \oplus D$

for some finite subset $\{1, 2, \ldots, s\} \subseteq \theta$. *Then there exists a finite set of indices* $\{1, 2, \ldots, t\} \subseteq \mathcal{L}$ *such that*

(b) $M = N_1 \oplus \cdots \oplus N_t \oplus D$

For any expressions (a) *and* (b) *as above*

(i) $s = t$.

Furthermore, there exists a permutation ρ *of* $\{1, \ldots, s\}$ *such that* (ii) *and* (iii) *hold:*

(ii) $M_i \cong N_{\rho i}$;

(iii) $M = N_{\rho 1} \oplus \cdots \oplus N_{\rho r} \oplus M_{r+1} \oplus \cdots \oplus M_s \oplus D \; 0 \leqslant r \leqslant s$; ρ *is independent of r.*

11-4.9 **Corollary to Theorem 11-4.8.** *If* ρ, (a) *and* (b) *are as above with* $t = s$ *and if* $\pi_j^N : M \to N_j, j = 1, \ldots, s$; *are the projections given by* (b), *then each of these projections* $\pi_{\rho i}^N$ *induces the isomorphism* $M_i \cong N_{\rho i}$, *i.e.*

(iv) $(\pi_{\rho i}^N \mid M_i) : M_i \to N_{\rho i}$ $i = 1, \ldots, s$

is an isomorphism.

Remark. Another way of stating (ii), (iii) and (iv) above is that we can reindex the N_j so that ρ becomes the identity permutation and $N_1 = N_{\rho 1}$, $N_2 = N_{\rho 2}, \ldots, N_s = N_{\rho s}$. This is the version that will be used.

Proof of 11-4.8 *and* 11-4.9. By 11-4.5, for any β, $N_\beta \cong M_\alpha$ for some α. Thus all the rings $\mathrm{End}_R N_\beta$ are local, and the hypotheses on the M_α's and the N_β's are symmetric.

(ii), (iii) and (iv). For $r = 0$, statement (iii) is the same as hypothesis (a). By induction assume that we have already selected and renumbered the first $r - 1$ of the N_j's such that $N_j \cong M_j$ for all $i = 1, \ldots, r - 1$ and for some $0 \leqslant r < t$, and so that (iii) holds with r replaced by $r - 1$:

(1) $M = N_1 \oplus \cdots \oplus N_{r-1} \oplus M_r \oplus \cdots \oplus M_s \oplus D.$

Let the projection maps and the inclusion maps given by (b) and then (1) be as follows:

(b) $M = N_1 \oplus \cdots \oplus N_t \oplus D$

projections: $\pi_1^N, \ldots, \pi_t^N, \pi_D^N$
inclusions: $\phi_1, \ldots, \phi_t, \phi_D$

$1_M = \phi_1 \pi_1^N + \cdots + \phi_t \pi_t^N + \phi_D \pi_D^N$

(1) $M = N_1 \oplus \cdots \oplus N_{r-1} \oplus M_r \oplus \cdots \oplus M_s \oplus D$

rth projection: $p : M \to M_r$
rth inclusion: $\eta : M_r \to M.$

The reader should think of the $\phi_j \pi_j^N$ and ηp as projection maps, except that their images are in M, e.g.

$\phi_j \pi_j^N : M \twoheadrightarrow N_j \subset M$ and $\eta p : M \twoheadrightarrow M_r \subset M.$

Also, a map preceded by η is just the restriction to M_r, e.g. $\pi_r^N \eta = \pi_r^N \mid M_r : M_r \to M_r \subset M.$

Let 1_r denote the identity endomorphism of M_r. The maps $p\phi_{r+i}\pi_{r+i}^N$ are endomorphisms of M_r, where $p\eta = 1_r$, and

$$M_r \xrightarrow{\;\;1_r = p\eta = p\phi_r \pi_r^N + \cdots + p\phi_t \pi_t^N\;\;} M_r$$

Since $\mathrm{End}_R M_r$ is a local ring, there must exist at least one map $p\phi_{r+i}\pi_{r+i}\eta$ (for some $i = 0, \ldots, t - r$) which is an automorphism of M_r. Now change notation and renumber at most two of the $N_r, \ldots, N_t$ so that the above old N_{r+i} becomes N_r, i.e. $p\phi_r\pi_r^N$ induces an automorphism of M_r.

Set $B = N_1 \oplus \cdots \oplus N_{r-1} \oplus M_{r+1} \oplus \cdots \oplus M_s \oplus D$. From (1) it follows that $B = \ker p$. Our objective now is to prove (iii), i.e. that $N_r + B = N_r \oplus B = M$.

Although the next diagram logically is not needed in the proof, it does show how and where various submodules are mapped by π_r^N and p, even though some restrictions and corestrictions of maps have been suppressed; all submodules are viewed as submodules of M.

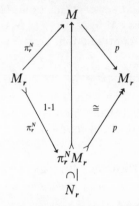

In rigorous notation, our previous diagram becomes the following commutative diagram, where the composite map $p(\phi_r\pi_r^N) = (p\phi_r)(\pi_r^N\eta)$ along either the top or the bottom of the diamond is an automorphism of M_r.

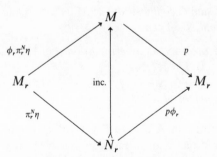

Our next step is to show that the inclusion $\pi_r^N\eta M_r = \pi_r^N M_r \subseteq \pi_r^N M = N_r$ is an equality. Now apply Lemma 11-4.2 twice. First along the bottom of the diamond (with $P = M_r = Q$, $M \leftrightarrow N_r$, $f = \pi_r^N\eta$, and $g = p\phi_r$) to conclude that $N_r = \pi_r^N M_r \oplus (N_r \cap \ker p) = \pi_r^N M_r \oplus (N_r \cap B)$. Since $p\phi_r\pi_r^N\eta$ is an automorphism of M_r, surely $\pi_r^N M_r \neq 0$. The indecomposability of N_r now requires that $N_r \cap B = 0$. But then $N_r = \pi_r^N M_r$ and $N_r + B = N_r \oplus B \leqslant M$.

A second application of 11-4.2 now along the top of the diamond (with $P = M_r = Q$, $M = M$, $f = \phi_r \pi_r^N$, and $g = p$) yields conclusion (iii) $M = \pi_r^N M_r \oplus \ker\, p = N_r \oplus B$. It now follows immediately from $M = M_r \oplus B$ that $M_r \cong N_r$. However, we just saw that this isomorphism is induced by the restriction of π_r^N to M_r, $(\pi_r^N \,|\, M_r) = \pi_r^N \pi : M_r \to N_r$ as required in (iv).

(i) Suppose that $s \leqslant t - 1$. For $r = s$, (iii) is $M = N_1 \oplus \cdots \oplus N_s \oplus D$. Thus by (b), $N_t \cap M = 0$, a contradiction. Next, if $t + 1 \leqslant s$, then for $r = t$, (iii) becomes $M = N_1 \oplus \cdots \oplus N_t \oplus M_{t+1} \oplus \cdots \oplus M_s \oplus D$. Again by (b), $M_s = 0$, a contradiction. So $s = t$. The theorem has been proved.

Next hypotheses are imposed on M which force it to satisfy the conditions in the last theorem with $D = 0$, and we obtain as a corollary a well known result. Recall that a module M has a composition series if and only if it satisfies both chain conditions.

11-4.9 Krull–Schmidt Theorem. *Any module M of finite composition length is a finite direct sum of indecomposable modules. Suppose that*

 (a) $M = M_1 \oplus \cdots \oplus M_s$

and

 (b) $M = N_1 \oplus \cdots \oplus N_t$

are any two direct sum decompositions of M as a direct sum of indecomposable modules M_i and N_j. Then

 (i) *$s = t$, and there exists a permutation ρ of $\{1, \ldots, s\}$ such that*
 (ii) *$M_i \cong N_{\rho i}$ $i = 1, \ldots, s$;*
 (iii) *$M = N_{\rho 1} \oplus \cdots \oplus N_{\rho r} \oplus M_{r+1} \oplus \cdots \oplus M_s$ for all $r = 0, 1, \ldots, s$;*
    ρ is independent of r.

Proof. By 11-3.5, either the A.C.C. or D.C.C. alone is enough to guarantee that M be a finite direct sum of indecomposables. Any indecomposable direct summand of M by Fitting's Lemma 11-3.8 has a local endomorphism ring. The above conclusions (i), (ii) and (iii) follow from the last theorem with $D = 0$.

Our next objective will be to show that if M satisfies the hypotheses of 11-4.8, that then any two direct sum decompositions of M as a direct sum of indecomposable are equivalent in the sense of 11-4.13. This proof will not use Theorem 11-4.8 and its corollaries.

11-4.10 Lemma. *Assume that*

$$M = \oplus\{M_\alpha \,|\, \alpha \in \theta\} \quad and \quad M = \oplus\{N_\beta \,|\, \beta \in \mathscr{L}\}$$

are two direct sum decompositions of M into indecomposable modules M_α and N_β all having local endomorphism rings. Suppose that C is any indecomposable summand $M = C \oplus B$ of M. Define sets X and Y by $X = \{\alpha \in \theta \,|\, M_\alpha \cong C\}$ and $Y = \{\beta \in \mathscr{L} \,|\, N_\beta \cong C\}$. Then $|X| = |Y|$.

Proof. Case 1. X is finite. Take any finite subset $F = \{\beta(1), \ldots, \beta(t)\} \subseteq Y$. By 11-4.6, $M = M_{\alpha(1)} \oplus \cdots \oplus M_{\alpha(t)} \oplus (\oplus\{N_\beta \,|\, \beta \in \mathscr{L}\backslash F\})$ with $M_{\alpha(i)} \cong N_{\beta(i)}$ for all i. Thus $\{\alpha(1), \ldots, \alpha(t)\} \subseteq X$, and $t \leqslant |X|$. Since X is finite, also Y must be finite, and furthermore $|Y| \leqslant |X|$. But now the hypotheses on X and Y are symmetric, and hence $|X| = |Y|$.

Case 2. X is infinite. Let $\pi_\beta^N : M \to N_\beta$ be the projections defined by $M = \oplus N_\beta$. For each $\alpha \in X$ first define $(\pi_\beta^N \,|\, M_\alpha) : M_\alpha \to N_\beta$ to be the restriction of π_β^N to M_α, and then define $F(\alpha) = \{\beta \in \mathscr{L} \,|\, \pi_\beta^N \,|\, M_\alpha$ is an isomorphism$\}$. For each $\beta \in Y$, use 11-4.5(i) and (ii) (with $C = N_\beta$, with $A_\alpha = M_\alpha$, and $B = \ker \pi_\beta^N$) to conclude that there is an $\alpha \in X$ with $\beta \in F(\alpha)$. Thus $Y = \cup\{F(\alpha) \,|\, \alpha \in X\}$. By 11-4.5(iii) used in the same way, each $F(\alpha)$ is finite.

Let $\mathbb{N} = 0, 1, 2, \ldots$. Since $Y = \cup\{F(\alpha) \,|\, \alpha \in X\}$ and each $|F(\alpha)| < |\mathbb{N}|$, there is a monic may $Y \to X \times \mathbb{N}$. By a known result of set theory $|X \times \mathbb{N}| = |X|$ if X is infinite. Thus $|Y| \leqslant |X|$ and by the symmetry of our hypotheses, $|Y| = |X|$. However, we do not want to assume that $|X \times \mathbb{N}| = |X|$, and next give a proof.

Let us form the disjoint union Z of the $F(\alpha)$:

$$Z = \bigcup_{\alpha \in X} \{\{\alpha\} \times F(\alpha)\} = \bigcup \{(\alpha, \beta) \,|\, \alpha \in X, \, \pi_\beta^N \,|\, M_\alpha \text{ is an isomorphism}\}.$$

If $g : Z \to X$ is the first projection $g((\alpha, \beta)) = \alpha$, then the equivalence classes of elements of Z identified by g are exactly the $F(\alpha)$. More precisely, define $(\alpha_1, \beta_1) \sim (\alpha_2, \beta_2)$ if and only if $g((\alpha_1, \beta_1)) = g((\alpha_2, \beta_2))$, in which case $\alpha_1 = \alpha_1 \equiv \alpha$, and $\beta_1, \beta_2 \in F(\alpha)$. Hence if $\pi : Z \to Z/\!\sim$ is the natural projection of Z onto the set of equivalence classes $Z/\!\sim$ of elements identified by g, then there is a commutative diagram with g being epic, and where $\bar{g}$ is one to one onto. Thus $|Z/\sim| = |X|$. Since the equivalence classes are finite but Z is not finite, by 0-1.12, $|Z| = |Z/\sim|$. The second projection $Z \to Y = \cup\{F(\alpha) \mid \alpha \in \theta\}$ is epic. Thus $|Y| \leqslant |Z| = |Z/\sim| = |X|$. By the symmetry of hypotheses, $|X| = |Y|$.

11-4.11 Krull–Remak–Schmidt Theorem. *Suppose that a module M is a direct sum $M = \oplus\{M_\alpha \mid \alpha \in \theta\}$ of indecomposable modules M_α each with a local endomorphism ring $\mathrm{End}_R M_\alpha$, and suppose that also $M = \oplus\{N_\beta \mid \beta \in \mathscr{L}\}$ where each N_β is indecomposable. Then there exists a bijection*

    (i) $\rho : \theta \to \mathscr{L}$ *such that*
    (ii) $M_\alpha \cong N_{\rho\alpha}$

for all α.

Proof. (i) and (ii). For $\alpha(1)$, $\alpha(2) \in \theta$, define an equivalence relation on θ by $\alpha(1) \sim \alpha(2)$ if and only if $M_{\alpha(1)} \cong M_{\alpha(2)}$. Let $\theta/\sim = \{X,\ldots\}$ be the set of equivalence classes. Similarly for N_β, $\beta \in \mathscr{L}$, let $\mathscr{L}/\sim = \{Y,\ldots\}$ be the corresponding equivalence classes. Define a function $h : \theta/\sim \to \mathscr{L}/\sim$ by $h(X) = Y$ if $M_\alpha \cong N_\beta$ for some $\alpha \in X$ and some $\beta \in Y$. By 11-4.5, the map h is onto. Suppose that $h(X_1) = h(X_2) = Y$ for some $X_1, X_2 \in \theta/\sim$. Then there are $\alpha(1) \in X_1$, $\alpha(2) \in X_2$, and a single $\beta \in Y$ such that $M_{\alpha(1)} \cong N_\beta \cong M_{\alpha(2)}$. Thus $\alpha(1) \sim \alpha(2)$, and hence $X_1 = X_2$. So $h : \theta/\sim \to \mathscr{L}/\sim$ is a bijection. Furthermore for any $X \in \theta/\sim$, if $Y = h(X)$ then $|X| = |Y|$ by the last lemma.

To recapitulate, there is a bijection $h : \theta/\sim \to \mathscr{L}/\sim$ such that moreover $|X| = |h(X)|$ for all $X \in \theta/\sim$. Now it is easy to construct a bijection $\rho : \theta \to \mathscr{L}$ which maps $\rho X = hX$ for all equivalence classes $X \subset \theta$. Thus there is a commutative diagram

$$
\begin{array}{ccc}
\theta = \displaystyle\bigcup_{X \in \theta/\sim} X & \xrightarrow[\cong]{\;\rho\;} & \displaystyle\bigcup_{Y \in \mathscr{L}/\sim} Y = \mathscr{L} \\[2em]
{\scriptstyle \mathrm{proj.}}\Big\downarrow & & \Big\downarrow{\scriptstyle \mathrm{proj.}} \\[1.5em]
\theta/\sim & \xrightarrow[h]{\;\cong\;} & \mathscr{L}/\sim
\end{array}
$$

In particular, $|\theta| = |\mathscr{L}|$, and since $hX = \{\rho\alpha \mid \alpha \in X\}$, $M_\alpha \cong N_{\rho\alpha}$ for all $\alpha \in \theta$.

11-4.12 Corollary 1 to Theorem 11-4.11. *Let $M = \{M_\alpha \mid \alpha \in \theta\}$ with*

$\text{End}_R M_\alpha$ *local for all* α, *and* N *be a module with* $N = \oplus\{N_\beta \mid \beta \in \mathscr{L}\}$ *where each* N_β *is indecomposable. If* $M \cong N$, *then there exists a bijection* $\rho : \theta \to \mathscr{L}$ *such that* $M_\alpha \cong N_{\rho\alpha}$.

The next lemma and corollary show that the Krull–Remak–Schmidt Theorem is applicable to wider classes of modules than merely those which satisfy the chain conditions.

Recall that every injective module is also quasi-injective (6-6.1).

11-4.13 Lemma. *If* G *is an indecomposable quasi-injective module, then its endomorphism ring* $\text{End}_R G$ *is local.*

Proof. If $g : G \to G$ is a monic map, then the semi-endomorphism $gG \to G$, $gx \to x$, extends to $f : G \to G$ by the quasi-injectivity of G. If $z = gy \in gG \cap \ker f$ for some $y \in G$, then $0 = fz = fgy = y$, and also $z = 0$. Thus $\ker f + gG = \ker f \oplus gG \leqslant G$. For any $x \in G$, $x = (x - gfx) + gfx$ with $x - gfx \in \ker f$ and $gfx \in gG$. Thus $G = \ker f \oplus gG$. Since G is indecomposable, $\ker f = 0$, and $G = gG$. Hence all monic endomorphisms of G are invertible.

Thus if $f, g \in \text{End}_R G$ are not invertible, then $\ker f \neq 0$, $\ker g \neq 0$, and hence $0 \neq \ker f \cap \ker g \subseteq \ker(f + g)$, because any indecomposable quasi-injective is uniform. (See Exercises 30–32.) Consequently $f + g$ is also noninvertible, and $\text{End}_R G$ is a local ring.

11-4.14 Corollary 2 to Theorem 11-4.11. *Assume that a module* G *is a direct sum* $G = \oplus\{G_\alpha \mid \alpha \in \theta\}$ *of indecomposable quasi-injective modules* G_α *and assume also that* $G = \oplus\{H_\beta \mid \beta \in \mathscr{L}\}$ *is another decomposition of* G *as a direct sum of indecomposable modules* H_β. *Then*

(i)   *all the* H_β *are quasi-injective, and*
(ii) $G_\alpha \cong H_{\rho\alpha}$ *for all* α *for some bijection* $\rho : \theta \to \mathscr{L}$ *of the index sets.*

11-5   Singular submodule

Every right R-module M over any ring R has certain intrinsic submodules. We have already encountered the socle soc M, and the radical $J(M)$. This section briefly introduces another such submodule.

11-5.1 Definition. For any right R-module M, the *singular submodule* $Z(M)$ is defined as $Z(M) = ZM = \{m \in M \mid m^\perp \leqslant R \text{ is essential}\}$. Sometimes the module M is called *singular* if $M = ZM$, and nonsingular if $ZM = 0$. Here we avoid the latter term 'nonsingular' because it does not mean 'not singular'. Here we define a module M to be *torsion free* if $ZM = 0$.

If $M = R_R$ is the ring viewed as a right R-module then $Z(R_R) = Z(R)$ is called the *right singular ideal* of R. When also left ideals are considered and when confusion is possible, it is frequently written as $Z_r(R) = Z(R_R)$. For left modules the left singular ideal $Z_l(R)$ is defined similarly.

A submodule $K \leqslant M$ of a module M is *fully invariant* in M if every endomorphism of M maps K back into itself, i.e. if $\mathrm{Hom}_R(M, M)K \subseteq K$.

11-5.2 Lemma. *For any right R-module M,*

   (i)   *ZM is a fully invariant submodule of M,*
   (ii) *If $M = R_R$, then $Z(R) \lhd R$ is an ideal.*

Proof. (i) If $x, y \in ZM$, then $(x - y)^{\perp} \supseteq x^{\perp} \cap y^{\perp} \leqslant R$ is large. For $x \in ZM$, and $r \in R$, if $xr \notin ZM$, then $(xr)^{\perp} \oplus C \leqslant R$ for some $0 \neq C \leqslant R$. Let $0 \neq ra \in C \cap x^{\perp}$, $a \in C$. Then $0 = xra$, and $0 \neq a \in (xr)^{\perp} \cap C = 0$ is a contradiction.

For $\varphi \in \mathrm{Hom}_R(M, M)$ and $x \in ZM$, $(\varphi x)^{\perp} \supseteq x^{\perp}$, and hence also $\varphi x \in ZM$. (ii) A fully invariant right ideal ZR of R is always an ideal $ZR \lhd R$.

11-5.3 Remarks. For any right R-modules M and N, and any module homomorphism $f : M \to N$, the following hold.

   (1) $f(ZM) \leqslant ZN$; and
   (2) $f(\mathrm{soc}\ M) \leqslant \mathrm{soc}(N)$.
   (3) There exist modules M and N such that $Z(M/(ZM)) \neq 0$, and $\mathrm{soc}(N/\mathrm{soc}(N)) \neq 0$.
   (4) In general $Z_r(R) \neq Z_l(R)$ and right $\mathrm{socle}(R_R) \neq$ left $\mathrm{socle}(_R R)$.

The next remarkable example (analysed in Storrer [72]) is a counter-example to several properties that rings with chain conditions have, as well as illustrates some concepts of this and previous chapters.

11-5.4 Example. For a field F, let R be the ring of polynomials in a countable number of commuting indeterminates $x_1, x_2, \ldots$, subject to the relations

$$x_1^2 = 0, x_{i+1}^2 = x_i;$$

and hence

$$x_i^{2^{i-m}} = x_m, 1 \leqslant i, 1 \leqslant m < 2^i.$$

Any arbitrary $r \in R$ can be written as $r = c_0 + c_1 x_i + c_2 x_i^2 + \cdots + c_n x_i^n$, $c_k \in F$, for some single x_i. If $c_0 \neq 0$, then r is a unit, because

$$1 + c_1 x_i + \cdots + c_n x_i^n = 1 - \lambda, \, (1 - \lambda)^{-1} = 1 + \lambda + \lambda^2 + \cdots + \lambda^{2^i}.$$

In general, $r = x_i^k(c_k + c_{k+1} x_i + \cdots + c_n x_i^{n-k}) = x_i^k u$, where k is smallest such that $c_k \neq 0$, and u is a unit.

The ideal $J = \Sigma \{x_i R \mid i = 1, 2, \ldots\} \lhd R$ is nil. Since all

$$x_i = x_{i+1}^2 \in J^2 \subseteq J, \, J = J^2$$

is not nilpotent. The elements r of any ideal of R are polynomials in some x_i as above with zero constant term $c_0 = 0$. Therefore J is the unique maximal ideal of R.

Since $J = J(R)$, J annihilates any minimal ideal V of R, $VJ = 0$. If $0 \neq r = x_i^k u \in V$, then $x_i^k J = 0$, $k \leqslant 2^i - 1$. Then for any $m > i$,

$$x_i^k x_m = x_m^{k 2^{m-i}+1}, \, k 2^{m-i} < (2^i - 1) 2^{m-i} + 1 \leqslant 2^m - 1.$$

Hence $x_i^k x_m \neq 0$ for $m > i$. This is a contradiction. Thus R contains no minimal ideals.

As a module over itself R is uniform. For if $x_i^k u$, $x_m^j v \in R \setminus \{0\}$ with u, v units and $m \leqslant i$, then x_m is a power of x_i as above. Then $x_i^k uR$ and $x_m^j vR$ are ordered by inclusion. If

$$k \leqslant j 2^{i-m} \quad \text{then} \quad x_m^j vR = x_i^{j 2^{i-m}} R \subseteq x_i^k uR;$$

or $j 2^{i-m} < k$, and the last inclusion is reversed. Thus all ideals of R are linearly ordered. Therefore the injective hull $E(R)$ of R is an indecomposable module.

If $P \lhd R$ is a prime ideal, then $P \subseteq J$. Since $J/P \lhd R/P$ is a nil ideal, $J = P$. That is J is the unique prime ideal of R. Every simple R-module is isomorphic $R/J \cong F$. Since R has no minimal ideals, $E(R)$ is an indecomposable injective which contains no simple submodules. In particular $E(R)$ contains no submodule isomorphic to R/P, $P = J \lhd R$ prime.

It is known that over a commutative Noetherian ring R, every indecomposable injective module is isomorphic to the R-injective hull $E(R/P)$ of the R-module R/P for some prime ideal $P \lhd R$. It is also known that over a noncommutative right Noetherian ring R, for $P \lhd R$ prime, the right R-module $E(R/P)$ is a finite direct sum of isomorphic indecomposable modules, and every indecomposable injective module is of this form. (See Storrer [72] and Lambek and Michler [73].)

Clearly, $Z(R) \subseteq J$. If $r = x_i^k u \in J$, $1 \leqslant k$ and $u \in R$ a unit, then

$$r^\perp = (x_i^k)^\perp = x_i^{2^i - k} R \neq 0.$$

Since R is uniform, $r^\perp < R$ is essential, and hence $Z(R) = J$.

11-6   Exercises

Since many of the problems use injective modules, assume for the
exercises that $1 \in R$.

1. For a prime $p \in \mathbf{Z}$, $\langle p^{-\infty}x \rangle$ is defined as the abelian group
    generated by the set of generators $p^{-\infty}x$ defined as $p^{-\infty}x = \{p^{-1}x, \ldots, p^{-n}x, \ldots\}$ subject to the relations $p^j(p^{-n}x) = p^{-n+j}x$,
    $j = 1, 2, \ldots$; $p(p^{-1}x) = x$. Verify that $\langle p^{-\infty}x \rangle$ is isomorphic to
    the additive group of rational numbers

    $$\{m/p^j \mid m \in \mathbf{Z}; j = 0, 1, 2, \ldots\}$$

    whose denominators are powers of p under an isomorphism that
    $x \to 1$, and that $\langle p^{-\infty}x \rangle$ is indecomposable.

2*. For $R = \mathbf{Z}$, let G be an additive subgroup of the p-adic integers
    J_p such that $G \cap pJ_p = pG$. Show that G is indecomposable, and
    in particular that J_p is an indecomposable abelian group of
    cardinality $|J_p| = 2^{\aleph_0}$. Show that every countable subgroup H of
    J_p can be embedded in a countable group G as above with
    $H \subseteq G \subseteq J_p$.

3*. Let q, $p_1, \ldots, p_n, \ldots$ be a finite or infinite sequence of distinct
    primes. Over the ring $\mathbf{Z}$, prove that the additive subgroup
    $G \subset \oplus_i \mathbf{Q}x_i \cong \oplus_i \mathbf{Q}$ is indecomposable, where

    $$G = \bigoplus_i \langle p_i^{-\infty}x_i \rangle + \sum_i \mathbf{Z}q^{-1}(x_1 + x_i)$$

    (See Fuchs [73]; vol. II, p. 123.)

4*. For distinct primes $p, q, p_1, \ldots, p_n$ for $n \geqslant 2$, let G be the abelian
    group

    $$G = \langle p_1^{-\infty}x_1 \rangle \oplus \cdots \oplus \langle p_n^{-1}x_n \rangle$$

    $$\oplus \left\langle p_1^{-\infty}y_1, \ldots, p_n^{-\infty}y_n; \frac{y_1+y_2}{pq}, \frac{y_1+y_3}{pq}, \ldots, \frac{y_1+y_n}{pq} \right\rangle$$

    (i) Show that the latter is a direct sum of $n + 1$ indecomposable
    abelian groups. Use Exercise 3 to conclude that the last direct
    summand is indecomposable.
       Let s, $t \in \mathbf{Z}$ satisfy $ps - qt = 1$. Set $z_i = -px_i + ty_i$ and
    $w_i = qx_i + sy_i$, and define

    $$A = \langle p_1^{-\infty}z_1, \ldots, p_n^{-\infty}z_n; p^{-1}(z_1+z_2),$$
    $$p^{-1}(z_1+z_3), \ldots, p^{-1}(z_1+z_n) \rangle$$
    $$B = \langle p_1^{-\infty}w_1, \ldots, p_n^{-\infty}w_n; q^{-1}(w_1+w_2),$$
    $$q^{-1}(w_1+w_3), \ldots, q^{-1}(w_1+w_n) \rangle.$$

Use Exercise 3 to conclude that A and B are indecomposable. Prove that $A, B \subseteq G$ and $G = A \oplus B$. (See Fuchs [73]; vol. II, p. 134.)

Remark. The above examples are intended to suggest to the reader that for moderately general rings, so far, the following two problems have not been solved: (i) To classify all indecomposable modules. (ii) To classify all possible indecomposable direct sum decompositions of a single arbitrary module.

5. (i) Find various hypotheses which guarantee that any two direct sum decompositions of a module M as

$$M = \oplus \{M_i \mid i \in I\} = \oplus \{N_j \mid j \in J\}$$

into indecomposables M_i, N_j are equivalent. (ii) Find hypotheses on direct sum decompositions of a module so that they complement direct summands. (See 11-4.1.) Show, in particular, that if $M = \oplus \{M_i \mid i \in I\}$ complements all direct summands, then all the M_i are indecomposable.

6. For $R = \mathbf{Z}$, explain why any indecomposable direct summand of the abelian group $M = \mathbf{Z} \oplus \cdots \oplus \mathbf{Z} = \mathbf{Z}^n$ is isomorphic to $\mathbf{Z}$ and why any two direct sum decompositions of M as a direct sum of indecomposable groups are equivalent. (The same holds for an arbitrary free abelian group M.)

7. With $R = \mathbf{Z}$ and $M = \mathbf{Z} \oplus \mathbf{Z}$, show that (i) $M = \mathbf{Z}(2, 5) \oplus \mathbf{Z}(1, 3)$. (ii) Show that the decomposition $M = \mathbf{Z} \oplus \mathbf{Z}$ does not complement $\mathbf{Z}(2, 5)$.

8. Suppose that M is an injective R-module with $ZM = 0$. Show that for any $0 \neq N < M$, M contains a unique injective hull of N.

9. If U is a uniform module and $0 \neq V < U$ a submodule, then $EU = EV$.

10. For a direct sum $M = \oplus \{M_i \mid i \in I\}$ of modules,

$$\text{soc } M = \oplus \{\text{soc } M_i \mid i \in I\} \quad \text{and} \quad JM = \oplus \{JM_i \mid i \in I\}.$$

11. For a projective right R-module P, $J(P) = PJ(R)$.

The following definition gives two classes of matrix-like rings, and the subsequent exercises describe their ideal structure and properties. For a very complete treatment, see Goodearl [76]; pp. 103–114, and for an abbreviated treatment see Rowen [88]; pp. 33–35.

Definition. Let S and T be rings with identity elements, and $_T B_S$ and $_S C_T$ bimodules. Define rings R_1 and R_2 by

$$R_1 = \begin{bmatrix} S & 0 \\ B & T \end{bmatrix}, \begin{bmatrix} s & 0 \\ b & t \end{bmatrix} \begin{bmatrix} \bar{s} & 0 \\ \bar{b} & \bar{t} \end{bmatrix} = \begin{bmatrix} s\bar{s} & 0 \\ b\bar{s} + t\bar{b} & t\bar{t} \end{bmatrix}$$ (1)

$s, \bar{s} \in S; t, \bar{t} \in T; b, \bar{b} \in B; c, \bar{c} \in C.$

$$R_2 = \begin{bmatrix} S & C \\ 0 & T \end{bmatrix}, \begin{bmatrix} s & c \\ 0 & t \end{bmatrix} \begin{bmatrix} \bar{s} & \bar{c} \\ 0 & \bar{t} \end{bmatrix} = \begin{bmatrix} s\bar{s} & s\bar{c} + c\bar{t} \\ 0 & t\bar{t} \end{bmatrix}$$ (2)

Let V be any right S-submodule of $S \oplus B$, with elements of V being viewed as column vectors; and similarly W a left T-submodule of $C \oplus T$, again elements of W being column vectors. Suppose that $T_1 \subseteq T_r$ and $S_2 \subseteq S_s$ are right ideals. Assume that the pairs (V, T_1) and (W, S_2) satisfy (3) $T_1 B \subseteq V$, and (4) $CS_2 \subseteq W$. Only if (3) and (4) hold, define

(5) $L_1 = \begin{bmatrix} \vdots & 0 \\ V & \\ \vdots & T_1 \end{bmatrix}$ and (6) $L_2 = \begin{bmatrix} S_2 & \vdots \\ & W \\ 0 & \vdots \end{bmatrix}$,

where the typical element of L_1 has its first column in V, and lower right-hand entry in T_1.

In Exercises 12 to 19, let $R = R_1$.
12. Prove that L_1 is a right ideal of R.
13. Conversely prove that every right ideal of R is of the form L_1.
    Let $S_1 \subseteq_S S$ be a left ideal, $D_1 \subseteq_S (B \oplus T)$ a left S-submodule with $BS_1 \subseteq D_1$. Then $\begin{bmatrix} S_1 & 0 \\ \cdots & D_1 & \cdots \end{bmatrix}$ is a left ideal of R.
14. Prove that every left ideal of R is of the above form.
15. Now let $S_1^* \subseteq S$ and $T_1^* \subseteq T$ be two-sided ideals, and $B_1 \subseteq B$ an $S-T$-bisubmodule of B such that $BS_1^* + T_1^* B \subseteq B_1$. Then $\begin{bmatrix} S_1^* & 0 \\ B_1 & T_1^* \end{bmatrix} \lhd R$ is an ideal of R.
16. Prove that every ideal of R is of the above form.
17. Let $G_S \subseteq B$ be any S-submodule. Then $\begin{bmatrix} 0 & 0 \\ G & 0 \end{bmatrix} < R$ is a right ideal. Hence conclude that if B_S fails to be right Artinian, Noetherian, or of finite Goldie dimension then so does R.
18. Show that the ring $\begin{bmatrix} \mathbf{Z} & 0 \\ \mathbf{Z}(p^\infty) & \mathbf{Z} \end{bmatrix}$ is neither right Artinian, nor right Noetherian.
19. Prove that L_2 is a right ideal of R.
20. Prove that every right ideal of R is of the form L_2.

Let $T_2 \subseteq_T T$ be a left ideal, and $D_2 \subseteq_T (S \oplus C)$ a left T-submodule with $CT_2 \subseteq D_2$. Then $\begin{bmatrix} \cdots & D_2 & \cdots \\ 0 & & T_2 \end{bmatrix}$ is a left ideal of R.

21. Prove that every left ideal of R is of the above form.
22. Now let $S_2^* \subseteq S$ and $T_2^* \subseteq T$ be two-sided ideals, and $C_2 \subseteq C$ a $T - S$-bisubmodule such that $T_2^*C + CS_2^* \subseteq C_2$. Then

$$\begin{bmatrix} S_2^* & C_2 \\ 0 & T_2^* \end{bmatrix} \triangleleft R$$

is an ideal of R.
23. Prove that every ideal of R is of the above form.
24. Consider subrings $S_0 \subset S$, $T_0 \subset T$ and ideals $I \triangleleft S$, $J \triangleleft T$. Develop the above theory, i.e. do the above exercises for the following rings:

$$\begin{bmatrix} S & 0 \\ S & S_0 \end{bmatrix} \begin{bmatrix} S_0 & 0 \\ S & S \end{bmatrix}, \begin{bmatrix} S & 0 \\ S/I & S \end{bmatrix} \begin{bmatrix} S & 0 \\ S/I & S/I \end{bmatrix} \begin{bmatrix} T_0 & T \\ 0 & T \end{bmatrix} \begin{bmatrix} T & T \\ 0 & T_0 \end{bmatrix},$$

$$\begin{bmatrix} T & T/J \\ 0 & T \end{bmatrix}, \begin{bmatrix} T & T/J \\ 0 & T/J \end{bmatrix}.$$

In Exercises 25 to 27, let $R = R_1$ with $T = B = S$.
25. Show that the right ideal L_1 of R in (1) is large in R if and only if $V_1, V_2 \subseteq S$ are large right S-ideals of S where

$$\begin{bmatrix} \vdots & 0 \\ V & \\ \vdots & 0 \end{bmatrix} \cap \begin{bmatrix} R & 0 \\ 0 & 0 \end{bmatrix} \equiv \begin{bmatrix} V_1 & 0 \\ 0 & 0 \end{bmatrix}, \begin{bmatrix} \vdots & 0 \\ V & \\ \vdots & 0 \end{bmatrix} \cap \begin{bmatrix} 0 & 0 \\ R & 0 \end{bmatrix} \equiv \begin{bmatrix} 0 & 0 \\ V_2 & 0 \end{bmatrix}.$$

26. For R as in the previous exercise, let $Z_r(S) = Z_n(S_S)$ be the right singular submodule of S_S. Prove that $Z(R) = \begin{bmatrix} Z_r(S) & 0 \\ Z_r(S) & Z_r(S) \end{bmatrix}$.
27. Show that the socle of R_R is $\mathrm{soc}(R_R) = \begin{bmatrix} \mathrm{soc}(S_S) & 0 \\ \mathrm{soc}(S_S) & 0 \end{bmatrix}$.
28. For the integers $\mathbf{Z}$ and rationals $\mathbf{Q}$, set

$$R = \mathbf{Q}e_{11} + \mathbf{Q}e_{21} + \mathbf{Z}e_{22} = \begin{pmatrix} \mathbf{Q} & 0 \\ \mathbf{Q} & \mathbf{Z} \end{pmatrix}.$$

The only proper right ideals are: $\mathbf{Q}e_{11}$; $J(R) = \mathbf{Q}e_{21}$; $\mathbf{Q}e_{11} + \mathbf{Q}e_{21}$; $\mathbf{Q}e_{21} + \mathbf{Z}n$, $n \in \mathbf{Z}$; and the pairwise disjoint family

$$(e_{11} + \lambda e_{21})R = \{qe_{11} + q\lambda e_{21} \mid q \in \mathbf{Q}\}, \lambda \in \mathbf{Q}.$$

Let $G \subset \mathbf{Q}$ be any additive subgroup. (i) Show that $\begin{pmatrix} 0 & 0 \\ G & 0 \end{pmatrix}$ is a left ideal of R. (ii) Deduce that R is right Noetherian, but not left Noetherian.

29. Give a discussion parallel to the last exercise for left ideals and the ring $R^{op} \cong \begin{pmatrix} \mathbf{Q} & \mathbf{Q} \\ 0 & \mathbf{Z} \end{pmatrix}$.

30. For any R-module G, G is quasi-injective $\Leftrightarrow G \leqslant EG$ is fully invariant in EG. (Hint: $\Rightarrow$: Let $\lambda \in \text{End}_R EG$. Extend $\lambda: G \cap \lambda^{-1}G \to G$ to $\psi: G \to G$. Show that $(\lambda - \psi)^{-1}G = \lambda^{-1}G$, that $G \cap (\lambda - \psi)^{-1}G = G \cap \lambda^{-1}G \leqslant \ker(\lambda - \psi)$, or that $(\lambda - \psi)G \cap G = 0$. Then $\lambda G = \psi G \leqslant G$.)

31. For any R-module M, define $\tilde{M}$ in $M \leqslant \tilde{M} \leqslant EM$ by $\tilde{M} = (\text{End}_R EM)M$. (i) Show that $\tilde{M}$ is the unique smallest quasi-injective submodule of EM containing M. ($\tilde{M}$ is called the quasi-injective hull of M.)

32. For any quasi-injective G, any direct sum decomposition of $EG = \oplus_\alpha H_\alpha$ induces one of $G = \oplus_\alpha (G \cap H_\alpha)$.

33. If G is quasi-injective, so is $G^n = G \oplus \cdots \oplus G$ for any finite $n < \infty$.

34. For $R = \mathbf{Z}$, any prime p, and $0 < n$, $\mathbf{Z}/p^n\mathbf{Z}$ and $\mathbf{Q}$ are both quasi-injective, but $\mathbf{Q} \oplus (\mathbf{Z}/p^n\mathbf{Z})$ is not.

CHAPTER 12

Simple Algebras

Introduction

It seems to be a formidable open problem in ring theory to find an acceptably inclusive theoretical framework for describing, and also classifying and thus unifying the wide variety of simple rings that are known.

In applications of ring theory, algebras over a field are so frequently encountered that they cannot be ignored. Some of our previous purely ring theoretic concepts can be extended to algebras over a field of scalars F. This amounts to assuming in addition that all our submodules, subrings, and right ideals are F-subspaces while module homomorphisms are F-linear. Not only is the material in Section 1 necessary equipment of every ring theorist, but it will allow us to ignore all of these additional technical complications later when we will deal with simple modules and simple algebras.

The second section shows that a simple algebra as well as every simple ring, with or without identity, is an algebra over a field called its centroid. As the name suggests, when the ring has an identity, the centroid is essentially the center. All of this holds for arbitrary nonassociative algebras, which is the natural framework for establishing these facts. It is only the second section which briefly touches upon non-associative rings. Aside from studying our main object – simple rings, section two gives the reader a brief glimpse of one of the last remaining unexplored important frontiers of ring theory – nonassociative rings. Through all of this it should be remembered that every theorem established for nonassociative rings holds a fortiori for associative ones. A better name for a nonassociative ring would be a 'not necessarily associative ring'. The only reason that the latter is not used is because of its clumsy length.

Every nonassociative ring A gives rise to two associative rings – the

239

multiplication algebra and the centroid of A. This process is important for two reasons. (i) The multiplication algebra is an indispensable instrument for studying both associative and nonassociative rings. (ii) The forming of multiplication algebras is a method of constructing new rings out of old. Thus if A is a simple nonassociative ring, its centroid is a field, and its multiplication algebra is a primitive ring.

12-1   Algebra modules

The topic of this section is the relationship between algebra modules and algebra homomorphisms and their ordinary ring theoretic counterparts. In one of our later applications, the algebra B of this section will be the multiplication algebra of a nonassociative algebra A.

12-1.1   Let B be an associative algebra over a field F (with $1 \in B$ or $1 \notin B$); and let $V = {}_F V$ be a left F-vector space which is simultaneously a right B-module $V = V_B$ over B as a ring only, satisfying the additional associativity hypothesis

$$(fv)b = f(vb) = v(fb) \qquad f \in F, v \in V, b \in B.$$

Such a V is called an F-algebra right B-module, which is a more restrictive concept than that ${}_F V_B$ is a left F, right B-bimodule, because in addition, the logical possibility that $(fv)b - v(fb) \neq 0$ has been explicitly ruled out by assumption only.

Next, five rings are defined. For $f \in F$, define a map $f_L: V \to V$ by $f_L v = fv$. Set $F_L = \{f_L \mid f \in F\}$. Thus $F \cong F_L$. Define $D = \operatorname{Hom}_B(V, V) = \operatorname{End}_B V$, and $E = \operatorname{Hom}_F(V, V) = \operatorname{End}_F V$. The ring $\Delta = D \cap E$ consists of those endomorphisms of V which are F-linear. Thus F_L, D, E, $\Delta \subseteq \operatorname{Hom}_{\mathbb{Z}}(V, V)$, where $\mathbb{Z} = 0, \pm 1, \pm 2, \dots$ and $1 = 1_L \in F_L \subset D \cap E \cap \Delta$. Furthermore, $F_L \subseteq$ center Δ.

In the next diagram, an ascending line means that the ring below is included in the ring above at the other end of the line.

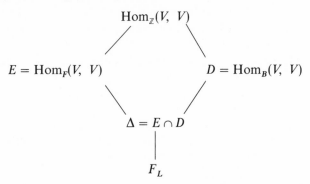

12-1.2   Lemma. *For F_L, B, V, D, and Δ as above, assume in addition that $V = VB$. Then*

(i) $D = \Delta$; *and hence*

(ii) $F_L \subseteq$ center D.

Proof. (i) and (ii). For any $f \in F$, $d \in D$, $w \in V$, and $b \in B$, set $v = wb$. Then

$$d(fv) - f(dv) = d[fwb] - f[d(wb)]$$
$$= d[w(fb) - f[(dw)b] = (dw)(fb) - (dw)(fb) = 0.$$

Since every element of V is a finite sum of elements of the form v, $df_L = f_L d$, and $D = \Delta$.

The next corollary says that the simple F-algebra B-modules are simple B-modules and hence we need only consider the latter.

12-1.3   Corollary. *For F_L, B, V, D, and Δ as in 12-1.1 assume in addition that $VB \neq 0$, and that V has no B-invariant F-subspaces aside from (0) and V. Then*

(i) V *is a simple B-module*;
(ii) $D = \Delta$; *and*
(iii) $F_L \subseteq$ *center D*.

Proof. (ii) and (iii). These conclusions follow from the last lemma, because $FVB = VB \neq 0$. (i) Set $T = \{v \in V \mid vB = 0\}$. Since $FT = T$, but $TB = 0$, $T \neq V$ – and hence necessarily $T = 0$. Consequently for any $0 \neq v \in V$, since $FvB = vB \neq 0$, $vB = V$ is simple.

12-2   Multiplication algebra

The reader might wish to recall the definition of non-associative rings, their subrings and ideals (0-2.3 and 0-2.4).

12-2.1   Definition. For elements x, y, z of a non-associative ring A, their *associator* is the element

$$(x, y, z) = (xy)z - x(yz).$$

The *nucleus* of A is the subring consisting of all $g \in A$ such that $(g, y, z) = (x, g, z) = (x, y, g) = 0$ for all x, y, $z \in A$. The *center* of A is the subring of A consisting of all those g in the nucleus such that $gw = wg$ for all $w \in A$. Thus center $A \subseteq$ nucleus A.

A nonassociative ring A is simple if $A^2 \neq 0$, and if A has only the two trivial ideals (0) and A. In this case $A^2 = A$.

12-2.2 Definition. Consider an associative and commutative ring F, and a nonassociative ring A that is a left F-module. Then A is a *nonassociative F-algebra* if for all k, k_1, $k_2 \in F$ and all x, $y \in A$ the following hold:

(0) $F \neq 0$; if $\exists 1 \in F$, then $1x = x$;
(i) $(k_1 k_2)x = k_1(k_2 x)$;
(ii) $k(xy) = (kx)y = x(ky)$.

Define "xk" by $xk = kx$.

An F-algebra ideal I of A is an ideal $I \lhd A$ of A as a nonassociative ring which also is an F-submodule of A, $FI \subseteq I$.

12-2.3 Remarks. (1) Every nonassociative ring is in a natural way an algebra over the ring of integers.

(2) Our previous definition 4-2.1 of an associative algebra over a field is a special case of this more general definition.

(3) In the next definition we could have allowed F to be noncommutative. Then an example of such an algebra would be the ring of 2×2 matrices over a non-commutative division ring F. Also, assumption (0) could have been omitted, in which case every nonassociative ring A would have been an algebra over its center (or even the nucleus). However, for us the only important cases are the case of a field F (by definition always commutative), or the ordinary algebraic case when $F = \mathbb{Z}$. For this reason only the next definition is not as general as it could be.

12-2.4 Definitions and notation. Let A be an F-algebra. Form the left F-module $_F A$, and the endomorphism ring $E = E(A) = \mathrm{Hom}_F(A, A)$ of the module $_F A$. For x, y, $a \in A$, define R_x, $L_y \in E$ to be the right left multiplication maps R_x, $L_y : A \to A$, $aR_x = ax$, $aL_y = ya$. These maps are written on the right side of A. Thus E is an F-algebra. The *multiplication algebra* $M = M(A)$ is the (associative) subalgebra of E generated by all the R_x, L_y for all x, $y \in A$. When $F = \mathbb{Z}$, M is also called the multiplier ring. The elements of M are of the form $\Sigma T_1 T_2 \cdots T_h$ where each $T_i = R_x$ or L_y. Note that $y(L_x R_z - R_z L_x) = (x, y, z)$ for all x, y, $z \in A$. For any subset $B \subseteq A$, $B^* \subseteq M$ is defined as the subalgebra of M generated by the set $\{R_b, L_b \mid b \in B\}$. In particular, $A^* = M$.

The *centroid* $C = C(A)$ consists of all those F-endomorphisms $T: A \to A$ which commute with all elements of M, i.e. $C = \{T \in E \mid \forall S \in M, TS = ST\} = \{T \in E \mid \forall x, y \in A, TR_x = R_x T, TL_y = L_y T\}$. Another frequently used name for the centroid is the *multiplication centralizer*. Let $Z = Z(A)$ be the center of A, $Z = \{z \in A \mid \forall x, y \in A, zy = yx, x(yz) = (xy)z\}$. Thus $Z^* \subseteq C$ is a subring.

Let $V = A$ as an additive abelian group. Then $V = {}_F V$ is a left F-module, and a right M-module $V = V_M$. (When F is a field V satisfies the associativity condition in 12-1.1 with $B = M$.) When A is regarded merely as a nonassociative ring, then an ideal $I \lhd A$ is the same as a right M-submodule of $V = A_M$, that is $IM \subseteq I \leqslant V$. The F-algebra ideals $J \lhd A$ are exactly the left F, right M-bisubmodules of V, i.e. $FJM \subseteq J \leqslant V$.

For a nonassociative ring A without an explicit mention of F, it is understood that $F = \mathbb{Z}$.

The next lemma follows immediately by applying 12-1.1, 1.2, and 1.3 to the preceeding with $B = M$. It interprets the five rings in 12-1.1 in this context.

12-2.5 Lemma. *Let A be a nonassociative algebra over a field F, and as before let $V = (A, +)$ be the underlying additive group of A. Recall that $F_L \subseteq C = \operatorname{Hom}_F(V, V) \cap \operatorname{Hom}_M(V, V)$ and set $D = \operatorname{Hom}_M(V, V)$.*

(1) *If $A^2 = A$, then $C = D$.*
(2) *The following are all equivalent:*
   (i) *A is a simple F-algebra;*
   (ii) *V is a simple F-algebra right M-module;*
   (iii) *A is a simple nonassociative ring;*
   (iv) *V is a simple right M-module.*

12-2.6 Lemma. *Let A be an arbitrary nonassociative ring or algebra and let $T \in E$. Then $T \in C$ if and only if for all $x, y \in A$*

$$(xT)y = (xy)T = x(yT).$$

Proof. By definition, for an element $T \in C$, $T \in C$ if and only if for all x, $y \in A$

$$R_y T = TR_y, \ TL_x = L_x T \Leftrightarrow xR_y T = xTR_y, \ yTL_x = yL_x T$$

$$\Leftrightarrow (xy)T = (xT)y, \ x(yT) = (xy)T.$$

12-2.7   Corollary. *Above for* $T \in E$,

$$T \in C \Leftrightarrow \begin{cases} \forall \; x, \; y \in A, \quad R_y T = TR_y = R_{yT} \\ \qquad\qquad L_x T = TL_x = L_{xT}. \end{cases}$$

Although all the subsequent results here will hold for F-algebras, for ease of reading, in view of Lemma 12-2.5, frequently they will be stated and proved only for $F = \mathbb{Z}$.

12-2.8   Lemma. *If* A *is a nonassociative ring with* $A^2 = A$, *then its centroid is commutative.*

Proof. All α, $\beta \in C$ and a, $b \in A$ satisfy

$$(ab)\alpha\beta = aR_b\alpha\beta = a\alpha R_b\beta = [(a\alpha)b]\beta.$$

Use of 12-2.6 gives

$$a b\alpha\beta = (a\alpha)(b\beta); \qquad \text{and}$$

$$(ab)\beta\alpha = [a(b\beta)]\alpha = aR_{b\beta}\alpha = a\alpha R_{b\beta} = (a\alpha)(b\beta).$$

Thus $\alpha\beta = \beta\alpha$ since $A^2 = A$.

Use of Lemma 12-2.5 gives the following corollary.

12-2.9   Corollary. *If* A *is a simple nonassociative ring or algebra, then its centroid* C *is a field.*

12-2.10   Lemma. *Let* A *be a nonassociative ring with center* Z *and centroid* C. *Then*

(i) $Z^* = \{R_z \mid z \in Z\}$, *and*
(ii) $Z^* \lhd C$.

Proof. (i) and (ii). An element $z \in A$ belongs to $z \in Z$ if and only if for all x, $y \in A$ equations (1) and (2) hold:

(1) $zy = yz$, or $L_z = R_z$; and
(2) $(xy)z = x(yz)$, or $R_y R_z = R_{yz}$.

In view of the latter, the subalgebra Z^* generated by the set $\{R_z \mid z \in Z\}$ already is $Z^* = \{R_z \mid z \in Z\}$ this set itself. The map $z \to R_z$ is a homomorphism of Z onto Z^*, where $Z^* \subseteq C$. It is asserted that $Z^* \lhd C$. Let x,

$y \in A$ and $T \in C$ be arbitrary. Use of 12-2.6 twice shows that zT satisfies (1):

$$y(zT) = (yz)T = (zy)T = (zT)y, \quad \text{or} \quad R_{zT} = L_{zT}$$

For the proof of (2), the fact that z is in the nucleus as well as 12-2.6 are used as follows

$$xR_{y(zT)} = x[y(zT)] = x[(yT)z] = [x(yT)]z$$
$$= [(xy)T]z = (xy)(zT) = xR_yR_{zT}, \quad \text{or} \quad R_{y(zT)} = R_yR_{zT}.$$

Thus $zT \in Z$, $R_zT = R_{zT} \in Z^*$, and $Z^*C \subseteq Z^*$. To prove that $CZ^* \subseteq Z^*$, it suffices to show that $TL_z \in Z^*$ if $z \in Z$ and $T \in C$ because $L_z = R_z$. In addition to 12-2.6, the already proved fact that $zT \in Z$ is now used in

$$yTL_z = z(yT) = (zy)T = (yz)T = y(zT), \quad \text{or} \quad TL_z = L_{zT} = R_{zT} \in Z^*.$$

Thus $Z^* \lhd C$.

12-2.11 Corollary. *If A contains an identity element $1 = e \in A$, then $C = Z^* \cong Z$.*

Proof. If $\beta \in C$ and $a \in A$, then

$$a(R_{e\beta} - \beta) = a(e\beta) - a\beta = (ae)\beta - a\beta = a\beta - a\beta = 0.$$

Since $e \in Z$, also $e\beta \in Z$, and hence $\beta = R_{e\beta} \in Z^*$.

12-2.13 Theorem. *Let A be a simple nonassociative ring with enter Z and centroid C.*

(1) *Then C is a field and A is an algebra over C.*
(2) (i) *EITHER $Z = 0$,*
      (ii) *OR $\exists 1 \in A$, Z is a field, and $C = Z^* = \{R_z \mid z \in Z\}$.*

Proof. (1) By 12-2.9, C is a field; and thus A is an algebra over C by 12-2.6.

(2) Since the right and left annihilators $\{z \in A \mid Az = 0\} \lhd A$, $\{z \mid zA = 0\} \lhd A$ are ideals of A, both of them must be zero. Thus $z \to R_z$ is an isomorphism of $Z \cong Z^*$. Let $Z \neq 0$. Since $Z^* \lhd C$, $Z^* = C$. Hence Z contains an identity element $e \in Z$. For any $a \in A$, $A(ae - a) = 0$ and $(ea - a)A = 0$. Hence $ea = a$, and $e = 1 \in A$.

12-2.14 Corollary to theorem 2.13. *Let A be a simple nonassociative ring* *or algebra with centroid C. Let B denote the algebra A over the field C. (I.e. B = A as a ring.) Then the entroid of B is also C.*

Proof. This is immediate from 12-2.5(1).

12-2.15 Definition. To say that a nonassociative algebra A over a field F is *central simple* means that A is simple, and that center $A = F$. The abbreviation 'central' is sometimes used for 'central simple'.

It follows from the definition that a central simple algebra has an identity element. Consequently, every simple nonassociative algebra with identity is central over its center.

12-3 Tensor products of simple rings

From now on all of our rings will be associative. Several notations are sometimes introduced for the same object because in different contexts, the various different notations have distinct advantages over others.

12-3.1 Notation and definitions. Let R be any ring, $R^1 = R$ if $1 \in R$ and $R^1 = \mathbb{Z} \times R$ if $1 \notin R$ where $\mathbb{Z} = 0, \pm 1, \pm 2, \ldots$. For a nonempty subset $Y \subseteq R$, define the *centralizer* of Y in R to be the subring $C_R(Y) = C(Y:R) = C(Y) = \{r \in R \mid \forall y \in y, ry = yr\} = R^Y$. (In formulas involving tensor products, and possibly several different R's and Y's, the notation "R^Y" will be used. The $C_R(Y) = C(Y)$ notation has been widely accepted in group and ring theory for a long time.) Note that always $Y \subseteq C_R(C_R(Y))$.

The *opposite ring* R^{op} has the same additive group $R^{op} = (R, +)$ as R but the new product $*$ in R^{op} is $a * b = ba$, where juxtaposition denotes the product in R.

If D is a division ring and F_n the $n \times n$ matrix ring over a field, then there are ring isomorphisms

$$D \cong D^{op} : d \to d^{-1} \qquad d \in D;$$

$$F_n \cong F_n^{op} : a \to a^T = \text{transpose} \qquad a \in F_n.$$

More generally, if $r \to r^c$ is a multiplication reversing isomorphism of R (an anti-automorphism), with $(ab)^c = b^c a^c$ for $r, a, b \in R$, then $R \cong R^{op}$.

Until further notice, "$- \otimes -$" denotes the tensor product $\_ \otimes_{\mathbb{Z}} \_$ over $\mathbb{Z}$. (If R were an algebra over F, $\_ \otimes_F \_$ could be used here just as well.) For rings A and B recall that $A \otimes B \cong B \otimes A$ is a ring under $(a \otimes b)(a' \otimes b') = aa' \otimes bb'$. Note that $(A \otimes B)^{op} = A^{op} \otimes B^{op}$. Then $R \otimes R$,

$R^{\mathrm{op}} \otimes R$, and $R \otimes R^{\mathrm{op}}$ are three rings with the same underlying additive subgroup $R \otimes R$.

Let $t = \Sigma a \otimes b$ and $s = \Sigma s \otimes d$, where summation indices have been surprised. Then R is a module over two rings, where for $x \in R$:

right $R^{\mathrm{op}} \otimes R: xt = \Sigma\ axb$, $ts = \Sigma\ a * c \otimes bd$;

left $R \otimes R^{\mathrm{op}}: tx = \Sigma axb$, $ts = \Sigma\ ca \otimes d * b$.

The same holds for $R' \otimes R''$ where R' and R'' are subrings and/or over rings of R, like R^1. Subsequently t will be thought of as a function $t: R \to R$, and $t[x]$ or $t(x)$ will denote the value of the function t at $x \in R$.

From the multiplier ring $M(R)$ of R. Frequently $t \in R^1 \otimes R + R \otimes R^1$ and then any such t can be written in the form $t = 1 \otimes d + c \otimes 1 + \Sigma a \otimes b$; $a, b, c, d \in R$. If R is viewed as a right $M(R)$-module $R_{M(R)}$ then $R_x L_y = L_y R_x$, $R_x R_y = R_{xy}$, and $L_x L_y = L_{yx}$ for all $x, y \in R$. Then $a \otimes 1 \to L_a$ and $1 \otimes b \to R_b$ defines a ring homomorphism $R^1 \otimes R^{\mathrm{op}} + R \otimes (R^1)^{\mathrm{op}} \to M(R)$. Thus $t \to R_d + L_c + \Sigma L_a R_b \in M(R)$; every element of $M(R)$ is of this latter form. Thus $t[x] =$

$$x(R_d + L_c + \Sigma L_a R_b) = xd + cx + \Sigma axb.$$

12-3.2   Definition. If $S \subseteq R$ is a subring, then elements $x_1, \ldots, x_n \in R$ are (right) *weakly independent* with respect to S if whenever $x_1 k_1 + \cdots + x_n k_n = 0$ with all $k_i \in S$, then $x_i k_i = 0$ for all $i = 1, \ldots, n$.

12-3.3   Conditions. Let $C \subset R$ be a subring such that

[a] $\forall\ 0 \neq b \in R$, $RbC = R$;

[b] $\forall\ 0 \neq x \in R$, $\exists\ u \in R^C$ such that $xu \neq 0$.

12-3.4   Consequences. If $C \subset R$ is a subring, then

(1) [a] $\Leftrightarrow$ R is a simple left $R \otimes C^{\mathrm{op}}$-module.
(2) [a] $\Leftrightarrow$ $\forall\ 0 \neq b \in R$, $\forall y \in R$, $\exists\ q \in R \otimes C$ such that $q[b] = y$.

12-3.5   Theorem. *Let R satisfy conditions* [a] *and* [b]. *Then for any integer n, for any elements $x_1, \ldots, x_n \in R$ which are weakly R^C-independent, and for any completely arbitrary $y_1, \ldots, y_n \in R$, there exists a $t \in R \otimes C$ such that $t[x_i] = y_i$ for all $i = 1, \ldots, n$.*

Proof. The theorem is true for $n = 1$ by [a]. By induction assume that it is true for $1, \ldots, n-1$. Given are x_i, y_j for $i, j \leqslant n$ as above. By adding a finite number of t's, it follows that without loss of generality it may be assumed that $y_1 = \cdots = y_{n-1} = 0$ and $y_n \neq 0$. By [b] select a $u \in R^C$ such that $x_n u \neq 0$. By induction applied $n-1$ times, there exist $t_1, \ldots,$ $t_{n-1} \in R \otimes C$ such that $t_i[x_j] = \delta_{ij} u$, $1 \leqslant i \leqslant n-1$, $i \neq n$.

Case 1. All $t_i[x_n] = k_i \in R^C$ for $i \leqslant n-1$. Set $\gamma = (x_1 \otimes 1)t_1 + \cdots + (x_{n-1} \otimes 1)t_{n-1} - 1 \otimes u \in R \otimes C^1 + R^1 \otimes C$. For $i \leqslant n-1$, $t_j[x_i] = \delta_{ji} u$, and hence $\gamma[x_i] = x_i u - x_i u = 0$. If $0 \neq \gamma[x_n] \in R$, then by 12-3.4(2) we are done. I.e. there is a $q \in R \otimes C$ such that $q[\gamma(x_n)] = y_n$. Take $t = q\gamma$. Even if $\gamma \notin R \otimes C^{op}$, nevertheless $t \in R \otimes C^{op}$. Otherwise $\gamma[x_n] = x_1 k_1 + \cdots + x_{n-1} k_{n-1} - x_n u = 0$. By weak independence, all $x_i k_i = 0$ and $x_n u = 0$, a contradiction.

Case 2. There exists an i such that $t_i[x_n] \notin R^C$. Thus there exists a $0 \neq c \in C$ such that $0 \neq b = ct_i[x_n] - t_i[x_n]c = p[x_n]$, where $p = (c \otimes 1)t_i - (1 \otimes c)t_i \in R \otimes C$. (Here $c \otimes 1 \in R \otimes C^1$ and $1 \otimes c \in R^1 \otimes C$.) Since $RbC = R$, it follows that there exists a $q \in R \otimes C$ such that $q[b] = y_n$. Thus $q[b] = q[p(x_n)] = y_n$. Set $t = qp \in R \otimes C^{op}$. If $j \leqslant n-1$, $t[x_j] = q[p(x_j)]$, where $p(x_j) = ct_i[x_j] - t[x_j]c = c\delta_{ij} u - \delta_{ij} uc = 0$. The theorem has been proved.

If above R were an algebra over a field F, clearly we could replace $\_ \otimes_Z \_$ with $\_ \otimes_F \_$ in the previous theorem. From now on in this section F is a given fixed field and $\_ \otimes_F \_$ will be written as $\_ \otimes \_$ unless stated otherwise.

12-3.6 Corollary 1 to theorem 3.5. *Consider a simple algebra A with identity with center $A = F$. Then*

> $\forall$ n, $\forall$ F − *linearly independent* $x_1, \ldots, x_n \in A$; *and*
>
> $\forall$ *arbitrary* $y_1, \ldots, y_n \in A$

there exists a $t \in A \otimes A$ *such that* $t[x_i] = y_i$ *for all* $i = 1, \ldots, n$.

Proof. In the theorem, take $R = C = A$. Then $A^A = F$.

12-3.7 Corollary 2 to theorem 3.5. *For a central simple algebra A over F as in the last corollary assume in addition that the dimension of A over F is $[A:F] = n$ is finite. For any ring such as F, let F_n denote the ring of all $n \times n$ matrices over F. Then $A \otimes A^{op} \cong F_n$.*

Proof. There is a ring homomorphism $A \otimes A^{op} \to \operatorname{Hom}_F(A, A) \cong F_n$ because each $t \in A \otimes A$ induces a map $A \to A$, $a \to t[a]$. It follows from $[A \otimes A^{\mathrm{op}}:F] = [A:F][A^{\mathrm{op}}:F] = [A:F]^2 = n^2$ that this map is an isomorphism.

12-3.8 Corollary 3 to Theorem 3.5. *Let A be a central simple algebra over F and let B be any ring with $F \subseteq$ center B. Then every ideal of the ring $A \otimes B$ is of the form $A \otimes \mathscr{L}$ for some $\mathscr{L} \lhd B$.*

Proof. If $I \lhd A \otimes_F B$, define $\mathscr{L} = I \cap (1 \otimes B)$. Clearly $\{b \mid 1 \otimes b \in \mathscr{L}\} \lhd B$. Then $(A \otimes 1)\mathscr{L} = A \otimes \mathscr{L} \subseteq I$. Any $\xi \in I$ is of the form $\xi = \Sigma_{k=1}^n a_k \otimes b_k$ with $b_k \in B$ and where the $a_k \in A$ are linearly independent. In the previous theorem if $R = C = A$, then $R^C = A^A =$ center $A = F$. Hence there exist for all $1 \leqslant i \leqslant n$ elements $t_i \in A \otimes A$ such that $t_i[a_k] = \delta_{ik}$. If $t_i = \Sigma_\lambda c_\lambda \otimes d_\lambda$, then $t_i[a_k] \otimes b_k = \Sigma_\lambda (c_\lambda \otimes 1)(a_k \otimes b_k)(d_\lambda \otimes 1)$. Hence $\Sigma_\lambda (c_\lambda \otimes 1)\xi(d_\lambda \otimes 1) = \Sigma_{k=1}^n t_i[a_k] \otimes b_k = 1 \otimes b_k \in I$. Thus $\xi \in A \otimes \mathscr{L}$ and $I = A \otimes \mathscr{L}$.

12-3.9 Corollary 4 to theorem 3.5. *If A and B are simple rings with identity with $F =$ center $A \subseteq$ center B, then $A \otimes B$ is simple.*

12-3.10 Corollary 5 to theorem 3.5. *Suppose that $A, B \subset R$ are subrings of a ring R with $F =$ center $A \subseteq$ center B and suppose that both A and B have the same identity element $e = 1_A = 1_B \in A \cap B$ (which need not be the identity element of R). If A is simple and $ab = ba$ for all $a \in A$, $b \in B$, then the subring AB has the structure $AB \cong A \otimes B$.*

Proof. Let $\bar{A} \cong A$ and $\bar{B} \cong B$ be new rings disjoint from R and from each other. Then the map $\bar{A} \otimes \bar{B} \to AB$, $\Sigma \bar{a} \otimes \bar{b} \to \Sigma ab$ is a ring homomorphism. By Corollary 3 the kernel is of the form $I = \bar{A} \otimes \bar{\mathscr{L}} \lhd \bar{A} \otimes \bar{B}$. For any $\bar{b} \in \bar{\mathscr{L}}$, $1 \otimes \bar{b} \in I$ and $1 \otimes \bar{b} \to b = 0$. Thus $\bar{\mathscr{L}} = 0$ and $A \otimes B \cong \bar{A} \otimes \bar{B} \cong AB$.

12-3.11 Consequence. In the last corollary, A and B are linearly disjoint subspaces of R (see 4-3.2, 3.3, and 3.4).

12-3.12 Proposition. *If A and B are central simple algebras over F then $A \otimes B$ is also central simple over F.*

Proof. Corollary 4 shows that $A \otimes B$ is simple. Take any $z =$

$\Sigma a_i \otimes b_i \in \text{center}(A \otimes B)$ with the b_i independent over F. Then for any $a \in A$, $0 = (a \otimes 1)z - z(a \otimes 1) = \Sigma(aa_i - a_i a) \otimes b_i$. So $aa_i - a_i a = 0$ for all i. Thus $a_i \in \text{center } A = F$; set $k_i = a_i \in F$. Hence $z = 1 \otimes b$ where $b = k_1 b_1 + \cdots + k_n b_n$. For any $x \in B$, $z(1 \otimes x) - (1 \otimes x)z = 1 \otimes (bx - xb) = 0$. So also $b \in \text{center } B = F$.

12-3.13 Proposition. *Suppose that D is a division ring with center $D = F$ and which is of finite dimension $[D:F]$ over F. Then $[D:F]$ is a perfect square.*

Proof. Form the algebraic closure $F \subseteq \bar{F}$ of F. Then $[D \otimes \bar{F} : \bar{F}] = [D:F]$. For if $\{x_i\}$ is a basis for D over F then $\{x_i \otimes 1\}$ is a basis for $D \otimes \bar{F}$ over $\bar{F}$. Since D is central simple over F by 12-3.9, $D \otimes \bar{F}$ is simple. It now follows that $D \otimes \bar{F} \cong \bar{F}_m$ for some m. (See 6-5.10(ii).) Consequently $[D:F] = [\bar{F}_m : \bar{F}] = m^2$.

12-3.14 Corollary. *If A is a central simple algebra over F of finite dimension $[A:F]$ over F, then $[A:F]$ is a perfect square.*

Proof. By Wedderburn's theorem $A \cong D_n$ where D is a division ring. Since center $D_n = (\text{center } D)I_n \supseteq FI_n$ where I_n is the $n \times n$ identity matrix, and since center $A = F1_A$ is mapped onto FI_n by the isomorphism $A \cong D_n$ it follows that center $D = F$. Thus $[A:F] = n^2[D:F]$.

12-4   Centralizers

12-4.1   Notation. Suppose that R and S are rings with identity both containing the field F. For any subsets $X \subset R$ and $Y \subset S$, define $X \otimes Y = \{\Sigma x \otimes y \mid x \in X, y \in Y\}$ to be the set of all finite sums of elements $x \otimes y$.

12-4.2   Lemma. *If in the above sets $1 \in X$ and $1 \in Y$, then*

(i) $(R \otimes S)^{X \otimes Y} = R^X \otimes S^Y$;
(ii) $(R \otimes S)^{X \otimes 1} = R^X \otimes S$.

Proof. (i) Trivially, $R^X \otimes S^Y \subseteq (R \otimes S)^{X \otimes Y}$. If $\xi \in (R \otimes S)^{X \otimes Y}$, then $\xi = \Sigma r_i \otimes s_i$ where s_i are independent over F. For any $x \in X$, $(x \otimes 1)\xi - \xi(x \otimes 1) = \Sigma(xr_i - r_i x) \otimes s_i = 0$. Hence for all i, $xr_i - r_i x = 0$, and

$r_i \in R^X$. Similarly, also all $s_i \in S^Y$. So $\xi \in R^X \otimes S^Y$, and $(R \otimes S)^{X \otimes Y} = R^X \otimes S^Y$. (ii) Conclusion (ii) is a special case of (i).

12-4.3 Consequence. If above R is central simple over F, then $(R \otimes S)^{R \otimes 1} = 1 \otimes S = F \otimes S$.

12-4.4 Lemma. *Let R be any ring ($1 \in R$ or $1 \notin R$), $Y \subset R$ any subset, and $\phi: R \to R$ an automorphism. Then $\phi C(Y) = C(\phi Y)$.*

12-4.5 Corollary. *If R is a ring with 1, t, $t^{-1} \in R$ and $Y \subset R$ any subset then $t^{-1} C(Y) t = C(t^{-1} Y t)$. In other words $t^{-1}(Y)^R t = (t^{-1} Y t)^R$; conjugation commutes with exponentiation.*

12-4.6 Proposition. *Suppose that R is any ring ($1 \in R$ or $1 \notin R$), and that $K \subset R$ is a commutative subring. Then*

$$K \text{ is a maximal commutative subring} \Leftrightarrow K = C(K).$$

Proof. $\Leftarrow$: If $K = C(K)$, and L is a commutative subring of R with $K \subseteq L \subseteq R$, then $L \subseteq C(K) = K$, and so $L = K$.

$\Rightarrow$: In general, for a subring $A \subset R$, $A \nsubseteq C(A)$. If however A is commutative, then $A \subseteq C(A)$. Thus $K \subseteq C(K)$. Let $a \in C(K)$. Form $\langle K, a \rangle = K + K^1 a + K^1 a^2 + \cdots = K + Ka + Ka^2 + \cdots + \mathbb{Z}a + \mathbb{Z}a^2 + \cdots \subseteq C(K)$. Since $\langle K, a \rangle$ is a commutative subring, $\langle K, a \rangle = K$ by the maximality of K. Thus $a \in K$ and $K = C(K)$.

12-4.7 Corollary. *If D is a division ring with center $D = F$, and $F \subseteq K \subseteq D$ is any subring, then the following are all equivalent*

  (i)   *K is a maximal subfield;*
  (ii) *K is a maximal commutative subring;*
  (iii) *$K = C(K)$.*

Proof. By the previous lemma (ii) $\Leftrightarrow$ (iii). If an element $a \in D$ commutes with all elements of K, then so does its inverse a^{-1}. (ii) $\Rightarrow$ (i). For any $a \in K$, $\langle K, a^{-1} \rangle = K + K^1 a^{-1} + K^1 a^{-2} + \cdots$ is a commutative subring; hence $a^{-1} \in K$. (i) $\Rightarrow$ (ii). If $a \in D$ such that $K[a] = K + Ka + Ka^2 + \cdots$ is a commutative subring, then the subfield $K(a)$ generated by K and a is commutative (because it is obtained by adjoining to $K[a]$ all inverses of elements of $K[a]$).

The next theorem is a special case of the following useful general problem. Given a simple algebra A over a field F, to find another F-algebra B such that $A \otimes B$ is a dense ring of linear transformations on A viewed as a vector space over some field K, where $F \subseteq K \subset A$.

Note that in the next theorem D need not be finite dimensional.

12-4.8   Theorem. *Let D be a division ring with center $D = F$ and K any maximal subfield with $F \subseteq K$, then $D \otimes K$ is a dense ring on D_K regarded as a right or left K-vector space.*

Proof. Use Theorem 12-3.5 with $R = D$ and $C = K$. Thus $R^C = D^K = K$. The map $D \to D^{\mathrm{op}}$, $d \to d^{-1}$ is a ring isomorphism. Thus D is a left $D \otimes K^{\mathrm{op}} = D \otimes K$-module and a right $D^{\mathrm{op}} \otimes K \cong D \otimes K$-module (where $v(d \otimes k) = d^{-1}vk$, $v \in D_K$, $d \in D$, $k \in K$). Weakly independent (12-3.2) now means simply right linearly K-independent. By 12-3.5, for any n, $D \otimes K$ acts n-transitively on D_K:

$\forall\ x_1, \ldots, x_n \in D$ right independent over K, and

$\forall\ y_1, \ldots, y_n \in D$

$\exists\ t \in D \otimes K$ such that $t[x_i] = y_i$ for $i = 1, \ldots, n$.

In the next corollary $[D:F]$ is not assumed to be finite.

12-4.9   Corollary 1 to theorem 4.8. *If D is a division ring and K a maximal subfield in $F = $ center $D \subseteq K \subseteq D$ such that D is of finite dimension $[D:K] = m$ over K as a right K-vector space, then $D \otimes K \cong K_m$.*

The next corollary extends 12-3.13. It gives an easy criterion for deciding whether a subfield is maximal or not.

12-4.11   Corollary 3 to theorem 4.8. *Let D be a division algebra of finite dimension $[D:F]$ over its center F, and let $K \subset D$ be any maximal subfield with $F \subset K$. Let $[D:K]$ be the dimension of D over K as a right K vector space. Then*

(i) $[D:K] = [K:F] = m$;
(ii) $[D:F] = m^2$;
(iii)   *the right dimension equals the left dimension of D over K.*

Proof. (i) and (ii). Set $m = [D:K]$. Then $[D \otimes K:K] = [K_m:K] = m^2$. But also $[D \otimes K:K] = [D:F]$. It now follows from $m^2 = [D:F] =$

$[D:K][K:F]$ that $[D:K] = [K:F] = m$. (iii) This follows from the fact that the hypotheses on F, K, and D are completely left right symmetric. An alternate way of seeing this is that for any $\Sigma d \otimes k \in D \otimes K$ and any $c \in K$, $(1 \otimes c)\Sigma d \otimes k = \Sigma d \otimes ck = (\Sigma d \otimes k)(1 \otimes c)$. Thus $D \otimes K$ as a right vector space coincides with $D \otimes K$ as a left vector space. The same is true for K_m. In the previous theorem we had $m^2 = [K_m:K] = [D \otimes K:K]$. In the left-handed analogue of 12-4.8 we would have $m^2 = [K_m:K] = [K \otimes D:K]$, which is the same.

12-4.12 Facts. (1) If the right R-submodules $V_1, \ldots, V_m$ of some R-module each has the descending chain condition, then so does also $V_1 + \cdots + V_m$ (See 11-3.3).

(2) If R is a ring and $S \subset R$ a subring, and if $R = V_1 + \cdots + V_m$ where each V_i is a right S-module with the descending chain condition on S-submodules, then R also satisfies the descending chain condition on right S-submodules. Hence automatically R has the descending chain condition as a right R-module (abbreviation: D.C.C.).

12-4.13 Lemma. *Suppose that A and B are algebras with identity over F, and that A has the right descending chain condition. Then $A \otimes B$ has the descending chain condition on right $A \otimes 1$-submodules if and only if the dimension of B over F is finite, that is $[B:F] \nleqq \infty$. In this case, in particular $A \otimes B$ has the D.C.C. on right ideals.*

Proof. Let $\{b_i | i \in I\} \subset B$ be any F-vector space basis of B. The tensor product commutes with direct sums

$$A \otimes B = A \otimes \left(\bigoplus_{i \in I} Fb_i \right) \cong \bigoplus_{i \in I} A \otimes b_i,$$

where the last isomorphism is a natural obvious isomorphism, and therefore must be replaced by an equality.

Each $A \otimes b_i$ is a right $A \otimes 1$-submodule of $A \otimes B$ and as such each $A \otimes b_i$ has the descending chain condition. The latter fact has two consequences. First, obviously, if I is infinite, then $A \otimes B$ fails to have the right $A \otimes 1$-descending chain condition. Secondly, if $I = \{1, \ldots, m\}$ is finite, then by 12-4.12(1), $A \otimes B = A \otimes b_1 + \cdots + A \otimes b_m$ has the descending chain condition on right $A \otimes 1$-submodules.

12-4.14 Theorem (Noether Skolem). *Consider a simple ring R with the*

D.C.C. and center $R = F$, and containing subalgebras $A, B \subset R$ satisfying the following conditions:

    (a)   *A and B are simple rings;*
    (b)   *$1_R \in F \subseteq A \cap B$; and*
    (c)   *$[A:F] < \infty$, $[B:F] < \infty$ are finite.*

If $\phi: A \to B$ is a ring isomorphism which is the identity on F, i.e. $\phi | F \cdot 1_R = 1$, then there exists a $c \in R$ such that $\phi a = c^{-1}ac$ for all $a \in A$. In particular, any F-isomorphism $A \cong B$ extends to an inner automorphism of R.

Proof. For any subalgebras such as $A, B \subset R$, R is a right module over $R^{op} \otimes A$ and $R^{op} \otimes B$. Since $A \cong B$, this makes R a right $R^{op} \otimes A$-module in two ways:

    (1) $x(r \otimes a) = rxa$ $x \in R, r \in R, a \in A$;
    (2) $(x, r \otimes a) = rx\phi a$.

First, $R^{op} \otimes A$ is a simple ring (12-3.9). Second, $R^{op} \otimes A$ has the D.C.C. (12-4.13). Third, any simple ring with the D.C.C. is a finite direct sum of simple right submodules (=minimal right ideals) all of which are isomorphic (10-1.6(ii)). Thus

    (1) $R = V_1 \oplus \cdots \oplus V_s$ V_i are simple right $R^{op} \otimes A$-modules;
                                 $V_1 \cong \cdots \cong V_s$;
    (2) $R = U_1 \oplus \cdots \oplus U_t$ U_j are simple right $R^{op} \otimes B = R^{op} \otimes (\phi A)$-
                                 modules; $U_1 \cong \cdots \cong U_t$.

Without loss of generality $s \leqslant t$. Since ϕ is an isomorphism each U_j is in effect a simple $R^{op} \otimes A$-module. Hence there exists an $R^{op} \otimes A$-isomorphism $\sigma_i: U_i \to V_i$; set $\sigma = \sigma_1 \oplus \cdots \oplus \sigma_s: R \to R$, a monic $R^{op} \otimes A$-homomorphism, where $\sigma(v_1 + \cdots + v_s) = \sigma_1 v_1 + \cdots + \sigma_s v_s$, $v_i \in V_i$; and where $\sigma[v(r \otimes a)] = r(\sigma v)\phi a$ for $r \in R, v \in R$, and $a \in A$. Thus $\sigma[rva] = r(\sigma v)\phi a$. Define $c = \sigma 1$.

If $v = 1, a = 1$, this becomes $\sigma r = rc$. In particular, if $r = a$,

$$\sigma a = ac.$$

Alternatively setting $r = 1, v = 1$ gives $\sigma a = (\sigma 1)\phi a$ or

$$\sigma a = c\phi a.$$

Since σ is one to one, $\ker \sigma = \{r \in R \mid rc = 0\} = 0$, i.e. c is not a right zero divisor. By Wedderburn's theorem R is isomorphic to a matrix ring over a division ring. Hence c has an inverse $c^{-1}c = cc^{-1} = 1$. But then the two expressions for σa give $\phi a = c^{-1}ac$ for any $a \in A$.

For the next corollary, observe that any simple ring with the right D.C.C. has an identity.

12-4.15 Corollary to theorem 4.14. *Any automorphism of a finite dimensional central simple algebra which leaves F elementwise fixed is inner.*

Proof. If $\phi: A \to A$ is the given automorphism, take $R = A = B$ in the theorem. Then $\phi a = c^{-1}ac$ for every $a \in A$.

12-4.16 Definition. A *derivation* δ of a ring R is an additive group homomorphism $\delta: R \to R$ such that $\delta(ab) = a(\delta b) + (\delta a)b$, for all $a, b \in R$. The notation $r^\delta = \delta r$, $r \in R$ is widespread and useful. The derivation δ is *inner* if there exists a $c \in R$ such that $\delta r = cr - rc \equiv [c, r]$, the latter is called a commutator.

Now, assume $1_R \in R$. Then $\delta 1_R = \delta(1_R \cdot 1_R) = 2\delta 1_R$ shows that $\delta(\mathbb{Z} \cdot 1_R) = 0$.

Next, assume that R is an F-algebra with identity $F = F \cdot 1_R \subset R$. Then

$$\delta \text{ is } F\text{-linear} \Leftrightarrow \delta(F \cdot 1_R) = 0.$$

In this latter case δ is called an F-algebra derivation.

12-4.17 Corollary to theorem 4.15. *Any derivation δ of a central simple finite dimensional algebra A over F which vanishes on F, $\delta(F \cdot 1_A) = 0$, is inner.*

Proof. The ring A_2 of all 2×2 matrices over A is simple by 7-2.16, and hence finite dimensional and central simple over F. It contains two subalgebras $\theta, \mathscr{L} \subset A_2$ and an isomorphism $\phi: \theta \to \mathscr{L}$, where

$$\theta = \left\{ \begin{vmatrix} a & 0 \\ 0 & a \end{vmatrix} \middle| a \in A \right\}, \qquad \mathscr{L} = \left\{ \begin{vmatrix} a & 0 \\ \delta a & a \end{vmatrix} \middle| a \in A \right\}$$

$$\phi \begin{vmatrix} a & 0 \\ 0 & a \end{vmatrix} = \begin{vmatrix} a & 0 \\ \delta a & a \end{vmatrix}.$$

It follows from $\delta F = 0$, that $\phi | F = 1$. Thus there exists an invertible matrix such that for all $a \in A$

$$\begin{vmatrix} a & 0 \\ \delta a & a \end{vmatrix} \begin{vmatrix} x_1 & x_2 \\ y_1 & y_2 \end{vmatrix} = \begin{vmatrix} x_1 & x_2 \\ y_1 & y_2 \end{vmatrix} \begin{vmatrix} a & 0 \\ 0 & a \end{vmatrix} \qquad x_1, x_2, y_1, y_2 \in A \\ x_1 y_2 - x_2 y_1 \neq 0;$$

$$ax_1 = x_1 a \qquad\qquad ax_2 = x_2 a$$
$$(\delta a)x_1 + ay_1 = y_1 a \qquad (\delta a)x_2 + ay_2 = y_2 a$$

Thus $x_1, x_2 \in F$ not both zero. If $x_1 \neq 0$, then $\delta a = x_1^{-1}y_1 a - ax_1^{-1}y_1$; if $x_2 \neq 0$, then $\delta a = [x_2^{-1}y_2, a]$. Thus δ is inner.

It is surprising that in the next theorem there are no restrictive assumptions on the algebra R whatever.

12-4.18 Theorem. *Let $F \cdot 1_R \subset A \subset R$ be algebras over F both having the same unit element and with A central simple over F of finite dimension $[A:F] = n \not\leqq \infty$ over F. If $C(A) \subset R$ is the centralizer of A in R, then R has the structure of the tensor product $R = AC(A) \cong A \otimes C(A)$.*

Proof. First assume that $A = F_n \subset R$. In this case there exist the usual matrix units $e_{ij} \in A$. In this proof only the tensor calculus summation convention will be used, where all repeated indices are summed. For any $r \in R$, $r_{ij} \in R$ is defined as $r_{ij} = e_{ki}re_{jk}$. (The notation "r_{ij}" means that there exists some single $r \in R$ which defines the r_{ij} this way.) To show that actually $r_{ij} \in C_R(F_n)$, it suffices to show that $r_{ij}e_{pq} = e_{pq}r_{ij}$ for any p, q. Both sides of the latter may be equated to the same expression as follows.

$$r_{ij}e_{pq} = e_{ki}r(e_{jk}e_{pq}) = e_{ki}re_{jq}\delta_{kp} = e_{pi}re_{jq}; \qquad \text{and}$$
$$e_{pq}r_{ij} = (e_{pq}e_{ki})re_{jk} = \delta_{qk}e_{pi}re_{jk} = e_{pi}re_{jq}.$$

Thus $r_{ij} \in C_R(F_n)$.

Let $1_R = \mathrm{diag}(1, \ldots, 1) = e_{kk} \in F_n$ be the identity element of R which now is the identity matrix. If some elements $c^{\alpha\beta} \in C_R(F_n)$ are given and an element $d \in R$ is defined by $d = c^{\alpha\beta}e_{\alpha\beta}$, then

$$d_{ij} = e_{ki}c^{\alpha\beta}e_{\alpha\beta}e_{jk} = c^{\alpha\beta}\delta_{i\alpha}e_{k\beta}e_{jk} = c^{i\beta}\delta_{\beta j}e_{kk}$$
$$= c^{ij}(e_{11} + \cdots + e_{nn}) = c^{ij} \cdot 1_R = c^{ij}.$$

Then $(c^{\alpha\beta}e_{\alpha\beta})_{ij} = c^{ij}$.

Conversely, for any $r \in R$, first calculate the $\{r_{ij}\}$, and then form

$$r_{ij}e_{ij} = e_{ki}re_{jk}e_{ij} = e_{ki}re_{jj}\delta_{ki} = e_{ii}re_{jj}$$
$$= (e_{11} + \cdots + e_{nn})r(e_{11} + \cdots + e_{nn}) = 1_R \cdot r \cdot 1_R = r.$$

Thus for any $r \in R$, $r = r_{ij}e_{ij} \in C_R(F_n)F_n = F_nC_R(F_n)$, or $R = C_R(F_n)F_n$.

Our previous result 12-3.10 (with $A \to F_n$ and $B \to C(F_n)$) immediately becomes $R \cong F_n \otimes C_R(F_n)$. However, in keeping with the spirit of this direct computational first part of the proof, an illuminating elementary proof of the latter is given. Let $C_R(F_n)_n$ denote the ring of $n \times n$ matrices over $C_R(F_n)$. The equality $R = C_R(F_n)F_n$ implies that there is a ring homomorphism $C_R(F_n)_n \to R$ obtained by identifying the matrix units of $C_R(F_n)_n$ with the $e_{ij} \in F_n \subset R$. This homomorphism is evidently onto. To show that the kernel is zero it suffices to show that if an element d of the form $d = c^{\alpha\beta}e_{\alpha\beta}$ with all $c^{\alpha\beta} \in C_R(F_n)$ is zero $d = 0$, that then necessarily all $c^{ij} = 0$. But our previous calculation showed that in general $(c^{\alpha\beta}e_{\alpha\beta})_{ij} = c^{ij}$, and hence in particular $c^{ij} = e_{ki}de_{jk} = 0$ for all i, j. Hence $R \cong F_n \otimes C_R(F_n)$.

Now turn to the general case. By 12-3.7, for simplicity, we will identify $A \otimes A^{\text{op}} \equiv F_n$. Define T by

(1) $1_T \in F_n = A \otimes A^{\text{op}} \subset R \otimes A^{\text{op}} = T.$

By the previous case, since $F_n \subset T$,

(2) $T = C_T(F_n) \otimes F_n = C_T(F_n) \otimes A \otimes A^{\text{op}}.$

Note that the theorem would follow at once if only we could cancel the "A^{op}" in (1) and (2), and then equate the remainders. By 12-4.3, the centralizer of the central simple algebra $A^{\text{op}} = 1 \otimes A^{\text{op}}$ in T from (1) is

(3) $(1 \otimes A^{\text{op}})^{R \otimes A^{\text{op}}} = R \otimes 1.$

However the centralizer of the same exact algebra $A^{\text{op}} = 1 \otimes 1 \otimes A^{\text{op}}$ in the same big algebra T is

(4) $[1 \otimes 1 \otimes A^{\text{op}}]^{C(F_n; T) \otimes A \otimes A^{\text{op}}} = C_T(F_n) \otimes A \otimes 1.$

Comparison of (3) and (4) shows that

$$R \otimes 1 = C_T(F_n) \otimes A \otimes 1, \quad \text{and} \quad R = C_T(F_n) \otimes A.$$

The previous proof was computational and indirect in the sense that $R \otimes A^{\text{op}}$ was formed to obtain information about R alone. The next lemma will give us an alternate conceptual proof that $R = A \otimes C(A)$. It is also of independent interest, and useful in other contexts.

12-4.19 Lemma. *Consider a central simple algebra A over F which is a subring $F = F \cdot 1_R \subset A \subset R$ in any ring R whatever with $F \subseteq$ center R.*

View R either as a left $A \otimes A^{op}$ or right $A^{op} \otimes A$-module. Suppose that p, $q \in A \otimes A$ such that $q[A] = \{0\}$ and $0 \neq p[A] = F$. Then

(i) $q = 0$ in $A \otimes A$; and hence
(ii) $q[r] = 0$ for all $r \in R$.
(iii) $R^A = \{p(x) \mid x \in R\}$.

Proof. (i) and (ii). Define $I = \{q \in A \otimes A^{op} \mid \forall a \in A, \ q[a] = 0\}$. For any $t \in A \otimes A^{op}$, $(qt)[a] = q[t(a)] \in q[A] = \{0\}$, and hence $qt \in I$. It is even simpler to see that $tq \in I$, and hence $I \lhd A \otimes A^{op}$ is an ideal. Since $1 \otimes 1 \notin I$, and since $A \otimes A^{op}$ is a simple ring, $I = 0$. But $q \in I$, and thus $q = 0$.

(iii) For any $a \in A$, define $q = (a \otimes 1)p - p(1 \otimes a)$, where the multiplication is in $A \otimes A$, and hence $q[x] = ap(x) - p(x)a$ for all $x \in R$. Since $q[A] = 0$, $q = 0$ by (i). Hence $q[x] = 0$ for all $x \in R$. Therefore $ap(x) = p(x)a$ for all $a \in A$. Hence $p(x) \in R^A$.

For the converse first take any $d \in A$ with $0 \neq p[d] = f \in F$. Set $c = f^{-1}d$. Then $p[c] = 1$. Now for any $z \in R^A$, $p[cz] = p[c]z = zp[c] = z$ because z can be interchanged with all the elements of A forming p. Thus $R^A = p[R]$.

An alternate proof of a somewhat different formulation of Theorem 12-4.19 is given next.

12-4.20 Theorem. *Let A be a central simple algebra over F of dimension $[A:F] = n$, and let $\{a_1, \ldots, a_n\} \subset A$ be any F-vector space basis of A. Suppose that R is any ring with $F \cdot 1 = F \cdot 1_R \subseteq \text{center } R$, and with $1 = 1_R \in A \subset R$. Then as a right R^A-module, R is free on $\{a_1, \ldots, a_n\}$; i.e.*

(i)   *every $x \in R$ is of the form $x = a_1 b_1 + \cdots + a_n b_n$, where all the $b_i \in R^A$ are unique.*
(ii) $R = A \cdot R^A \cong A \otimes R^A$.
(iii) $\text{center } R^A = \text{center } R \cong F \otimes (\text{center } R)$.

Proof. (i) There exist elements $t_i \in A \otimes A$ with $t_i[a_j] = \delta_{ij} \cdot 1$ for i, $j = 1, \ldots, n$. Set $p = (a_1 \otimes 1)t_1 + \cdots + (a_n \otimes 1)t_n - 1 \otimes 1 \in A \otimes A$ where the products $(a_i \otimes 1)t_i$ are formed in $A \otimes A$ also. For any j,

$$p[a_j] = \sum_{i=1}^{n} a_i t_i[a_j] - 1 \cdot a_j \cdot 1 = 0.$$

Each t_i maps A onto $t_i[A] = F$, and so $t_i[R] \in R^A$ by the previous lemma. But $p[A] = 0$. Since A is central simple, also $p = 0$. Then for any

$x \in R$, $p[x] = 0$ of course, but this means that $\Sigma a_i t_i[x] - x = 0$, or $x = \Sigma a_i t_i[x] \in a_1 R^A + \cdots + a_n R^A$ with all $t_i[x] \in R^A$.

Suppose that $x = a_1 b_1 + \cdots + a_n b_n \in R$ for some $b_i \in R^A$. The b_j commute with all the elements of A forming the t_i. Thus each b_i is uniquely determined by x alone by

$$t_i[x] = t_i[a_1]b_1 + \cdots + t_i[a_n]b_n = b_i.$$

(ii) This follows immediately from general principles (12-3.10). However, conclusion (i) allows us to construct an explicit concrete isomorphism $R \cong A \otimes R^A$ by mapping $x = a_1 b_1 + \cdots + a_n b_n \to a_1 \otimes b_1 + \cdots + a_n \otimes b_n$, $b_i \in R^A$. By (i), the map is well defined and it is easy to see that it is an onto ring homomorphism. If x is in the kernel of this map, then $\Sigma a_i \otimes b_i = 0$. The linear independence of the a_i forces each $b_i = 0$ also, and hence $x = 0$. Thus $R \cong A \otimes R^A$.

(iii) Set $Z = \text{center } R$. Always for any $A \subset R$, $Z \subseteq R^A$, and hence $Z \subseteq \text{center } R^A$. Since $R = AR^A$, center $R^A \subseteq Z$. Hence $Z = \text{center } R^A$. Since $F \cdot 1 \subseteq Z \cap A \subseteq \text{center } A = F \cdot 1$, $Z \cap A = F$. Now select a more special basis of A than was necessary for (i), and (ii), namely with $a_1 = 1 \in R$. Then under the isomorphism $R \cong A \otimes R^A$ in (ii), $Z = a_1 Z$ is mapped onto $a_1 \otimes Z = 1 \otimes Z$ as was to be shown.

12-5   Double centralizers

To begin with the dimension $[R:S]$ of a ring R over a simple Artinian subring $S \subset R$ is constructed. It then becomes clear that $[R:S]$ behaves in many ways like the more familiar dimension of a ring over a subfield.

The main application of this notion is in the double centralizer theorem. However, a large part of the content of this theorem and almost all of its corollaries do not use dimensions over simple subrings. The statement and proof of these deliberately have been so formulated that they are understandable to someone unfamiliar with this more complicated dimension.

12-5.1   Lemma. *Let S be a simple Artinian ring which is a subring $S \subset R$ of a ring R. Suppose that P_1, $P_2 \subset S$ are any minimal right S-ideals and that $\phi: P_1 \to P_2$ is any isomorphism of right S-modules. Then ϕ extends to a unique right R-module isomorphism $\bar{\phi}: P_1 R \to P_2 R$ of the right R-ideals $P_1 R$, $P_2 R \leqslant R$.*

Proof. It is possible to express $S = e_1 S \oplus e_2 S \oplus \cdots \oplus e_m S$, where $1_S = e_1 + e_2 + \cdots + e_m$ is a decomposition of the identity element of S into orthogonal idempotents $e_i e_j = \delta_{ij} e_i \in S$, such that two additional conditions hold. First, $P_1 = e_1 S$, and secondly, if $P_1 \neq P_2$, then $P_2 = e_2 S$. If $P_1 = P_2$, then of course $P_2 = e_1 S$ also. Due to the orthogonality of these idempotents, $e_1 R \oplus e_2 R \oplus \cdots \oplus e_m R \leqslant R$ is also a direct sum of right ideals of R with $P_1 R = e_1 R$ and $P_2 R = e_2 R$.

Irrespective of whether $P_1 = P_2$, or $P_1 \neq P_2$ there is some $t \in S$ such that $\phi e_1 = e_2 t e_1$, and $\phi(e_1 s) = e_2 t e_1 s$ for all $s \in S$. Define $\bar{\phi} : e_1 R \to e_2 R$ the same way by $\bar{\phi}(e_1 r) = e_2 t e_1 r$ for $r \in R$. The uniqueness of $\bar{\phi}$ follows from the abelian group isomorphisms and subgroup inclusion

$$\operatorname{Hom}_S(e_1 S, e_2 S) \cong e_2 S e_1 \subset e_2 R_1 \cong \operatorname{Hom}_R(e_1 R, e_2 R).$$

The structure of two important types of S-submodules P and Q of R is determined next.

12-5.2 For $S \subset R$ as in the previous lemma, suppose that $P \subset R$ is a subgroup which is a simple right S-submodule of R, and that $Q \subset R$ is a right S-submodule that is isomorphic to S as a right S-module.

(i) Let $eS \subset S$ with $e^2 = e \in S$ be any minimal right ideal of S whatever, and $\phi : eS \to P$ any right S-module isomorphism. Set $y = \phi e \in R$. Thus $y = ye$, and $P = yeS$ and $\phi(es) = ys$, where $ys = 0$ for $s \in S$ if and only if $es = 0$.

(ii) For any isomorphism $\phi : S \to Q$ of right S-modules, set $x = \phi 1_s$. Then $S \to xS = Q$, $s \to xs$, $s \in S$, is an isomorphism with $x^\perp \cap S = 0$.

(iii) As before, let $1_s = e_1 + \cdots + e_m$ be orthogonal idempotents so that $S = e_1 S \oplus \cdots \oplus e_m S$ is a direct sum of minimal right ideals. Next, assume that $P_1, \ldots, P_m \subset R$ are completely arbitrary simple S-submodules of R subject only to the restriction that their sum $Q = P_1 \oplus \cdots \oplus P_m \subset R$ be direct. It will be shown that $Q_S = xS \cong S_S$ for some $x \in R$.

For each i, and some $y_i \in R$, there is a right S-isomorphism $e_i S \to y_i S = P_i$ with $e_i \to y_i = y_i e_i$. Thus the sum of these gives an isomorphism of right S-modules $S = e_1 S \oplus \cdots \oplus e_m S \to Q$, where for $s_i \in S$, $e_1 s_1 + \cdots + e_m s_m \to y_1 s_1 + \cdots + y_m s_m$. In particular, $1_s = e_1 + \cdots + e_m \to x \equiv y_1 e_1 + \cdots + y_m e_m = y_1 + \cdots + y_m$. Thus $S \cong xS = P_1 \oplus \cdots \oplus P_m$ as right S-modules.

12-5.3 For $S \subset R$ as before suppose that eS and fS are minimal right ideals of S with $e^2 = e$, $f^2 = f \in S$. Since every S-module is a direct sum of simple S-modules, in view of 12-5.2(i), $fR = \oplus\{y_\alpha S \,|\, \alpha \in \theta\}$ for some $y_\alpha = y_\alpha e \in R$ which depend on our choice of e.

12-5.4   Dimension of a ring over simple subring. For rings $1 \in S \subset R$ now both having the same joint identity element 1, suppose that S is simple Artinian. Let $1 = e_1 + \cdots + e_m$, $e_i e_j = \delta_{ij} e_i \in S$ be arbitrarily chosen primitive orthogonal idempotents so that $S = e_1 S \oplus \cdots \oplus e_m S$ is a direct sum of minimal right S-ideals $e_i S$ of S. In particular, if $eS \subset S$ is a minimal right S-ideal with $e^2 = e \in S$, then the e_i may be so chosen that $e_1 = e$.

By 12-5.1, $R = e_1 R \oplus \cdots \oplus e_m R$ is a direct sum of isomorphic $e_1 R \cong \cdots \cong e_m R$ right R-ideals of R; and hence they are automatically isomorphic as right S-modules.

Since every S-module is a direct sum of simple S-modules, in view of 12-5.3, each $e_i R = \oplus\{y_{\alpha i} S \,|\, \alpha \in \theta\}$ for some $y_{\alpha i} \in R$ with $y_{\alpha i} = y_{\alpha i} e_i$. Since all the $e_i R$ are isomorphic as S-modules, we may take the indexing set θ to be the same for all i.

For each $\alpha \in \theta$, define $x_\alpha = y_{\alpha 1} + \cdots + y_{\alpha m}$. As seen in 12-5.2(iii), $S \cong x_\alpha S$ as right S-modules. However, now $R = \oplus\{x_\alpha S \,|\, \alpha \in \theta\}$. Thus R is a free right S-module with basis $\{x_\alpha \,|\, \alpha \in \theta\}$. In particular, for elements $s_\alpha \in S$, $\Sigma x_\alpha s_\alpha = 0$ if and only if all $s_\alpha = 0$ individually.

Define the right dimension $[R:S]$ of R over S to be the cardinal number $[R:S] = |\theta|$.

There are several ways to see that $|\theta|$ is unique and independent of the various arbitrary choices of elements made in its definition. Since $R = \oplus x_\alpha S$, R can be written as a direct sum of $m|\theta|$ simple right S-modules. By the Krull-Remark-Schmidt theorem, $m|\theta|$ is unique, and hence so is $|\theta|$.

Another way of seeing this is to use the fact that the rank such as $|\theta|$ of a free module is unique, provided that it is infinite. If θ is finite (or infinite for that matter), then it follows from Wedderburn's theorem applied to S that R is a left vecctor space over a division ring. Again, it could be argued from the uniqueness of this vector space dimension that also $|\theta|$ is unique.

If A is an algebra with identity over F containing a subfield K with $1 \in F \subset K \subset A$, then $[A:F] = [A:K][K:F]$. The latter will next be generalized from fields $F \subset K$ to simple Artinian rings $T \subset S$.

Now suppose that in addition $1 = 1_R \in T \subset S \subset R$ where S and T are simple subrings of R both with the D.C.C. Express as above $R = \oplus\{x_\alpha S \,|\, \alpha \in \theta\}$ and $S = \oplus\{y_\beta T \,|\, \beta \in \mathscr{L}\}$, where $\{x_\alpha\}$ is a free S-basis of R, while $\{y_\beta\}$ is a free T-basis of S. Suppose that $x_\alpha y_\beta t = 0$ for some $t \in T$ and some α, β. Then $x_\alpha s = 0$ with $s = y_\beta t \in S$ implies that $s = y_\beta t = 0$, which in turn implies that $t = 0$. This shows that each $x_\alpha y_\beta T$ is a free right T-module. From this it now follows that R is free on $\{x_\alpha y_\beta\}$ as a right T-module,

$$R = \bigoplus_{\alpha \in \theta} \bigoplus_{\beta \in \mathscr{L}} x_\alpha y_\beta T,$$

$$[R:T] = |\theta \times \mathscr{L}| = |\theta||\mathscr{L}| = [R:S][S:T].$$

12-5.4   Lemma. *Let R be any ring and $A \subset R$ and $B \subset R$ two simple subrings both with the D.C.C. Suppose that A is mapped isomorphically onto B by some ring automorphism of R. Then $[R:A] = [R:B]$.*

Proof. Let $\phi: R \to R$ with $B = \phi A$ be the ring automorphism. Express $R = \oplus\{x_\alpha A \mid \alpha \in \theta\}$ where $\{x_\alpha\}$ is a free basis for the right A-module R with $[R:A] = |\theta|$. Then

$$R = \phi R = \bigoplus_{\alpha \in \theta}(\phi x_\alpha)\phi A = \bigoplus_{\alpha \in \theta}(\phi x_\alpha)B.$$

Since $B \to (\phi x_\alpha)B$, $b \to (\phi x_\alpha)b$ is a right B-module isomorphism, R is free on $\{\phi x_\alpha\}$ as a right B-module.

12-5.5   Lemma. *Suppose that $1 = 1_R \in S \subset R$ and T are three algebras over F with identity where S and T are simple rings with the D.C.C. Assume also that $S \otimes T$ is simple with the D.C.C. (This will hold if one of S and T is central over F, and one of $[S:F]$ or $[T:F]$ is finite.) Then*

$$[R \otimes T : S \otimes T] = [R:S].$$

Proof. Let $\{x_\alpha\}$ be a free right S-basis of $R = \oplus\{x_\alpha S \mid \alpha \in \theta\}$ with $[R:S] = |\theta|$. Each $(x_\alpha S) \otimes T = (x_\alpha \otimes 1)(S \otimes T)$ is a right $S \otimes T$-submodule of $R \otimes T$. Suppose that $(x_\alpha \otimes 1)\Sigma s_i \otimes t_i = 0$. Take the t_i linearly independent over F. Then $\Sigma x_\alpha s_i \otimes t_i = 0$ requires that each $x_\alpha s_i = 0$, and hence that all $s_i = 0$. Hence $(x_\alpha \otimes 1)(S \otimes T) \cong S \otimes T$ are naturally isomorphic. Since tensor products commute with direct sums, we get

$$R \otimes T = \left[\bigoplus_\alpha x_\alpha S\right] \otimes T = \bigoplus_\alpha[(x_\alpha S) \otimes T] = \bigoplus_\alpha(x_\alpha \otimes 1)(S \otimes T).$$

Thus $\{x_\alpha \otimes 1\}$ is a free right $S \otimes T$-basis of $R \otimes T$, and $[R \otimes T : S \otimes T] = |\theta| = [R:S]$.

If the algebra A below were of finite dimension $[A:F] = n$, then by selecting a vector space basis of A over F, the next construction could have been done more concretely in the ring F_n as opposed to the more abstract ring $\text{End}_F A$.

12-5.6 Construction. Suppose that A is any algebra with identity over F. View $_F A$ as an F-vector space. Each element $a \in A$ defines linear transformations L_a, $R_a : {}_F A \to {}_F A$ written on the right $vL_a = av$ and $vR_a = va$ for $v \in {}_F A$. The map $a \to R_a$ maps A isomorphically onto a subalgebra $\mathscr{R} A \subset \text{End}_F A$, $A \cong \mathscr{R} A$, where $\text{End}_F A$ is the ring of all linear transformations of $_F A$, written on the right. Similarly $A \to \mathscr{L} A$, $a \to L_a$ is an F-algebra anti-isomorphism $A \cong (\mathscr{L} A)^{op}$ of A onto the subalgebra $\mathscr{L} A \subset \text{End}_F A$. (If the linear transformations in $\text{End}_F A$ were written on the left, then we would have $A \cong \mathscr{L} A \cong (\mathscr{R} A)^{op}$.) It is useful to realize that $(\mathscr{L} A)^{op}$ is not a subring of $\text{End}_F A$, but that $A \cong \mathscr{R} A$ and $A^{op} \cong \mathscr{L} A$ are two subrings of the ring $\text{End}_F A$ with $F \subseteq \mathscr{R} A \cap \mathscr{L} A = \{R_z = L_z \,|\, z \in \text{center } A\}$. Next, it is shown that the inclusions $\mathscr{R} A \subseteq C(\mathscr{L} A; \text{End}_F A) \equiv C(\mathscr{L} A)$ and $\mathscr{L} A \subseteq C(\mathscr{R} A; \text{End}_F A) \equiv C(\mathscr{R} A)$ are equalities.

Let $T : {}_F A \to {}_F A$ be a linear transformation. Set $c = 1T \in A$.

If $T \in C(\mathscr{R} A)$, then $vR_a T = vTR_a$ for all $v \in {}_F A$ and $a \in A$. Thus $vT = (1v)T = (1R_v)T = (1T)R_v = cv = vL_c$. Hence $T = L_c \in \mathscr{L} A$.

On the other hand, if $T \in C(\mathscr{L} A)$, then $vT = (v1)T = (1L_v)T = (1T)L_v = cL_v = vc$, and $T = R_c \in \mathscr{R} A$. Thus

$$C(\mathscr{R} A; \text{End}_F A) = \mathscr{L} A \quad \text{and} \quad C(\mathscr{L} A; \text{End}_F A) = \mathscr{R} A.$$

12-5.7 Remark. If in the previous construction in addition $[A:F] = n < \infty$ and A is central simple over F, then $A \otimes A^{op} \cong F_n$, and $A \otimes A^{op} \cong \mathscr{R} A \otimes \mathscr{L} A \cong \mathscr{R} A \mathscr{L} A = \text{End}_F A$.

The type of situation to be encountered in the double centralizer theorem is isolated and then put in a more general framework below, in order to obtain a broader perspective.

12-5.8 Suppose that (a) R and E are central simple algebras over F, that (b) R has the D.C.C. while $[E:F] \not\gtreqless \infty$ is finite. Suppose further that (c) $F \subseteq A \subset R$ and $F \subseteq B \subset E$ are isomorphic under an isomorphism leaving F elementwise fixed.

These hypotheses guarantee that $R \otimes E$ is central simple over F with the D.C.C., and that $A \otimes 1 = t^{-1}(1 \otimes B)t$ for some element $t \in R \otimes E$.

The commutators of these two algebras are also conjugate under the same inner automorphism.

(1) $(R \otimes E)^{A \otimes 1} = R^A \otimes E$, $(R \otimes E)^{1 \otimes B} = R \otimes E^B$,

   $R^A \otimes E = t^{-1}(R \otimes E^B)t$.

Once more verbatim by the same argument, since now $R^A \otimes E$ and $R \otimes E^B$ are conjugate, so also are their commutators in turn:

(2) $(R \otimes E)^{R^A \otimes E} = C_R(R^A) \otimes 1$, $(R \otimes E)^{R \otimes E^B} = 1 \otimes C_E(E^B)$,

   $C_R(R^A) \otimes 1 = t^{-1}(1 \otimes C_E(E^B))t$

With the same notation and hypotheses as above, now assume in addition that (d) R^A, E^B, $R^A \otimes E$, and $R \otimes E^B$ are all simple with the D.C.C.

Then first, 12-5.4 shows that

$$[R \otimes E : R^A \otimes E] = [R \otimes E : R \otimes E^B].$$

Then by 12-5.5, $[R \otimes E : R^A \otimes E] = [R : R^A]$ and $[R \otimes E : R \otimes E^B] = [E : E^B]$. Consequently

(3) $[R : R^A] = [E : E^B]$.

12-5.9   Double centralizer theorem. Suppose that R is central simple over F, and suppose that $A \subset R$ is a simple subalgebra containing F in its center $F \subseteq A \subset R$. Assume that R has the D.C.C. while the dimension $[A : F]$ of A over F is finite. Let $E = \text{End}_F A$ be the ring of all F-linear transformations of A, and $A \to \mathcal{R}A \subset E$, $a \to R_a$, and $A^{\text{op}} \to \mathcal{L}A \subset E$, $a \to L_a$ be the isomorphisms as in 12-5.6. Then for some invertible element $t \in R \otimes E$, the following hold:

(i) R^A is simple;
(ii) $C_R(R^A) = A$;
(iii) $R^A \otimes E = t^{-1}(R \otimes \mathcal{L}A)t$;
(iv) R^A has the D.C.C.;
(v) $[R : R^A] = [A : F] = [E : \mathcal{L}A] = [E : \mathcal{R}A]$.

Proof. If it were the case that $A = \mathcal{R}A$ while $R = E$, then (i) and (ii) of the theorem have already been proved in the last construction 12-5.6. For in this case, (i) $C(\mathcal{R}A : E) = \mathcal{L}A \cong A^{\text{op}}$ is simple; and (ii) $C(C(\mathcal{R}A : E) : E) = C(\mathcal{L}A : E) = \mathcal{R}A$.

(i), (ii), (iii). Set $n = [A : F]$. Since R and E are central simple over F, and since R has the D.C.C. while $[E : F] = n^2$ is finite, $R \otimes E$ is simple with the D.C.C., and with center $R \otimes E = F$. The two subalgebras $A \otimes 1$, $1 \otimes \mathscr{R}A \subset R \otimes E$ are isomorphic $A \otimes 1 \cong 1 \otimes \mathscr{R}A$ under $a \otimes 1 \to 1 \otimes R_a$, $a \in A$; with $F \subseteq (A \otimes 1) \cap (1 \otimes \mathscr{R}A)$ (where the latter actually is an equality). By 12-4.14, there exists a $t \in R \otimes E$ such that

(1) $\quad A \otimes 1 = t^{-1}(1 \otimes \mathscr{R}A)t$

Now the idea of the proof is to take double commutators in the big algebra $R \otimes E$ of both sides of (1) by using 12-4.2 and 12-4.4. The latter just says that inner automorphisms by t commute with the operation of taking the commutator, while 12-4.2 tells us that we can take the commutators of tensor products by taking commutators of their factors separately. Thus

(2) (a) $\quad (R \otimes E)^{A \otimes 1} = R^A \otimes E$, (b) $(R \otimes E)^{R^A \otimes E} = C_R(R^A) \otimes 1$;

(3) (a) $\quad (R \otimes E)^{t^{-1}(1 \otimes \mathscr{R}A)t} = t^{-1}[(R \otimes E)^{1 \otimes \mathscr{R}A}]t = t^{-1}(R \otimes \mathscr{L}A)t$,

(b) $\quad (R \otimes E)^{t^{-1}(R \otimes \mathscr{L}A)t} = t^{-1}[(R \otimes E)^{R \otimes \mathscr{L}A}]t = t^{-1}(1 \otimes \mathscr{R}A)t$.

Two important facts must now be put together. First, since (2) and (3) were computing commutators of the same algebra written in two ways $A \otimes 1 = t^{-1}(1 \otimes \mathscr{R}A)t$, the end results 2(b) and 3(b) are equal. Secondly, due to the very special nature of the algebra $1 \otimes \mathscr{R}A$, it turned out that it was equal to its own second commutator, which in turn by (1) is equal to $A \otimes 1$. Thus

$$C_R(R^A) \otimes 1 = A \otimes 1, \quad \text{or} \quad C_R(R^A) = A.$$

However, 2(a) and 3(a) are also equal, and hence $R^A \otimes E = t^{-1}(R \otimes \mathscr{L}A)t \cong R \otimes A^{\mathrm{op}}$. Since $R \otimes A^{\mathrm{op}}$ is simple, so is also R^A by 12-3.9.

(iv) Since R has the D.C.C. and $[\mathscr{L}A:F] = [A:F]$ is finite, so does also $R \otimes \mathscr{L}A$, and hence also $R \otimes \mathscr{L}A \cong R^A \otimes E$. For any two distinct right ideals $L_1 \neq L_2$ of R^A, $L_1 \otimes E \neq L_2 \otimes E$ are two distinct right ideals of $R^A \otimes E$. Thus R^A also has the right D.C.C.

(v) In view of (iii), first 12-5.4 shows that

$$[R \otimes E : R^A \otimes E] = [R \otimes E : R \otimes \mathscr{L}A]$$

Then by 12-5.5, $[R \otimes E : R^A \otimes E] = [R : R^A]$, and $[R \otimes E : R \otimes \mathscr{L}A] = [E : \mathscr{L}A]$. Thus $[R : R^A] = [E : \mathscr{L}A]$.

Now use the multiplicative property of dimensions as in 12-5.4 to get

$$n^2 = [E:F] = [E:\mathscr{L}A][\mathscr{L}A:F] = [E:\mathscr{L}A]n, \quad \text{and}$$

$$n^2 = [E:\mathscr{R}A][\mathscr{R}A:F] = [E:\mathscr{R}A]n.$$

Consequently

$$[R:R^A] = [E:\mathscr{L}A] = [E:\mathscr{R}A] = n.$$

12-5.10 Corollary 1 to theorem 5.9. *Under the hypotheses and with the notation of the last theorem, if* $n = [A:F]$, *then* $E \cong F_n$, *and in particular* $R^A \otimes F_n \cong R \otimes A^{\mathrm{op}}$.

In the next corollary every algebra that appears in the double centralizer theorem is central simple.

12-5.11 Corollary 2 to theorem 5.9. *If for* $F \subset A \subset R$ *as in the theorem, in addition center* $A = F$, *then also center* $R^A = F$.

Proof. Since now both R and A are central simple over F, also center$(R \otimes A^{\mathrm{op}}) = F$. By the last corollary, $[\text{center } R^A] \otimes F \subseteq$ center$[R^A \otimes F_n] = F$. The latter requires that center $R^A = F$.

12-5.12 Corollary 3 to theorem 5.9. *In the theorem, assume in addition that* $[R:F] \neq \infty$ *is finite. Then* $[R:F] = [R^A:F][A:F]$.

Proof. Equating the dimensions of both sides of $R^A \otimes F_n \cong R \otimes A^{\mathrm{op}}$ gives $[R^A:F]n^2 = [R:F][A:F]$. Since $[A:F] = n$, the result follows.

The next corollary generalizes 12-4.11.

12-5.13 Corollary 4 to theorem 5.9. *Suppose that under the conditions of the theorem on* $F \subset A \subset R$, *in addition* A *is a subfield having the property that* $[R:F] = [A:F]^2$. *Then*

   (i) $A = R^A$;

  (ii) A *is maximal among commutative subrings of* R *(and hence a fortiori* A *is a maximal commutative subfield of* R).

Proof. The previous corollary combined with the special assumptions now becomes

$$[R:F] = [A:F]^2 = [R^A:F][A:F].$$

Thus $[R^A:F] = [A:F]$. Because A is commutative $A \subseteq R^A$, and $[A:F] \leqslant [R^A:F]$. Hence $A = R^A$.

12-6   Exercises

Prove the following assertions

1. For a field F, let $F(x)$, $F(y)$, and $F(x, y)$ be rational function fields. (i) Show that there is an embedding $F(x) \otimes_F F(y) \subset F(x, y)$

under the identification $\alpha \otimes \beta \equiv \alpha\beta$ for $\alpha \in F(x)$, $\beta \in F(y)$. (ii) Show that the latter inclusion is proper. (iii) Prove that $F(x) \otimes_F F(y)$ does not satisfy the descending chain condition by use of 12-4.13(iv) Give an alternate proof of this using the fact that a ring with the D.C.C. and without zero divisors is a division ring.

2. If $\mathbf{H}$ are the quaternions over the reals $\mathbf{R}$, then $\mathbf{H} \otimes_{\mathbf{R}} \mathbf{H} \cong \mathbf{R}_4$ where the latter is the ring of 4×4 matrices over the reals $\mathbf{R}$.

3. Let $F[x, y]$ be the ring of noncommutative polynomials in indeterminates x and y subject only to the relation $xy - yx = 1$. (i) Then $F[x, y]$ is a simple ring. (ii) Prove that the right ideal $x^2F[x, y] + (xy + 1)F[x, y] = x^2F[x, y] + (xy + 1)F[y]$ is non-principal. ([Rinehart 62; p. 343]).

The next definition will give a class of simple finite dimensional algebras for some of the following exercises.

Definition. Suppose that $F \subset K$ is a finite, normal, separable extension of fields with a cyclic Galois group $G(K/F) = \{\sigma, \ldots, \sigma^{m-1}, \sigma^m = 1\}$ of order m. Write $\sigma(n) = \sigma^n$ and $k\sigma(n) = k^{\sigma(n)}$ for $k \in K$. Take any $0 \neq g \in F$. A *cyclic algebra* is an algebra over F denoted by $(K/F, \sigma, g)$ which is generated by K and an element $y \notin K$ according to the following multiplication extended by linearity and distributivity:

$$(K/F, \sigma, g) = K + yK + \cdots + y^{m-1}K; \quad ky = yk^{\sigma}, \ y^m = g, \ k \in K.$$

4. $(K/F, \sigma, g)$ is an associative algebra over F satisfying $ky^i = y^ik^{\sigma(i)}$ for $i \leqslant m - 1$, $y^{-1} = y^{m-1}g^{-1}$, and $y^{-1}ky = k^{\sigma}$ for $k \in K$.

5. If $\Delta = (K/F, \sigma, g)$, then center $\Delta = F$; Δ is simple; the centralizer of K in Δ is K itself; $K \subset \Delta$ is a maximal subfield; and $\dim_F \Delta = [\Delta : K][K : F] = m^2$. (See [Dauns 82; p. 48].)

6. For $(K/F, \sigma, g)$ suppose that $m = 2$ and $g = cc^{\sigma} \in F$ for some $c \in K$. Show that $(K/F, \sigma, g)$ has zero divisors.

7. The real quaternions $\mathbf{H}_{\mathbf{R}}$ are a cyclic algebra.

8. Generalize the construction of a cyclic algebra as follows. Let B be an F-algebra with $1 \in B$, and $\sigma: B \to B$ an algebra endomorphism. Let $g \in B$ with $g\sigma = g$, and $bg = gb^{\sigma(m)}$ for all $b \in B$. Define $A = B + yB + \cdots + y^{m-1}B$; $y^m = g$, $by = yb^{\sigma}$, $b \in B$.
   (i) Verify that A is an associative algebra over F.
   (ii) Explain how a cyclic algebra $(K/F, \sigma, g)$ is a special case of the above A.

Remark. General hypotheses on B can be found so that A is simple, see [Dauns 82a; p. 46, Theorem 111-1.15].

9. For an automorphism $\theta: K \to K$ of a skew field K, let $K[x, x^{-1}; \theta]$ be the subring of $K(x; \theta)$ generated by $K[x; \theta]$ and x^{-1}. Define $\theta(-1) = \theta^{-1}$ and $\theta(-n) = \theta^{-n}$ for $n = 0, 1, 2, \ldots$.
    (i) Show that $K[x, x^{-1}; \theta]$ can alternatively be defined as the ring of all polynomials in x and x^{-1} with right side K-coefficients subject to

    $$xx^{-1} = x^{-1}x = 1, \ kx = xk^\theta, \ kx^{-1} = x^{-1}k^{\theta(-1)}, \ k \in K$$

    (ii) Show that $K[x, x^{-1}; \theta]$ is not a division ring.

10. For $K[x, x^{-1}; \theta]$ as above assume that for some integer $1 \leqslant p$, $\theta(p)$ is inner, i.e. there exists a fixed $d \in K$ such that $k\theta(p) = d^{-1}kd$ for all $k \in K$. Next, suppose that $c \in \operatorname{center} K$ such that $(d^{-1}c)\theta = d^{-1}c$. Then
    (i) $x^p d^{-1} c K[x, x^{-1}; \theta] \lhd K[x, x^{-1}; \theta]$.
    (ii) Suppose that for an integer $m = \pm 1, \pm 2, \ldots, c_0 \in \operatorname{center} K$ such that $(d^{-m}c_0)\theta = d^{-m}c_0$. Then $x^{mp}d^{-m}c_0 K[x, x^{-1}; \theta] \lhd K[x, x^{-1}; \theta]$.
    (iii) Guess what is the form of the most general ideal of $K[x, x^{-1}; \theta]$.

11. For $K[x, x^{-1}; \theta]$ as in exercise 10, assume that for no nonzero integer $1 \leqslant p$ is the p-th power $\theta(p)$ of θ an inner automorphism of K by an element of K. Then $K[x, x^{-1}; \theta]$ is a central simple algebra over the field $F = \{c \in \operatorname{center} K \mid c\theta = c\}$.

12. Suppose that K is a skew field and $\delta: K \to K$ a derivation. (i) Show that $K_\delta = \{k \in K \mid \delta k = 0\}$ is a skew subfield of K. (ii) If in addition the characteristic of K is zero, then K_δ contains the rational numbers. (iii) If $K[x; \theta]$ is the skew polynomial ring, then center $K[x; \theta] = F$, where F is the field $F = K_\delta \cap \operatorname{center} K$.

13. Suppose that K is a skew field and $\delta: K \to K$ is a derivation which is not the inner derivation by any element of K. Then $K[x; \delta]$ is a central simple algebra over the field F of exercise 12. (See [Dauns 82a; p. 224].)

CHAPTER 13

Hereditary rings, free and projective modules

Introduction

Thus far the following classes of rings have been studied or mentioned: prime, primitive, simple, semiprimitive (right) Artin, division ring, field, and commutative principal right ideal domain (3-2.4). The last four are special instances of hereditary rings. Finitely generated modules are discussed and then applied to semihereditary rings.

Aside from their own intrinsic interest, the semiprimitive Artin rings, and the right hereditary rings are fundamental building blocks in terms of which other rings are understood. Similarly, modules with a satisfactory structure theory, or modules which appear frequently are standard concepts needed to understand more general modules. Certainly, finitely generated modules fall in this class. Just as the word 'Artinian' refers ambiguously to either the right or left chain condition, so also by a hereditary ring usually is meant either a left or right hereditary one, and here the latter.

13-1 Hereditary rings

Throughout this chapter R is a ring with an identity element.

13-1.1 Definition. A ring R is right *hereditary* if $1 \in R$ and provided each of its right ideals is projective. A ring is *right semihereditary* if each finitely generated right ideal is projective.

The next lemma holds for any ring R.

13-1.2 Lemma. *Let $C < N \oplus B$ be modules with $(C + N)/N$ projective.*

269

Then there exists a submodule $A \leqslant N \oplus B$ *such that* $C = (N \cap C) \oplus A$ *and,* $A \cong (C + N)/N$.

Equivalently, if $\pi \colon N \oplus B \to B$ *is the natural projection, then* $C = (N \cap C) \oplus A$ *where* $\pi(C) \cong A < N \oplus B$.

Proof. Since any surjective map $C \to (C + N)/N$ with a projective image splits, $C = (C \cap N) \oplus A$ for some $A \leqslant C$ with $A \cong (C + N)/N$.

13-1.3   Theorem. *Let R be a right hereditary ring. Then any submodule of a free right R-module is isomorphic to a direct sum of right ideals of R.*

I.e., if $F = \oplus \{R_\alpha \mid 0 \leqslant \alpha < \tau\}$ *is free with all* $R_\alpha \cong R$ *for some ordinal* τ, *and* $C < F$ *is arbitrary, then* $C \cong \oplus \{L_\alpha \mid 0 \leqslant \alpha < \tau\}$ *where* $L_\alpha \leqslant R$, $\alpha < \tau$.

Proof. For any $0 \leqslant \beta < \tau$, define $C_\beta = C \cap \oplus \{R_\alpha \mid \alpha < \beta\}$. Since $\oplus \{R_\alpha \mid \alpha < 0\}$ is the empty sum, $C_0 = 0$. Thus $C_\beta \subseteq C_{\beta+1} = C \cap \oplus \{R_\alpha \mid \alpha \leqslant \beta\}$ and $C_\tau = C$. Let $\pi_\beta \colon F \to R_\beta$ be the natural projection, and $p_\beta = \pi_\beta \mid C_{\beta+1} \colon C_{\beta+1} \to R_\beta$ be restriction of π_β to $C_{\beta+1}$.

Since $L_\beta = p_\beta(C_{\beta+1}) < R$, by hypothesis the image of p_β is projective. Hence the surjective map $p_\beta \colon C_{\beta+1} \to p_\beta(C_{\beta+1})$ splits. The kernel of p_β is $\ker p_\beta = C_\beta$. By 13-1.2, $C = C_\beta \oplus A_\beta$ where $A_\beta \cong p_\beta(C_{\beta+1}) = L_\beta$. For any β and $\alpha \leqslant \beta$, $A_\alpha \leqslant C_\beta$.

If $\beta = 1$, then $C_1 = C_0 \oplus A_0 = A_0 = C \cap R_0 < R$ as $C_0 = 0$. If $\beta = 2$, $C_2 = C_1 \otimes A_1 = A_0 \oplus A_1$, etc. for an integer n, $C_{n+1} = C_n \oplus A_n = C_{n-1} \oplus A_{n-1} \oplus A_n = A_0 \oplus A_1 \oplus \cdots \oplus A_n$, where $A_i \cong L_i \leqslant R$, $0 \leqslant i \leqslant n$. Thus for any β, $C_\beta \supseteq A_\alpha$ for all $\alpha < \beta$.

It will be shown that C is the direct sum of all the A_α's. First, if ΣA_α is not direct, then for some integer n, and some $0 \neq a_{\alpha(j)} \in A_{\alpha(j)}$, $a_{\alpha(1)} + \cdots + a_{\alpha(n)} = 0$ where $\alpha(1) < \cdots < \alpha(n)$. But then

$$0 \neq a_{\alpha(n)} = -a_{\alpha(1)} - \cdots - a_{\alpha(n-1)} \in C_{\alpha(n)} \cap A_{\alpha(n)} = 0,$$

a contradiction.

If $C \neq \oplus A_\alpha$, then since $C = \cup C_\alpha$, there exists a smallest ordinal $\lambda \leqslant \tau$ such that $C_\lambda \nsubseteq \otimes A_\alpha$. Since $C_n \subseteq \oplus A_\alpha$ for all finite n, $\omega \leqslant \lambda$, where ω is the first finite ordinal.

If λ has a predecessor $\lambda - 1$, then $C_{\lambda-1} \subseteq \oplus A_\alpha$ and also $C_\lambda = C_{\lambda-1} \oplus A_{\lambda-1} \subseteq \oplus A_\alpha$, a contradiction.

Thus λ is a limit ordinal. Since each element has only a finite number of nonzero components it follows that each element of $C \cap \oplus \{R_\alpha \mid \alpha < \lambda\}$

is contained in some $C \cap \oplus \{R_\alpha | \alpha < \beta\}$ where $\beta < \lambda$. (The latter statement is false if $\lambda - 1$ exits.) Thus

$$C_\lambda \equiv C \cap \bigoplus_{\alpha < \lambda} R_\alpha = \bigcup_{\beta < \lambda} [C \cap \oplus \{R_\alpha | \alpha < \beta\}] = \bigcup_{\beta < \lambda} C_\beta.$$

Since λ is the least ordinal with $C_\lambda \nsubseteq \oplus A_\alpha$, for any $\beta < \lambda$, we have $C_\beta \subseteq \oplus A_\alpha$. Hence $C_\lambda = \cup\{C_\beta | \beta < \lambda\} \subseteq \oplus A_\alpha$ a contradiction. Hence $C = \oplus A_\alpha$, and $C \cong \oplus L_\alpha$.

13-1.4 Corollary 1. *Over a right hereditary ring R, any projective module is isomorphic to a direct sum of right ideals of R.*

Proof. If P is any projective module, then $F = P \oplus Q$ for some free module F and $Q \leqslant F$.

13-1.5 Corollary 2. *A ring R is right hereditary* $\Leftrightarrow$

$\forall$ *free* $F = F_R$, $\forall C \leqslant F$, *C is projective.*

Proof. $\Rightarrow$: By the last theorem. $\Leftarrow$: If $F = R$, then a submodule of F is a right ideal.

13-1.6 Corollary 3. *The ring R is right hereditary* $\Leftrightarrow$ *each submodule of any projective module inherits the property of being projective.*

Proof. Let $C \leqslant P$ be any submodule of any projective P. Then there exists a free module $F = P \oplus Q$ and hence $C \leqslant F$. By Corollary 2, any such

C is projective $\Leftrightarrow$ R is right hereditary.

13-1.7 Corollary 4. *The ring R has the property that each submodule of each free module is free* $\Leftrightarrow$ *all right ideals of R are free.*

Proof. $\Rightarrow$: Since $F = R_R$ is free, any $L < F$ is free by hypothesis.
$\Leftarrow$: For any free module F, for any submodule $C < F$, by the last theorem, $C = \oplus A_\alpha$, where each $A_\alpha \cong L_\alpha \leqslant R$. Since the L_α are all assumed free, so is also their sum C.

13-1.8   Commutative rings. *Let C be a commutative ring with identity.*

(1)   *For $0 \neq x \in C$ a nonzero divisor, xC is free. Conversely every nonzero free ideal of C is of this form.*

Proof. If x is a nonzero divisor then $xC \cong C/x^{\perp} \cong C$ is free. Conversely, if $I \lhd C$ is free but not principal, then I is free on a basis $\{x, y, \ldots\} \subset I$ where $xc = 0$ implies $c = 0$. Thus $I = xC \oplus yC \oplus F$ where $F \lhd C$ is free. But then $xy = yx \in xC \cap yC = 0$ implies that $y = 0$, a contradiction. Thus $I = xC$.

(2)   *If each principal ideal of C is generated by a nonzero divisor then C is a domain.*

Proof. Let $0 = ab \in C$. Then $bC = xC$ where x is not a zero divisor in C. Thus $x = bd$ for some $d \in C$. But then $ax = abd = 0$ is a contradiction. Hence C is a domain.

(3)   *The ring C has the property that each submodule of each free module is free $\Leftrightarrow$ C is a principal ideal domain.*

Proof. $\Leftarrow$: This follows from first 13-1.8(1) and then 13-1.7. $\Rightarrow$: Now every ideal of C is of the form xC where $x \in C$ is a nonzero divisor. Hence C is a principal ideal ring which is a domain by (2).

13-2   Injectivity and projectivity

The first two propositions give improved better criteria for a module to be projective or injective. Then these are applied to dualize 13-1.6 in terms of injectivity.

13-2.1   Proposition. *A right R-module P is projective. $\Leftrightarrow$ For any surjective map $\pi: I \to J$ with I injective, and any $f: P \to J$, there exists a $g: P \to I$ such that $f = \pi g$ factors through I.*
I.e. P is projective $\Leftrightarrow$

$$\forall \qquad \begin{array}{c} P \\ \downarrow f \\ I \xrightarrow{\pi} J \longrightarrow 0 \end{array} \qquad \Rightarrow \exists \qquad \begin{array}{c} P \\ {}^{g}\nearrow \ \downarrow f \\ I \xrightarrow{\pi} J \longrightarrow 0 \end{array}$$

where I is injective, the rows are exact, and diagrams commute.

Proof. $\Rightarrow$: Trivial. $\Leftarrow$: Given $A < B$ and any map $f: P \to B/A$. Take any injective $I \supset B$ and let f' be f followed by inclusion $f' : P \to B/A \subset I/A$; let $\pi : I \to I/A$. By hypothesis there exists $g : P \to I$ such that $\pi g = f'$. I.e. there are commutative diagrams with exact rows:

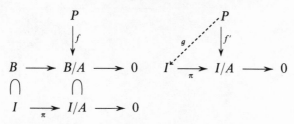

It is claimed that $g(P) \subseteq B$; for any subset $X \subset I$,

$$X \subseteq B \Leftrightarrow \pi(X) \subseteq B/A.$$

If $X = g(P)$, then $\pi X = \pi g(P) = f'(P) = f(P) \subseteq B/A$. Identify g with its corestriction to $g : P \to B$. Then $f = \pi g$ shows that P is projective.

The proof of the dual of the last proposition is noticeably longer because the dual of inclusions are quotients which are harder to handle.

13-2.2 Proposition. *A right R-module I is injective.* $\Leftrightarrow$ *For any monic map* $i : Q \to P$ *with* P *projective, and any* $f : Q \to I$, *there exists a map* $g : P \to I$ *with* $f = gi$.

I.e., I is injective $\Leftrightarrow$

$$
\begin{array}{ccc}
0 \longrightarrow Q \xrightarrow{\ i\ } P & & 0 \longrightarrow Q \xrightarrow{\ i\ } P \\
\forall \qquad f\big\downarrow \qquad \Rightarrow \exists & & \qquad f\big\downarrow \quad \swarrow g \\
\qquad I & & \qquad I
\end{array}
$$

where P is projective, the rows are exact, and diagrams commute.

Proof. $\Rightarrow$: Trivial. $\Leftarrow$: Given are a monic map $j : A \to B$ and $f : A \to I$. Then B is the quotient $\pi : P \to B$ of some free module P. Define $C = \pi^{-1} jA \subset P$ and let $i : C \to P$ be the inclusion. Then define ϕ to be the composite map below

$$
\begin{array}{ccc}
C & \xrightarrow{\ i\ } & P \\
& & \big\downarrow \pi \\
A & \xleftarrow{\cong} & jA \longleftarrow 0 \\
& a \longleftarrow ja &
\end{array}
$$

By hypothesis there is a map $g: P \to I$ such that $f\phi = gi$. There are commutative diagrams with exact rows and columns as indicated.

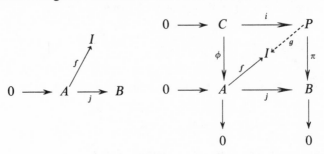

Let $D = \phi^{-1}0$ be the kernel of ϕ, and $k: D \to C$ the inclusion map. Thus $gikD = f\phi kD = 0$ because $\phi kD = 0$. Since $ikD = \pi^{-1}0$ is the kernel of π, and $g(\pi^{-1}0) = 0$, g factors through π. There is a map $h: B \to I$ with $g = h\pi$. Thus the new commutative diagram is:

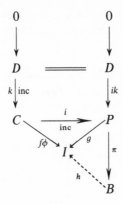

It is asserted that $f = hj$. Since $h\pi = g$, also $h\pi i = gi$. But $gi = f\phi$, and thus $h\pi i = f\phi$. Now from $\pi i = j\phi$ it follows that $hj\phi = f\phi$. Since ϕ is epic, $hj = f$. Thus I is injective.

13-2.3 Theorem. *For a ring R, the following are equivalent.*

(a) *R is right hereditary.*
(b) *Each submodule of a projective is projective.*
(c) *Each quotient of an injective module is injective.*

Proof. To prove that (b) $\Leftrightarrow$ (c), consider the following diagram with P projective and I injective.

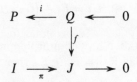

(b) $\Rightarrow$ (c): To show that J is injective, given are the data Q, i, P, and f as in 13-2.2, except for I. By the assumed projectivity of Q, there is a map $f^1: Q \to I$ with $f = \pi f^1$. Then since I is injective, there is a map $f^2: P \to I$ with $f^2 i = f^1$. The commutative diagrams are:

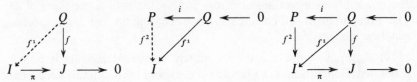

The last diagram must commute because it is pasted together from the previous two commutative diagrams. Hence $f = (\pi f^2)i$ shows that J is injective.

(c) $\Rightarrow$ (b). To prove that Q is projective, take the data I, π, J and f as in 13-2.1. Since by hypothesis J is injective, there is a map $f^1: P \to J$ with $f^1 i = f$. Then the projectivity of P guarantees an $f^2: P \to I$ with $f^1 = \pi f^2$. The diagrams are as follows, where the last diagram commutes because it is pieced together from the first two commutative ones:

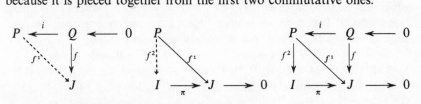

Thus $f = \pi(f^2 i)$ shows that Q is projective.

(a) $\Leftrightarrow$ (b). By 13-1.6, (a) and (b) are equivalent.

13-3   Finitely generated modules

The basic facts about direct sums of finitely generated modules are not only of independent interest, but are needed for obtaining the main properties of right semihereditary rings. First, some elementary facts about modules which are used over and over again are singled out once and for all.

13-3.1   Facts. (1) Let $M = F_1 \oplus H_1 = G \oplus H_2$ with $F_1 \subset G$ be any modules. Then $G = F_1 \oplus F_2$ for some $F_2 < M$.

Proof. By the modular law, $G = G \cap (F_1 \oplus H_1) = F_1 \oplus (G \cap H_1)$. Set $F_2 = G \cap H_1$.

(2) If a module G is finitely generated then any epimorphic image of G is likewise finitely generated. In particular, if $G = F_1 \oplus F_2$, then F_1 and F_2 are also finitely generated.

(3) A finitely generated submodule G of a free module M is always contained in a finitely generated free module F, with $G \subset F \leqslant M$. Furthermore, F is a direct summand of M.

Proof. Simply express the generators of G in terms of a free basis of M. The finite number of basis elements appearing in the latter process generate the required F.

(4) An epimorphic image of a finitely generated module is finitely generated. Conclusions (2), (3), and (4) also hold for countably generated.

13-3.2 Lemma. *A countably generated R-module M is a direct sum of finitely generated submodules of M* $\Leftrightarrow$ *Each finitely generated submodule of M is contained in a finitely generated direct summand.*

Proof. $\Rightarrow$: This is easy and is omitted. $\Leftarrow$: If M is finitely generated we are done, so let $M = \Sigma_1^\infty x_i R = \langle x_1, x_2, \ldots \rangle$. Since $\langle x_1 \rangle$ is a finitely generated submodule of M, $\langle x_1 \rangle \subset F_1$ with $M = F_1 \oplus H_1$ where F_1 is finitely generated. But then so is also $\langle F_1 \cup \{x_2\} \rangle$ and hence $\langle F_1 \cup \{x_2\} \rangle \subset G_2 < M = G_2 \oplus H_2$, where G_2 is finitely generated. Since $F_1 \subset G_2$, 13-3.1(1), $G_2 = F_1 \oplus F_2$ for some $F_2 < M$ which is finitely generated by 13-3.1(2). Thus $M = G_2 \oplus H_2 = F_1 \oplus F_2 \oplus H_2$ with $\{x_1, x_2\} \subset F_1 + F_2$.

Then for some finitely generated $G_3 < M$, we have $\langle F_1 + F_2, \{x_3\} \rangle \subset G_3 < M = G_3 \oplus H_3$.

Again since $G_3 \supset F_1 \oplus F_2$ and the latter is a direct summand of M, $G_3 = F_1 \oplus F_2 \oplus F_3$ for some finitely generated $F_3 < M$. Thus $M = F_1 \oplus F_2 \oplus F_3 \oplus H_3$ with $\{x_1, x_2, x_3\} \subset F_1 + F_2 + F_3$.

For each finite integer n, $M = F_1 \oplus \cdots \oplus F_n \oplus H_n$ where all the F_i are finitely generated and $\{x_1, \ldots, x_n\} \subset F_1 + \cdots + F_n$. Then the sum $\Sigma_1^\infty F_i = \oplus_1^\infty F_i$ is direct, and contains all the generators. Hence $M = \oplus_1^\infty F_i$.

The hypotheses of the last lemma are now weakened to get an even more useful criterion.

13-3.3 Lemma. *For a countably generated module M, assume that for* *any direct summand H of M, every element of H is contained in a finitely generated direct summand of H. Then M is a direct sum of finitely generated submodules.*

Proof. Let $M = \Sigma_1^\infty x_i R$ be generated by $\{x_1, x_2, \dots\} \subset M$. With $x_1 \in H = M$, the hypotheses imply that $M = F_1 \oplus H_1$ with $x_1 \in F_1$, and with F_1 finitely generated. Write $x_2 = f + h$, $f \in F_1$ and $h \in H_1$. By hypothesis $H_1 = F_2 \oplus H_2$ where $h \in F_2$ and F_2 is finitely generated. Then $M = F_1 \oplus F_2 \oplus H_2$ with $\{x_1, x_2\} \subset F_1 + F_2$.

Next, $x_3 = f' + h'$ where $f' \in F_1 + F_2$ and $h' \in H_2$. Again $H_2 = F_3 \oplus H_3$ with F_3 finitely generated and $h' \in F_3$. Since $f' \in F_1 + F_2$, $\{x_1, x_2, x_3\} \subset F_1 + F_2 + F_3$, and $M = F_1 \oplus F_2 \oplus F_3 \oplus H_3$. The rest is now clear.

13-3.4 Theorem. *Let R be a right semihereditary ring. Then any finitely generated submodule of a free module is isomorphic to a finite direct sum of finitely generated right ideals of R.*

Proof. Let G be a finitely generated submodule of a free module. By 13-3.1(3) it may be assumed that $G \subset R_1 \oplus \cdots \oplus R_n$ where all $R_i = R$. If $n = 1$, then $G < R_1 = R$ is already a finitely generated right ideal.

Induction hypothesis: For any finitely generated submodule $H < R_1 \oplus \cdots \oplus R_{n-1}$, $H = H_1 \oplus \cdots \oplus H_{n-1}$ where $H_i \cong L_i \leqslant R$, and each L_i is finitely generated.

Let $\pi: R_1 \oplus \cdots \oplus R_n \to R_n$ be the natural projection. Since $L_n = \pi(G) \leqslant R$ is a finitely generated right ideal of a right semihereditary ring, L_n is projective. Thus $\pi \mid G: G \to L_n$ splits; $G = (G \cap \ker \pi) \oplus G_n$ where $L_n \cong G_n < R_1 \oplus \cdots \oplus R_n$. Since $G \cap \ker \pi$ is a finitely generated submodule of $R_1 \oplus \cdots R_{n-1}$ as seen from 13-3.1(2), by induction $G \cap \operatorname{Ker} \pi = G_1 \oplus \cdots \oplus G_{n-1}$ where $G_i \leqslant R_1 \oplus \cdots R_{n-1}$ and $G_i \cong L_i \leqslant R$ with all L_i finitely generated. Thus $G = G_1 \oplus \cdots \oplus G_n \cong L_1 \oplus \cdots \oplus L_{n-1} \oplus L_n$ as required.

The following is immediate from the previous proof.

13-3.5 Corollary 1. *For a right semihereditary ring, suppose that G is a finitely generated submodule of the form $G < R_1 \oplus \cdots \oplus R_n$, with all $R_i \cong R$. Then there are finitely generated right ideals $L_j \leqslant R$ such that*

$$G = H_1 \oplus \cdots \oplus H_n \quad where$$

$$H_j \leqslant R_1 \oplus \cdots \oplus R_j, H_j \cong L_j, \qquad j = 1, \dots, n.$$

13-3.6   Corollary 2. *Over a right semihereditary ring, any finitely gener-ated projective module is isomorphic to a finite direct sum of finitely generated right ideals.*

Proof. Let G be a finitely generated projective module. Then there exists a free module F with $F = G \oplus H$, and the previous theorem gives the corollary.

13-3.7   Main Corollary 3. *For a ring R the following are all equivalent.*

(a)   *R is right semihereditary.*
(b)   *Each finitely generated submodule of any free module is projective.*
(c)   *Any finitely generated submodule of any projective module is also projective.*

Proof. (a) $\Rightarrow$ (b): This follows from the previous theorem because finitely generated right ideals are projective.

(b) $\Rightarrow$ (a): Any finitely generated right ideal is a submodule of the free module R.

(c) $\Rightarrow$ (a): Since R is projective, by assumption, finitely generated right ideals are now projective.

(a) $\Rightarrow$ (c): Let G be a finitely generated submodule $G < P$ of a projec-tive P. There exists a free module F such that $F = P \oplus Q$. Thus $G < F$, and the last theorem shows (c).

13-3.8   Corollary 4. *The ring R has the property that any finitely gener-ated submodule of any free module is free. $\Leftrightarrow$ Each finitely generated right ideal of R is free.*

13-3.9   Corollary 5. *Let C be a commutative ring with $1 \in C$. Every finitely generated submodule of any free C-module is free $\Leftrightarrow$ C is a domain all of whose finitely generated ideals are cyclic.*

Proof. The basic fact that is to be used here is 13-1.8(1) that an ideal of C is free if and only if it is a principal one generated by a nonzero divisor.

$\Rightarrow$: Any finitely generated ideal $L \lhd C$ is a finitely generated sub-module of the free module C. Now by hypothesis L is free, and consequently principal. By 13-1.8(2), C is a domain.

$\Leftarrow$: By hypothesis, C is certainly semihereditary. Thus by the last theorem any finitely generated submodule G of some free C-module is isomorphic to $G \cong L_1 \oplus \cdots \oplus L_n$, where each L_i is a finitely generated ideal $L_i \lhd C$. By assumption, each L_i is cyclic. Since C is a domain, the free ideals of C are the cyclic ones.

The next theorem remains true if the restrictive hypothesis that the projective module be finitely generated is omitted altogether. (See [Anderson and Fuller 74; p 297, Corollary 26.2], or [Head 74; p. 133, Theorem 7], or [Kasch 82; p. 355, Lemma 13.6.3]). The usual proofs of the more general result are based on the next theorem.

13-3.10 Theorem. *Let R be a right semihereditary ring. Then any countably generated projective module is isomorphic to a direct sum of finitely generated right ideals of R.*

Proof. Let M be a countably generated projective module. Since a direct summand of a projective is a projective, in view of Corollary 2, 13-3.6, it suffices to show that M is a direct sum of finitely generated modules. For the latter it suffices to verify that any direct summand P of $M = P \oplus P'$ satisfies the criterion in 13-3.3. Thus it suffices to show the following. If P is any projective module and $w \in P$ any element, then w is contained in a finitely generated direct summand of P. There exists a free module $F = P \oplus Q$. There is a second decomposition of $F = G \oplus H$, where $w \in G$ and G is finitely generated and free. (Simply express w in terms of any basis of F.) Thus $w \in G \cap P$. It will now be shown that $G \cap P$ is a finitely generated direct summand of F, and hence also of P (13-3.1(1)).

Let $\pi: F = P \oplus Q \to Q$ be the natural projection and let $\phi = (\pi \mid G): G \to \phi(G)$ be the restriction and corestriction to G and $\pi(G) = (G + P) \cap Q = \phi(G)$. The kernel of ϕ is $\ker \phi = G \cap P$. The image $\phi(G)$ is finitely generated because G is. Since R is right semihereditary and $\phi(G) < F$ is a finitely generated submodule of a free module, by 13-3.7(b) also $\phi(G)$ is projective. Hence the epimorphism $\phi: G \to \phi(G)$ splits with $G = (G \cap P) \oplus I$ where $I < G$ and $I \cong \phi(G)$. Since G is finitely generated, it follows from 13-3.1(2) that $G \cap P$ is finitely generated. Thus $P = (G \cap P) \oplus [P \cap (I \oplus H)]$.

13-4 Examples

For commutative domains, 2-3.19 gives a criterion for being hereditary and a ring satisfying this criterion is called a Dedekind domain. Thus $\mathbf{Z} = 0, \pm 1, \pm 2, \ldots$ is hereditary. By 10-5.5, every semiprimitive ring with the right D.C.C. is hereditary.

Principal right ideal domains are right hereditary, and in particular the class of rings of the form $K[x; \theta; \delta]$ as in 3-6.1, 6.5, and 6.6.

13-4.1   Example. For a field F, let $F(t)$ be the field of rational functions in an indeterminate t, and let $\theta: F(t) \to F(t)$ be the endomorphism induced by $t\theta = t^2$; i.e. $[a(t)/b(t)]\theta = a(t^2)/b(t^2)$ for $a(t)$, $b(t) \in F[t]$. The skew polynomial ring $F(t)[x; \theta]$ is right hereditary.

Note that it contains as a subring the ring $F[t, x]$ of all non-commutative polynomials in t and x subject to the one relation $tx = xt^2$.

13-4.2   Example. Let R be the ring $R = Ke_{11} + Ke_{21} + Ke_{22}$ of two by two lower triangular matrices over a skew field K. The proper right ideals of R are projective: $e_{11}K = e_{11}R$, $e_{21}K + e_{22}K = e_{22}R$, and for $0 \ne \lambda \in K$, $\{e_{11}k + e_{21}\lambda k + \bar{k}e_{22} \mid k, \bar{k} \in K\} = fR$, $f^2 = f = e_{11} + \lambda e_{12}$. Thus R is hereditary.

Next, a ring that is semihereditary but not hereditary is given.

13-4.3   Rings of sets. For any set X, the set $\mathscr{P}(X)$ of all subsets of X is a ring under the multiplication $A \cdot B = A \cap B$ and addition $A + B = (A \cup B) \backslash (A \cap B)$ as symmetric difference for A, $B \subseteq X$; $0 = \varnothing$ and $1 = X$. The ring $\mathscr{P}(X)$ contains two distinct classes of ideals. For any infinite cardinal number $\aleph$, the set of all subsets of X of cardinality $\leqslant \aleph$ is an ideal. Each subset $Y \subset X$ defines an ideal that is a direct summand: $\mathscr{P}(Y) \lhd \mathscr{P}(X) = \mathscr{P}(Y) \oplus \mathscr{P}(X \backslash Y)$.

Each finitely generated ideal $L = Y_1 \cdot \mathscr{P}(X) + \cdots + Y_n \cdot \mathscr{P}(X) \lhd \mathscr{P}(X)$ is of this latter type $L = \mathscr{P}(Y)$, where $Y = Y_1 \cup \cdots \cup Y_n$. Therefore the ring $\mathscr{P}(X)$ is always semihereditary.

Let X be uncountable, and $L = \{Y \in \mathscr{P}(X) \mid Y \text{ is countable}\}$. Then L is not projective.

Proof. Assume L is projective. Then L is a direct sum of finitely generated ideals $L = \oplus\{\mathscr{P}(Y_i) \mid i \in I\}$, $Y_i \cap Y_j = \varnothing$ for $i \ne j \in I$. For any $x_0 \in X$, $\{x_0\} \in L$. Hence $X = \cup\{Y_i \mid i \in I\}$ is a partition of X into countable subsets X_i. We will use the known fact that a countable union of countable sets is countable. Thus I is uncountable. For a countably infinite subset $J \subset I$, $W = \cup\{Y_j \mid j \in J\}$ is countable, and hence $W \in L = \oplus_\alpha \mathscr{P}(Y_\alpha)$. Thus $W \in \mathscr{P}(Y_{\alpha(1)}) \oplus \cdots \oplus \mathscr{P}(Y_{\alpha(n)})$ for a finite number of summands. But $W \backslash [Y_{\alpha(1)} \cup \cdots \cup Y_{\alpha(1)}]$ is infinite, a contradiction.

The next proposition shows that regular rings (Definition 6-7.6) are semihereditary.

13-4.4 Proposition. *For a regular ring R, every finitely generated right ideal $L < R$ is of the form $L = eR$, $e = e^2 \in R$.*

Proof. It suffices to let $L = aR + bR$, $a, b \in R$. Let $axa = a$ for $x \in R$, and set $e = ax$. Then $e^2 = e$ and $aR = eR$. Since $bR \subseteq ebR + (1 - e)bR$, $aR + bR = eR + (1 - e)bR$. Write $(1 - e)bR = fR$, $f^2 = f \in R$. Then $e(1 - e)bR = efR = 0$, or $ef = 0$, and $aR + bR = eR + fR$. Set $g = f(1 - e)$. Then $eg = ge = 0$; $g^2 = f(1 - e)f(1 - e) = f^2(1 - e) = g$; and lastly, $gf = f(1 - e)f = f^2 = f$. Thus $f \in gR$ and $g = f(1 - e) \in fR$, or $gR = fR$. Consequently $aR + bR = eR + gR \supseteq (e + g)R$.

In general, if e and g are two orthogonal idempotents then always $eR + gR = (e + g)R$. For if $ex + gy \in eR + gR$, then $ex + gy = (e + g)(ex + gy)$. Thus $L = (e + g)R$.

13-4.5 Definition. A ring R is *Boolean* if $r^2 = r$ for every $r \in R$.

13-4.6 Facts. (1) It can be shown that every Boolean ring is isomorphic to a subring of the Boolean ring $\mathscr{P}(X)$, where X is the maximal ideal space of R.

(2) Any finite $n \times n$ lower triangular matrix ring over a skew field is both right and left hereditary ([Goodearl 76; p. 112]). There are rings R hereditary on the right but not left, for example the subring $R = Qe_{11} + Qe_{21} + Ze_{22}$ of the two by two lower triangular matrix ring over the rationals, where Z are the integers. (See [Small 66] or [Kaplansky 58b].)

13-5 Exercises

Definition. For an arbitrary ring R with identity, a right R-module M is a *hereditary module* if every submodule of M is projective. The module M is *semihereditary* if every finitely generated submodule of M is projective. Thus a ring R is hereditary or semihereditary if and only if R_R is likewise.

Prove the assertions below.

1. Generalize as much as possible the results of Chapter 13 from the case $M = R_R$ to an arbitrary module M.
2*. For $R = Z$ the ring of integers, the hereditary modules coincide with the free modules ([Fuchs 70; vol. I, p. 75]).
3*. As a Z-module, $\Pi_1^\infty Z$ is semihereditary but not hereditary ([Fuchs 70; vol. I, p. 94]).

4. For an automorphism $\theta: K \to K$ of a skew field K, let $K[[x; \theta]]$
    be a formal skew power series ring (see Chapter 1, exercise 23).
    (i) Prove that any right ideal of $K[[x; \theta]]$ is of the form
    $x^n K[[x; \theta]] \lhd K[[x; \theta]]$, $n = 0, 1, 2, \ldots$.
    (ii) Deduce that $K[[x; \theta]]$ is (right and left) hereditary.

5. For a skew field K, let $\delta: K \to K$ be a derivation. Abbreviate
    $\delta(n) = \delta^n$, $n = 1, 2, \ldots$, and $\delta(0) = 1$. Let $K[[y; \delta]]$ be the abelian
    group of all formal power series with right side coefficients
    $\alpha = a_0 + ya_1 + y^2 a_2 + \cdots$, $\beta = b_0 + yb_1 + y^2 b_2 + \cdots$ a_i, $b_i \in K$.
    (i) Show that $K[[y; \delta]]$ becomes an associative ring under
    formal power series multiplication and subject to the commuta-
    tion rules:

    $$\alpha\beta = \left(\sum_i y^i a_i \right)\left(\sum_j y^j b_j \right) = \sum_{i,j} y^i a_i y^j b_j$$

    $$ky = yk + yk^\delta y = yk + y^2 k^\delta + y^3 k^{\delta(2)} + \cdots + y^n k^{\delta(n-1)} + y^n k^{\delta(n)} y$$

    $$= \sum_{n=1}^{\infty} y^n k^{\delta(n-1)}.$$

    (ii) If $\alpha = \Sigma x^i a_i$ with $a_0 \neq 0$, then α has an inverse in $K[[y; \delta]]$.
    (iii) Every right ideal in $K[[y; \delta]]$ is of the form
    $y^n K[[y; \delta]] \lhd K[[y; \delta]]$. (iv) $K[[y; \delta]]$ is (right and left) here-
    ditary.

6. For a Boolean ring R ($1 \in R$) prove that (i) R is commutative; (ii)
    $r + r = 2r = r - r = 0$ for all $r \in R$. (iii) For any maximal ideal
    $M \lhd R$, $R/M \cong \mathbf{Z}/2\mathbf{Z} = \mathbf{Z}_2$.

7. Let R be a Boolean ring. (i) For any $0 \neq b \in R$, there exists a
    maximal ideal $M \lhd R$ with $b \notin M$. (Hint: $b \notin (1 - b)R$.) (ii) For
    any $b \in R$ and any maximal ideal $M \lhd R$, $b \notin M \Leftrightarrow 1 - b \in M$. (iii)
    Let X be the set of maximal ideals of R. For $b \in R$, define
    $\hat{b}: X \to \mathbf{Z}_2$ by $\hat{b}(M) = b + M \in R/M$ and let $\hat{R} = \{\hat{b} \mid b \in R\}$.
    Map $\psi: R \to \mathscr{P}(X)$ by $\psi(b) = \{M \mid \hat{b}(M) = 1\}$. Show that ψ is a
    monic ring homomorphism. (iv) Prove that $R \cong \hat{R} \cong \psi(R) \subset$
    $\mathscr{P}(X)$.

8*. (Continuation of 7). Regarding $\mathbf{Z}_2$ as a discrete topological space
    (ring), endow X with the smallest topology making all the
    functions $\hat{b}$ continuous. (i) The X is a compact Hausdorff space
    having a basis for its topology consisting of closed and open sets.
    (Such a space is called a *Boolean space*.) (ii) $\{\{M \mid b \notin M\} \mid b \in R\}$
    is a basis for the topology of X.

CHAPTER 14

Module constructions

Introduction

Although we are not prepared to define the terms 'universal property' and 'categorical proof' rigorously, nevertheless an attempt to describe these might be more informative. A module is said to satisfy a universal property if for any variable module satisfying some restrictive hypotheses, there exists a module map making a certain diagram commute. The map whose existence is given connects the variable module with the one having the universal property. Roughly speaking, theorems and their proofs which do not use elements of modules are called categorical. Such proofs proceed by manipulating modules, maps, and commutative diagrams, using the associativity of maps, and invoking either assumed hypotheses and universal properties to get the existence of new maps. Indeed, it is the abundance of such proofs which will lead us to a formal study of categories.

This chapter does not give new structure theories for rings, nor does it study special classes of rings. However, it does something just as important; it gives the basic fundamental module constructions which is standard equipment for all ring theorists. Also these module constructions and manipulations is what creates a general theorem proving ability. The emphasis is on details of proofs, not general results. Some proofs will be given elementwise as well as categorically a second time. The objective will be to develop the ability to give proofs in either of the two modes, so as to be able to exploit the strengths and weaknesses of either modes.

14-1   Pullbacks

Here R is any ring with or without identity. Right R-modules are in use, and maps, functions, or arrows will automatically be assumed to be module homomorphisms unless otherwise stated.

14-1.1   Definition. For given module homomorphism $\alpha: A \to C$ and $\beta: B \to C$ a *pullback* of α and β is a triple $\langle D, \gamma, \delta \rangle$ where $\gamma: D \to A$, $\delta: D \to B$ with $\alpha\gamma = \beta\delta$. Furthermore, for any module E for any $f: E \to A$ and $g: E \to B$ there exists a unique map $\varepsilon: E \to D$ such that $f = \gamma\varepsilon$ and $g = \delta\varepsilon$. The definition is summarized in the following commutative diagrams

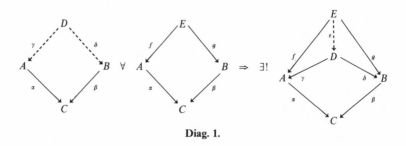

Diag. 1.

The next paragraph is the basis of the common abuse of language when one talks about the pullback instead of a pullback of two given maps.

14-1.2.   In the last definition, $\langle E, f, g \rangle$ is another pullback of α and $\beta \Leftrightarrow \varepsilon$ is an isomorphism.

Proof. $\Rightarrow$: In diagram 2, the pullback property of E is used to obtain a (unique) map η. Thus $f\eta = \gamma$, and $g\eta = \delta$, and hence $(\delta\varepsilon)\eta = g\eta = \delta$, and similarly $(\gamma\varepsilon)\eta = \gamma$. This shows that diagram 3 commutes with either $\varepsilon\eta$ or with the identity map 1_D on D. Hence the uniqueness property of the pullback $\langle D, \gamma, \delta \rangle$ implies that $\varepsilon\eta = 1_D$. Similarly in diagram 4, $\eta\varepsilon = 1_E$.

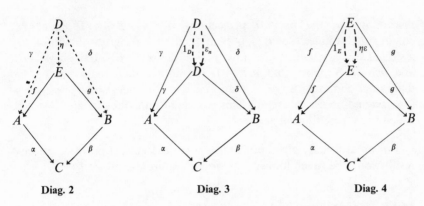

Diag. 2   **Diag. 3**   **Diag. 4**

$\Leftarrow$: Let $f_1: E_1 \to A$ and $g_1: E_1 \to B$ be given with $\alpha f_1 = \beta g_1$. Use of the pullback property of D gives a unique $\varepsilon_1: E_1 \to D$ with $f_1 = \gamma \varepsilon_1$ and $g_1 = \delta \varepsilon_1$. From $\gamma \varepsilon = f$ and $\delta \varepsilon = g$, one gets that $\gamma = f \varepsilon^{-1}$ and $\delta = g \varepsilon^{-1}$. Set $\rho = \varepsilon^{-1} \varepsilon_1$. Then $f\rho = (f\varepsilon^{-1})\varepsilon_1 = \gamma \varepsilon_1 = f_1$ and $g\rho = (g\varepsilon^{-1})\varepsilon_1 = \delta \varepsilon_1 = g_1$. Thus there is a commutative diagram.

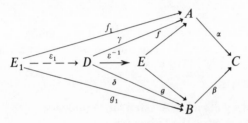

Suppose that $\lambda: E_1 \to E$ is another map with $f_1 = f\lambda$ and $g_1 = g\lambda$. Then $\eta \equiv \varepsilon\lambda: E_1 \to D$ with $f_1 = f\varepsilon^{-1}\eta = \gamma\eta$ and $g_1 = g\varepsilon^{-1}\eta = \delta\eta$. By the uniqueness of ε_1 it follows that $\varepsilon_1 = \eta = \varepsilon\lambda$ and $\lambda = \varepsilon^{-1}\varepsilon_1 = \rho$.

14-1.3 Corollary. *Suppose that $\langle D, \gamma, \delta \rangle$ and $\langle E, \gamma, \delta \rangle$ are two pullbacks of α and β with $E \subseteq D$ (and where the restrictions of γ and δ to E are also written as γ and δ). Then $E = D$.*

Proof. In the last lemma take $f = \gamma$, $g = \delta$, and $\varepsilon: E \to D$ the inclusion map. Thus $\eta: D \to E$ with $\varepsilon\eta = 1_D$. Hence $D = \varepsilon\eta D \subseteq \varepsilon E = E$.

14-1.4 Construction. *Every pair of maps $\alpha: A \to C$ and $\beta: B \to C$ has a pullback $\langle D, \gamma, \delta \rangle$.*

Proof. Set $D = \{(a, b) \in A \oplus B \,|\, \alpha a = \beta b\}$. Define γ and δ to be the restrictions of the natural projections on $A \oplus B$ to D: $\gamma(a, b) = a$, $\delta(a, b) = b$. Suppose that E, f, g as in diagram 1 are given. Then for any $x \in E$, $\alpha f x = \beta g x$; hence $\varepsilon x = (fx, gx) \in D$ defines $\varepsilon \colon E \to D$ with $\gamma\varepsilon = \partial f$ and $\delta\varepsilon = g$. Suppose that $\eta \colon E \to D$ is another such map making the diagram 1 commute by $\gamma\eta = f$ and $\delta\eta = g$. But then $\eta x = (\gamma\eta x, \delta\eta x) = (fx, gx)$ and necessarily $\eta = \varepsilon$. This particular pullback of α and β will be called the standard pullback.

For an abstractly given pullback a more categorical proof of the next result would be much longer and would require the use of 14-1.3.

14-1.5 For the standard pullback $\langle D, \gamma, \delta \rangle$ as in 14-1.3 of α and β, assume in addition that there exists a $\lambda \colon A \to B$ such that $\alpha = \beta\lambda$. Define $K = \beta^{-1}0 < B$, $\tilde{K} = \{(0, k) \mid k \in K\} < D$, and $\tilde{A} = \{(a, \lambda a) \mid a \in A\} < D$. Then

$$D = \tilde{A} \oplus \tilde{K} \cong A \oplus \ker \beta.$$

Proof. See 2-4.2. (It must not be assumed here that the pullback diagram with λ inserted commutes. Not all diagrams commute. However, since $\beta(\lambda\gamma - \delta) = \alpha\gamma - \beta\delta = 0$, $(\lambda\gamma - \delta)D \subseteq K$.)

14-1.6 Properties of pullbacks. A standard pullback $\langle D, \gamma, \delta \rangle$ of $\alpha \colon A \to C$ and $\beta \colon B \to C$ satisfies the following:

(1) $\ker \gamma = \{0\} \times \ker \beta \subset D$, $\ker \delta = \ker \alpha \times \{0\} \subset D$, $\ker \gamma + \ker \delta = \ker \alpha \times \ker \beta \subset D$. Hence

δ is monic $\Leftrightarrow$ α is monic; (γ monic $\Leftrightarrow$ β monic).

(2) α is epic $\Rightarrow$ δ is epic; (β epic $\Rightarrow$ γ epic).

Proof of (2). Given any $b \in B$, since $\beta b \in C = \alpha A$, there is an $a \in A$ with $\alpha a = \beta b$. Thus $(a, b) \in D$ and $b = \delta(a, b)$.

Next, a frequently occuring useful construction is discussed.

14-1.7 Pullback of a pullback is a pullback. Given are any module maps $C_2 \xrightarrow{\gamma_2} C_1 \xrightarrow{\gamma_1} C$ and $B \xrightarrow{\sigma} C$. For the pullback $\langle B_1, \beta_1, \sigma_1 \rangle$ of σ and γ_1, and then the pullback $\langle B_2, \beta_2, \sigma_2 \rangle$ of σ_1 and γ_2. Then $\langle B_2, \beta_1\beta_2, \sigma_2 \rangle$ is a pullback of σ and $\gamma_1\gamma_2$.

1st proof. In the first proof all the given and constructed pullbacks are standard. Let $\langle B_{21}, \beta, \rho \rangle$ be the pullback of σ and $\gamma_1 \gamma_2$. There are two comutative diagrams.

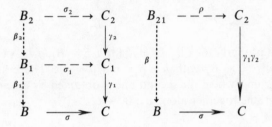

If $b \in B$, $c_1 \in C_1$, and $c_2 \in C_2$ represent arbitrary elements, then $B_1 = \{(b, c_1) \mid \sigma b = \gamma_1 c_1\}$ with $\beta_1(b, c_1) = b$ and $\sigma_1(b, c_1) = c_1$. Then

$$B_2 = \{((b, c_1), c_2) \mid \sigma_1(b, c_1) = \gamma_2 c_2, \sigma b = \gamma_1 c_1\}$$
$$= \{((b, \gamma_2 c_2), c_2) \mid \sigma b = \gamma_1 \gamma_2 c_2\};$$
$$B_{21} = \{(b, c_2) \mid \sigma b = \gamma_1 \gamma_2 c_2\}.$$

The map $\phi: B_2 \to B_{21}$, $\phi x = (b, c_2)$ is an isomorphism, where $x = ((b, \gamma_2 c_2), c_2) \in B_2$. Furthermore $\rho \phi x = \rho(b, c_2) = c_2 = \sigma_2 x$ and hence $\rho \phi = \sigma_2$. Next, $\beta \phi x = \beta(b, c_2) = b$, while $\beta_1 \beta_2 x = \beta_1(b, \gamma_2 c_2) = b$. Thus $\beta \phi = \beta_1 \beta_2$. From 14-1.2 and the commutative diagram below with ϕ an isomorphism it follows that $\langle B_2, \beta_1 \beta_2, \sigma_2 \rangle$ is a pullback of σ and $\gamma_1 \gamma_2$.

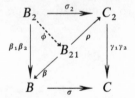

2nd proof. Now only the universal property of pullbacks will be used. Start with $C_2 \xrightarrow{\gamma_2} C_1 \xrightarrow{\gamma_1} C$, $B \xrightarrow{\sigma} C$, $\langle B_1, \beta_1, \sigma_1 \rangle$, and $\langle B_2, \beta_2, \sigma_2 \rangle$ as before. Let $f: E \to B$ and $g: E \to C_2$ with $\sigma f = \gamma_1 \gamma_2 g$ be given. But then $\gamma_2 g: E \to C_1$, and the universal property of $\langle B_1, \beta_1, \sigma_1 \rangle$ gives a unique $h: E \to B_1$ with $f = \beta_1 h$ and $\gamma_2 g = \sigma_1 h$. There are two commutative diagrams.

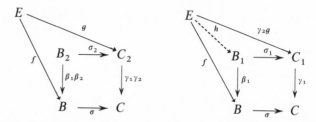

Now the universal property for $\langle B_2, \beta_2, \sigma_2 \rangle$ for $h: E \to B_1$, $g: E \to C_2$ with $\sigma_1 h = \gamma_2 g$ gives an $\varepsilon: E \to B_2$ with $h = \beta_2 \varepsilon$ and $g = \sigma_2 \varepsilon$. There are two commutative diagrams, where the second one is obtained by glueing two commutative diagrams together along $\sigma_1: B_1 \to C_1$.

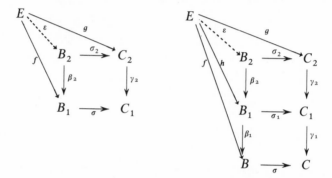

In particular, $f = \beta_1 \beta_2 \varepsilon$ and $g = \sigma_2 \varepsilon$, and the existence part of the proof has been accomplished. Suppose now that $\eta: E \to B_2$ is another map with $f = \beta_1 \beta_2 \eta$ and $g = \sigma_2 \eta$. Then both $\beta_2 \varepsilon$, $\beta_2 \eta: E \to B_1$. Furthermore, the commutativity of the last diagram gives that $f = \beta_1(\beta_2 \varepsilon) = \beta_1(\beta_2 \eta)$ and $\gamma_2 g = (\gamma_2 \sigma_2)\varepsilon = \sigma_1(\beta_2 \varepsilon) = \sigma_1(\beta_2 \eta)$. By the uniqueness property for the pullback B_1, $\beta_2 \varepsilon = \beta_2 \eta$. Thus there are two commutative diagrams.

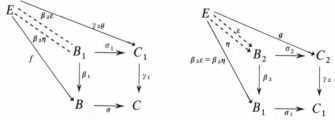

Finally the uniqueness for the pullback B_2 in the second diagram yields that $\eta = \varepsilon$. Thus $\langle B_2, \beta_1 \beta_2, \sigma_2 \rangle$ is the pullback of σ and $\gamma_1 \gamma_2$.

The above construction can be iterated for any finite and in fact countable number of C_n's.

14-2   Pushouts

Reversal of arrows in 14-1.1, 1.2, and 1.3 give the dual results below.

14-2.1   Definition. A *pushout* of two module homomorphisms $\alpha: C \to A$ and $\beta: C \to B$ is a triple $\langle D, \gamma, \delta \rangle$ where $\gamma: A \to D$ and $\delta: B \to D$ with $\alpha\gamma = \beta\delta$, and having the following universal property. For any module E and any $f: A \to E$ and $g: B \to E$, it follows that there exists a unique $\varepsilon: D \to E$ with $\varepsilon\gamma = f$ and $\varepsilon\delta = g$. The following commutative diagrams summarize the definition

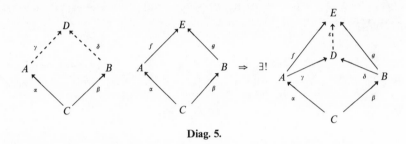

Diag. 5.

14-2.2 In the last definition of a pushout, ε is an isomorphism $\Leftrightarrow$ $\langle E, f, g \rangle$ is another pushout of α and β.

As already seen in the last section the implication "$\Rightarrow$" above in particular simplifies many proofs.

14-2.3   Corollary. *If* $\langle D, \gamma, \delta \rangle$ *and* $\langle E, \gamma, \delta \rangle$ *are two pushouts of* α *and* β *with* $D \subseteq E$, *then* $D = E$.

14-2.4   Construction. *Every pair of maps* $\alpha: C \to A$ *and* $\beta: C \to B$ *has a pushout* $\langle D, \gamma, \delta \rangle$.

Proof. Set $K = \{(\alpha c, -\beta c) \mid c \in C\} \subseteq A \oplus B$, and define $D = (A \oplus B)/K$. Then define $\gamma: A \to D$ and $\delta: B \to D$ by $\gamma a = (a, 0) + K$ and $\delta b = (0, b) + K$; $a \in A$, $b \in B$. From $(\alpha c, 0) + K = (0, \beta c) + K$ it follows that $\gamma\alpha = \delta\beta$.

Suppose that E, f, and g are as in diagram 5. Define $\varepsilon: D \to E$ by $\varepsilon[(a, b) + K]] = fa + gb$; ε is well defined because for $(\alpha c, -\beta c) \in K$, $f(\alpha c) + g(-\beta c) = 0$. Next, $\varepsilon\gamma\alpha = \varepsilon[(a, 0) + K] = fa$ and $\varepsilon\delta\beta b = gb$ show that $f = \gamma\varepsilon$ and $g = \delta\varepsilon$.

Suppose that $\eta: D \to E$ is another such map making the diagram 5 commute with $\eta\gamma = f$ and $\eta\delta = g$. Thus

$$\eta\gamma a = \eta[(a, 0) + K] = fa = \varepsilon\gamma a,$$

$$\eta\delta b = \eta[(0, b) + K] = gb = \varepsilon\delta b, \quad \text{and}$$

$$\eta[(a, b) + K] = \eta(\gamma a + \delta b) = fa + gb = \varepsilon(\gamma a + \delta b) = \varepsilon[(a, b) + K],$$

$$a \in A, \ b \in B.$$

Hence $\eta = \varepsilon$ is unique. This particular pushout of α and β will be referred to as the standard pushout.

A categorical proof of the next result would not show exactly how D is equal to an internal direct sum.

14-2.5 For the standard pushout $\langle D, \gamma, \delta \rangle$ given by 13-2.4 of α and β, suppose in addition that $\lambda: B \to A$ such that $\alpha = \lambda\beta$. Define $I = \{(-\lambda b, b) + K \mid b \in B\}$. Then $D = \gamma A \oplus I$ where $\gamma A \cong A$ and $I \cong B/\beta C = \operatorname{coker} \beta$.

Proof. If $(-\lambda b, b) + K = K$, then $(-\lambda b, b) = (-\alpha c, \beta c)$ for some $c \in C$. But then $b = \beta c$ and $(-\lambda b, b) = (-\lambda \beta c, \beta c) = (-\alpha c, \beta c) \in K$. Thus $I \cong B/\beta C$. Every $(a, b) + K \in D$ is of the form

$$(a, b) + K = (a + \lambda b, 0) + K + (-\lambda b, b) + K \in \gamma A + I.$$

Thus $D = \gamma A + I$. If $z \in \gamma A \cap I$, then $z = (a, 0) + K = (-\lambda b, b) + K$ for some $a \in A$, $b \in B$. Thus in $A \oplus B$, the equation $(a, 0) = (-\lambda b, b) + (\alpha c, -\beta c)$ holds for some $c \in C$. Consequently,

$$b = \beta c, \quad \text{and} \quad (-\lambda b, b) = (-\lambda \beta c, \beta c) = (-\alpha c, \beta c) \in K$$

also. Hence $(a, 0) \in K$, $z = 0$, and $D = \gamma\beta A \oplus I$. If $\gamma a = (a, 0) + K = K$, then $(a, 0) = (\alpha c, -\beta c)$ for some $c \in C$. Then $\beta c = 0$. But $\alpha = \lambda\beta$ and also $a = \lambda\beta c = 0$. Thus $\gamma A \cong A$.

The next paragraph identifies certain submodules of $A \oplus B$ which always must be factored out in forming the pushout $D = (A \oplus B)/K$.

14-2.6   Properties of pushouts. If $\langle D, \gamma, \delta \rangle$ is a standard pushout of $\alpha: C \to A$ and $\beta: C \to B$, then

    (1) $\alpha(\ker \beta) \times \{0\} \subset K, \ \{0\} \times \beta(\ker \alpha) \subset K, \ \alpha(\ker \beta) \times \beta(\ker \alpha) \subset K;$

(2) α is monic $\Rightarrow \delta$ is monic. (β monic $\Rightarrow \gamma$ monic.)

(3) $A/\alpha C \to D/\delta B$, $a + \alpha C \to \gamma a + \delta B$ is an isomorphism; $(B/\beta C \cong D/\gamma A)$. Hence δ is epic $\Leftrightarrow \alpha$ is epic; (γ epic $\Leftrightarrow \beta$ epic).

Proof. (2): If $\delta b = (0, b) + K = K$ for some $b \in B$, then $(0, b) = (\alpha c, -\beta c) = (0, -\beta c)$ for some $c \in \ker \alpha \subset C$. Since $\ker \alpha = 0$, $c = 0$, and $b = \beta c = 0$.

(3): The map is well defined and onto because $\gamma A + \delta B = D$. If $\gamma a \in \delta B$, then $(a, 0) = (\alpha c, -\beta c) + (0, b)$ for some $c \in C$ and $b \in B$. Hence $a = \alpha c$, and $A/\alpha C \cong D/\delta B$.

14-2.7 Pushout of a pushout is a pushout. Start with module maps $C \xrightarrow{\sigma} B$ and $C \xrightarrow{\gamma_1} C_1 \xrightarrow{\gamma_2} C_2$. Form the pushout $\langle B_1, \beta_1, \sigma_1 \rangle$ of σ and γ_1, and then the pushout $\langle B_2, \beta_2, \sigma_2 \rangle$ of σ_1 and γ_2. Then $\langle B_2, \beta_2 \beta_1, \sigma_2 \rangle$ is a pushout of β and $\gamma_2 \gamma_1$.

1st Proof. In the first proof the given and yet to be constructed pushouts are standard. In one step, let $\langle B_{21}, \beta, \rho \rangle$ be the pushout of σ and $\gamma_2 \gamma_1$. Thus there are two commutative diagrams

Here $b \in B$, $c \in C$, $c_1 \in C_1$, and $c_2 \in C_2$ will represent arbitrary elements. First, $B_1 = (B \oplus C_1)/N_1$, $N_1 = \{(\sigma c_1 - \gamma_1 c) \mid c \in C\}$ and $\beta_1 b = (b, 0) + N_1$, $\sigma_1 c_1 = (0, c_1) + N_1$. Second,

$$B_2 = (B_1 \oplus C_2)/N_2, \quad \text{and} \quad N_2 = \{[\sigma_1 c_1, -\gamma_2 c_1] \mid c_1 \in C_1\}.$$

Thirdly,

$$N_2 = \{[(0, c_1) + N_1, -\gamma_2 c_1] \mid c_1 \in C_1\}.$$

Fourthly,

$$B_{21} = (B \oplus C_2)/N,$$

where $N = \{(\sigma c, -\gamma_2 \gamma_1 c) \mid c \in C\}$. Map $\phi: B_{21} \to B_2$ by

$$\phi\{(b, c_2) + N\} = [\beta_1 b, c_2] + N_2 = [(b, 0) + N_1, c_2] + N_2.$$

We verify that ϕ is well defined and independent of the choice of coset representative, that it is onto, and one to one.

Suppose that $N = (\sigma c, -\gamma_2\gamma_1 c) + N$. Then by use of $\beta_1\sigma = \sigma_1\gamma_1$, it follows that $\phi\{N\} = [B_1\sigma c, -\gamma_2\gamma_1 c] + N_2 = [\sigma_1\gamma_1 c, -\gamma_2\gamma_1 c] + N_2$. If $c^* = \gamma_1 c$, then the latter coset representative is $[\sigma_1 c^*, -\gamma_2 c^*] \in N_2$. Thus $\phi\{N\} = N_2 = 0$, and ϕ is well defined.

Let $y = [(b, c_1) + N_1, c_2] + N_2 \in B_2$ be arbitrary. Define $x = (b, c_2 + \gamma_2 c_1) + N \in B_{21}$. Then $\phi x = [(b, 0) + N_1, c_2 + \gamma_2 c_1] + N_2$. Since $[\sigma_1 c_1, -\gamma_2 c_1] = [(0, c_1) + N_1, -\gamma_2 c_1] \in N_2$, $\phi x = [(b, c_1) + N_1, c_2] + N_2 = y$. So ϕ is onto.

Suppose that $z = (b, c_2) + N \in B_{21}$ with $0 = \phi z = (\beta_1 b, c_2) + N_2$. Thus $(\beta_1 b, c_2) = [\sigma_1 c_1, -\gamma_2 c_1]$ for some $c_1 \in C_1$. Upon equating components we get $c_2 = -\gamma_2 c_1$, and that $(b, 0) + N_1 = \beta_1 b = \sigma_1 c = (0, c_1) + N_1$. Consequently $(b, 0)$ and $(0, c_1)$ as elements of $B \oplus C_1$ differ by an element of N_1, i.e. $(b, 0) = (0, c_1) + (\sigma c^*, -\gamma_1 c^*)$ for some $c^* \in C$. Thus $b = \sigma c^*$, $c_1 = \gamma_1 c^*$ and $(b, c_2) = (\sigma c^*, -\gamma_2 c_1) = (\sigma c^*, -\gamma_2\gamma_1 c^*) \in N$. Hence $z = 0$, and ϕ is monic.

For any $b \in B$, $\phi\beta b = \phi\{(b, 0) + N\} = (\beta_1 b, 0) + N_2 = \beta_2\beta_1 b$. Thus $\phi B = \beta_2\beta_1$. For any $c_2 \in C_2$, $\phi\rho c_1 = \phi\{(0, c_2) + N\} = (0, c_2) + N_2 = \sigma_2 c_2$. Hence also $\phi\rho = \sigma_2$. The commutative diagram below with ϕ an isomorphism by 13-2.2 implies that $\langle B_2, \beta_2\beta_1, \sigma_2 \rangle$ is a pushout of σ_1 and $\gamma_2\gamma_1$.

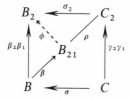

2nd Proof. This is obtained by dualizing the second proof in 14-1.7.

Just as before, this construction may be iterated for any finite hence countable number of C_n's.

The next simple observation gives another way of looking at pushouts, and its dual–at pullbacks.

14-2.8 Given the commutative square of module maps below, from it form the sequence $C \to A \oplus B \to D$ as follows:

$$C \xrightarrow{\beta} B \qquad C \xrightarrow{\alpha \oplus (-\beta)} A \oplus B \xrightarrow{\text{so}(\gamma \oplus \delta)} D$$

$$\alpha \downarrow \qquad \downarrow \delta \qquad c \longrightarrow (\alpha c, -\beta c)$$

$$A \xrightarrow{\gamma} D \qquad \qquad (a, b) \longrightarrow \gamma a + \delta b$$

$$\qquad \qquad \qquad c \in C;\ a \in A,\, b \in B.$$

Then $\langle D, \gamma, \delta \rangle$ is a pushout of α and $\beta \Leftrightarrow C \to A \oplus B \to D \to 0$ is exact.

Proof. Form the standard pushout $(A \oplus B)/K$ of α and β. In both cases there will be a well defined isomorphism $\varepsilon: D \to (A \oplus B)/K$ such that $\varepsilon \gamma a = (a, 0) + K$ and $\varepsilon \delta b = (0, b) + K$ for all $a \in A$, $b \in B$. The rest of the proof is omitted.

14-3   Pushout application

There are at least two distinct ways of using module constructs. Sometimes it helps to recognize that a module that we have constructed during a proof is of a certain known type, say a pullback. Or in attempting to construct a module satisfying certain constraints, the missing module just might be a pushout. The key to one such application of the latter kind is 14-2.6(2).

14-3.1   Problem. Suppose that given are modules $C \subset A$, B and any homomorphism $\beta: C \to B$; and that the problem is to (i) embedd B in any bigger module $B \subset D$ such that (ii) β extends to $\gamma: A \to D$ so that the restriction $\gamma \mid C$ of γ to C is $\gamma \mid C = \beta$.

Solution. Let $\alpha: C \to A$ be the inclusion and $\langle D, \gamma, \delta \rangle$ the standard pushout of α and β. Then $D = (A \oplus B)/K$ where $K = \{(\alpha c, -\beta c) \mid c \in C\} = \{(c, -\beta c) \mid c \in C\}$. The latter is the graph of $(-\beta): C \to B$. Write $K = \text{gr}(-\beta)$. The advantage of this more explicit notation for K is that it emphasizes that $\delta: B \to D$ is monic, for whenever $(0, b) \in \text{gr}(-\beta)$ then necessarily $b = 0$. Hence identify $B \equiv [\{0\} \times B + \text{gr}(-\beta)]/\text{gr}(-\beta) \subset D$.

The commutativity $\gamma \alpha = \delta \beta$ of a square simply says that $\gamma \mid C = \beta: \gamma \alpha c = \gamma c = (c, 0) + \text{gr}(-\beta) = (c, 0) + (-c, \beta c) + \text{gr}(-\beta) = (0, \beta c) + \text{gr}(-\beta) = \delta \beta c$. The commutative diagrams below indicate two equivalent ways to picture the present solutions of our problems.

$$C \underset{1-1}{\overset{\alpha}{\rightarrowtail}} A \qquad c \qquad C \qquad \qquad \subset \quad A \qquad a$$

$$\beta \downarrow \qquad \downarrow \gamma \qquad \Big| \qquad \Big| \beta \qquad\qquad\qquad\qquad \gamma \downarrow \qquad\qquad \Big|$$

$$B \underset{\delta}{\overset{1-1}{\rightarrowtail}} D \qquad \Big| \qquad\qquad B \equiv \frac{\{0\} \times B + \mathrm{gr}(j-\beta)}{\mathrm{gr}(-\beta)} \subset \frac{A \oplus B}{\mathrm{gr}(-\beta)} \qquad \Big|$$

$$\downarrow \qquad\qquad\qquad\qquad\qquad\qquad\qquad\qquad\qquad\qquad\qquad\qquad \downarrow$$

$$(0, \beta c) + \mathrm{gr}(-\beta) \qquad\qquad\qquad\qquad\qquad\qquad\qquad (a,\ 0) + \mathrm{gr}(-\beta)$$

14-3.2 Problem. Suppose that $\mathcal{F} = \{f\}$ is an indexed family of module homomorphisms $f : C_f \to B$ where all $C_f \leqslant A$ for some module A. It is required to construct a module D which (i) contains B, and such that (ii) every $f \in \mathcal{F}$ extends to a map $\gamma_f : A \to D$.

Solution. Set $A_f = A$ for all $f \in \mathcal{F}$ and $H = \oplus \{A_f \mid f \in \mathcal{F}\}$. Let $f, g \in \mathcal{F}$, $f \neq g$. We could regard $C_f,\ C_g \leqslant A$, in which case $C_f \cap C_g \neq 0$ could hold. However, we will not do this. Instead we view $C_f \leqslant A_f < H$. In so doing we have disjointified the C_f's, i.e. $C_f + C_g = C_f \oplus C_g \leqslant A_f \oplus A_g < H$ with $A_f \cap A_g = 0$. Then $\mathrm{gr}(-f) = \{(c, -f(c)) \mid c \in C_f\} < C_f \oplus B \leqslant A_f \oplus B < H \oplus B$, and $\mathrm{gr}(-f)$ is a right R-submodule of $H \oplus B$. Hence $K \equiv \Sigma\{\mathrm{gr}(-f) \mid f \in \mathcal{F}\} < H \oplus B$, because K is a sum of right R-submodules. More explicitly, every $0 \neq k \in K$ is a unique sum $k = \Sigma_f(c_f, -f(c_f))$, where $0 \neq c_f \in C_f$.

 Define $D = (H \oplus B)/K$ and $\delta : B \to D$ by $\delta b = (0, b) + K$. Then δ is monic. For if $\bar{0} = K = \delta b = (0, b) + K$, then $(0, b) = \Sigma_f(c_f, -f(c)) \in K$. By uniqueness, $b = 0$. Identity $B \cong \delta B = [\{0\} \oplus B + K]/K$.

 Given any $f \in \mathcal{F}$, define $\gamma_f : A \to D$ by identifying $A = A_f$, and $\gamma_f(a) = (a, 0) + K$ for $a \in A$, where $(a, 0) \in A_f \oplus B < H \oplus B$. If $c \in C_f < A$, then

$$\gamma_f(c) = (c, 0) + K = (c, 0) + (-c, f(c)) + K = (0, f(c)) + K = \delta(f(c)).$$

Thus $\gamma_f = \delta f$. (Note that in general for $f \neq g \in \mathcal{F}$, $\gamma_f \neq \gamma_g$.)

$$C_f \qquad\qquad \subset \qquad\qquad A$$

$$\downarrow f$$

$$B \qquad\qquad\qquad\qquad \vdots\, \gamma_f$$

$$\downarrow \delta$$

$$B \cong \frac{\{0\} \oplus (B + K)}{K} \subset \frac{[\oplus A_f] \oplus B}{K}$$

The crucial step in showing that every module M has an injective hull was to show that M could be embedded in an injective module. A totally different alternate proof of this will now be given by use of pushouts. We simply will keep enlarging M until the resulting module satisfies Baer's criterion.

14-3.3. Here we only need to assume that $1 \in R$. Consider right ideals $L < R$ and the set $\mathscr{F} = \{f\}$ of all homomorphisms $f : L \to M$. In the notation of 14-3.2, set $C_f = L$, $A = R$, and $B = M$. Thus for every $f \in \mathscr{F}$, there exists an embedding $M = B \subset D$, and $\gamma : R \to D$ such that $\gamma \mid L = f$. The module D need not satisfy Baer's criterion, because there might be a right ideal $L < R$ and a homomorphism $L \to D$ which is not given by any $f \in \mathscr{F}$, and hence which might not lift to R.

So set $M_0 = M$, $M_1 = D$ and repeat the above construction with M replaced by M_1 yielding a new $D = M_2$ with $M_0 \subseteq M_1 \subseteq M_2$. We continue and obtain a countable sequence $M_0 \subseteq \cdots \subseteq M_n \subseteq M_{n+1} \subseteq \cdots$ for all integers n. Set $M_\omega = \cup M_n$. Now for each nonlimit ordinal like $\omega + 1$, repeat the above construction. If we reach a limit ordinal β, define $M_\beta = \cup \{M_\alpha \mid \alpha < \beta\}$ to be the ascending union. Let θ be the (unique) smallest ordinal whose cardinality $|\theta|$ is equal to the successor cardinal $|R|^+$ of the cardinality $|R|$ of R, i.e. $|\theta| = |R|^+ > |R|$. For every ordinal $\lambda < \theta$, $|\lambda| < |\theta|$; θ is an initial ordinal, and $M_\theta = \cup \{M_\lambda \mid \lambda < \theta\}$ is a union of an ascending chain of submodules $M_\lambda < M_\theta$.

To see that M_θ is injective, let $L < R$ and $f : L \to M_\theta$ be any R-map. For every $r \in L$, let $\lambda(r)$ be the smallest ordinal such that $fr \in M_{\lambda(r)}$. Then $\lambda(r) < \theta$, and $|\lambda(r)| < |\theta| = |R|^+$. Hence $|\lambda(r)| \leqslant |R|$. Every set of ordinal numbers has a supremum; in particular, define

$$\tau = \text{supremum}\{\lambda(r) \mid r \in R\}.$$

Since each $|\lambda(r)| \leqslant |R|$ also $|\tau| \leqslant |R| < |\theta|$. Hence $\tau < \theta$. Since θ is a limit ordinal also $\tau + 1 < \theta$. Consequently, for every $r \in R$,

$$r \in M_{\lambda(r)} \subseteq M_\tau \subset M_{\tau+1} \subset M_\theta.$$

Therefore, $f(L) \subseteq M_\tau$. By the construction of $M_{\tau+1}$, the map $f : L \to M_\tau$ can be enlarged at both ends to a map $\gamma : R \to M_{\tau+1}$, with $\gamma \mid L = f$. However, view γ as a map $\gamma : R \to M_\theta$. By Baer's criterion, M_θ must be an injective module with $M \subset M_\theta$.

14-4 Exercises

Prove the following assertions.

1. Let $\langle D, \gamma, \delta \rangle$ be a pullback of $\alpha : A \to C$ and $\beta : B \to C$. Then α is

monic $\Leftrightarrow \delta$ is monic. (*Hint:* use 14-1.4.)

2. For a pushout $\langle D, \gamma, \delta \rangle$ of $\alpha : C \to A$ and $\beta : C \to B$, the map α is onto $\Leftrightarrow \delta$ is onto.

3. Given is a commutative diagram with an exact bottom row.

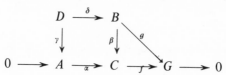

Then $\langle D, \gamma, \delta \rangle$ is a pullback of α and $\beta \Leftrightarrow 0 \to D \xrightarrow{\delta} B \xrightarrow{g} G$ is exact.

4. Show that in the row exact commutative diagram below the dotted arrow h can be filled in to yield a commutative diagram. Then dualize this by reversing all arrows.

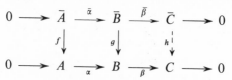

5. *Nine lemma.* In the commutative diagram below with exact rows and exact middle column, the left column is exact if and only if the right one is exact.

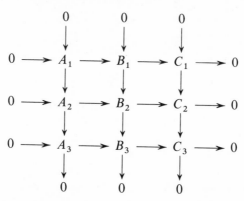

6. Consider the row exact commutative diagram below and prove the following.

$$\begin{array}{ccccccccc}
A & \xrightarrow{\alpha} & B & \xrightarrow{\beta} & C & \xrightarrow{\gamma} & D & \xrightarrow{\delta} & E \\
\downarrow{\scriptstyle f} & & \downarrow{\scriptstyle g} & & \downarrow{\scriptstyle h} & & \downarrow{\scriptstyle i} & & \downarrow{\scriptstyle j} \\
\bar{A} & \xrightarrow{\bar{\alpha}} & \bar{B} & \xrightarrow{\bar{\beta}} & \bar{C} & \xrightarrow{\bar{\gamma}} & \bar{D} & \xrightarrow{\bar{\delta}} & \bar{E}
\end{array}$$

(i) $fA = \bar{A}$, ker $g = 0$, ker $i = 0 \Rightarrow$ ker $h = 0$;

(ii) $fA = \bar{A}$, ker $j = 0$, ker $g = 0$, ker $i = 0 \Rightarrow h$ is an isomorphism;

(iii) ker $j = 0$, $gB = \bar{B}$, $iD = \bar{D} \Rightarrow hC = \bar{C}$.

7. Given in the diagram below are the exact bottom row and $\gamma : \bar{C} \to C$. Show that there exist a commutative diagram, with an exact top row where $\alpha = 1$ is the identity map on A.

$$
\begin{array}{ccccccccc}
0 & \longrightarrow & \bar{A} & \xrightarrow{\bar{f}} & \bar{B} & \xrightarrow{\bar{g}} & \bar{C} & \longrightarrow & 0 \\
& & \big\| {\scriptstyle\alpha = 1_A} & & \Big| {\scriptstyle\beta} & & \Big\downarrow {\scriptstyle\gamma} & & \\
0 & \longrightarrow & A & \xrightarrow[f]{} & B & \xrightarrow[g]{} & C & \longrightarrow & 0
\end{array}
$$

(*Hint*: Let $\bar{B}$ be the pullback of g and γ, $\bar{B} \subseteq B \oplus \bar{C}$. Define $\bar{f}$ by $\bar{f}a = (fa, 0) \in \bar{B}$.)

8. Given in the diagram below is the exact top row and $\alpha : A \to \bar{A}$. Prove that there exists a commutative diagram with an exact bottom row where $\gamma = 1_C$ is the identity on C.

$$
\begin{array}{ccccccccc}
0 & \longrightarrow & A & \xrightarrow{f} & B & \xrightarrow{g} & C & \longrightarrow & 0 \\
& & \Big\downarrow {\scriptstyle\alpha} & & \Big\downarrow {\scriptstyle\beta} & & \big\| {\scriptstyle\gamma = 1_C} & & \\
0 & \dashrightarrow & \bar{A} & \dashrightarrow[\bar{f}] & \bar{B} & \dashrightarrow[\bar{g}] & C & \dashrightarrow & 0
\end{array}
$$

(*Hint*: Let $\bar{B}$ be the pushout of α and f, $\bar{B} = (\bar{A} \oplus B)/K$. Show that $\bar{g}$ is well defined by $\bar{g}[(\bar{a}, b) + K] = gb$.)

9. In the row exact commutative diagram below, all the columns are given to be exact, except column 3. Prove also that column 3 is exact.

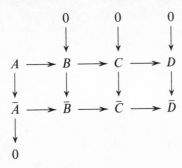

CHAPTER 15

Categories and functors

Introduction

The first section contains categories, morphisms, functors, natural transformations of functors, the hom functors, and bifunctors. Only in the second section the objects, subobjects, and quotient objects of categories make their appearance.

In order for the concepts of additive and exact functor to even make sense, pre-additive, exact, and abelian categories have to be at least defined in Section 3. Section 4 develops adjoint functors without imposing any restrictive hypotheses on the underlying categories.

It is not the intention of this chapter to develop an axiomatically exhaustive treatment of abelian categories. For later purposes it suffices to be aware of the fact that some proofs carried out in module categories could be done in an abelian category, and that some proofs done in terms of elements could be done categorically.

However since category theory requires a large amount of notation and terminology, some brief comments are made along the way on alternative notation and definitions which the reader will encounter in the literature.

15-1 Basics of categories

The blanket assumption is made throughout that the so-called morphisms between any two objects of a category form a set, as opposed to a class, although this assumption is not used until Section 3, and even then it could be avoided.

15-1.1 Category. Consider a class $\mathrm{Ob}\,\mathscr{C}$ whose elements are called *objects*, and another class $\mathrm{Morph}\,\mathscr{C}$ whose elements are called *morphisms*, or arrows, and sometimes even maps. Suppose that for any pair $A, B \in \mathrm{Ob}\,\mathscr{C}$, there is a definite subset $\mathrm{Morph}_{\mathscr{C}}(A, B) \subset \mathrm{Morph}\,\mathscr{C}$, whose elements are morphisms. Abbreviate $\text{Morph}_{\mathscr{C}}(A, B) = \mathscr{C}(A, B)$. Suppose further that for each triple $A, B, C \in \text{Ob}\,\mathscr{C}$ there is a function, or a composition law '$\circ$', where for $f \in \mathscr{C}(A, B)$ and $g \in \mathscr{C}(B, C)$, the composite or *product* $g \circ f = gf \in \mathscr{C}(A, C)$:

$$\circ : \mathscr{C}(B, C) \times \mathscr{C}(A, B) \longrightarrow \mathscr{C}(A, C)$$

$$(g, f) \longrightarrow g \circ f = gf$$

$$A \xrightarrow{\;f\;} B \quad B \xrightarrow{\;g\;} C.$$

Let $\mathscr{C}(\cdot, \cdot)$ denote the above function on $\text{Ob}\,\mathscr{C} \times \text{Ob}\,\mathscr{C}$. Then the four truple $\langle \text{Ob}\,\mathscr{C}, \text{Morph}\,\mathscr{C}, \mathscr{C}(\cdot, \cdot), \circ \rangle$ is a *category* provided it satisfies the following three axioms.

   (i) **Identity.** For any object $B \in \text{Ob}\,\mathscr{C}$ there exists a unique morphism $1_B : B \to B$ called the *identity morphism* such that for any other morphisms $f : A \to B$ and $g : B \to C$ the equations $1_B \circ f = f$ and $g \circ 1_B = g$ hold. Then 1_B is called the identity or sometimes the identity map.
   (ii) **Associativity.** Every triple of morphisms of the form $f : A \to B$, $g : B \to C$, and $h : C \to D$ associate, $(f \circ g) \circ h = f \circ (g \circ h)$.
   (iii) For any $A, B, C, D \in \text{Ob}\,\mathscr{C}$, if

$$(A, B) \neq (C, D), \quad \text{then } \mathscr{C}(A, B) \cap \mathscr{C}(C, D) = \varnothing.$$

So far the symbol '$\mathscr{C}$' by itself has not been defined.

15-1.2. Definition. Define $\mathscr{C}$ to be the disjoint union

$$\mathscr{C} = \text{Ob}\,\mathscr{C} \cup \text{Morph}\,\mathscr{C}.$$

(Thus $\text{Ob}\,\mathscr{C} \cap \text{Morph}\,\mathscr{C} = \varnothing$.) In place of $A \in \text{Ob}\,\mathscr{C}$ and $f \in \text{Morph}\,\mathscr{C}$, the notation $A \in \mathscr{C}$ and $f \in \mathscr{C}$ will sometimes be used. The previously defined category in 15-1.1 will be simply referred to as $\mathscr{C}$.

Below are given some notations which are not used here, but are used by other authors elsewhere.

15-1.3 Remark. (1) $\text{Ob}\,\mathscr{C} = \text{Obj}\,\mathscr{C}$; $\text{Morph}\,\mathscr{C} = \text{Mor}\,\mathscr{C} = \text{Map}\,\mathscr{C}$.
   (2) $\text{Morph}_{\mathscr{C}}(A, B) = \text{Mor}_{\mathscr{C}}(A, B) = \text{Mor}(A, B) = \text{Hom}_{\mathscr{C}}(A, B) = \text{Hom}(A, B) = (A, B)$.
   (3) Some authors identify each object $A \in \text{Ob}\,\mathscr{C}$ with the unique

morphism $1_A \in$ Morph $\mathscr{C}$, and then dispense with the objects Ob $\mathscr{C}$ altogether (Freyd [64]).

15-1.4 Subcategory. For some category $\mathscr{C}$ consider a subsystem

$$\mathscr{D} = \langle \text{Ob } \mathscr{D}, \text{ Morph } \mathscr{D}, \mathscr{D}(\cdot, \cdot), \circ \rangle$$

of $\mathscr{C}$, where Ob $\mathscr{D} \subseteq$ Ob $\mathscr{C}$; Morph $\mathscr{D} \subseteq$ Morph $\mathscr{C}$; and for any $A, B \in$ Ob $\mathscr{D}$, $\mathscr{D}(A, b) \subseteq \mathscr{C}(A, B)$. If under the same composition law '$\circ$' as in $\mathscr{C}$, $\mathscr{D}$ is a category (i.e. satisfies 15-1.1(i) and (ii)), then $\mathscr{D}$ is a *subcategory* of $\mathscr{C}$.

Furthermore, $\mathscr{D}$ is a *full* subcategory of $\mathscr{C}$ if $\mathscr{D}(A, B) = \mathscr{C}(A, B)$ for all $A, B \in$ Ob $\mathscr{C}$. Thus a subcategory $\mathscr{D}$ is full in $\mathscr{C}$ if and only if $\mathscr{D}(\cdot, \cdot)$ is the restriction of $\mathscr{C}(\cdot, \cdot)$ to Ob $\mathscr{D} \times$ Ob $\mathscr{D}$.

Subsequently, symbols of the form A, B, C, D, or '$\alpha : A \to B$', mean that A, B, C, D are objects and that α is actually a morphism in some fixed underlying, category without repeating the words 'object', 'morphism', or naming the category.

15-1.5 Morphisms. Let $f : A \to B$ be any morphism in some category.

(i) Then f is an *isomorphism* if there exists a morphism $g : B \to A$ such that $gf = 1_A$ and $fg = 1_B$ are the identity maps. Suppose that also $g_2 : B \to A$ with $g_2 f = 1_A$ and $fg_2 = 1_B$. From $gf = g_2 f$ it follows that $gfg = g_2 fg$; but $g = g1_B = gfg$ and $g_2 = g_2 1_B = g_2 fg$. Thus $g = g_2$. The unique morphism g is called the *inverse* of f and written also as $g = f^{-1}$. In this case we say that A is isomorphic to B, and write $A \cong B$, or $B \cong A$.

(ii) The morphism f is a *monomorphism* if for any other morphisms $g, h : C \to A$ with $fg = fh$, it follows that $g = h$. Monomorphisms may be cancelled from the left.

(iii) The morphism f is an *epimorphism* if whenever $g, h : B \to C$ with $gf = hf$, then it is the case that $g = h$. Epimorphisms are right cancellable.

An isomorphism is both a monomorphism and an epimorphism, whereas the converse is false in general.

(iv) We say that A can be mapped to B if there exists a morphism from A to B. If every object A of the category can be mapped by a unique morphism to B, then B is called a *terminal object*. A fixed object A is an *initial object*, if for any object B, A can be mapped uniquely to B.

A *zero object* is an object $(0) \in$ Ob $\mathscr{C}$ such that for any objects $A, B \in$ Ob $\mathscr{C}$, there exist unique morphisms

$$0_{(0),A} : A \to (0) \quad \text{and} \quad 0_{B,(0)} : (0) \to B.$$

Define the zero morphism $0_{A,B}$ from A to B to be the composite $0_{A,B} = 0_{B,(0)} 0_{(0)A}$ of $A \to (0)$ and $(0) \to B$. First, any two zero objects of $\mathscr{C}$ are isomorphic. Secondly, the above defined morphism $0_{A,B}$ is independent of the choice of zero object. In view of these two reasons, we refer to any zero object as the zero object, and frequently abbreviate this zero object, and all zero maps by $(0) = 0$ and $0 = 0_{A,B}$. A zero object is also called a null object. The composition of any zero morphism $0 = 0_{A,B} : A \to B$ with any other morphism $f : A' \to A$ or $g : B \to B'$ is zero, i.e. $0_{A,B}f = 0$ and $g0_{A,B} = 0$.

In all our examples of categories, the axiom 15-1.1(iii) is automatically invoked.

15-1.6 Notation. Some standard categories are listed.

(1) For any ring R with or without an identity there is a category whose objects are right R-module and whose morphisms are module homomorphisms, denoted by $\mathscr{M}od_R$. To avoid trivialities, in case $1 \in R$, it will be assumed that $\mathscr{M}od_R$ denotes the category of unital right R-modules. The same conventions apply to the category of left R-modules $_R\mathscr{M}od$.

(2) If $R = \mathbb{Z}$, the $\mathscr{A}b = \mathscr{M}od_{\mathbb{Z}} = {_\mathbb{Z}}\text{Mod}$ denotes the category of abelian groups.

(3) $\mathscr{S}ets$ is the category whose objects are sets and morphisms are functions of sets.

(4) $\mathscr{Q}uoset$ is the category whose objects are quasi-ordered sets and morphisms are order preserving functions. (See 0-1.3.)

(5) $\mathscr{P}oset$: partially ordered sets and order preserving functions.

(6) $\mathscr{S}grp$: semigroups and semigroup homomorphisms.

(7) $\mathscr{M}on$: monoids and identity preserving semigroup homomorphisms.

(8) $\mathscr{G}rp$: groups and group homomorphisms.

(9) $\mathscr{R}ing$: rings R ($1 \in R$ or $1 \notin R$) and ring homomorphisms.

15-1.7 Examples. (i) In $\mathscr{M}od_R$, monomorphisms are one to one functions (injective, monic); epimorphisms are onto functions (surjective, epic); isomorphisms are surjective monics. The same holds for $\mathscr{A}b$ and $\mathscr{S}et$.

(ii) In the category $\mathscr{R}ing$, those with identity and identity preserving ring homomorphism form a subcategory $\mathscr{R}ing^{(1)}$, but not a full one. The inclusion map $\mathbb{Z} \to Q$ of the integers into the rationals is an epimorphism in $\mathscr{R}ing$ and hence automatically in $\mathscr{R}ing^{(1)}$, but it is not an isomorphism.

(iii) In the category of topological Hausdorff spaces and continuous functions, the inclusion map $Q \to \mathbb{R}$ of the rationals into the reals is both an epimorphism and a monomorphism which is not an isomorphism.

15-1.8 Functor. Let $\mathscr{C}$ and $\mathscr{D}$ be two categories and suppose that there is a rule F which maps $\mathrm{Ob}\,\mathscr{C} \to \mathrm{Ob}\,\mathscr{D}$, and $\mathrm{Morph}\,\mathscr{C} \to \mathrm{Morph}\,\mathscr{D}$ such that $F(1_A) = 1_{F(A)}$ for any object A in $\mathscr{C}$. Let $f : A \to B$ and $g : B \to C$ in $\mathscr{C}$ be arbitrary.

Then F is a *covariant functor* if the composite of the two morphisms $F(f) : F(A) \to F(B)$ and $F(g) : F(B) \to F(C)$ in $\mathscr{D}$ is the same as $\mathscr{F}$ applied to the composite gf of f and g in $\mathscr{C}$, i.e. if $F(gf) = (Fg) \circ (Ff)$.

An F is a *contravariant functor* if F reverses all arrows, $F(f) : F(B) \to F(A)$ in going from $\mathscr{C}$ to $\mathscr{D}$, and $F(gf) = (Ff) \circ (Fg)$.

The term 'covariant functor' will be abbreviated to just *functor*.

A functor $F : \mathscr{C} \to \mathscr{D}$ induces a function

$$f \to Ff : \mathscr{C}(A, B) \to \mathscr{D}(FA, FB);$$

and F is *faithful* provided that this new induced map of sets is injective (i.e. one to one) for all $A, B \in \mathrm{Ob}\,\mathscr{C}$. (The term '$F$ is an embedding' is also used in place of faithful, but this may be misleading. A faithful functor need not be one to one on objects.) The functor F is *full* if all of the above maps are surjective (onto), i.e. $\mathscr{D}(FA, FB) = \{Ff \mid f \in \mathscr{C}(A, B)\}$.

The identity function on all objects and maps of $\mathscr{D}$ is called the *identity functor* $1_{\mathscr{D}} : \mathscr{D} \to \mathscr{D}$ of $\mathscr{D}$.

If $\mathscr{C} \subset \mathscr{D}$ is a subcategory, then evidently the inclusion function of objects and maps of $\mathscr{C}$ and $\mathscr{D}$ is a functor. It is called the *inclusion functor*. Thus $\mathscr{C}$ is a full subcategory of $\mathscr{D}$ if and only if the inclusion functor is a full functor.

If $F : \mathscr{C} \to \mathscr{D}$ and $G : \mathscr{D} \to \mathscr{E}$ are functors, then their composite $GF : \mathscr{C} \to \mathscr{E}$ is one too.

For any category whose objects are sets and whose morphisms are functions of sets (with 15-1.1(iii)), the functor which maps this category into the category $\mathscr{S}et_{\mathfrak{s}}$ is called the *forgetful functor*. More generally the functor which forgets or discards some of the additional structure on a category is called the forgetful functor. For example, the functor $F : \mathscr{M}od_R \to \mathscr{A}b$ which regards right R-module and R-homomorphisms as simply abelian groups and abelian group homomorphisms is called a forgetful functor.

15-1.9 Example. The image of a category under a functor need not be a category. Let $\mathrm{Ob}\,\mathscr{C} = \{1, 2, 3, 4\}$ and aside from the four identity morphisms, let $\mathscr{C}$ have the two morphisms $(2, 1) : 1 \to 2$ and $(4, 3) : 3 \to 4$. Let $\mathrm{Ob}\,\mathscr{D} = \{5, 6, 7\}$ and let the nonidentity morphisms of $\mathscr{D}$ be $(6, 5) : 5 \to 6$, $(7, 6) : 6 \to 7$, and their composite $(7, 6) \circ (6, 5) = (7, 5) : 5 \to 7$.

Define a functor $F : \mathscr{C} \to \mathscr{D}$ by:

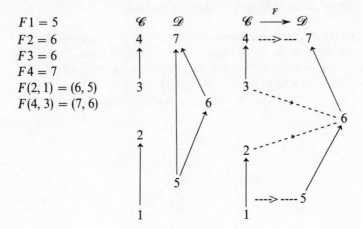

$$F1 = 5$$
$$F2 = 6$$
$$F3 = 6$$
$$F4 = 7$$
$$F(2, 1) = (6, 5)$$
$$F(4, 3) = (7, 6)$$

The composite $F(4, 3) \circ F(2, 1) = (7, 6) \circ (6, 5) = (7, 5)$ is not in the image of F. Hence the image of F is not a category.

Clearly, for any functor $F : \mathscr{C} \to \mathscr{D}$ of any categories, the image of F will be a subcategory in two cases. (i) if F is one to one on objects; or (ii) if F is a full functor. If so, write $F(\mathscr{C}) \subset \mathscr{D}$ for the image subcategory.

15-1.10 Opposite category. Let $\mathscr{C} = \langle \mathrm{Ob}\, \mathscr{C}, \mathrm{Morph}\, \mathscr{C}, \mathscr{C}(\circ, \circ), \circ \rangle$ be any category. The objects of the opposite category $\mathscr{C}^{\mathrm{op}}$ are the same as those of $\mathscr{C}$. With each map $f \in \mathrm{Morph}\, \mathscr{C}$, associate a symbol f° so that there is a bijective $\mathrm{Morph}\, \mathscr{C} \to \mathrm{Morph}\, \mathscr{C}^{\mathrm{op}} = \{f^\circ \mid f \in \mathrm{Morph}\, \mathscr{C}\}$, $f \to f^\circ$, but where $\mathrm{Morph}\, \mathscr{C} \cap \mathrm{Morph}\, \mathscr{C}^{\mathrm{op}} = \varnothing$. For arbitrary $f : A \to B$ and $g : B \to C$ in $\mathscr{C}$, form their composite $gf : A \to C$. Then the corresponding arrows are reversed in $\mathscr{C}^{\mathrm{op}}$ as follows

$$C \xrightarrow{g^\circ} B \xrightarrow{f^\circ} A, \quad C \xrightarrow{(gf)^\circ} A, \quad (gf)^\circ = f^\circ \circ g^\circ.$$

The new resulting category $\mathscr{C}^{\mathrm{op}} = \langle \mathrm{Ob}\, \mathscr{C}^{\mathrm{op}}, \mathrm{Morph}\, \mathscr{C}^{\mathrm{op}}, \mathscr{C}^{\mathrm{op}}(\cdot, \cdot), \cdot \rangle$ is called the *opposite category* of $\mathscr{C}$. (In other places it is also sometimes called the dual category of $\mathscr{C}$.) The function $\mathrm{op} : \mathscr{C} \to \mathscr{C}^{\mathrm{op}}$, $\mathrm{op}(A) = A$, and $\mathrm{op}(f) = f^\circ$, is a contravariant functor.

If $\mathscr{C}$ and $\mathscr{D}$ are categories and F is a function mapping $\mathrm{Ob}\, \mathscr{C}$ into $\mathrm{Ob}\, \mathscr{D}$, and $\mathrm{Morph}\, \mathscr{C}$ into $\mathrm{Morph}\, \mathscr{D}$, then there always is a composite function

$$(\mathrm{op}) \circ F : \mathscr{C} \to \mathscr{D} \to \mathscr{D}^{\mathrm{op}}, \quad ((\mathrm{op}) \circ F)(A) = F(A) \quad \text{and} \quad ((\mathrm{op}) \circ F)(f) = (Ff)^\circ.$$

Then $F : \mathscr{C} \to \mathscr{D}$ is a contravariant functor if and only if $(\mathrm{op}) \circ F : \mathscr{C} \to \mathscr{D}^{\mathrm{op}}$ is a covariant functor, and similarly for covariant.

It is clear that $\mathscr{C}$ and $(\mathscr{C}^{\mathrm{op}})^{\mathrm{op}}$ are categories which are isomorphic, in a sense which will now be defined precisely.

15-1.11 Natural transformations. Consider two functors F, $G:\mathscr{C} \to \mathscr{D}$ both from $\mathscr{C}$ to $\mathscr{D}$. A *natural transformation* from F to G is a function $\phi:\mathrm{Ob}\,C \to \mathrm{Morph}\,\mathscr{D}$ satisfying the following two axioms:

(i) $\forall A \in \mathrm{Ob}\,\mathscr{C}$, $\phi(A) \in \mathscr{D}(F(A), G(A))$; and

(ii) $\forall h:A \to B$ in $\mathscr{C}$, there is a commutative diagram:

$$\text{in } \mathscr{C}: \quad A \xrightarrow{\ h\ } B$$

$$\text{in } \mathscr{D}: \quad \begin{array}{ccc} F(A) & \xrightarrow{F(h)} & F(B) \\ {\scriptstyle \phi(A)}\downarrow & & \downarrow{\scriptstyle \phi(B)} \\ G(A) & \xrightarrow[G(h)]{} & G(B) \end{array}$$

If (i) and (ii) hold the abbreviation $\phi:F \to G$ is used. Also the notation $\phi_A = \phi(A)$ and $\phi_B = \phi(B)$ will be used (to emphasize that ϕ_A is a morphism). And if no confusion is possible, sometimes both maps $\phi(A)$ and $\phi(B)$ are replaced by the single ϕ, even in the same formula.

If all $\phi(A)$, $A \in \mathrm{Ob}\,\mathscr{C}$ are isomorphisms, then ϕ is called a *natural isomorphism* of F and G (and by some other writers, natural equivalence, a term never used here). Then there is a natural transformation $\phi^{-1}:G \to F$ which is defined by $\phi^{-1}(A) = \phi(A)^{-1}$. In this case F and G are said to be *isomorphic*; write $F \cong G$. Then '$\cong$' is an equivalence relation on the class of all functors from $\mathscr{C}$ to $\mathscr{D}$.

If F, G, $H:\mathscr{C} \to \mathscr{D}$ are any functors, and $\phi:F \to G$ and $\psi:G \to H$ are any natural transformations, then so is their composite $\psi\phi:F \to H$, where $(\psi\phi)(A) = \psi(A) \circ \phi(A)$.

All of the above in 15-1.11 holds verbatim equally well if all the functors F, G, $H:\mathscr{C} \to \mathscr{D}$ are contravariant, except that the arrows $F(h)$ and $G(h)$ above in (ii) have to be reversed.

15-1.12 Observation. If $H:\mathscr{C} \to \mathscr{C}$ is a functor such that $1_{\mathscr{C}} \cong H$, then H is faithful and full. That is, equivalently, H induces a bijection $\mathscr{C}(A, B) \to \mathscr{C}(HA, HB)$ for all A, $B \in \mathrm{Ob}\,\mathscr{C}$.

Proof. Let $\phi:1_{\mathscr{C}} \to H$ be a natural isomorphism of functors and $f:A \to B$ an arbitrary morphism in $\mathscr{C}$. Then $f = \phi(B)^{-1}Hf\phi(A)$ shows that H is faithful. Given $\gamma:HA \to HB$ in $\mathscr{C}$, define $\delta = \phi(B)^{-1}\gamma\phi(A):A \to B$. Then $\delta = \phi(B)^{-1}(H\delta)\phi(A)$, and hence the last two equations imply that $\gamma = H\delta$.

15-1.13 Two categories $\mathscr{C}$ and $\mathscr{D}$ are *isomorphic* if there exist functors $F:\mathscr{C} \to \mathscr{D}$ and $G:\mathscr{D} \to \mathscr{C}$ such that $GF = 1_{\mathscr{C}}$ and $FG = 1_{\mathscr{D}}$, where $1_{\mathscr{C}}$ and $1_{\mathscr{D}}$ are identity functors. In particular, F and G are bijections between $\text{Ob}\,\mathscr{C}$ and $\text{Ob}\,\mathscr{D}$ as well as between $\text{Morph}\,\mathscr{C}$ and $\text{Morph}\,\mathscr{D}$.

A functor $F_1:\mathscr{C} \to \mathscr{D}$ is a *category equivalence* if there exists some functor $F_2:\mathscr{D} \to \mathscr{C}$ such that $1_{\mathscr{C}} \cong F_2F_1$ and $1_{\mathscr{D}} \cong F_1F_2$. To be more explicit, there exist two natural isomorphisms of functors $\phi_{21}:1_{\mathscr{C}} \to F_2F_1$ and $\phi_{12}:1_{\mathscr{D}} \to F_1F_2$. In this case F_2 is also a category equivalence (by definition) and the categories $\mathscr{C}$ and $\mathscr{D}$ are said to be equivalent, written as $\mathscr{C} \approx \mathscr{D}$.

Let F_1 and F_2 be any pair of category equivalences. It will be shown that F_1 (and F_2) is a faithful and full functor. It follows from $1_{\mathscr{C}} \cong F_2F_1$ that F_1 is faithful, by 15-1.12. For A, $B \in \text{Ob}\,\mathscr{C}$, let $\gamma \in \mathscr{D}(F_1A, F_1B)$. Since by 15-1.12, F_2F_1 induces a surjective function $\mathscr{C}(A, B) \to \mathscr{C}(F_2F_1A, F_2F_1B)$, there is an $f:A \to B$ in $\mathscr{C}$ such that $F_2\gamma = F_2F_1f$. Since F_2 is faithful, $\gamma = F_1f$.

Isomorphic categories are equivalent, but not conversely as the next example shows.

15-1.14 Example. In $\mathscr{M}od_R$ with $R = \mathbb{Z}/2\mathbb{Z} = \{\bar{0}, \bar{1}\}$, let $\mathscr{D}$ be the full subcategory generated by all modules isomorphic to $\mathbb{Z}/2\mathbb{Z}$, $\text{Ob}\,\mathscr{D} = \{Z_2^{\alpha}, Z_2^{\beta}, \ldots, Z_2^{*}, \ldots\}$ is indexed by a class $\{\alpha, \beta, \ldots, *, \ldots\}$ where $Z_2^{\alpha} \cap Z_2^{\beta} = \varnothing$ if $\alpha \neq \beta$, and where $*$ is a distinguished index, with also $Z_2^{*} \cap Z_2^{\alpha} = \varnothing$ except in the one case when $\alpha = *$. Write $\mathscr{D}(Z_2^{\alpha}, Z_2^{\beta}) = \{0^{\alpha\beta}, 1^{\alpha\beta}\}$. Thus $1^{\alpha\alpha}$ and 1^{**} are the identity maps of Z_2^{α} and Z_2^{*}. Let $\mathscr{C}$ be the full subcategory containing the one object Z_2^{*} and thus $\text{Morph}\,\mathscr{C} = \{0^{**}, 1^{**}\}$. Let $F:\mathscr{C} \to \mathscr{D}$ be the inclusion functor. Define $G:\mathscr{D} \to \mathscr{C}$ by $GZ_2^{\alpha} = Z_2^{*}$, $G0^{\alpha\beta} = 0^{**}$, $G1^{\alpha\beta} = 1^{**}$ for all α, β. Then $GF = 1_{\mathscr{C}}$. The next step in showing that $FG \cong 1_{\mathscr{D}}$ will be to define $\phi:FG \to 1_{\mathscr{D}}$. Since FG, $1_{\mathscr{D}}:\mathscr{D} \to \mathscr{D}$, we need $\phi:\text{Ob}\,\mathscr{D} \to \text{Morph}\,\mathscr{D}$ with $\phi(Z_2^{\alpha}) \in \mathscr{D}(FGZ_2^{\alpha}, 1_{\mathscr{D}}Z_2^{\alpha})$. Set $\phi(Z_2^{\alpha}) = 1^{*\alpha}$. If $f^{\alpha\beta} = 1^{\alpha\beta}$ or $0^{\alpha\beta}$ for some α, β, then $Gf^{\alpha\beta} = f^{**}$ where f is the symbol 1 or 0. The following two equal diagrams verify that ϕ is a natural transformation.

$$
\begin{array}{ccc}
FG(Z_2^{\alpha}) & \xrightarrow{\;FG(f^{\alpha\beta})\;} & FG(Z_2^{\beta}) \\
{\scriptstyle\phi(Z_2^{\alpha})}\downarrow & & \downarrow{\scriptstyle\phi(Z_2^{\beta})} \\
1_{\mathscr{D}}(Z_2^{\alpha}) & \xrightarrow[\;1_{\mathscr{D}}(f^{\alpha\beta})\;]{} & 1_{\mathscr{D}}(Z_2^{\beta})
\end{array}
\qquad
\begin{array}{ccc}
Z_2^{*} & \xrightarrow{\;f^{**}\;} & Z_2^{*} \\
{\scriptstyle 1^{*\alpha}}\downarrow & & \downarrow{\scriptstyle 1^{*\beta}} \\
Z_2^{\alpha} & \xrightarrow[\;f^{\alpha\beta}\;]{} & Z_2^{\beta}
\end{array}
$$

Thus $FG \cong 1_{\mathscr{D}}$. Hence $\mathscr{C} \approx \mathscr{D}$, but $\mathscr{C} \not\cong \mathscr{D}$ because there can be no bijection between the singleton set $\text{Ob}\,\mathscr{C}$ and the proper class $\text{Ob}\,\mathscr{D}$.

The above example easily generalizes.

15-1.15 Definition. A category $\mathscr{C}$ is *skeletal* if any two isomorphic objects in $\mathscr{C}$ are equal. Now let $\mathscr{C} \subset \mathscr{D}$ be a subcategory of some category $\mathscr{D}$; $\mathscr{C}$ is *representative* in $\mathscr{D}$ if for any $B \in \text{Ob}\,\mathscr{D}$, there exists $A \in \text{Ob}\,\mathscr{C}$ with $A \cong B$. A *skeleton* of $\mathscr{D}$ is a subcategory $C \subset \mathscr{D}$ which is (i) representative, (ii) skeletal, and moreover (iii) a full subcategory of $\mathscr{D}$.

The proof in the last example can be modified to show the following.

15-1.16 Facts. (1) Any category contains a skeleton, provided that the axiom of choice can be applied to the class of isomorphy classes of objects of this category. (This includes 15-1.6 (1)–(9).)

(2) All skeletons of a category are isomorphic.

(3) The inclusion functor of a skeleton into its original category is a category equivalence. Any category is equivalent to any one of its skeletons.

(4) Let $F_1 : \mathscr{C} \to \mathscr{D}$ be any functor. Then F_1 is a category equivalence $\Leftrightarrow$

(i) F_1 is faithful;

(ii) F_1 is full;

(iii) Every object of $\mathscr{D}$ is isomorphic, in $\mathscr{D}$, to the F-image of some object of $\mathscr{C}$. That is $\forall X \in \text{Ob}\,\mathscr{D}$, $\exists A \in \text{Ob}\,\mathscr{C}$ and there exists an isomorphism $\gamma : FA \to X$ with $\gamma \in \text{Morph}\,\mathscr{D}$.

(5) Let $\mathscr{D}$ be a category with a skeleton $\mathscr{C}$, and $F_1 : \mathscr{C} \to \mathscr{D}$ the inclusion functor. Then there is a category equivalence $F_2 : \mathscr{D} \to \mathscr{C}$ associated to F_1 with $F_1 F_2 \cong 1_{\mathscr{D}}$ but with $F_2 F_1 = 1_{\mathscr{C}}$.

(6) Two categories are equivalent if and only if they have isomorphic skeletons. Hence

(7) category equivalence is an equivalence relation on the class of all categories.

The so-called hom functors discussed below are a generalization of $\text{Hom}_R(\cdot, \cdot)$ on $\mathscr{M}\!od_R$. They are important because they exist on every category.

15-1.16 Hom functors. As before, the category whose objects are sets and morphisms are functions of sets will be called $\mathscr{S}\!ets$. For a completely arbitrary category $\mathscr{C}$, suppose that $X \in \text{Ob}\,\mathscr{C}$ is fixed but $f : A \to B$, $\alpha : X \to A$, and $\beta : B \to X$ are arbitrary. Then a functor $\mathscr{C}(X, \cdot) : \mathscr{C} \to \mathscr{S}\!ets$ and a similar contravariant functor $\mathscr{C}(\cdot, X)$, called the hom functors are defined as follows:

$$X \xrightarrow{\alpha} A \xrightarrow{f} B \qquad\qquad A \xrightarrow{f} B \xrightarrow{\beta} X$$

$\mathscr{C}$:

$$A \xrightarrow{f} B \qquad\qquad\qquad A \xrightarrow{f} B$$

$$\Big\downarrow \mathscr{C}(X,\,\cdot\,) \qquad\qquad\qquad \Big\uparrow \mathscr{C}(\,\cdot\,,X)$$

$\mathscr{Sets}$: $\quad \mathscr{C}(X, A) \xrightarrow[\mathscr{C}(X, f)]{} \mathscr{C}(X, B) \qquad \mathscr{C}(A, X) \xleftarrow[\mathscr{C}(f, X)]{} \mathscr{C}(B, X)$

$$\alpha \longrightarrow f\alpha \qquad\qquad\qquad \beta f \longleftarrow \beta$$

Any morphism $h: Y \to X$ in $\mathscr{C}$ induces a natural transformation $\rho: \mathscr{C}(X, \cdot) \to \mathscr{C}(Y, \cdot)$ which is easy to describe — it is simply right multiplication by h. Officially, ρ is defined by

$$\rho(A): \mathscr{C}(X, A) \to \mathscr{C}(Y, A), \ \rho(A)\alpha = \alpha h.$$

Similarly there is a natural transformation $\lambda: \mathscr{C}(\cdot, Y) \to \mathscr{C}(\cdot, X)$ which is left multiplication by h. Note that

$$\rho(A) = \mathscr{C}(h, A) \qquad \text{and} \qquad \lambda(B) = \mathscr{C}(B, h).$$

15-1.17 Generators and cogenerators. For any category $\mathscr{C}$, let $f_1 \ne f_2: X \to Y$ denote generically any two unequal morphisms whatever.

An object $G \in \mathrm{Ob}\,\mathscr{C}$ is a *generator* in $\mathscr{C}$ if there exists in $\mathscr{C}$ a morphism $g: G \to X$ such that $f_1 g \ne f_2 g$.

An object $C \in \mathrm{Ob}\,\mathscr{C}$ is a *cogenerator* in $\mathscr{C}$ if there exists in $\mathscr{C}$ a morphism $h: Y \to C$ such that $hf_1 \ne hf_2$.

These concepts will now be reinterpreted. Suppose that $G \in \mathrm{Ob}\,\mathscr{C}$ is any object whatever, and apply the functor $\mathscr{C}(G, \cdot): \mathscr{C} \to \mathscr{Sets}$ to $f_i: X \to Y, i = 1, 2$.

$$X \overset{f_1}{\underset{f_2}{\rightrightarrows}} Y$$

$$\mathscr{C}(G, X) \overset{\mathscr{C}(G, f_1)}{\underset{\mathscr{C}(G, f_2)}{\rightrightarrows}} \mathscr{C}(G, Y)$$

Thus $\mathscr{C}(G, f_1)$ and $\mathscr{C}(G, f_2)$ are two elements of

$$\mathrm{Morph}(\mathscr{C}(G, X), \mathscr{C}(G, Y))$$

in the category $\mathscr{S}ets$. The functor $\mathscr{C}(G, \cdot)$ is faithful if and only if $f_1 \neq f_2$ implies that $\mathscr{C}(G, f_1) \neq \mathscr{C}(G, f_2)$, i.e. that there exists some $g \in \mathscr{C}(G, X)$ such that

$$f_1 g = \mathscr{C}(G, f_1)g \neq \mathscr{C}(G, f_2)g = f_2 g.$$

Thus an arbitrary object $G \in \mathrm{Ob}\,\mathscr{C}$ is a generator if and only if $\mathscr{C}(G, \cdot)$ is a faithful functor.

Similarly $C \in \mathrm{Ob}\,\mathscr{C}$ is a cogenerator if and only if $\mathscr{C}(\cdot, C) : \mathscr{C} \to \mathscr{S}ets$ is a faithful contravariant functor.

Later we will have to show that loosely speaking an entity of the above type '$\mathscr{C}(\cdot, \cdot)$', which is a functor in each variable separately, is naturally isomorphic — in both arguments jointly — to another such gadget. Next, a framework is developed which handles this and more complicated cases with ease, equally well from two different viewpoints, which ever is more convenient.

15-1.18 Product categories and bifunctors. Let $\mathscr{C}_1$, $\mathscr{C}_2$, and $\mathscr{D}$ be any categories; and let $\alpha : A \to B$, $\beta : B \to C$ in $\mathscr{C}_1$, and $f : X \to Y$, $g : Y \to Z$ in $\mathscr{C}_2$ be arbitrary. The *product category* $\mathscr{C}_1 \times \mathscr{C}_2$ has ordered pairs as its objects $\mathrm{Ob}(\mathscr{C}_1 \times \mathscr{C}_1) = \mathrm{Ob}\,\mathscr{C}_1 \times \mathrm{Ob}\,\mathscr{C}_2$ and morphisms

$$\mathrm{Morph}(\mathscr{C}_1 \times \mathscr{C}_2) = \mathrm{Morph}\,\mathscr{C}_1 \times \mathrm{Morph}\,\mathscr{C}_2.$$

The two morphisms $(\alpha, f) : (A, X) \to (B, Y)$ and $(\beta, g) : (B, Y) \to (C, Z)$ of $\mathscr{C}_1 \times \mathscr{C}_2$ compose componentwise

$$(\beta, g) \circ (\alpha, f) = (\beta\alpha, gf) : (A, X) \to (C, Z).$$

(In view of 15-1.2, $\mathscr{C}_1 \times \mathscr{C}_2$ cannot be an ordinary set theoretic product of the two classes $\mathscr{C}_1$ and $\mathscr{C}_2$.)

An ordinary functor $\mathscr{C}_1 \times \mathscr{C}_2 \to \mathscr{D}$ is called a *bifunctor*. It is said to be covariant in both $\mathscr{C}_1$ and $\mathscr{C}_2$. A bifunctor, that is a functor $\mathscr{C}_1^{\mathrm{op}} \times \mathscr{C}_2 \to \mathscr{D}$ is called contravariant in $\mathscr{C}_1$ and covariant in $\mathscr{C}_2$. Similarly, a bifunctor $\mathscr{C}_1^{\mathrm{op}} \times \mathscr{C}_2^{\mathrm{op}} \to \mathscr{D}$ is said to be contravariant in both $\mathscr{C}_1$ and $\mathscr{C}_2$. (Thus a bifunctor $\mathscr{C}_1 \times \mathscr{C}_2^{\mathrm{op}} \to \mathscr{D}$ could be said to be contravariant in $\mathscr{C}_1^{\mathrm{op}}$ and covariant in $\mathscr{C}_2^{\mathrm{op}}$.) What is important is that the term bifunctor as used here is always a functor in the strict ordinary sense on a product category, which may or may not involve opposite categories (see also MacLane (1971). For a bifunctor on any one of the four categories $\mathscr{C}_1 \times \mathscr{C}_2$, $\mathscr{C}_1^{\mathrm{op}} \times \mathscr{C}_2$, $\mathscr{C}_1 \times \mathscr{C}_2^{\mathrm{op}}$, or $\mathscr{C}_1^{\mathrm{op}} \times \mathscr{C}_2^{\mathrm{op}}$, somewhat ambiguously, $\mathscr{C}_1$ is called its first argument or variable, while its second argument refers to $\mathscr{C}_2$.

In an obvious way, all of the above definitions extend to the direct product of any indexed family of categories indexed by a finite or infinite index set.

Consider a function

$$H : \mathscr{C}_1 \times \mathscr{C}_2 = \mathrm{Ob}\, \mathscr{C}_1 \times \mathscr{C}_2 \cup \mathrm{Morph}\, \mathscr{C}_1 \times \mathscr{C}_2 \to \mathscr{D}$$

mapping objects to objects and morphism to morphisms. Fix $X \in \mathrm{Ob}\, C_2$. Next, a function $H(\cdot, X) : C_1 \to D$ is defined which maps objects to objects and sends morphisms to morphisms by $H(\cdot, X)(A) = H(A, X)$, and $H(\cdot, X)(\alpha) = H(\cdot, X)\alpha = H(\alpha, X) \equiv H(\alpha, 1_X)$. Similar notation and definitions apply to $H(A, \cdot) : \mathscr{C}_2 \to \mathscr{D}$. We also have $H(\cdot, \cdot) = H$.

15-1.19 Remarks. (1) Here is a nonintrinsic unsymmetry in the notation. In the symbol '$H(\cdot, X)$' the 'X' in the second variable is just an arbitrary tag which could be written as well in some other way. If in place of '$H(\cdot, X)$' we instead had tried to use '$H(\cdot, 1_X)$' we would have been forced to define $H(\cdot, 1_X)A = H(A, 1_X) = H(A, X) \in \mathrm{Ob}\, \mathscr{D}$. This, however, would be incompatible with our definition according to which $H(A, 1_X)$ by definition is the morphism $H(A, 1_X) = H(1_A, 1_X) = 1_{H(A,X)} \in \mathrm{Morph}\, \mathscr{D}$.

(2) Starting with only H as defined above in 15-1.18, here we extended its definition as a function to hybrid entities

$$H[(\mathrm{Ob}\, \mathscr{C}_1 \times \mathrm{Morph}\, \mathscr{C}_2) \cup (\mathrm{Morph}\, \mathscr{C}_1 \times \mathrm{Ob}\, \mathscr{C}_2)] \subseteq \mathrm{Morph}\, \mathscr{D}.$$

15-1.20. Return again to the hom functors on a category $\mathscr{C}$. Let $\gamma : X_2 \to X_1$, $f : X_1 \to Y_1$, and $\delta : Y_1 \to Y_2$ be arbitrary in $\mathscr{C}$. Define $\mathscr{C}(\cdot, \cdot) : \mathscr{C}^{\mathrm{op}} \times \mathscr{C} \to \mathscr{Sets}$ by:

in $\mathscr{C}$:

$$X_1 \xleftarrow{\gamma} X_2$$

$$\begin{array}{l} f \searrow \\ \quad Y_1 \xrightarrow{\delta} Y_2 \end{array}$$

in $\mathscr{C}^{\mathrm{op}} \times \mathscr{C}$:

$$(X_1, Y_1) \xrightarrow{(\gamma^\circ, \delta)} (X_2, Y_2)$$

$$\Big\downarrow \mathscr{C}(\cdot, \cdot)$$

in $\mathscr{Sets}$ $\qquad \mathscr{C}(X_1, Y_1) \xrightarrow[C(\gamma^\circ, \delta)]{} \mathscr{C}(X_2, Y_2)$

$$C(\gamma^\circ, \delta)f = \delta f \gamma$$

Thus $\mathscr{C}(\cdot, \cdot)$ is covariant in the first argument and contravariant in the second, and that it actually is a bifunctor will follow from the more general result 15-1.21 proven next.

Next, return to the situation as in 15-1.18.

15-1.21 Proposition. *For categories* $\mathscr{C}_1$, $\mathscr{C}_2$, *and* $\mathscr{D}$, *let* H *be a function mapping*

$$H[\mathrm{Ob}(\mathscr{C}_1 \times \mathscr{C}_2)] \subseteq \mathrm{Ob}\,\mathscr{D} \quad and \quad H[\mathrm{Morph}(\mathscr{C}_1 \times \mathscr{C}_2)] \subseteq \mathrm{Morph}\,\mathscr{D}.$$

Let $\alpha: A \to B$ *in* $\mathscr{C}_1$, *and* $f: X \to Y$ *in* $\mathscr{C}_2$ *be arbitrary. Then* $H: \mathscr{C}_1 \times \mathscr{C}_2 \to \mathscr{D}$ *is a functor* $\Leftrightarrow$ (i), (ii), *and* (iii) *hold for all* α, A, B, f, X, *and* Y.

$H(\cdot, X): \mathscr{C}_1 \to \mathscr{D}$ *is a functor.*   (i)
$H(A, \cdot): \mathscr{C}_2 \to \mathscr{D}$ *is a functor.*   (ii)
The following diagram commutes: (iii)

Proof. Let $\beta: B \to C$ and $g: Y \to Z$ be arbitrary.

$\Rightarrow$: (i) Since H is a functor, $H(1_A, 1_X) = 1_{H(A,X)}$. By definition of $H(\cdot, X)$, $H(\cdot, X)(1_A) = H(1_A, 1_X) = 1_{H(\cdot,X)A}$. It remains to show that

$$H(\beta, X) \circ H(\alpha, X) = H(\beta\alpha, X).$$

Since H is a functor, we get the second equality in

$$H(\beta, X) \circ H(\alpha, X) = H(\beta, 1_X) \circ H(\alpha, 1_X) = H(\beta\alpha, 1_X) = H(\beta\alpha, X).$$

(ii) This conclusion is entirely similar.

$\Rightarrow$ (iii): Again, by definition $H(\alpha, Y) = H(\alpha, 1_Y)$, and thus by the functoriality of H

$$H(\alpha, Y) \circ H(A, f) = H(\alpha, 1_Y) \circ H(1_A, f) = H(\alpha 1_A, 1_Y f) = H(\alpha, f).$$

Similarly, the bottom triangle also commutes.

$\Leftarrow$: By (i), $H(1_A, 1_X) = H(1_A, X) = 1_{H(\cdot,X)A} = 1_{H(A,X)}$. It remains to show that $H(\beta, g)H(\alpha, f) = H(\beta\alpha, gf)$. From (iii) alone the following diagram commutes.

Go clockwise from (A, X) to (C, Z) around the outer edge, and equate this to going along the diagonal. The result is

$$H(\beta, Z)H(\alpha, Z)H(A, g)H(A, f) = H(\beta, g)H(\alpha, f).$$

By (i), $H(\beta, Z)H(\alpha, Z) = H(\beta\alpha, Z)$, and by (ii) $H(A, g)H(A, f) = H(A, gf)$. Hence

$$H(\beta\alpha, Z)H(A, gf) = H(\beta, g)H(\alpha, f).$$

Now by a second application of (iii), the following diagram commutes:

$$
\begin{array}{ccc}
H(A, X) & \xrightarrow{H(X, gf)} & H(A, Z) \\
& \searrow{\scriptstyle H(\beta\alpha, gf)} & \downarrow{\scriptstyle H(\beta\alpha, Z)} \\
& & H(C, Z)
\end{array}
$$

Thus

$$H(\beta\alpha, Z)H(A, gf) = H(\beta\alpha, gf).$$

Comparison with a previous equation finally gives $H(\beta, g)H(\alpha, f) = H(\beta\alpha, gf)$. Thus H is a functor.

15-1.22 Corollary 1. *Suppose that we are given only a function*

$$H[\text{Ob}(\mathscr{C}_1 \times \mathscr{C}_2)] \subseteq \text{Ob}$$

and

$$H[\text{Ob }\mathscr{C}_1 \times \text{Morph }\mathscr{C}_2 \cup \text{Morph }\mathscr{C}_1 \times \text{Ob }\mathscr{C}_2] \subseteq \text{Morph }\mathscr{D}$$

so that all the $H(\alpha, X)$, $H(A, f)$, $H(A, X)$ etc., satisfy 15-1.21(i), (ii), and (iii)—except that in (iii) the diagonal arrow is removed.

In this situation, define $H(\alpha, f) = H(\alpha, Y)H(A, g) = H(B, f)H(\alpha, X)$. Then this extended H is a functor $H: \mathscr{C}_1 \times \mathscr{C}_2 \to \mathscr{D}$.

15-1.22 Corollary 2. *Let $E: \mathscr{A} \to \mathscr{B}$ and $F: \mathscr{B} \to \mathscr{A}$ be any functors of categories $\mathscr{A}$ and $\mathscr{B}$. Then*

$$\mathscr{B}(E\cdot, \cdot): \mathscr{A}^{\text{op}} \times \mathscr{B} \to \mathscr{Sets} \quad \text{and} \quad \mathscr{A}(\cdot, F\cdot): \mathscr{A}^{\text{op}} \times \mathscr{B} \to \mathscr{Sets}$$

are functors.

15-1.23 Observation. Let $H, T: \mathscr{C} \to \mathscr{D}$ be any contravariant functors. Define $H^\circ(A) = H(A)$ and $H^\circ(f^\circ) = H(f)$ for $A \in \text{Ob }\mathscr{C}$ and $f \in \text{Morph }\mathscr{C}$. Then $H^\circ, T^\circ: \mathscr{C}^{\text{op}} \to \mathscr{D}$ are covariant functors. A function $\phi: \text{Ob }\mathscr{C}_1 \times \text{Ob }\mathscr{C}_1^{\text{op}} \to \text{Morph }\mathscr{D}$ is a natural transformation of $\phi: H \to T$ if and only if it is simultaneously a natural transformation of $\phi: H^\circ \to T^\circ$.

15-1.24 Proposition. *Suppose that* $H, T : \mathscr{C}_1 \times \mathscr{C}_2 \to \mathscr{D}$ *are bifunctors and that* $\phi : \mathrm{Ob}(\mathscr{C}_1 \times \mathscr{C}_2) \to \mathrm{Morph}\, \mathscr{D}$ *is a function such that*

$$\phi(A, X) \in \mathscr{D}(H(A, X), T(A, X))$$

for all $A \in \mathrm{Ob}\, \mathscr{C}_1$, *and* $X \in \mathrm{Ob}\, \mathscr{C}_2$. *Then* $\phi : H \to T$ *is a natural transformation* $\Leftrightarrow$ (a) *and* (b) *hold for all* A, X.

(a) $\phi(A, \cdot) : H(A, \cdot) \to T(A, \cdot)$ *is a natural transformation of the functors* $H(A, \cdot), T(A, \cdot) : \mathscr{C}_2 \to \mathscr{D}$.

(b) $\phi(\cdot, X) : H(\cdot, X) \to T(\cdot, X)$ *is a natural transformation of* $H(\cdot, X)$, $T(\cdot, X) : \mathscr{C}_1 \to \mathscr{D}$.

Proof. $\Rightarrow$: omitted. $\Leftarrow$: Let $\alpha : A \to B$ in $\mathscr{C}_1$ and $f : X \to Y$ in $\mathscr{C}_2$. The left side of the following 3-dimensional cube commutes by (b), the right side by (a), the front because T is a functor, and the back because H is a functor.

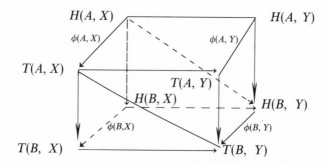

Now the diagram in the plane of the front and back diagonals commutes, which is the statement that $\phi : H \to T$ is a natural transformation.

15-2   Objects

In our abstract categorical setting, an object C of a category need not have any elements. Hence a subobject of C cannot be defined as a subset of C.

15-2.1   Definition. Two monomorphisms $\alpha : A \to C$ and $\beta : B \to C$ in a category $\mathscr{C}$ are *equivalent* if there exist morphism $f : A \to B$ and $g : B \to A$ in $\mathscr{C}$ such that the first two diagrams following commute:

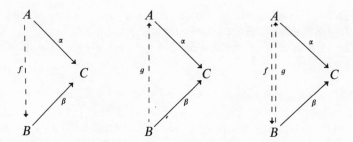

It follows that the third one commutes also. It can be shown that f and g are isomorphisms, and that f and g are unique.

An equivalence class $\{\alpha, \beta, \ldots\} \subseteq \text{Morph } \mathscr{C}$ consisting of the class of all equivalent monomorphisms into C is defined to be a subobject of C. However, any member of the equivalence class, say $\alpha: A \to C$, will also be referred to as a subobject, provided that whatever is done with α could equally well have been carried out with any other monomorphism β euivalent to α.

Suppose that $\alpha: A \to C$ and $\beta: B \to C$ represent two subobjects of C. Then the subobject α is said to be contained in the subobject β if there is a morphism $i: A \to B$ such that $\beta i = \alpha$. Necessarily, i is a unique monomorphism. In general it is *not* the case that the equivalence class of α is contained in that of β. The subobjects of $\mathscr{C}$ form a partially ordered class under containment in this non-set theoretic sense. If in addition the subobject β is also contained in the subject α, then α and β are equivalent and are (or represent) the same subobject.

Two epimorphisms $\alpha_1: C \to A$ and $\beta_1: C \to B$ and *equivalent* if there exist morphisms $g: A \to B$ and $f: B \to A$ such that the first two (and hence all) diagrams commute:

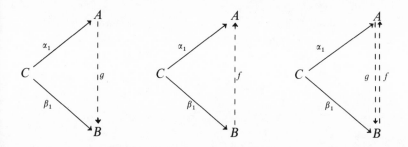

Then g and f have to be isomorphisms, and they are unique. A quotient object of C is the equivalence class $\{\alpha_1, \beta_1, \ldots\}$ of all equivalent epimorphisms emanating from C.

Subobjects and quotient objects are dual; the contravariant functor op: $\mathscr{C} \to \mathscr{C}^{\mathrm{op}}$ sends one into the other.

The various morphisms next defined need not always exist in arbitrary categories.

15-2.2 Kernel and cokernel. Assume that the category $\mathscr{C}$ has a zero object $(0) \in \mathrm{Ob}\,\mathscr{C}$. (See 15-1.5.)

A *kernel* of f is a morphism $k: K \to A$ such that (i) $fk = 0$; and (ii) any $g: X \to A$ with $fg = 0$ factors uniquely through k; there exists a unique map $\phi: X \to K$ such that $g = k\phi$.

That is k is a kernel of f if there are commutative diagrams as follows

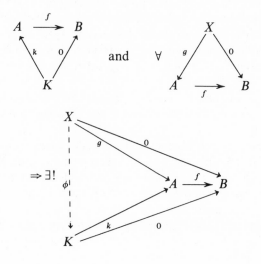

The notation

$$\ker(f) = \ker f = K \qquad \text{and} \qquad \mathrm{Ker}(f) = \mathrm{Ker}\, f = k,$$

or

$$\mathrm{Ker}(A \to B) = (K \to A)$$

will be used.

A cokernel of f is a morphism $c: B \to C$ such that (i) $cf = 0$; and (ii) any $g: B \to X$ such that $gf = 0$ factors uniquely through c, that is, there exists a unique morphism $\phi: C \to X$ such that $g = \phi c$. There is an associated commutative diagram

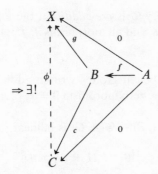

Write $\text{coker}(f) = \text{coker } f = X$ and $\text{Coker } f = c$; or $\text{Coker}(A \to B) = (B \to C)$. Again, note that the cokernel diagram is obtained by systematically reversing all arrows in the kernel diagram. Thus these two concepts are dual in the sense that the op functor maps one into the other.

The concepts of kernel and cokernel have been defined by means of a universal property, and the assertions below are immediate consequences.

15-2.3. Any kernel $k: K \to A$ of $f: A \to B$ is a monomorphism. Any two kernels of f are equivalent monomorphisms, so that all kernels of f form one and the same subobject of A. Furthermore k is a largest subobject of A among all subobjects $j: Y \to A$ such that $fj = 0$. Dual statements hold for cokernels.

15-2.4 Image and coimage. In any category, an *image* of a morphism $f: A \to B$ is defined as a smallest subobject $i: I \to B$ of B such that f factors through i, that is $f = ig$ for some $g: A \to I$. Since i is a monomorphism, g is unique. Write $\text{im } f = I$ and $\text{Im } f = i$. Any two images of f are equivalent.

Dually, the *coimage* of $f: A \to B$ is the smallest quotient object $q: A \to Q$ of A through which f factors, $f = hq$ for some $h: Q \to B$. Write $\text{coim } f = Q$ and $\text{Coim } f = q$. Thus $q = \text{Coim } f$ if and only if $q^\circ = \text{Im } f^\circ$ in the opposite category.

15-3   Pre-additive categories

The approach here is to postulate the minimal properties needed on a category in order to develop some concepts. Thus finite coproducts, products, biproducts, and additive functors only require a pre-additive category. At the end of this section a brief description of exact, additive, and abelian categories is given.

15-3.1 Pre-additive category. A category $\mathscr{C}$ is *pre-additive* if the following properties hold for all A, B, $C \in \mathrm{Ob}\,\mathscr{C}$ and all morphisms f, $f' : A \to B$ and g, $g' : B \to C$.

(i) $\mathscr{C}$ has a zero object $(0) \in \mathrm{Ob}\,\mathscr{C}$.

(ii) All $\mathscr{C}(A, B)$ (and $\mathscr{C}(B, C)$) are abelian groups written additively. The zero group element is the same as the zero morphism $0 = 0_{A,B} : A \to B$ (see 15-1.5(iv)).

(iii) The composition $\mathscr{C}(B, C) \times \mathscr{C}(A, B) \to \mathscr{C}(A, C)$ is bilinear,

$$g(f + f') = gf + gf' \qquad \text{and} \qquad (g + g')f = gf + g'f.$$

15-3.2 Additive functor. A functor $H : \mathscr{C} \to \mathscr{D}$ of two pre-additive categories $\mathscr{C}$ and $\mathscr{D}$ is *additive* if for any f, $g : A \to B$ in $\mathscr{C}$, $H(f + g) = Hf + Hg : HA \to HB$ in $\mathscr{D}$. That is, H is additive provided that the induced map $\mathscr{C}(A, B) \to \mathscr{D}(HA, HB)$ is an abelian group homomorphism.

15-3.3 Coproduct and product. In any category, for any family of objects, indexed by a set, their sum and product can be defined exactly as in 1-2.2 and 1-2.3 by their universal mapping properties — provided they exist. In category theory and frequently elsewhere, the sum is called the *coproduct*.

15-3.4 Definition. Let $\mathscr{C}$ be any pre-additive category and

$$A_1, \ldots, A_n \in \mathrm{Ob}\,\mathscr{C}$$

any finite indexed set of objects (with repetitions allowed). Let $1_i : A_i \to A_i$ denote the identity map, and let $f_{ij} : A_j \to A_i$ be $\delta_{ij} = 1_i$ if $i = j$, and the zero morphism $\delta_{ij} = 0 : A_j \to A_i$ when $i \neq j$.

A *biproduct* of $A_1, \ldots, A_n$ is a $(2n + 1)$-tuple $C \in \mathrm{Ob}\,\mathscr{C}$, and $\phi_1, \ldots, \phi_n$, $\pi_1, \ldots, \pi_n \in \mathrm{Morph}\,\mathscr{C}$ such that for all i and j

$$A_j \xrightarrow{\ \phi_j\ } C \xrightarrow{\ \pi_i\ } A_i, \ \pi_i \phi_j = \delta_{ij}; \ \sum_{i=1}^{n} \phi_i \pi_i = 1_C$$

It follows from $\pi_i \phi_i = 1_i$, that ϕ_i is monic, and π_i is epic for any biproduct.

The next lemma proves among other things that a biproduct is both a product and a coproduct.

15-3.5 Lemma. *Suppose that* $A_1, \ldots, A_n$, $C \in \mathrm{Ob}\,\mathscr{C}$ *are as above in a pre-additive category* $\mathscr{C}$. *Then the following hold.*

(1) *Morphisms* $\pi_i : C \to A_i$, $i = 1, \ldots, n$; *are a product of* $A_1, \ldots, A_n \Leftrightarrow \exists$ *morphisms* $\phi_j : A_j \to C$ *such that* C; $\phi_1, \ldots, \phi_n$; $\pi_1, \ldots, \pi_n$ *is a biproduct of* $A_1, \ldots, A_n$.

(2)   *Morphisms $\phi_j: A_j \to C, j = 1, \ldots, n$; form a coproduct (i.e. sum) of $A_1, \ldots, A_n \Leftrightarrow \exists \pi_i: A_i \to C$, $i = 1, \ldots, n$ such that C; $\phi_1, \ldots, \phi_n$, $\pi_1, \ldots, \pi_n$ is a biproduct.*

(3)   *In particular every (finite) biproduct is both a product and a coproduct.*

Proof. (1) $\Leftarrow$: It suffices to show that C satisfies the universal property which defines the product. So given any object $X \in \mathrm{Ob}\,\mathscr{C}$ and morphisms $f_i: X \to A_i$ for all i, simply define $f = \Sigma_1^n \phi_j f_j: X \to C$. Then invoking the bilinear property, for any i, $\pi_i f = \Sigma_1^n (\pi_i \phi_j) f_j = \Sigma_1^n \delta_{ij} f_j = f_i$. If $g: X \to C$ is another such map with $\pi_i g = f_i$ for all i, then

$$g = 1_C g = \left(\sum_{i=1}^n \phi_i \pi_i \right) g = \sum_{i=1}^n \phi_i (\pi_i g) = \sum_{i=1}^n \phi_i f_i = f.$$

Then $C, \pi_1, \ldots, \pi_n$ is a product of the $A_1, \ldots, A_n$.

(1) $\Rightarrow$: Fix j, and define morphisms $f_i: A_j \to A_i$ by $f_i = \delta_{ij}$ for all $i = 1, \ldots, n$. By the universal property of the product, there exists a (unique) morphism $\phi_j: A_j \to C$ with $\pi_i \phi_j = f_i = \delta_{ij}$. Then by bilinearity (15-3.1(i)), $\pi_i(\Sigma_1^n \phi_j \pi_j) = \Sigma_1^n (\pi_i \phi_j) \pi_j = \Sigma_1^n \delta_{ij} \pi_j = \pi_i = \pi_i 1_C$. Thus 1_C, $\Sigma_1^n \phi_j \pi_j: C \to C$ are two morphisms such that $\pi_i 1_C = \pi_i \Sigma_1^n \phi_j \pi_j$ for all i. By the universal property of the product (for the morphisms $\pi_i: C \to A_i$) it follows that $1_C = \Sigma_1^n \phi_j \pi_j$. The commutative diagrams below clarify this, where in the second diagram the universal product property is used for the top C.

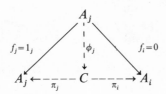

The proof of (2) is obtained by dualizing (1), and then (3) follows.

3.6.   A category is said to be with (finite) products, if every (finite) indexed set of objects has a product, and similarly for coproducts. Thus a pre-additive category is with finite products if and only if it is with finite coproducts, or more precisely with finite biproducts. Such a finite biproduct of objects $A_1, \ldots, A_n$ is frequently denoted by $A_1 \oplus \cdots \oplus A_n$ where the $2n$ morphisms have been surpressed. A pre-additive category with finite biproducts is called an additive category.

15-3.7 Proposition. *Suppose that* $H: \mathscr{C} \to \mathscr{D}$ *is an additive functor of pre-additive categories* $\mathscr{C}$ *and* $\mathscr{D}$. *Then for any product, coproduct, or biproduct* $A_1 \oplus \cdots \oplus A_n$ *in* $\mathscr{C}$, $HA_1 \oplus \cdots \oplus HA_n$ *is a product, coproduct, or biproduct in* $\mathscr{D}$.

Proof. Let $C = A_1 \oplus \cdots \oplus A_n$. If C is given only as a product or as a coproduct, first use Lemma 15-3.5 to produce a biproduct $C; \phi_1, \ldots, \phi_n$; $\pi_1, \ldots, \pi_n$ as in 15-3.4 with $\pi_i \phi_j = \delta_{ij}$ and $\Sigma_1^n \phi_i \pi_i = 1_C$. Since H preserves identity and zero morphisms, $(H\pi_i)(H\phi_j) = \delta_{ij}$ in $\mathscr{D}$. The additivity of H implies that $\Sigma_1^n (H\pi_i) = 1_{HC}$. Thus $HC; H\phi_1, \ldots, H\phi_n; H\pi_1, \ldots, H\pi_n$ satisfies Definition 15-3.4 of a biproduct.

Although property (iv) below is a consequence of the previous ones, we simply include it in the description of an exact category.

15-3.8 Exact category. A category is *exact* if it has a zero object and if the following five properties hold.

  (i)   Every morphism has a kernel and a
  (i)*   cokernel.
  (ii)   Every monomorphism is the kernel of a morphism. (Any category possessing this property (ii) is called a *normal* category.)
  (ii)* Every epimorphism is the cokernel of a morphism. (This is the *conormal* property.)
  (iii) Every morphism $f: A \to B$ can be factored as $g: A \to C$ and $h: C \to B$ with $f = hg$, where g is epic and h is monic.
  (iv)   Furthermore $h = \operatorname{Im} f$ is the image of f, $g = \operatorname{Coim} f$ is the coimage of f, and any such a factorization $f = hg$ as above is unique up to equivalent monics h and epics g.

15-3.9 A category is called *balanced* if every morphism that is both monic and epic is an isomorphism. Every exact category is known to be balanced.

15-3.10 Exact sequences. In an exact category, a sequence of morphisms $E: A \xrightarrow{\alpha} B \xrightarrow{\beta} C$ is *exact* at B if $\operatorname{Im} \alpha = \operatorname{Ker} \beta$, that is if they are equal as subobjects of B. Exact and short exact sequences are defined the same as in 1-4.1 and 1-4.2. Similarly, the definitions of '$0 \to A \to B$ splits' and '$B \to C \to 0$ splits also carry over from 1-4.3. Any finite or infinite exact sequence of morphisms having a zero object and a zero morphism on the left, i.e. $0 \to A \to B \to \cdots$, is called left exact. Similarly an exact sequence of the form $\cdots \to B \to C \to 0$ is called *right exact*.

15-3.11 Remarks. In an exact category the following hold.

(i) $A \to B$ is monic $\Leftrightarrow 0 \to A \to B$ is exact.

(ii) $A \to B$ is epic $\Leftrightarrow A \to B \to 0$ is exact.

(iii) $A \to B$ is an isomorphism $\Leftrightarrow A \to B$ is both monic and epic $\Leftrightarrow 0 \to A \to B \to 0$ is exact.

(iv) $\mathscr{C}$ is an exact category $\Leftrightarrow \mathscr{C}^{\mathrm{op}}$ is exact, in which case $A \xrightarrow{\alpha} B \xrightarrow{\beta} C$ is exact in $\mathscr{C} \Leftrightarrow A \xleftarrow{\alpha^{\circ}} B \xleftarrow{\beta^{\circ}} C$ is exact in $\mathscr{C}^{\mathrm{op}}$.

15-3.12 Exact functor. Let $\mathscr{C}$ and $\mathscr{D}$ be exact categories. A functor $F : \mathscr{C} \to \mathscr{D}$ is zero preserving if F maps the zero object 0 of $\mathscr{C}$ to the zero object $F(0)$ of $\mathscr{D}$, in which case F automatically maps zero morphisms to zero morphisms. A zero preserving functor $F : \mathscr{C} \to \mathscr{D}$ is *exact* if F preserves zero objects and zero morphisms, and if for any sequence $A \to B \to C$ in $\mathscr{C}$ which is exact at B, the sequence $FA \to FB \to FC$ in $\mathscr{D}$ is exact at FB. (In particular, A or C may equal 0.) The same definition also applies to a contravariant functor, except that all the arrows in $\mathscr{D}$ have to be reversed. Thus a zero preserving functor is exact if and only if it carries any finite or infinite exact sequence of $\mathscr{C}$ into an exact sequence of $\mathscr{D}$.

15-3.13 One-sided exact functors. Suppose that $F : \mathscr{C} \to \mathscr{D}$ is a (covariant) functor and $G : \mathscr{C} \to \mathscr{D}$ a contravariant one between exact categories $\mathscr{C}$ and $\mathscr{D}$, and assume both functors are zero preserving. Let $0 \to A \to B \to C$ and $P \to N \to Q \to 0$ represent arbitrary exact sequences in $\mathscr{C}$, each involving only four objects. Let us call such sequences *short* left exact and *short* right exact respectively. Then

F is *left* exact if $0 \to FA \to FB \to FC$ is left exact.

F is *right* exact if $FP \to FN \to FQ \to 0$ is right exact.

Thus F is left exact if and only if F preserves short left exact sequences, and similarly for right.

However, for a contravariant functor

G is *left* exact if $0 \to GQ \to GN \to GP$ is left exact.

G is *right* exact if $GC \to GB \to GA \to 0$ is right exact.

Thus a left exact contravariant functor turns a short right exact sequence into a left exact one. As already stated, a contravariant functor G composed with the functor $\mathrm{op} : \mathscr{D} \to \mathscr{D}^{\mathrm{op}}$ gives a covariant functor $(\mathrm{op}) \circ G : \mathscr{C} \to \mathscr{D}^{\mathrm{op}}$. Now G is left exact if and only if $(\mathrm{op}) \circ G$ is likewise; i.e. iff $(\mathrm{opp}) \circ G$ preserves short left exact sequences. The dual holds for right.

15-3.14 Observation. A zero preserving functor $F: \mathscr{C} \to \mathscr{D}$ between exact categories satisfies the following.

(i) F is left exact $\Leftrightarrow F$ preserves kernels.
(ii) F is right exact $\Leftrightarrow F$ preserves cokernels.
(iii) F is exact $\Leftrightarrow F$ is both left and right exact $\Leftrightarrow F$ preserves short exact sequences $\Leftrightarrow F$ preserves arbitrary exact sequences.

15-3.15. If $\mathscr{C} \subset \mathscr{D}$ is a subcategory of a category $\mathscr{D}$, then morphisms in $\mathscr{C}$ may possess one of the following properties with respect to $\mathscr{C}$, but not with respect to $\mathscr{D}$: monomorphism, epimorphism, isomorphism, kernel, cokernel, image, coimage, exact sequence, etc. Thus these properties should be prefixed (by either $\mathscr{C}$ or $\mathscr{D}$) to indicate the category in which they apply.

15-3.16. Exact subcategory. A subcategory $\mathscr{C}$ of an exact category $\mathscr{D}$ is an exact subcategory if the inclusion functor $\mathscr{C} \to \mathscr{D}$ is an exact zero preserving functor.

Given objects A and C in an exact category, we can not always construct a short exact splitting sequence $0 \to A \to B \to C \to 0$. This suggests the next definition.

15-3.17 Abelian category. A category $\mathscr{C}$ is *abelian* if it is pre-additive, exact, and with finite biproducts.

An equivalent definition due to Freyd (1964), is that a category $\mathscr{C}$ is *abelian* if the following hold

(0) $\mathscr{C}$ has a zero object.
(i)   Any two objects have a product and a
(i)*   coproduct (or sum).
(ii)   Every morphism has a kernel, and a
(ii)* cokernel.
(iii) Every monomorphism is a kernel of a morphism;
(iii)* every epimorphism — a cokernel.

To repeat, all of our previous concepts of exact sequences, short exact sequences, and splitting exact sequences can now be defined in an abelian category using only kernels, cokernels, images, coimages, a zero object, and zero morphisms. So-called homological algebra can be done in the more general framework of abelian categories. A category is pre-additive exact, or abelian if and only if the opposite category is likewise.

15-3.18 Remark. It is known that a left or right exact covariant or contravariant functor between two abelian categories is automatically additive (see Freyd [64] pp. 65–66). For the functors encountered here, the additivity is trivial to verify.

The next result says that on an abelian category, the hom functor is left exact in each variable separately.

15-3.19. Let $\mathscr{C}$ be an exact and a pre-additive category and $M \in \mathrm{Ob}\,\mathscr{C}$ any object. Then for any exact sequences in $\mathscr{C}$

$$0 \longrightarrow A \xrightarrow{\ \alpha\ } B \xrightarrow{\ \beta\ } C \quad \text{and} \quad P \xrightarrow{\ \gamma\ } Q \xrightarrow{\ \delta\ } N \longrightarrow 0,$$

there are induced exact sequences of abelian groups

(i) $\quad 0 \longrightarrow \mathscr{C}(M, A) \xrightarrow{\ \mathscr{C}(M,\alpha)\ } \mathscr{C}(M, B) \xrightarrow{\ \mathscr{C}(M,\beta)\ } \mathscr{C}(M, C)$ and

(ii) $\quad 0 \longrightarrow \mathscr{C}(N, M) \xrightarrow{\ C(\delta,M)\ } \mathscr{C}(Q, M) \xrightarrow{\ \mathscr{C}(\gamma,M)\ } \mathscr{C}(P, M).$

Proof. (i) Suppose that $f : M \to A$ and $0 = \mathscr{C}(M, \alpha)f = \alpha f \in \mathscr{C}(M, B)$. Since $\ker \alpha = 0$, by the universal property of the kernel, f factors through 0. Hence $f = 0$. Let $\alpha = iq$ as in 15-3.8(iii) where $q : A \to \ker \beta$ is epic and $i : \operatorname{im} \alpha \to B$ is monic. Since $\alpha(\operatorname{Ker} q) = iq(\operatorname{Ker} q) = 0$, $\operatorname{Ker} q$ factors through $\ker \alpha = 0$. Thus $\operatorname{Ker} q = 0$. Since q is both epic and monic it is an isomorphism with inverse $q^{-1} : \operatorname{im} \alpha \to A$. Next, suppose that $\gamma : M \to B$ and $0 = \mathscr{C}(M, \beta)\gamma = \beta\gamma$. Hence there is a unique morphism $h : M \to \ker \beta$ such that $\gamma = ih$. Then $q^{-1}h : M \to A$ and $\mathscr{C}(M, \alpha)q^{-1}h = -\alpha q^{-1}h = iqq^{-1}h = \gamma$. Thus the first sequence (i) is exact. In this and some later proofs, we place an explanatory diagram at the end of the proof.

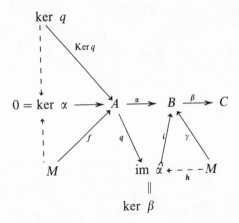

(ii) Applying the first part to the exact pre-additive category $\mathscr{C}^{\mathrm{op}}$ we get two exact sequences in $\mathscr{C}$: $0 \longrightarrow N \xrightarrow{\delta^\circ} Q \xrightarrow{\gamma^\circ} P$

in $\mathscr{Ab}$: $0 \longrightarrow \mathscr{C}^{\mathrm{op}}(M, N) \xrightarrow{\mathscr{C}^{\mathrm{op}}(M,\delta^\circ)} \mathscr{C}^{\mathrm{op}}(M, Q)$

$\xrightarrow{\mathscr{C}^{\mathrm{op}}(M,\gamma^\circ)} \mathscr{C}^{\mathrm{op}}(M, P).$

For

$$f \in \mathscr{C}(Q, M), \ f^\circ \in C^{\mathrm{op}\circ}(M, Q),$$

and

$$C^{\mathrm{op}}(M, \gamma^\circ)f^\circ = \gamma^\circ f^\circ = (f\gamma)^\circ = [\mathscr{C}(\gamma, M)f]^\circ.$$

Identify as sets and as abelian groups $C^{\mathrm{op}}(MP) \equiv C(P, M)$ via $f \equiv f^\circ$. Under this identification $C^{\mathrm{op}}(M, \gamma^\circ) \equiv C(\gamma, M)$ as functions of sets. In other words, the last sequence above is exact if and only if (ii) is exact.

Some facts about morphisms in an abelian category $\mathscr{C}$ are proved next which are needed later. Since also $\mathscr{C}^{\mathrm{op}}$ is abelian, the proofs of all the dual statements follow automatically. Many authors simply define the Image of a morphism f in an abelian category by the formula Im $f =$ Ker Coker f. The later fact is all that is needed to prove Proposition 15-3.23. Thus if the reader is willing to redefine the Image this way then he or she may skip the technicalities 15-3.20, 3.21 and 3.22, and proceed instead immediately to the main result — 15-3.23.

15-3.20. Lemma. *Let $f : A \to B$ be any morphism in an abelian category. Then*

    (i)   Ker $f =$ Ker(Coker(Ker f)), *and dually*
    (ii) Coker $f =$ Coker(Ker(Coker f)).

Proof. (i) Set $k =$ Ker f and $c =$ Coker k. It has to be shown that any $g : X \to A$ with $cg = 0$ factors uniquely through k. Since $fk = 0$, by 15-2.2, there is a morphism $\rho_{\mathrm{OUT}} : \operatorname{coker} k \to B$ with $f = \rho_{\mathrm{OUT}} c$. Since $fg = \rho_{\mathrm{OUT}} cg = 0$, there is a (unique) morphism $\rho_{\mathrm{IN}} : X \to \ker f$ with $g = k\rho_{\mathrm{IN}}$. Suppose that $h : X \to \ker f$ is another such morphism with $g = k \circ h$. Then $k(h - \rho_{\mathrm{IN}}) = 0$. Since k is monic, $h - \rho_{\mathrm{IN}} = 0$. Thus $k =$ Ker c.

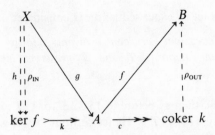

15-3.21 Corollary. *If $i: I \to A$ is any monomorphism in an abelian category, then $i = \mathrm{Ker}(\mathrm{Coker}\, i)$. Dually, any epimorphism is the Cokernel of its Kernel.*

Proof. By 15-3.17(iii), $i = \mathrm{Ker}\, f$ for some morphism $f: A \to B$. Now 15-3.20(i) says that $i = \mathrm{Ker}(\mathrm{Coker}\, i)$.

15-3.22 Lemma. *For any morphism $f: A \to B$ in an abelian category,*

  (i) $\mathrm{Ker\, Coker}\, f = \mathrm{Im}\, f$, *and dually*
  (ii) $\mathrm{Coker\, Ker}\, f = \mathrm{Coim}\, f$.

Proof. (i) Set $c = \mathrm{Coker}\, f$ and $k = \mathrm{Ker}\, c$. Since $cf = 0$, there is a (unique) morphism $\rho_{\mathrm{IN}}: A \to \ker c$ with $f = k\rho_{\mathrm{IN}}$. Thus f factors through k. It remains to show that for any monomorphism $i: I \to B$ through which f can be factored as $f = ig$, $g: A \to I$, there is a monomorphism $j: \ker c \to I$ such that $k = ij$ (see 15-2.4). Since $(\mathrm{Coker}\, i)f = (\mathrm{Coker}\, i)ig = 0$, there exist a (unique) $\rho_{\mathrm{OUT}}: \mathrm{coker}\, f \to \mathrm{coker}\, i$ with $\rho_{\mathrm{OUT}}c = \mathrm{Coker}\, i$.

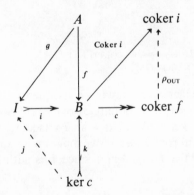

Then $ck = 0$ and hence also $(\mathrm{Coker}\, i)k = \rho_{\mathrm{OUT}}ck = 0$. By 15-3.21, $i = \mathrm{Ker}(\mathrm{Coker}\, i)$. Hence there exists a (unique) $j: \ker c \to I$ with $k = ij$. Thus $\mathrm{Im}\, f = k$ as required.

How functors map inexact or exact sequences is considered next.

15-3.23 Proposition. *A covariant or contravariant additive functor* $T : \mathcal{C} \to \mathcal{D}$ *between abelian categories* $\mathcal{C}$ *and* $\mathcal{D}$ *preserves inexact sequences* $\Leftrightarrow$ T *is faithful.*

Proof. If T is contravariant, then as before in 15-3.13, $(\text{op}) \circ T : \mathcal{C} \to \mathcal{D}^{\text{op}}$, and T is faithful if and only if $(\text{op}) \circ T$ is. Hence without loss of generality we may assume that T is covariant. $\Rightarrow$: Let $0 \neq \alpha : A \to B$ in $\mathcal{C}$. Consider the following two sequences

$$A \xrightarrow{\ 1_A\ } A \xrightarrow{\ \alpha\ } B, \quad TA \xrightarrow{\ T1_A = 1_{TA}\ } TA \xrightarrow{\ T\alpha\ } TB$$

The first is inexact because $\alpha \neq 0$, and hence the second one is inexact by hypothesis. The latter means that $T\alpha \neq 0$, and T is faithful.

$\Leftarrow$: It suffices to show that for any sequence

$$E : A \xrightarrow{\ \alpha\ } B \xrightarrow{\ \beta\ } C \quad \text{in } \mathcal{C} \quad \text{if } TE : TA \xrightarrow{\ TA\ } TB \xrightarrow{\ T\beta\ } TC$$

is exact in $\mathcal{D}$, that then the original sequence E was exact in $\mathcal{C}$.

If $k = \text{Ker }\beta$ and $c = \text{Coker }\alpha$, then $(Tc) \circ (T\alpha) = T(\text{Coker }\alpha \circ \alpha) = 0$ and $(T\beta) \circ (Tk) = T(\beta \circ \text{Ker }\beta) = 0$. Thus by the universal properties of the Kernel and Cokernel morphisms there is a commutative diagram

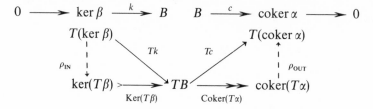

The fact that TE is exact has two consequences. First, $(T\beta) \circ (T\alpha) = T(\beta\alpha) = 0$, and since T is faithful, also $\beta\alpha = 0$. Secondly, $\text{im}(T\alpha) = \text{ker}(T\beta)$. But always $\text{ker Coker}(T\alpha) = \text{im } T\alpha$ (see 15-3.22(i)). The later implies that the bottom row of the above diagram is exact, and in particular that $0 = (Tc) \circ (Tk) = T(ck)$. Again, since T is faithful, also $ck = 0$.

(*Remark.* In module categories $\mathcal{C}$ and $\mathcal{D}$ the proof now would be immediate. In this case $\text{im }\alpha \subseteq B$ and $\text{ker }\beta \subseteq B$ are subsets, and $c : B \to \text{coker }\alpha = B/\text{im }\alpha$ the natural projection. Thus $\beta\alpha = 0$ implies that $\text{im }\alpha \subseteq \text{ker }\beta$, while $kc = 0$ gives the reverse $\text{ker }\beta \subseteq \text{im }\alpha$.) Let

$i = \operatorname{Im}\alpha : \operatorname{im}\alpha \to B$ and $\alpha = iq$ for some epic $q : A \to \operatorname{im}\alpha$. Since $0 = \beta\alpha = \beta iq$, and q is epic, also $\beta i = 0$. Hence there is a unique map $j : \operatorname{im}\alpha \to \ker\beta$ with $i = kj$. Since i is monic, j is necessarily also monic.

Conversely, since $kc = 0$, and since $i = \operatorname{Ker} c$ by 15-3.22(i), there is a unique morphism $h : \ker\beta \to \operatorname{im}\alpha$ with $k = ih$. Again h is monic since k is. Thus $i = kj = ihj$ and $k = ih = kjh$. Since i and k are monic, it follows that j and h are inverses of each other. Hence T is exact.

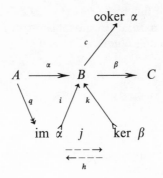

15-4 Adjoint functors

In this section $E : \mathscr{A} \to \mathscr{B}$, and $F : \mathscr{B} \to \mathscr{A}$ will be functors on arbitrary categories $\mathscr{A}$ and $\mathscr{B}$. Since adjoint functor situations occur in the most unexpected situations, here they are discussed in as general a context as possible. The important later applications are when $\mathscr{A}$ and $\mathscr{B}$ are categories of modules over rings R and S, where in many cases $S = \mathbb{Z}$ is the ring of integers. However, it is left to the reader to establish the additionally available structure in case of module categories.

Let A, $A' \in \operatorname{Ob}\mathscr{A}$; $f, g : A \to A'$ be morphism in $\mathscr{A}$, and $B \in \operatorname{Ob}\mathscr{B}$. In general, $\operatorname{Morph}_{\mathscr{A}}(EB, \cdot)$ is a functor $\operatorname{Morph}_{\mathscr{A}}(EB, \cdot) : \mathscr{A} \to \operatorname{Sets}$, where $\operatorname{Morph}_{\mathscr{A}}(EB, A)$, $\operatorname{Morph}_{\mathscr{A}}(EB, A')$ are only sets, and

$$\operatorname{Morph}_{\mathscr{A}}(EB, f) : \operatorname{Morph}_{\mathscr{A}}(EB, A) \to \operatorname{Morph}_{\mathscr{A}}(EB, A')$$

is only a function between two sets. However, if $\mathscr{A}$ and $\mathscr{B}$ are module categories, then the above becomes a homomorphism of two abelian groups. Moreover, the functor $\operatorname{Morph}_{\mathscr{A}}(EB, \cdot) : \mathscr{A} \to Ab$ into the category Ab of abelian groups now is additive:

$$\operatorname{Morph}_{\mathscr{A}}(EB, f + g) = \operatorname{Morph}_{\mathscr{A}}(EB, f) + \operatorname{Morph}_{\mathscr{A}}(EB, g)$$

15-4.1 Notation. Let A and B be categories. As before for A, $A' \in \operatorname{Ob}\mathscr{A}$ and B, $B' \in \operatorname{Ob}\mathscr{B}$, the morphism sets are abbreviated $\operatorname{Morph}_{\mathscr{A}}(A, A') = \mathscr{A}(A, A')$, and $\operatorname{Morph}_{\mathscr{B}}(B, B') = \mathscr{B}(B, B')$. Throughout, $E : \mathscr{A} \to \mathscr{B}$ and

$F: \mathcal{B} \rightarrow \mathcal{A}$ will be functors. There are induced bifunctors $\mathcal{A}(\cdot, F\cdot)$, $\mathcal{B}(E\cdot, \cdot): \mathcal{A}^{\mathrm{op}} \times \mathcal{B} \rightarrow \mathcal{S}ets$ (see 15-1.18). This bifunctor is contravariant in $\mathcal{A}$, covariant in $\mathcal{B}$. The identity functors on $\mathcal{A}$ and $\mathcal{B}$ are denoted by $1_{\mathcal{A}}$ and $1_{\mathcal{B}}$.

Generically, $\alpha: A' \rightarrow A$ and $\beta: B \rightarrow B'$ will refer to two arbitrary morphisms in $\mathcal{A}$ and $\mathcal{B}$; furthermore, c and n will denote morphisms $c: B \rightarrow FA$ in $\mathcal{B}$ and $n: EB \rightarrow A$ in $\mathcal{A}$.

15-4.2   Definition. For functors $\mathcal{A} \xrightarrow{E} \mathcal{B} \xrightarrow{F} \mathcal{A}$ as above, E is a *left adjoint* of F if there is a natural isomorphism of functors $\phi: \mathcal{B}(E\cdot, \cdot) \rightarrow \mathcal{A}(\cdot, F\cdot)$. This in particular entails that each

$$\phi(A, B) = \phi_{A,B}: \mathcal{B}(EA, B) \rightarrow \mathcal{A}(A, FB)$$

is an isomorphism.

15-4.3   Definition. For functors $E: \mathcal{A} \rightarrow \mathcal{B}$, and $F: \mathcal{B} \rightarrow \mathcal{A}$ the *front adjunction* is a natural transformation η of functor $\eta: 1_{\mathcal{A}} \rightarrow FE$ satisfying the following universal property:

$$\forall A, B, c \Rightarrow \exists! n \in \mathcal{B}(EA, B), \ c = \eta \circ F(n), \ \eta_A = \eta(A);$$

i.e. there is a commutative diagram:

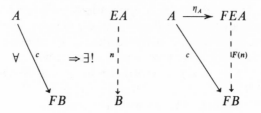

A dual notion called the back adjunction is obtained from the front one by (i) reversing all arrows, and then interchanging (ii) A and B, (iii) E and F, (iv) $\mathcal{A}$ and $\mathcal{B}$, and (v) c and n.

15-4.4   Definition. For functors $E: \mathcal{A} \rightarrow \mathcal{B}$ and $F: \mathcal{B} \rightarrow \mathcal{A}$, the *back adjunction* ψ is a natural transformation $\psi: EF \rightarrow 1_{\mathcal{B}}$ such that

$$\forall A, B, n \Rightarrow \exists! c \in \mathcal{A}(A, FB), \ n = \psi_B \circ E(C), \ \psi_B = \psi(B);$$

in other words, there is a commutative diagram:

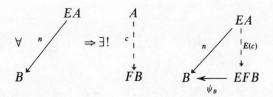

The next construction is a preliminary step in constructing adjunctions η and ψ out of a given pair of adjoint functors E and F. Even though the end result is stated first, its meaningfulness becomes evident only by going through the construction.

15-4.5 Construction. Assume that E is the left adjoint of F, and that $\phi = \phi_{A,B} : \mathscr{B}(EA, B) \to \mathscr{A}(A, FB)$ is an isomorphism of functors. Define $\eta(A) = \eta_A = \phi_{A,EA}(1_{EA})$ and $\psi(B) = \psi_B = \phi_{FB,B}^{-1}(1_{FB})$. Then:

(i) $\forall f : EA \to B$, $f \in \mathscr{B}(EA, B)$, $\phi(f) = (Ff) \circ \eta_A$; i.e. there is a commutative diagram:

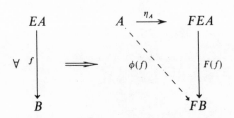

(ii) $\forall \; g : A \to FB$, $g \in \mathscr{A}(A, FB)$, $\phi^{-1}(g) = \psi_B \circ (Eg)$; i.e. there is a commutative diagram:

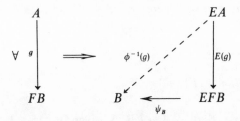

Proof. (i) In "$\phi_{A,B} : \mathscr{B}(EA, B) \to \mathscr{A}(A, FB)$" substitute $B = EA$ to get $\phi_{A,EA} : \mathscr{B}(EA, EA) \to \mathscr{A}(A, FEA)$. Therefore $\eta_A = \phi(1_{EA}) : A \to FEA$ is a morphism in $\mathscr{A}$. Set $f_* = \mathscr{B}(EA, f)$ and $(Ff)_* = \mathscr{A}(A, Ff)$. There are commutative diagrams:

$$\begin{array}{ccc}
\mathscr{B}(EA, EA) & \xrightarrow{\phi=\phi_{A,EA}} & \mathscr{A}(A, FEA) \\
\downarrow{f_*} & & \downarrow{(Ff)_*} \\
\mathscr{B}(EA, B) & \xrightarrow[\phi=\phi_{A,B}]{} & \mathscr{A}(A, FB)
\end{array}$$

$$\begin{array}{ccc}
1_{EA} & \dashrightarrow & \phi(1_{EA}) = \eta_A \\
\downarrow{f_*} & & \downarrow{(Ff)_*} \\
f \circ 1_{EA} & \dashrightarrow & \phi(f \circ 1_{EA}) = (Ff) \circ \eta_A
\end{array}$$

The second diagram shows how the identity morphism 1_{EA} is mapped around the first square. Thus

$$\forall f : EA \to B \Rightarrow \phi(f) = (Ff) \circ \eta_A$$

(ii) Set $g^* = \mathscr{A}(g, FB)$ and $(Eg)^* = -\mathscr{B}(Eg, B)$. In "$\phi_{A,B}:\mathscr{B}(EA, B) \to \mathscr{A}(A, FB)$" replace A with FB and obtain $\phi^{-1} = \phi^{-1}_{FB,B}: \mathscr{A}(FB, FB) \to \mathscr{B}(EFB, B)$. Consequently $\psi_B = \phi^{-1}(1_{FB}): EFB \to B$ is a well defined morphism of B. As before there are commutative diagrams:

$$\begin{array}{ccc}
\mathscr{B}(EFB, B) & \xrightarrow{\phi^{-1}=\phi^{-1}_{FB,B}} & \mathscr{A}(FB, FB) \\
(Eg)^* \downarrow & & \downarrow g^* \\
\\
\mathscr{B}(EA, B) & \xrightarrow[\phi\mu^{-1}=\phi^{-1}_{A,B}]{} & \mathscr{A}(A, FB)
\end{array}$$

$$\begin{array}{ccc}
\psi_B = \phi^{-1}(1_{FB}) & \longleftarrow\!\dashleftarrow & 1_{FB} \\
(Eg)^* \downarrow & & \downarrow g^* \\
\psi_B \circ (Eg) = \phi^{-1}(1_{FB} \circ g) & \dashleftarrow & 1_{FB} \circ g
\end{array}$$

Therefore

$$\forall g : A \to FB \Rightarrow \phi^{-1}(g) = \psi_B \circ (Eg)$$

15-4.6   Proposition. *For any two functors $E: \mathscr{A} \to \mathscr{B}$ and $F: \mathscr{B} \to \mathscr{A}$ for any categories $\mathscr{A}$ and $\mathscr{B}$, the following are equivalent.*

(1)   *E is a left adjoint of F.*
(2)   *$\exists$ a front adjunction $\eta : 1_{\mathscr{A}} \to FE$.*
(3)   *$\exists$ a back adjunction $\psi : EF \to 1_{\mathscr{B}}$.*

Proof. $(1) \Rightarrow (2)$. Suppose that $c : A \to FB$ is given in 15-4.3. Since $\phi^{-1}: \mathscr{A}(A, FB) \to \mathscr{B}(EA, B)$ is an isomorphism, define

$$n = \phi^{-1}(c) : EA \to B.$$

In 15-4.5(i) substitute $f = n$ in $\phi(f) = (Ff) \circ \eta_A$ to yield $c = (Fn) \circ \eta_A$. Suppose that $m : EA \to B$ is any other morphism such that also $c = (Fm) \circ \eta_A$. By 15-4.5(i), $c = \phi(m) = (Fm) \circ \eta_A$. Hence

$$m = \phi^{-1}(\phi m) = \phi^{-1}(c) = n$$

is unique. Thus η is a front adjunction.

(2) $\Rightarrow$ (1). The first step in defining a natural isomorphism

$$\phi : \mathscr{B}(E\cdot, \cdot) \to \mathscr{A}(\cdot, F\cdot)$$

is to define an isomorphism $\phi_{A,B} : \mathscr{B}(EA, B) \to \mathscr{A}(A, FB)$ for any objects A and B. Given any $n \in \mathscr{B}(EA, B)$, define $\phi_{A,B}$ by

$$\phi_{AB}(n) = (Fn) \circ \eta_A : A \to FB.$$

For any $c \in \mathscr{A}(A, FB)$, by hypothesis there exists a unique $n \in \mathscr{B}(EA, B)$ such that $c = (Fn) \circ \eta_A = \phi_{A,B}(n)$. Hence $\phi(A, {}^+B) = \phi_{A,B}$ is an isomorphism.

Let $\alpha : A' \to A$ and $\beta : B \to B'$ be any morphisms of $\mathscr{A}$ and $\mathscr{B}$. In order to complete the proof that ϕ is a natural isomorphism of functors, it only remains to verify that ϕ is a natural transformation of the two functors $\phi : \mathscr{B}(E\cdot, \cdot) \to \mathscr{A}(\cdot, F\cdot)$ defined on the category $\mathscr{A}^{\mathrm{op}} \times \mathscr{B}$. Then $(\alpha, \beta) : A \times B \dashrightarrow A' \times B'$ is a typical morphism in $\mathscr{A}^{\mathrm{op}} \times \mathscr{B}$. Thus it suffices to verify that the following diagram commutes:

$$
\begin{array}{ccc}
\mathscr{B}(EA, B) & \xrightarrow{\ \phi = \phi(A,\,B)\ } & \mathscr{A}(A, FB) \\
{\scriptstyle B(E\alpha,\,\beta)} \downarrow & & \downarrow {\scriptstyle \mathscr{A}(\alpha,\,{}^+ F\beta)} \\
B(EA', B') & \xrightarrow[\ \phi = \phi(A',B')\ ]{} & \mathscr{A}(A', FB')
\end{array}
$$

First, since η is a natural transformation, it follows that $\eta(A) \circ \alpha = (FE\alpha) \circ \eta(A')$ from the following commutative diagram:

$$
\begin{array}{ccc}
A' & \xrightarrow{\ \eta(A')\ } & FEA' \\
{\scriptstyle \alpha} \downarrow & & \downarrow {\scriptstyle FE\alpha} \\
A & \xrightarrow[\ \eta(A)\ ]{} & FEA
\end{array}
$$

Given $n \in \mathscr{B}(EA, B)$, and using that F is a functor in $F(\beta \circ n \circ E\alpha) = (F\beta) \circ (Fn) \circ (FE\alpha)$, we get along the bottom of the square

$$\phi(A', B')\mathscr{B}(E\alpha, \beta)n = \phi(A', B')(\beta \circ n \circ E\alpha)$$
$$= [F(\beta \circ n \circ E\alpha)] \circ \eta(A')$$
$$= (F\beta) \circ (Fn) \circ (FE\alpha) \circ \eta(A')$$
$$= (F\beta) \circ (Fn) \circ \eta(A) \circ \alpha.$$

Mapping around the top of the square produces

$$\mathscr{A}(\alpha, F\beta) \circ \phi(A, B)n = \mathscr{A}(\alpha, F\beta)[(Fn) \circ \eta(A)]$$
$$= (F\beta) \circ (Fn) \circ \eta(A) \circ \alpha.$$

Thus $\phi: \mathscr{B}(E\cdot, \cdot) \to \mathscr{A}(\cdot, F\cdot)$ is a natural equivalence of functors.

(1) $\Rightarrow$ (3). By hypothesis $\phi: \mathscr{B}(EA, B) \to \mathscr{A}(A, FB)$ is an isomorphism. Hence for $n \in \mathscr{B}(EA, B)$ as in 15-4.4, define $c = \phi(n): A \to FB$. In 15-4.5(ii) $\phi^{-1}(g) = \psi_B \circ (Eg)$, substitution of $g = \phi n = c$ gives $n = \psi_B \circ (Ec)$. If $g: A \to FB$ were another morphism with $n = \psi_B \circ (Eg)$, then a second application of 15-4.5(ii) shows that $\phi^{-1}(g) = = \psi_B \circ (Eg) = n$. Hence $g = \phi(\phi^{-1}g) = \phi n = c$ is unique. Thus ψ is a back adjunction.

(3) $\Rightarrow$ (1). Since ϕ is an equivalence if and only if ϕ^{-1} is, it suffices to define $\phi^{-1}: \mathscr{A}(\cdot, F\cdot) \to \mathscr{B}(E\cdot, \cdot)$. Given $c \in \mathscr{A}(A, FB)$, define $\phi^{-1}(c) = \psi_B \circ (EC)$. For any $n \in (EA, B)$, by hypothesis there exists a unique $c \in \mathscr{A}(A, FB)$ such that $n = \psi_B \circ (Ec) = \phi_{A,B}^{-1}(c)$. Thus $\phi^{-1}(A, B) = \phi(A, B)^{-1} = \phi_{A,B}^{-1}$ is an isomorphism.

Let $\alpha: A' \to A$ and $\beta: B \to B'$ be as before. The fact that ψ is a natural transformation implies that $\beta \circ \psi(B) = \psi(B') \circ (EF\beta)$. It suffices now to verify that the diagram below commutes

$$
\begin{array}{ccc}
\mathscr{B}(EA, B) & \xleftarrow{\phi^{-1} = \phi^{-1}(A, B)} & \mathscr{A}(A, FB) \\
{\scriptstyle \mathscr{B}(E\alpha, \beta)}\downarrow & & \downarrow{\scriptstyle \mathscr{A}(\alpha, F\beta)} \\
\mathscr{B}(EA', B') & \xleftarrow{\phi^{-1} = \phi^{-1}(A', B')} & \mathscr{A}(A', FB')
\end{array}
$$

An element $c \in \mathscr{A}(A, FB)$ is mapped along the top of the square as follows:

$$B(E\alpha, \beta)\phi^{-1}(A, B)c = \mathscr{B}(E\alpha, \beta)[\psi(B) \circ (Ec)]$$
$$= \beta \circ \psi(B) \circ (Ec) \circ (E\alpha).$$

Along the bottom route, $E[(F\beta) \circ c \circ \alpha] = (EF\beta) \circ (Ec) \circ (E\alpha)$ because E is a functor, and

$$\phi^{-1}(A' \cdot B') \mathscr{A}(\alpha, F\beta)c = \phi^{-1}(A', B')[(F\beta) \circ c \circ \alpha]$$
$$= \psi(B') \circ E((F\beta) \circ c \circ \alpha)$$
$$= \psi(B') \circ (EF\beta) \circ (Ec) \circ (E\alpha)$$
$$= \beta \circ \psi(B) \circ (Ec) \circ (E\alpha).$$

Hence $\mathscr{B}(E\cdot, \cdot) \cong \mathscr{A}(\cdot, F\cdot)$.

15-4.7. Examples. Let $\mathscr{B}$ be a category with the property that for any set X, there is a free object EX on that set with the usual universal property. Let $\mathscr{A} = \mathscr{S}ets$ and $F: \mathscr{B} \to \mathscr{A}$ the forgetful functor, whereby for $B \in \text{Ob } \mathscr{B}$, $FB = B \in \text{Ob } \mathscr{A}$ is viewed as a set only. Then E is the left adjoint of F.

Here are some examples of B. (i) The category of groups and group homomorphisms. (ii) The category of algebras over a field F and identity preserving ring homomorphisms. (iii) $\mathscr{B} = \mathscr{M}od_R$.

In the latter, let X be a set, EX the free module on X, and $f: EX \to M$ any module homomorphism. Then

$$\phi(X, M): \text{Hom}_R(EX, M) \to \mathscr{S}ets(X, FM),$$

where $\phi(X, M)f = f \mid X: X \to FM = M$ is the restriction of f to X viewed as a set map.

The similarity in notation between the next example and the Adjoint Associativity Theorem is not accidental. This is so because in many categories the tensor product is the left adjoint of the hom functor, which is also the case below.

15-4.8 Example. Here $\mathscr{A} = \mathscr{B} = \mathscr{S}ets$. For sets A, B, C define functors $E, F: \mathscr{S}ets \to \mathscr{S}ets$ by $F = \text{Morph}_{\mathscr{S}ets}(B, \cdot) = \mathscr{S}ets(B, \cdot)$, and $E = (\cdot) \times B$, where $E(A) = A \times B$ is the direct product. For any sets like A and C, C^A denotes as usual the set of all functions from A to C. Next will be defined a natural equivalence ϕ of the functors $\phi: \mathscr{S}((\cdot) \times B, \cdot) \to \mathscr{S}(\cdot, \mathscr{S}(B, \cdot))$. Define $\phi(A, B) = \phi: \mathscr{S}(A \times B, C) \to \mathscr{S}(A, \mathscr{S}(B, C))$ by the following:

$$f \in C^{A \times B}, \ \phi f \in (C^B)^A; \quad \text{for } a \in A, b \in B,$$

$$(\phi f)(a) \in C^B, (\phi f)(a)(b) = f(a, b).$$

Thus $(\cdot) \times B$ is a left adjoint of the functor $\mathscr{S}ets(B, \cdot)$.

Thus far in our treatment of adjoint functors no restrictive hypotheses have been invoked on the underlying categories $\mathscr{A}$ and $\mathscr{B}$.

15-4.9 Proposition. *Suppose that* $E: \mathscr{A} \to \mathscr{B}$ *and* $F: \mathscr{B} \to \mathscr{A}$ *are zero preserving additive functors of two abelian categories* $\mathscr{A}$ *and* $\mathscr{B}$, *and that*

E is the left adjoint of F. Then if

(i)   *$\mathscr{A}$ has a generator $\Rightarrow$ F is left exact. If*
(ii) *$\mathscr{B}$ has a cogenerator $\Rightarrow$ E is right exact.*

Proof. (i) Given an exact sequence (1) in $\mathscr{B}$, we have to show that (2) is exact in $\mathscr{A}$:

$$0 \to B_1 \to B_2 \to B_3 \quad \text{in } \mathscr{B}; \tag{1}$$

$$0 \to FB_1 \to FB_2 \to FB_3 \quad \text{in } \mathscr{A}. \tag{2}$$

There is a natural equivalence of functors $\phi : \mathscr{B}(E\cdot, \cdot) \to \mathscr{A}(\cdot, F\cdot)$ because E and F are an adjoint pair. Let $G \in \mathrm{Ob}\,\mathscr{A}$ be a generator of $\mathscr{A}$.

By 15-3.19(i) the functor $\mathscr{B}(EG, \cdot)$ is left exact. Hence the sequence (3) below is an exact sequence of abelian groups.

$$0 \to \mathscr{B}(EG, B_1) \to \mathscr{B}(EG, B_2) \to \mathscr{B}(EG, B_3) \tag{3}$$

$$\phi \downarrow \qquad\qquad \phi \downarrow \qquad\qquad \phi \downarrow$$

$$0 \to \mathscr{A}(G, FB_1) \to \mathscr{A}(G, FB_2) \to \mathscr{A}(G, FB_3) \tag{4}$$

The above diagram (3) and (4) commutes because the three vertical morphisms are isomorphism given by the natural equivalence of functors $\phi : \mathscr{B}(E\cdot, \cdot) \to \mathscr{A}(\cdot, F\cdot)$. Consequently also the bottom row (4) is exact. Since G is a generator of $\mathscr{A}$, by 15-1.17 the functor $\mathscr{A}(G, \cdot)$ is faithful. However we know from Proposition 15-3.23 that a faithful functor preserves inexact sequences. Now assume that (2) is inexact. But then (4) would be inexact, which is a contradiction. Hence sequence (2) must be exact, and F is a left exact functor.

(ii) Let $C \in \mathrm{Ob}\,\mathscr{B}$ be a cogenerator of $\mathscr{B}$. Now starting with an arbitrary exact sequence (1') in $\mathscr{A}$, we will show that 2' is exact in $\mathscr{B}$.

$$A_1 \to A_2 \to A_3 \to 0 \quad \text{in } \mathscr{A}; \tag{1'}$$

$$EA_1 \to EA_2 \to EA_3 \to 0 \quad \text{in } \mathscr{B}. \tag{2'}$$

Since by 15-3.19(ii) the contravariant functor $\mathscr{A}(\cdot, FC)$ is also left exact, it turns the right exact sequence (1') into the left exact sequence (3') where below the diagram (3'), (4') commutes and sequence (4') is exact for the same reasons as in part (i).

$$0 \to \mathscr{A}(A_1, FC) \to \mathscr{A}(A_2, FC) \to \mathscr{A}(A_3, FC) \tag{3'}$$

$$\phi^{-1} \downarrow \qquad\qquad \phi^{-1} \downarrow \qquad\qquad \phi^{-1} \downarrow$$

$$0 \to \mathscr{B}(EA_1, C) \to \mathscr{B}(EA_2, C_2) \to \mathscr{B}(EA_3, C) \tag{4'}$$

Use of the contravariant parts of 15-1.17 and 15-3.23 shows that now the functor $\mathscr{B}(\cdot, C)$ also preserves inexact sequences. Hence (2') and E are right exact.

15-5   Exercises

Prove the following

1. $\mathbf{Z} \to \mathbf{Q} \times \mathbf{Z}/(2) \times \mathbf{Z}/(3)$, $n \to (n, n + (2),\ n + (3))$ is epic but not onto in the category Ring[1].

2*. (Burgess and Stewart [89], p. 484). For the following rings S, there is an identity preserving epic ring homomorphism $v: \mathbf{Z} \to S$.
   (a) $S \cong \mathbf{Z}/I, I \lhd \mathbf{Z}$.
   (b) For a finite set of primes $p_1, \ldots, p_k$ and any natural numbers $n(i) \geqslant 1$, take any ring D in $\mathbf{Z} \subseteq D \subseteq \mathbf{Q}$ such that every element of D is divisible by all the $p_1, \ldots, p_k$. Then

$$S \cong D \times \mathbf{Z}/(p_1^{n(1)}) \times \cdots \times \mathbf{Z}/(p_k^{n(k)}).$$

   (c) For an infinite set of primes $\{p_i \,|\, i = 1, 2, \ldots\}$ and any integers $n(i) \geqslant 1$, let D be any ring in $\mathbf{Z} \subseteq D \subseteq \mathbf{Q}$, where D is divisible by all the p_i. Show that for any $a/b \in D$, there exists an m such that eventually for all $i \geqslant m$, the element $\bar{b} = b + (p_i^{n(i)})$ has an inverse $1/\bar{b} \in \mathbf{Z}/(p_i^{n(i)})$. Let $S \subset \Pi_{i=1}^{\infty} \mathbf{Z}(p_i^{n(i)})$ be the subring

$$S = \{(r_i) \,|\, \exists\, a/b \in D, \text{ for almost all } i,\ r_i = \bar{a}/\bar{b} \in \mathbf{Z}/(p_i^{n(i)})\}.$$

   In all cases, find $v: \mathbf{Z} \to S$.

Remark. It can be shown that every epimorph S of $\mathbf{Z}$ in Ring[1] is of the above form. All of the above generalizes verbatim to commutative principal ideal domains.

3. The inclusion $2Z_4 \subset Z_4 = \mathbf{Z}/4\mathbf{Z}$ is an epimorphism in the category of rings.

4. Prove that in any category $\mathscr{C}$, Ob $\mathscr{C}$ is a set if and only if Morph $\mathscr{C}$ is a set. Such a category will be called *small*.

5. In a category $\mathscr{C}$ assume that for all ordered pairs of objects A, $B \in$ Ob $\mathscr{C}$, the morphisms $f, f' \in \mathscr{C}(A, B)$ are divided into equivalence classes $[f], [f']$. Assume that this equivalence relation respects composition, i.e. if $[f] = [f']$, then $[hf] = [hf']$ and $[fh] = [f'h]$ whenever composition with h makes sense. Define $\langle Ob\, \mathscr{D}, \text{Morph } \mathscr{D}, \mathscr{D}(\cdot, \cdot), \circ \rangle$ as follows: (i) Ob $\mathscr{D} =$ Ob $\mathscr{C}$; (ii) Morph $\mathscr{D} = \{[f] \,|\, A,\ B \in$ Ob $\mathscr{C}, f \in \mathscr{C}(A, B)\}$; (iii) composition is $[f][g] = [fg]$. Prove then that $\mathscr{D}$ is a category called the *quotient category* of $\mathscr{C}$ with respect to the equivalence relation.

6. In any category a monomorphism $A \xrightarrow{f} B$ is defined to be *essential* if for any morphism g with $A \xrightarrow{f} B \xrightarrow{g} C$, whenever gf is monic, g is monic. In Mod_R show that f is essential if and only if $f(A) < B$ is an essential submodule.

7. Let $\mathscr{C}$ and $\mathscr{D}$ be any two categories where $\mathscr{C}$ is a small category. Prove that the following so-called *functor category* $(\mathscr{C}, \mathscr{D})$ is actually a category. The objects are all functors from $\mathscr{C}$ to $\mathscr{D}$, and the morphisms are natural transformations of functors. If $\mathscr{D}$ is small then Functor $(\mathscr{C}, \mathscr{D})$ is a small category.

8*. If above $\mathscr{C}$ is a small abelian category, and $\mathscr{D} = Ab$, the class of all additive F, G functors from $\mathscr{C}$ to Ab is an abelian category where for $A \xrightarrow{\alpha} B$ in $\mathscr{C}$,

$$(F \oplus G)(A) \xrightarrow{(F \oplus G)(\alpha)} (F \oplus G)(B)$$

is

$$F(A) \oplus G(A) \xrightarrow{(F\alpha, G\alpha)} F(B) \oplus G(B).$$

If $\eta : F \to G$ is a natural transformation the "kernel of η" is a subfunctor K of F defined on objects by requiring $0 \to K(A) \to F(A) \xrightarrow{\eta(A)} G(A)$ to be exact.

9. Suppose that $\mathscr{D}$ is a category whose objects are sets and morphisms are functions, and let $F, G : \mathscr{C} \to \mathscr{D}$ be functors from any category $\mathscr{C}$. Define F to be a *subfunctor* of G, written as $F \leqslant G$, if for any object A of $\mathscr{C}$, $F(A) \subseteq G(A)$ is a subobject of GA, and for any morphism $A \xrightarrow{f} B$ in $\mathscr{C}$, $f(G(A)) \subseteq G(B)$, and Gf is the restriction and corestriction of $F(f)$ to $G(A) \xrightarrow{F(f)} G(A)$. Show that "$\leqslant$" is a partial order.

10. For a field F, let Mod_F be the category of F-vector spaces and $\text{Alg}_F^{(1)}$ the category of F-algebras with identity and identity preserving algebra homomorphism. For a vector space V, let TV be the tensor algebra on V. (i) Show that $T : \text{Mod}_F \to \text{Alg}_F^{(1)}$ is a functor. Prove that T is the left adjoint of the forgetful functor $\text{Alg}_F^{(1)} \to \text{Mod}_F$. Spell out clearly what the front and back adjunctions are.

11. If EV is the exterior algebra for $V \in \text{Mod}_F$, prove that E is a functor. Explain why E is not the left adjoint of a forgetful functor.

CHAPTER 16

Module categories

Introduction

The first section discusses generators and cogenerators in the category $\mathcal{M}od_R$. Then the basic properties of the hom functor $\mathrm{Hom}_R(\,\cdot\,,\,\cdot\,)$ and the tensor product functor $\underline{\ }\otimes_R\underline{\ }$ are derived. The right exactness of the latter is used to derive precise conditions for an element of the tensor product of a module to be zero. The Adjoint Associativity Theorem is proved. For almost all of sections 16-1, 2, 3, and 4 an identity element for the ring R is not needed. The only exception is Proposition 16-1.5 which uses injective hulls. In section 16-5 an identity element for the ring R is assumed.

In this chapter inverse and direct limits of modules and rings are covered. Inverse and direct limits are used in many areas in mathematics outside of algebra, frequently with inadequate proofs. They are particularly useful in category theory. For those who will want to go on with ring and module theory will find these limit modules and rings an interesting source of examples, and an indispensable tool later on in the book (18-3).

16-1   Generators and cogenerators

The category of right R-modules over any ring (with $1 \in R$ or with $1 \notin R$) is an abelian category. The special subobjects and morphisms defined previously (in 15-2.2, 2.3, and 2.4) can now be concretely identified as sets and functions.

16-1.1   Let $f \colon A \to B$ be a right R-module homomorphism in $\mathcal{M}od_R$ with kernel $\ker f < A$. Then

(i) $\mathrm{im}\, f = fA$, $\mathrm{coim}\, f = A/\ker f$, $\mathrm{coker}\, f = B/fA$;

(ii)   natural inclusion and projection are denoted by "inc." and "proj."

$$\text{Im } f : fA \xrightarrow{\text{inc.}} B \qquad \text{Coim } f : A \xrightarrow{\text{proj.}} A/\ker f$$

$$\text{Ker } f : \ker f \xrightarrow{\text{inc.}} A \quad \text{Coker } f : B \xrightarrow{\text{proj.}} B/fA$$

Thus

(iii)   Ker Coker $f = \text{Im } f$ ker Coker $f = \text{im } f$
(iv)   Coker Ker $f = \text{Coim } f$ coker Ker $f = \text{coim } f$.

16-1.2   Lemma. *For a module G, the following are all equivalent.*

(i)   *G is a generator of $\mathcal{M}od_R$ (i.e. $\forall N = N_R, \forall 0 \neq f \in \text{Hom}_R(M, N)$, $\exists g \in \text{Hom}_R(G, M)$ such that $fg \neq 0$).*
(ii) $\forall M = M_R, M = \text{Hom}_R(G, M)G = \Sigma\{gG \mid g \in \text{Hom}_R(G, M)\}$.
(iii) $\forall M = M_R, \exists$ *an index set I and an epimorphism $\oplus\{G \mid I\} \to M$.*

Proof. (iii) $\Rightarrow$ (ii) and (ii) $\Rightarrow$ (iii) are easy and are omitted. (For the latter, take $I = \text{Hom}_R(G, M)$.) (ii) $\Rightarrow$ (i). As in 15-1.7, let $0 \neq f : M \to N$ be given. Take $m \in M$ for which $fm \neq 0$. By hypothesis $m = g_1 x_1 + \cdots + g_n x_n$ for some $g_1, \ldots, g_n \in \text{Hom}_R(G, M)$ and $x_1, \ldots, x_n \in G$. Thus $fg_1 x_1 + \cdots + fg_n x_n \neq 0$. Without loss of generality, let $fg_1 x_1 \neq 0$. Thus $fg_1 \neq 0$ as required.
   (i) $\Rightarrow$ (ii). If not, let $0 \neq f : M \to M/K$ be the projection where $K = \text{Hom}_R(G, M)G < M$. By hyhpothesis $fg \neq 0$ for some $g \in \text{Hom}_R(G, M)$. This is a contradiction because $g(G) \subseteq K$. Hence $K = M$.

Property (ii) below says that homomorphisms into C separate the points of any module.

16-1.3   Lemma. *For a module C, the following are all equivalent.*

(i)   *C is a cogenerator of $\mathcal{M}od_R$ (i.e. $\forall N = N_R, \forall 0 \neq f \in \text{Hom}_R(N, M)$, $\exists h \in \text{Hom}_R(M, C)$ such that $hf \neq 0$).*
(ii) $\forall M = M_R, \cap\{\ker h \mid h \in \text{Hom}_R(M, C)\} = 0$.
(iii) $\forall M = M_R, \exists$ *index set I and a monomorphism $M \to \Pi\{C \mid I\}$.*

Proof. Again (iii) $\Rightarrow$ (ii) and (ii) $\Rightarrow$ (iii) are omitted. (For the latter, take $I = \text{Hom}_R(M, C)$.) (ii) $\Rightarrow$ (i). If $0 \neq f : N \to M$, then $0 \neq fN \nsubseteq \ker h$ for some $h \in \text{Hom}_R(M, C)$. Hence $hf \neq 0$.
   (i) $\Rightarrow$ (ii). For any $0 \neq m \in M$, if $f : mR^1 \to M$ is the inclusion, then by hypothesis $hf \neq 0$ for some $h : M \to C$. Hence $0 \neq hfm = hm$, $m \notin \ker h$, and (ii) holds.

16-1.4 Recall that $R^1 = R$ if $1 \in R$, but otherwise $R^1 = \mathbb{Z} \times R$ if $1 \notin R$. The right R-module R^1 is a generator of $\mathcal{M}od_R$.

Proof. For any right R-module M, and for any $0 \neq m \in M$, and for $e = 1 \in R^1$, the map $R^1 \to mR^1$, $e \to me = m$ defines a right R-module epimorphism. By 16-1.3(ii), the right R-module R^1 is a generator of $\mathcal{M}od_R$.

The module K below is called the *minimal cogenerator* of $\mathcal{M}od_R$. It is unique up to isomorphism.

16-1.5 Proposition. *Assume that the ring R has an identity. Let $\mathcal{S} = \{S\}$ be any irredundant set of representatives of isomorphism classes of simple R-modules, and define $K = \oplus\{\hat{S} \mid S \in \mathcal{S}\}$ where $\hat{S}$ denotes as before the injective hull of S. Then*

(i) *If C is any module such that for any $S \in \mathcal{S}$, C contains an isomorphic copy of $\hat{S}$, then C is a cogenerator. In particular,*

(ii) *K is a cogenerator of $\mathcal{M}od_R$. Moreover,*

(iii) *any module C is a cogenerator of $\mathcal{M}od_R$. if and only if C contains an isomorphic copy of K.*

Proof. (i) and (ii). Given any module M, and any element $0 \neq x \in M$, by Zorn's lemma there exists a submodule $N < M$ maximal with respect to $x \notin N$. Thus $S = (xR + N)/N$ is simple. The natural quotient map $h: xR + N \to S$ extends to $\hat{h}: M \to \hat{S}$ with $\hat{h}x = hx = x + N \neq 0$. By 16-1.3(ii), conclusions (i) and (ii) follow.

(iii) By 16-1.3(ii), any module C containing a cogenerator is itself a cogenerator. Conversely, if C is a cogenerator and S any simple module, then $\cap\{\ker h \mid h \in \operatorname{Hom}_R(\hat{S}, C)\} = 0$ by 16-1.3(ii). Hence there must exist a homomorphism $h: \hat{S} \to C$ such that $S \nsubseteq \ker h$. Since S is simple, $S \cap \ker h = 0$; and since $S \leqslant \hat{S}$ is large, h is monic. Thus $\hat{S} \cong h(S) \subset C$. For every $S \in \mathcal{S}$, select submodules $S \cong T_S \leqslant \hat{T}_S \leqslant C$. Assume $\Sigma\{T_S \mid S \in \mathcal{S}\}$ is not direct. Let n be minimal such that a finite sum $T_1 + \cdots + T_n$ of the T_S is not direct. Then $T_1 + \cdots + T_{n-1} = T_1 \oplus \cdots \oplus T_{n-1}$, and $(T_1 \oplus \cdots \oplus T_{n-1}) \cap T_n \neq 0$. Since the module T_n is simple, $T_n \subseteq T_1 \oplus \cdots \oplus T_{n-1} = W$. If $\pi_i: W \to T_i$, $1 \leqslant i \leqslant n-1$, is the natural projection, then without loss of generality we may assume that $0 \neq \pi_1(T_n) \leqslant T_1$. But then $T_n \cong \pi_1(T_n) = T_1$ is a contradiction of the irredundancy of $\mathcal{S}$. Thus $\Sigma\{T_S \mid S \in \mathcal{S}\} = \oplus\{T_S \mid S \in \mathcal{S}\}$. Since $T_S \leqslant \hat{T}_S \leqslant C$ is essential, we still have $\Sigma\{\hat{T}_S \mid S \in \mathcal{S}\} = \oplus\{\hat{T}_S \mid S \in \mathcal{S}\} \cong K$.

16-1.6   Remark. For the ring $R = \mathbb{Z} = 0,\ \pm 1,\ \pm 2, \ldots$ and Q the rationals, the injective module $Q/\mathbb{Z}$ is a cogenerator of the category of abelian groups by 16-1.5(i).

16-2   Hom functor

16-2.1   Let M be any right R-module, and let $\alpha\colon A \to B$, $g\colon B \to M$, and $f\colon M \to A$ be module maps. As before, $\mathbb{Z} = 0,\ \pm 1,\ \pm 2, \ldots$ and $\mathcal{M}\!od_\mathbb{Z} = \mathcal{A}\!b$. Then $\operatorname{Hom}_R(M,\ \cdot)\colon \mathcal{M}\!od_R \to \mathcal{M}\!od_\mathbb{Z}$ is a functor, where $\alpha_* = \operatorname{Hom}_R(M, \alpha)\colon \operatorname{Hom}_R(M, A) \to \operatorname{Hom}_R(M, B)$ by $\alpha_*(f) = \alpha_* f = \operatorname{Hom}_R(M, \alpha) f = \alpha f$. The functor $\operatorname{Hom}_R(\cdot, M)\colon \mathcal{M}\!od_R \to \mathcal{M}\!od_\mathbb{Z}$ is contravariant; where $\alpha^* = \operatorname{Hom}_R(\alpha, M)\colon \operatorname{Hom}_R(B, M) \to \operatorname{Hom}_R(A, M)$, $\alpha^*(g) = \alpha^* g = g\alpha$. Thus

$\qquad \alpha_* =$ left multiplication by α, and

$\qquad \alpha^* =$ right multiplication by α.

The following are straightforward to verify.

(i) $\operatorname{Hom}_R(M,\ \cdot)$ is a left exact functor, by 15-3.19;

(ii) $\operatorname{Hom}_R(\cdot, M)$ is a left exact contravariant functor. (See 15-3.12 and 15-3.19.)

(iii)   If $1 \in R$, then evaluation at $1 \in R$ is an R-isomorphism of right R-modules $\operatorname{Hom}_R(R, M) \to M$, $\phi \to \phi 1$; where $(\phi * r)(x) = \phi(rx)$ for $r, x \in R$.

(iv) $\operatorname{Hom}_R(M,\ \cdot)$ and $\operatorname{Hom}_R(\cdot, M)$ are additive functors. For any indexed set of modules $\{A_i \mid i \in I\}$ there are canonical isomorphisms

$$\operatorname{Hom}_R(M,\ \oplus A_i) \cong \oplus \operatorname{Hom}_R(M, A_i);$$

$$\operatorname{Hom}_R(\oplus A_i,\ M) \cong \Pi \operatorname{Hom}_R(A_i, M).$$

(v) M is projective $\Leftrightarrow \operatorname{Hom}_R(M,\ \cdot)$ is exact.

(vi) M is injective $\Leftrightarrow \operatorname{Hom}_R(\cdot,\ M)$ is exact (if $1 \in R$).

16-2.2   Example.   For the integers $\mathbb{Z}$ and rationals $\mathbb{Q}$, with $R = \mathbb{Z}$, form the exact sequence

$$0 \to \mathbb{Z} \to \mathbb{Q} \to \mathbb{Q}/\mathbb{Z} \to 0.$$

(i)   For $M = \mathbb{Z}/2\mathbb{Z}$, $\operatorname{Hom}_R(M,\ \cdot)$ is not right exact. When applied to the above sequence, it gives the nonexact sequence

$$0 \to 0 \to 0 \to \operatorname{Hom}_\mathbb{Z}(\mathbb{Z}/2\mathbb{Z},\ \mathbb{Q}/\mathbb{Z}) \to 0.$$

(ii) For $M = \mathbb{Z}$, the contravariant functor $\operatorname{Hom}_R(\,\cdot\,, M)$ is not right exact either. It turns the above sequence around into the nonexact sequence

$$0 \leftarrow \operatorname{Hom}_{\mathbb{Z}}(\mathbb{Z}, \mathbb{Z}) \leftarrow 0 \leftarrow 0 \leftarrow 0.$$

16-3 Tensor product functor

16-3.1 Tensor products with respect to R will be abbreviated as $\_\otimes\_ \equiv \_\otimes_R\_$ by omitting the R. Let M be a left R-module. So then $\_\otimes M = \_\otimes_R M$. If $\alpha: A \to B$ is a right R-module homomorphism, then $\alpha \otimes M$ is the abelian group homomorphism $\alpha \otimes M: A \otimes M \to B \otimes M$ defined on generators by $(\alpha \otimes M)(a \otimes m) = \alpha a \otimes m$, $a \in A$, $m \in M$.

If $f: M \to N$ is a homomorphism of left R-modules in the category $_R\mathscr{M}\!od$, then there is an induced abelian group homomorphism $\alpha \otimes f: A \otimes M \to B \otimes N$. In particular, if $M = N$ and $f = 1_M$ is the identity then $\alpha \otimes 1_M = \alpha \otimes M$. If M is fixed and understood, abbreviate $\alpha \otimes 1_M = \alpha \otimes 1$.

The value of the functor $\_\otimes M$ at A is $(\_\otimes M)A = A \otimes M$, and its value at α is $(\_\otimes M)\alpha = \alpha \otimes M = \alpha \otimes 1_M$. The notation $\_\otimes 1_M = \_\otimes M$ is also convenient for some purposes, where $(\_\otimes 1_M)\alpha = \alpha \otimes 1_M$, and $(\_\otimes 1_M)A = A \otimes M$ although the latter is infrequently used.

The map $R \otimes M \to M$, $x \otimes m \to xm$, $x \in R$, $m \in M$, is an isomorphism of left R-modules, where $r(x \otimes m) = rx \otimes m$ for $r \in R$, if $1 \in R$.

There are only two facts about the tensor product available for our use in the proof below – the universal property, and properties of generators $b \otimes m \in B \otimes M$, such as linearity in b and m.

16-3.2 The functor $\_\otimes M$ is right exact.

Proof. Assume that the first sequence is exact, and then form the second one:

$$A \xrightarrow{\ \alpha\ } B \xrightarrow{\ \beta\ } C \longrightarrow 0$$

$$A \otimes M \xrightarrow{\ \alpha \otimes 1\ } B \otimes M \xrightarrow{\ \beta \otimes 1\ } C \otimes M \longrightarrow 0.$$

The following have to be shown: (i) $\beta \otimes 1$ is epic; (ii) $\operatorname{im}(\alpha \otimes 1) \subset \ker(\beta \otimes 1)$; and (iii) $\ker(\beta \otimes 1) \subset \operatorname{im}(\alpha \otimes 1)$; (i) is obvious. Set $D = \operatorname{im}(\alpha \otimes 1) = (\alpha A) \otimes M \subset B \otimes M$, and let $\rho: B \otimes M \to (B \otimes M)/D$ be the projection map. Then let $\eta: C \times M \to C \otimes M$ be the R-bilinear map $\eta(c, m) = c \otimes m$.

(ii) Since $\beta \alpha = 0$, $(\beta \otimes 1)(\alpha \otimes 1) = \beta \alpha \otimes 1 = 0$, and $D \subseteq \ker(\beta \otimes 1)$.

(iii) Thus there is a commutative diagram of abelian group homomorphisms, where $\beta \otimes 1 = \bar{\beta}\rho$:

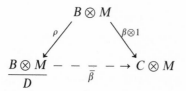

Define $f: C \times M \to (B \otimes M)/D$ by $f(c, m) = b \otimes m + D$ where $b \in B$ is any element such that $\beta b = c$. If also $\beta b = \beta b_1 = c$ for another $b_1 \in B$, then $\beta(b - b_1) = 0$, and $b - b_1 = \alpha a \in \ker \beta = \operatorname{im} \alpha$ for some $a \in A$. Thus $b \otimes m + D = b_1 \otimes m + \alpha a \otimes m + D = b_1 \otimes m + D$, and f is a well-defined R-bilinear map. Now by the universal property of $C \otimes M$ there exists a unique abelian group homomorphism $\bar{f}: C \otimes M \to (B \otimes M)/D$ such that $f = \bar{f}\eta$. There is a diagram which we know commutes if $\bar{\beta}$ is removed from it.

$$
\begin{array}{ccc}
C \times M & \xrightarrow{\ \eta\ } & C \otimes M \\
{\scriptstyle f} \downarrow & {\scriptstyle \bar{f}}\;{\scriptstyle \bar{\beta}} & \\
B \otimes M & & \\
\hline
D & &
\end{array}
$$

We show that $\bar{\beta}$ and $\bar{f}$ are inverses. For any generator $c \otimes m \in C \otimes M$, select an element $b \in B$ with $\beta b = c$, and then

$$
\begin{aligned}
\bar{\beta}\bar{f}(c \otimes m) &= \bar{\beta}[\bar{f}\eta(c, m)] = \bar{\beta}[f(c, m)] \\
&= \bar{\beta}[b \otimes m + D] = (\bar{\beta}\bar{\rho})(b \otimes m) \\
&= (\beta \otimes 1)(b \otimes m) = c \otimes m.
\end{aligned}
$$

Thus $\bar{\beta}\bar{f} = 1$. For any generator $b \otimes m + D \in (B \otimes M)/D$, use of the definition of f shows that

$$
\bar{f}\bar{\beta}[b \otimes m + D] = \bar{f}\bar{\beta}\rho(b \otimes m) = \bar{f}(\beta \otimes 1)(b \otimes m) = \bar{f}(\beta b \otimes m)
$$
$$
= \bar{f}[\eta(\beta b, m)] = f(\beta b, m) = b \otimes m + D.
$$

Thus also $\bar{f}\bar{\beta} = 1$, and $\bar{f}$ and $\bar{\beta}$ are inverses. In particular $\bar{\beta}$ is monic, and hence $\ker(\beta \otimes 1) = \ker \bar{\beta}\rho = \ker \rho = D = \operatorname{im}(\alpha \otimes 1)$. Thus $\_\otimes M$ is right exact.

16-3.3 Example. For $R = \mathbb{Z}$ and $_R M = \mathbb{Z}/2\mathbb{Z}$, $\_\otimes M$ is not left exact. When applied to the exact sequence $0 \to \mathbb{Z} \to \mathbb{Q} \to \mathbb{Q}/\mathbb{Z} \to 0$, it gives the nonexact sequence $0 \to \mathbb{Z}/2\mathbb{Z} \to 0 \to 0 \to 0$.

Here in 16-3.4 we assume 16-6.1 to 16-6.5. (See the second paragraph on page 1.

16-3.4 For a quasi-ordered set I (not assumed upper directed), suppose that $\{A_i, \phi_{ij}, i \leqslant j \in I\}$ is a direct system of right R-modules with direct limit $A = \varinjlim A_i$, $\{\alpha_i : A_i \to A \mid i \in I\}$. Let $1_M : M \to M$ be the identity map. Then

(i) $\{A_i \otimes M, \phi_{ij} \otimes 1_M, i \leqslant j \in I\}$ is a direct system of abelian groups with direct limit $A \otimes M$, $\{\alpha_i \otimes 1_M : A_i \otimes M \to A \otimes M \mid i \in I\}$.

(ii) In particular

$$(\varinjlim A_i) \otimes M = \varinjlim (A_i \otimes M).$$

Proof. (i) and (ii). We will use the notation of 16-5.5 for H, $A = (\oplus A_i)/H$ and $\alpha_i : A_i \to A$ which is the natural inclusion $e_i : A_i \to \oplus A_i$ followed by the projection modulo H. Let $1_M : M \to M$ be the identity map. If $\pi : \oplus(A_i \otimes M) \to [\oplus(A_i \otimes M)]/H \otimes M$ is the projection, then also $\pi(e_i \otimes 1_M) : A_i \otimes M \to [\otimes(A_i \otimes M)]/H \otimes M$ is the inclusion followed by the projection modulo $H \otimes M$. From 16-6.5(i) we conclude that $\varinjlim(A_i \otimes M) = [\otimes((A_i \otimes M)]/H \otimes M$ and $\{\pi(e_i \otimes 1_M), i \in I\}$. Let us identity $(\oplus A_i) \otimes M = \oplus(A_i \otimes M)$ under the obvious natural isomorphism. Since $\_ \otimes M$ is right exact, we get two exact sequences

$$H \longrightarrow \oplus A_i \xrightarrow{S \circ (\oplus \alpha_i)} A \longrightarrow 0$$
$$H \otimes M \longrightarrow \oplus(A_i \otimes M) \xrightarrow{\oplus(\alpha_i \otimes 1_M)} A \otimes M \longrightarrow 0$$

There is an isomorphism ρ and a commutative diagram

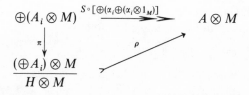

where S is the sum map as in 16-2.1. The latter gives us two facts.

First, for any index $j \in I, \rho\pi(e_j \otimes 1_M) = [\oplus_i(\alpha_i \otimes 1_M)](e_j \otimes 1_M) = (\oplus_i(\alpha_i e_j) \otimes 1_M = \alpha_j \otimes 1_M$. Secondly, also the following is a direct limit of the $A_i \otimes M$:

$$\rho \frac{\oplus(A_i \otimes M)}{H \otimes M} = A \otimes M, \quad \{\rho\pi(e_i \otimes 1_M) = \alpha_i \otimes 1_M, i \in I\}.$$

16-4   Adjoint associativity

16-4.1 Notation. For two given rings R and S consider arbitrary modules $A = A_R$ and $C = C_S$ and a bimodule $B = {}_R B_S$. The latter guarantees that $A \otimes B = A \otimes_R B$ is a right S-module in the obvious way, $(a \otimes b)s = a \otimes (bs)$; $a \in A$, $b \in B$, $s \in S$. Recall that $\text{Hom}_S(B, C)$ is a right R-module, where for $r \in R$, $f \in \text{Hom}_S(B, C)$, the function $f * r \in \text{Hom}_S(B, C)$ is defined at $b \in B$ by $(f * r)(b) = f(rb)$.

Regard B as given and fixed while A and C are variable. Then $\_\otimes B = \_\otimes_R B \colon \mathcal{M}od_R \to \mathcal{M}od_S$ is a functor. If $\alpha\colon A' \to A$ is a right R-homomorphism, then $\alpha \otimes B = \alpha \otimes 1 \colon A' \otimes B \to A \otimes B$ is an S-homomorphism, where on generators $(\alpha \otimes B)(a' \otimes b) = (\alpha \otimes 1)(a' \otimes b) = (\alpha a') \otimes b$ for $a' \in A'$, $b \in B$.

In the opposite direction, $\text{Hom}_S(B, \cdot)\colon \mathcal{M}od_S \to \mathcal{M}od_R$ is also a functor. For let $\gamma\colon C \to C'$ be a right S-homomorphism, and let $r \in R$, $b \in B$, and $f \in \text{Hom}_S(B, C)$. Then $\text{Hom}_S(B, \gamma) = \gamma_* \colon \text{Hom}_S(B, C) \to \text{Hom}_S(B, C')$, where $(\gamma * f)(b) = \gamma(f(b))$. Thus $\text{Hom}_S(B, \gamma)f = \gamma_* f = \gamma \circ f$ is simply the composite of f and then γ. Next, $[\gamma \circ (f * r)](b) = \gamma[(f * r)(b)] = \gamma[f(rb)] = (\gamma f)(rb) = [(\gamma f) * r](b)$. Thus $\gamma_*(f * r) = \gamma(f * r) = (\gamma f) * r = (\gamma_* f) * r$.

16-4.2 In order to show that $\_\otimes B$ is the left adjoint functor of $\text{Hom}_S(B, \cdot)$ first we have to define an abelian group isomorphism $\phi = \phi_{A,C} = \phi(A, C)\colon \text{Hom}_S(A \otimes B, C) \to \text{Hom}_R(A, \text{Hom}_S(B, C))$. Let $f \in \text{Hom}_S(A \otimes B, C)$ and $g \in \text{Hom}_R(A, \text{Hom}_S(B, C))$. This means that $f[(a \otimes b)s] = f(a \otimes bs) = [f(a \otimes b)]s = f[(a \otimes b)s]$ and that $g(a) = ga\colon B \to C$ with $[g(a)](b) = (ga)(b) \in C$. Then $g(ar) = (ga) * r$ translates into the condition $[g(ar)](b) = (ga)(rb) = garb$.

Next, ϕ depending on A and C is defined as follows:

$$\phi f\colon A \to \text{Hom}_S(B, C), \quad [\phi f](a)\colon B \to C,$$
$$[\phi f](a)(b) = f(a \otimes b).$$

An element $\psi g \in \text{Hom}_S(A \otimes B, C)$ is defined by specifying its value on generators of $A \otimes B$ by $(\psi g)(a \otimes b) = (ga)(b)$. Then ψg is an S-homomorphism because $(ga)(bs) = [(ga)(b)]s$, and hence $(\psi g)(a \otimes bs) = (ga)(bs) =$

$[(\psi g)(a \otimes b)](s)$.

It is next shown that the abelian group homomorphism

$$\psi: \mathrm{Hom}_R(A, \mathrm{Hom}_S(B, C)) \to \mathrm{Hom}_S(A \otimes B, C)$$

is the inverse of ϕ, $\psi = \phi^{-1}$. First, $(\psi \phi f)(a \otimes b) = ((\phi f)(a))(b) = f(a \otimes b)$, or $\psi \phi f = f$. Secondly $[\phi \psi g](a)(b) = (\psi g)(a \otimes b) = (ga)(b)$. Hence ϕ is an isomorphism.

16-4.3 Adjoint associativity theorem. Let R and S be any rings (with or without identity), $B = {}_R B_S$ a fixed given right S and left R-bimodule, and $A = A_R$, $C = C_S$ arbitrary right modules over R and S. Set $\_\otimes B = \_\otimes_R B$.

(i) There are four functors:

$$\mathcal{M}od_R \underset{\mathrm{Hom}_S(B,\,\cdot)}{\overset{\_\otimes B}{\rightleftarrows}} \mathcal{M}od_S$$

$\mathrm{Hom}_S(\_\otimes B, \cdot)$, $\mathrm{Hom}_R(\cdot, \mathrm{Hom}_S(B, \cdot))$: $\mathcal{M}od_R^{\mathrm{op}} \times \mathcal{M}od_S \to \mathcal{A}b$.

(ii) There are abelian group isomorphisms ϕ and ψ (depending upon A and C, $\phi = \phi_{A,C} = \phi(A, C)$; $\psi = \psi_{A,C} = \psi(A, C)$)

$$\mathrm{Hom}_S(A \otimes B, C) \underset{\psi}{\overset{\phi}{\rightleftarrows}} \mathrm{Hom}_R(A, \mathrm{Hom}_S(B, C))$$

defined by

$$[\phi f](a)(b) = f(a \otimes b) \qquad f \in \mathrm{Hom}_S(A \otimes B, C), \, a \in A, \, b \in B,$$
$$(\psi g)(a \otimes b) = (ga)(b) \qquad g \in \mathrm{Hom}_R(A, \mathrm{Hom}_S(B, C)).$$

Furthermore, ϕ and ψ are inverses of each other.

(iii) $\_\otimes B$ is a left adjoint of $\mathrm{Hom}_S(B, \cdot)$. Moreover, $\phi: \mathrm{Hom}_S(\_\otimes B, \cdot) \to \mathrm{Hom}_R(\cdot, \mathrm{Hom}_S(B, \cdot))$ is a natural isomorphism of functors.

Proof. Conclusion (ii) was established in 16-2.1, 2.2. (i) That the two homs are actually bifunctors follows from general principles valid in a more general framework and does not depend on the special choice of categories and functors used here (15-1.22).

In order to show that $\phi: \mathrm{Hom}_S(\_\otimes B, \cdot) \to \mathrm{Hom}_R(\cdot, \mathrm{Hom}_S(B, \cdot))$ is a natural transformation of bifunctors, according to 15-1.24 it suffices to verify this in each variable separately with the other variable held fixed. Let $\alpha: A \to A'$ in $\mathcal{M}od_S$ and $\gamma: C \to C'$ in $\mathcal{M}od_R$. It suffices to show that the following two diagrams commute:

$$\text{Hom}_S(A \otimes B,\ C) \xrightarrow{\ \phi\ } \text{Hom}_R(A, \text{Hom}_S(B,\ C)) \qquad \text{Hom}_S(B,\ C)$$

$$\downarrow{\scriptstyle \text{Hom}_S(A\otimes B,\gamma)} \qquad\qquad \downarrow{\scriptstyle \text{Hom}_R(A,\gamma^*)} \qquad\qquad\qquad \downarrow{\scriptstyle \gamma^* = \text{Hom}_S(B,\gamma)}$$

$$\text{Hom}_S(A \otimes B,\ C') \xrightarrow[\ \phi\ ]{} \text{Hom}_R(A, \text{Hom}_S(B, C')) \qquad \text{Hom}_S(B, C')$$

$$\text{Hom}_S(A \otimes B,\ C) \xrightarrow{\ \phi\ } \text{Hom}_R(A, \text{Hom}_S(B,\ C))$$

$$\downarrow{\scriptstyle \text{Hom}_S(\alpha \otimes B, C)} \qquad\qquad \downarrow{\scriptstyle \text{Hom}_R(\alpha, \text{Hom}_S(B,C))}$$

$$\text{Hom}_S(A' \otimes B,\ C) \xrightarrow[\ \phi\ ]{} \text{Hom}_R(A', \text{Hom}_S(B, C))$$

For f, g, a, a', and b as before in 16-2.1, 2.2, going along the top of the first diagram, we have $[\text{Hom}_R(A, \gamma^*)\phi f](a)(b) = \{\gamma[(\phi f)(a)]\}(b) = \gamma f(a \otimes b)$, while along the bottom, $[\phi \text{Hom}_S(A \otimes B, \gamma)f](a)(b) = [\phi(\gamma f)](a)(b) = \gamma f(a \otimes b)$. Similarly, in the second diagram,

$$(\text{Hom}_R(\alpha, \text{Hom}_S(B, C))\phi f)(a')(b) = (\phi f)(\alpha a')(b) = f(\alpha a' \otimes b);$$

and

$$[\phi(\text{Hom}_S(\alpha \otimes B, C)f)](a')(b) = [\phi(f \circ (\alpha \otimes 1_B))](a')(b)$$
$$= f(\alpha a' \otimes b).$$

Thus ϕ is a natural equivalence, and hence by 16-4.2 a natural isomorphism of functors.

The proof of the next corollary is similar, and is omitted.

16-4.4   Corollary 1 to theorem 16-4.3. *With the previous notation but in the situation $({}_S A_R,\ {}_R B,\ {}_S C)$ let ${}_R \mathcal{M}od$ and ${}_S \mathcal{M}od$ denote categories of left modules.*

(i) *There are four functors:*

$${}_R \mathcal{M}od \;\underset{\text{Hom}_s(A,\ \cdot)}{\overset{A \otimes \_}{\rightleftarrows}}\; {}_S \mathcal{M}od$$

$\text{Hom}_s(A \otimes \_\_,\ \cdot)$, $\text{Hom}_R(\cdot,\ \text{Hom}_s(A,\ \cdot))$: ${}_R \mathcal{M}od^{\text{op}} \times {}_S \mathcal{M}od \to \mathcal{A}\ell$.

(ii) *There are abelian group isomorphisms ϕ and ψ (depending upon B and C)*

$$\text{Hom}_S(A \otimes B,\ C) \;\underset{\psi}{\overset{\phi}{\rightleftarrows}}\; \text{Hom}_R(B, \text{Hom}_S(A,\ C))$$

defined by

$$[\phi f](b)(a) = f(a \otimes b) \quad f \in \text{Hom}_S(A \otimes B, C), \quad a \in A, b \in B,$$
$$(\psi g)(a \otimes b) = (gb)(a) \quad g \in \text{Hom}_R(B, \text{Hom}_S(A, C)).$$

The adjoint Associativity Theorem together with Proposition 15-4.9 can be used to give a very roundabout, but completely rigorous proof that the tensor product functor $\_\otimes_R B \colon \mathcal{M}od_R \to \mathcal{M}od_S$ above is right exact. The proof of 15-4.9 used the fact that the hom-functor is left exact.

16-4.5 Corollary 2 to theorem 16-4.3. *Suppose that $S = \mathbb{Z} = 0, \pm 1, \pm 2, \ldots$ and hence $B = {}_R B$ can be any left R-module. Then $\_\otimes_R B \colon \mathcal{M}od_R \to \mathcal{A}\ell$ is a right exact functor.*

Proof. Use 15-4.9 with $\mathcal{A} = \mathcal{M}od_R$ and $\mathcal{B} = \mathcal{A}\ell$. By 16-1.6, $\mathcal{A}\ell$ has a cogenerator and hence by 15-4.9, the tensor product is right exact.

16-5 Elements of tensor products

We next investigate the elements of the tensor product of two modules and in particular when they are zero, and when they can be written in some unique way. From now on to the end of this section, the ring R needs to have an identity element.

16-5.1 Lemma. *Let xR be a free module on one free generator x, and let $m \in M = {}_R M$. If $x \otimes m = 0$ in $xR \otimes M$, then necessarily $m = 0$.*

Proof. Define $\pi \colon xR \times M \to xR \otimes M$ and $\phi \colon xR \times M \to M$ by $\pi(xa, b) = xa \otimes b$, and $\phi(xa, b) = ab$ for $a \in R$, $b \in M$. Since $x^\perp = 0$, ϕ is a well defined R-bilinear map which extends to an abelian group homomorphism $\bar{\phi} \colon xR \otimes M \to M$ with $\bar{\phi}(x \otimes m) = \phi(x, m) = m$.

16-5.2 Corollary. *If $F = \oplus \{xR \mid x \in X\}$ is free on X, then every nonzero element of $F \otimes M$ is uniquely of the form $x_1 \otimes m_1 + \cdots + x_n \otimes m_n$ for distinct $x_1, \ldots, x_n \in X$ and $0 \neq m_i \in M$, $i = 1, \ldots, n$.*

Proof. Since $F \otimes M = \oplus(xR \otimes M) = \oplus(x \otimes M)$, this follows from the last lemma.

The proof of the next theorem uses the fact that the functor $\_\otimes B$ is right exact. The theorem illustrates how the question whether an element

$\Sigma_1^n a_i \otimes b_i \in A \otimes B$ is zero or not, is a nonlocal property of finite character depending on all of A and B.

16-5.3   Theorem. *Let* $1 \in R$ *and* $A = A_R$, $B = {}_RB$ *be any right and left* R-modules, and let $\Sigma_1^n a_i \otimes b_i \in A \otimes B$, $a_i \in A$, $b_i \in B$. The $\Sigma_1^n a_i \otimes b_i = 0$ in $A \otimes B$ if and only if there exists a finite number $p + s$ elements $a_{n+1}, \ldots,$ $a_{n+p} \in A$; $y_1, \ldots, y_s \in B$ and an $(n + p) \times s$ matrix $\|r_{ij}\|$ with $r_{ij} \in R$ such that for every $j = 1, \ldots, s$ the following hold:

(1) $\displaystyle\sum_{i=1}^{n+p} a_i r_{ij} = 0$

(2) $\displaystyle\sum_{j=1}^{s} r_{ij} y_j = \begin{cases} b_i & i = 1, \ldots, n; \\ 0 & i = n+1, \ldots, n+p. \end{cases}$

Proof. $\Leftarrow$: Substitution of (2) first, and then use of (1) gives

$$\sum_{i=1}^{n} a_i \otimes b_i = \sum_{i=1}^{n+p} a_i \otimes \left(\sum_{j=1}^{s} r_{ij} y_j\right) = \sum_{j=1}^{s} \left(\sum_{i=1}^{n+p} a_i r_{ij}\right) \otimes y_j = 0.$$

$\Rightarrow$: There exists a finitely generated submodule $A' = a_1 R + \cdots + a_n R + a_{n+1} R + \cdots + a_{n+p} R < A$ such that $\Sigma_1^n a_i \otimes b_i = 0$ already in $A' \otimes B$. Let $F = F_R$ be the free module on generators $x_1, \ldots, x_{n+p}$. Map $\phi: F \to A'$ by $\phi x_i = a_i$ for all i, and let $K = \ker \phi$. Then $0 \to K \to F \to A \to 0$ is exact, and since $\_\otimes B$ is right exact, we get an exact sequence $K \otimes B \to F \otimes B \to A \otimes B \to 0$. If $\xi = \Sigma_1^n x_i \otimes b_i \in \ker(\phi \otimes 1_B) = K \otimes B$, then there exist $2s$ elements $k_j \in K$ and $y_j \in B$ such that $\xi = \Sigma_1^s k_j \otimes y_j$. Since $k_j \in F = \oplus_1^{n+p} x_i R$, there exist $r_{ij} \in R$ such that $k_j = \Sigma_1^{n+p} x_i r_{ij}$. Hence

(1) $\quad 0 = \phi(k_j) = \displaystyle\sum_{i=1}^{n+p} a_i r_{ij} \qquad \forall j = 1, \ldots, s.$

Substitution for k_j in ξ gives

$$\xi = \sum_{j=1}^{s} \left(\sum_{i=1}^{n+p} x_i r_{ij}\right) \otimes y_j = \sum_{i=1}^{n+p} x_i \otimes \left(\sum_{j=1}^{s} r_{ij} y_j\right)$$

$$\Rightarrow \sum_{i=1}^{n} x_i \otimes b_i = \sum_{i=1}^{n+p} x_i \otimes \left(\sum_{j=1}^{s} r_{ij} y_j\right)$$

By 16-3.6,

(2) $\quad b_i = \displaystyle\sum_{j=1}^{s} r_{ij} y_j \qquad 1 \leqslant i \leqslant n;$

$$0 = \sum_{j=1}^{s} r_{ij} y_j \qquad n + 1 \leqslant i \leqslant n + p.$$

The p and s in the next corollary bear no relation to the previous ones. Its proof is the left analogue of the previous theorem, and is omitted.

16-5.4 Corollary. *With the hypotheses and notation of the previous theorem, $\sum_1^n a_i \otimes b_i = 0$ in $A \otimes B$ if and only if there exist a finite number $p + s$ elements $b_{n+1}, \ldots, b_{n+p} \in B$; $y_1, \ldots, y_s \in A$ and an $s \times (n + p)$ an R-matrix $\|r_{ij}\|$ such that for every $i = 1, \ldots, s$ the following hold:*

(1) $\displaystyle\sum_{j=1}^{n+p} r_{ij} b_j = 0$

(2) $\displaystyle\sum_{i=1}^{s} y_i r_{ij} = \begin{cases} a_j & j = 1, \ldots, n; \\ 0 & j = n + 1, \ldots, n + p. \end{cases}$

16-6   Direct and inverse limits

16-6.1   In this section, $(I, \leqslant)$ will be a quasi-ordered set (i.e. "$\leqslant$" is reflexive, and transitive, but not necessarily antisymmetric, see 0-1.3), and A_i, $i \in I$, will be an indexed family of either sets, modules, or rings. In the definitions below, in case of modules or rings, the word map should be interpreted as a module map or a ring homomorphism. The case when the A_i are abelian groups is included as the special case of $\mathbb{Z} = 0$, ± 1, $\pm 2, \ldots$ modules.

16-6.2   A direct system is a triple $\langle (I, \leqslant), \{A_i | i \in I\}, \{\phi_{ji} | i \leqslant j \in I\} \rangle$ where for any $i \leqslant j$ there exists a map $\phi_{ji} : A_i \to A_j$ such that for all $i \leqslant j \leqslant k$ and $m \in I$

(i) $\phi_{mm} = 1$, the identity on A_m; and
(ii) $\phi_{ki} = \phi_{kj} \phi_{ji}$

Next, let A be an object (i.e. a set, module, or ring) and $\alpha_i : A_i \to A$, $i \in I$, a family of maps commuting with the ϕ_{ij}'s, i.e. for any $i \leqslant j$, $\alpha_i = \alpha_j \phi_{ji}$.

That is, the diagram below commutes

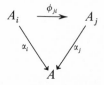

Then A is a *direct limit* of a direct system if it satisfies the following universal property. For any other object B and any family of maps $\beta_i: A_i \to B$, $i \in I$ commuting with the ϕ_{ij}'s (i.e. for $i \leqslant j$; $\beta_i = \beta_j \phi_{ji}$), there exists a unique map $\beta: A \to B$ such that for any $i \in I$, $\beta_i = \beta \alpha_i$.

An equivalent definition of a direct limit A is that for every B and β_i, $i \in I$ for which the lower right triangle below commutes for all $i \leqslant j$, there exists a unique β making the whole diagram to commute

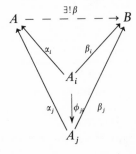

The abbreviation $A = \varinjlim A_i$ is often used.

The definition of an inverse system and limit is dual to that of a direct one and can be obtained by systematically and consistently reversing both all the arrows as well as all the inequalities '$\leqslant$' in the definition of a direct limit. However, here instead (i) always $i \leqslant j \leqslant k$; (ii) ϕ above is replaced by π below; and (iii) for convenience, the subscripts on doubly indexed maps are interchanged, so that always the second (right) index refers to the domain, while the first (left) one to the range.

16-6.3 An inverse system is a triple $\langle (I, \leqslant), \{A_i | i \in I\}, \{\pi_{ij} | i \leqslant j \in I\} \rangle$, where for any $i \leqslant j$, there exists a map $\pi_{ij}: A_j \to A_i$ satisfying the following for all $i \leqslant j \leqslant k$ and m in I

(i) $\pi_{mm} = 1$;
(ii) $\pi_{ik} = \pi_{ij}\pi_{jk}$, i.e. the diagram below commutes.

Suppose that A is an object and $\alpha_i \colon A \to A_i$, $i \in I$, are maps commuting with the π_{ij}'s, i.e. for all $i \leqslant j$, $\alpha_i = \pi_{ij}\alpha_j$. Then A is an *inverse limit* of the inverse system if for any other object B and maps $\beta_i \colon B \to A_i$ similarly commuting with the π_{ij}'s, there exists a unique map $\beta \colon B \to A$ making the diagram below to commute for all $i \leqslant j$.

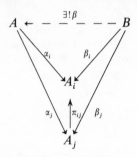

Again, write $A = \varprojlim A_i$.

Some authors use, 'colimit' for inverse limit and simply 'limit' for 'direct limit'. The uniqueness part in both definitions imply that direct and inverse limits are unique up to isomorphism.

16-6.4 Proposition. *The inverse limit A of an inverse system $\{A_i, \pi_{ij}, i \leqslant j \in I\}$ exists and $A = \{(a_i)_{i \in I} \in \Pi A_i \mid a_i = \pi_{ij}a_j, i \leqslant j \in I\}$, where $\alpha_i \colon A \to A_i$ is the restriction of the i-th projection $\Pi A_i \to A_i$ to A. Furthermore, A is a set, module, or ring if the A_i are.*

Proof. Given maps $\beta_i \colon B \to A_i$ that commute with the π_{ij}'s, for $b \in B$ set $\beta b = \{\beta_i b \mid i \in I\} \in \Pi A_i$. Actually $\beta b \in A$, for if $i \leqslant j$, then $\beta_i = \pi_{ij}\beta_j$, and hence $\beta_i b = \pi_{ij}\beta_j b$. Thus β is a map $\beta \colon B \to A$ (of the same kind as the β_i, α_i, and π_{ij}'s). For any $j \in I$, $\alpha_j[\beta b] = \alpha_j[(\beta_i b)_{i \in I}] = \beta_j b$, and $\alpha_j \beta = \beta_j$ as required.

If $\gamma \colon B \to A$ is another map such that $\beta_i = \alpha_i \gamma$ for all $i \in I$, and $b \in B$, then $\alpha_i(\gamma b) = \beta_i b = \alpha_i(\beta b)$. Thus $\gamma b = \beta b$, and $\gamma = \beta$.

When the A_i are R-modules, then for $r \in R$, and $(a_i) \in A \subseteq \Pi A_i$, also $(a_i)r = (a_i r) \in A$, because $a_i r = (\pi_{ij}a_j)r = \pi_{ij}(a_j r)$. If the A_i are rings, and (a_i), $(c_i) \in A$, then $a_i c_i = (\pi_{ij}a_j)(\pi_{ij}c_j) = \pi_{ij}(a_j c_j)$, and hence also $(a_i)(c_i) = (a_i c_i) \in A$. (Note that in general, A need not be an ideal in the ring ΠA_i.)

16-6.5 **Proposition.** *Let* $\{A_i, \phi_{ji}, i \leqslant j \in I\}$ *be a direct system of right R-modules, and let* $e_i: A_i \to \oplus A_i$, $a_i \to e_i a_i$, $i \in I$, *be the natural inclusion maps. Define* H *to be the abelian subgroup generated by the set* $\{-e_i a_i + e_j \phi_{ji} a_i \mid i \leqslant j \in I\}$. *Then*

    (i) $H \subseteq \oplus A_i$ *is a submodule and* $A = (\oplus A_i)/H$ *is the direct limit of the direct system, where* $\alpha_i: A_i \to A$, $\alpha_i a_i = e_i a_i + H$.

Now for (ii) *and* (iii) *assume in addition to* I *being quasi-ordered, that* I *is also an upper directed set. Then*

    (ii) $A = \{e_j a_j + H \mid a_j \in A_j, j \in I\}$ *and*
    (iii) $e_i a_i + H = 0 \Leftrightarrow \exists q \geqslant i, \phi_{qi} a_i = 0$.

Proof. (i) For $r \in R$, $(-e_i a_i + e_j \phi_{ji} a_i)r = -e_i(a_i r) + e_j \phi_{ji}(a_i r)$, and hence $H \leqslant \oplus A_i$ is a submodule. Let module maps $\beta_i: A_i \to B$ commuting with the ϕ_{ij}'s be given. Define $\beta: (\oplus A_i)/H \to B$ by defining for $c_i \in A_i$, $\beta[\Sigma e_i c_i + H] = \Sigma \beta_i c_i$. It then is a consequence of $\beta_i = \beta_j \phi_{ji}$, $i \leqslant j$, that β is a well defined R-map, for

$$\beta[-e_i a_i + e_j \phi_{ji} a_i] = -\beta_i a_i + \beta_j(\phi_{ji} a_i) = 0.$$

For any $i \in I$, $\beta \alpha_i a_i = \beta(e_i a_i + H) = \beta_i a_i$, and hence $\beta_i = \beta \alpha_i$ as required.

Suppose that $\gamma: (\oplus A_i)/H \to B$ is another map such that $\beta_i = \gamma \alpha_i$ for all $i \in I$. Then for any i and any $c_i \in A_i$, $\gamma(e_i c_i + H) = \gamma(\alpha_i c_i) = \beta_i c_i = \beta(e_i c_i + H)$. Hence $\gamma = \beta$.

(ii) For any $\xi = \Sigma e_i c_i + H \in A$, by the upper directed property of I, take any $j \in I$ such that $i \leqslant j$ for all i such that $c_i \neq 0$. Then

$$\xi = \Sigma(e_i c_i + H) = \Sigma(e_i c_i - e_i c_i + e_j \phi_{ji} c_i + H) = e_j a_j + H,$$

where $a_j = \Sigma \phi_{ji} c_i \in A_j$.

(iii) $\Leftarrow$: If $\phi_{qi} a_i = 0$, then $e_i a_i + H = e_i a_i - e_i a_i + \phi_{qi} a_i + H = H$. $\Rightarrow$: Conversely, if $e_i a_i + H = 0$, then $e_i a_i \in H$ is a finite sum $e_i a_i = \Sigma \rho_k \in \oplus A_i$ of relators $\rho_k \in H$ each of the form $\rho_k = -e_k c_k + e_{j(k)} \phi_{j(k)k} c_k$, $c_k \in A$, $k < j(k) \in I$. Take any $q > i$ and $q > j(k)$ for all k. Fix k and set $j = j(k)$. Then

$$\rho_k = [-e_k c_k + e_q \phi_{qk} c_k] + [-e_q \phi_{qj}(\phi_{jk} c_k) + e_j(\phi_{jk} c_k)].$$

Note that ρ_k is now a sum of two relators in H where the large index in both is q. Thus $e_i a_i$ can be expressed as a sum of relators with the same large index q in ech term. Any finite sum of relators having the same large and small index equals a single relator with the same indices. Next,

$$e_q \phi_{qi} a_i = [e_q \phi_{qi} a_i - e_i a_i] + e_i a_i, \quad \text{and} \quad e_i a_i = \Sigma \rho_k \in H$$

show that also $e_q \phi_{qi} a_i$ can be expressed as a finite sum of relators each of whom has the same largest index q, i.e.

$$e_q \phi_{qi} a_i = \Sigma_{p \leqslant q}(-e_p d_p + e_q \phi_{qp} d_p) \qquad d_p \in A_p.$$

The latter is a relation in $\oplus A_i$, where an element is zero if and only if each component is separately equal to zero. Thus for $p < q$, $-e_p d_p = 0$, or $d_p = 0$, and hence $\phi_{qp} d_p = 0$. If $p = q$, then $\phi_{qq} = 1$, and again $-e_q d_q + e_q \phi_{qq} d_q = 0$. Thus $e_q \phi_{qi} a_i = 0$.

16-6.6 Corollary 1. *If I is arbitrary* (not assumed upper directed) *and $\xi = e_p a_p + H \in A$, then*

(i) *for any $i \geqslant p$, $\xi = e_i \phi_{ip} a_p + H$. If $\xi = e_p a_p + H = e_q a_q + H \in A$, then*

(ii) *for any $i \geqslant p$ and $i \geqslant q$, $\phi_{ip} a_p = \phi_{iq} a_q$.*

Proof. (i) Clearly

$$e_p a_p + H = e_i \phi_{ip} a_p + (e_p a_p - e_i \phi_{ip} a_p) + H = e_i \phi_{ip} a_p + H.$$

(ii) Since by (i), $e_i \phi_{ip} a_p + H = e_i \phi_{iq} a_q + H$,

$$e_i(\phi_{ip} a_p - \phi_{iq} a_q) \in H \cap e_i A_i = (0),$$

or $\phi_{ip} a_p = \phi_{iq} a_q$.

16-6.7 Corollary 2. *If the quasi-ordered index set I is also upper directed, and if $\{A_i, \phi_{ij}, i \leqslant j \in I\}$ is a direct system of rings A_i (with or without an identity), and if all the ϕ_{ij} are ring homomorphisms, then their direct limit A is also a ring.*

Proof. Any two arbitrary α, $\beta \in A$ by 16-6.5(ii) may be taken to be of the form $\alpha = e_p a_p + H$ and $\beta = e_p b_p + H$ for the same (sufficiently big) index p. Simply define $\alpha\beta = e_p a_p b_p + H \in A$. Suppose that also $\alpha = e_q a_q + H$ and $\beta = e_q b_q + H$. Then take any $i \in I$, $p, q \leqslant i$. Now since ϕ_{ip} and ϕ_{iq} are ring homomorphisms, by 1-5.6,

$$\phi_{ip}(a_p b_p) = (\phi_{ip} a_p)(\phi_{ip} b_p) = (\phi_{iq} a_q)(\phi_{iq} b_q) = \phi_{iq}(a_q b_q).$$

Upon combining the above with

$$e_p a_p b_p + H = e_i \phi_{ip}(a_p b_p) + H,$$
$$e_q a_q b_q + H = e_i \phi_{iq}(a_q b_q) + H,$$

we conclude that $\alpha\beta = e_p a_p b_p + H = e_q a_q b_q + H$ gives a well defined ring multiplication on A.

16-6.8   Remark. The ring structure in 16-6.7 on $A = (\oplus A_i)/H$ is not that of a quotient ring of $\oplus A_i$. In general, H is not a subring in the ring direct sum $\oplus A_i$.

16-6.9   Alternate description of direct limits. Suppose that $\{A_i, \phi_{ji}, i \leqslant j \in I\}$ is a direct system of right R-modules over an upper directed index set I. By replacing some of the A_i's with isomorphic but unequal copies, we may assume that $A_i \cap A_j = \varnothing$ for all $i \neq j$. Then afterwards let us take their union. This is called the disjoint union, and is denoted by $\uplus A_i$.

Define an equivalence relation $\sim$ on $\uplus A_i$ by

$$a_i \sim a_j \quad \text{if} \quad \phi_{ki} a_i = \phi_{kj} a_j \quad \text{for some } k \geqslant i, j; \ a_i \in A_i, \ a_j \in A_j.$$

Let $[a_i]$ be the equivalence class of $a_i \in A_i$.

The set E of all equialence classes carries a natural R-module structure. Since multiplication by $r \in R$ respects equivalence classes, i.e. if $a_i \sim a_j$ then also $a_i r \sim a_j r$, we may define $[a_i]r = [a_i r]$.

Given any two equivalence classes $[a_f]$ and $[b_g]$, $a_f \in A_f$, $b_g \in A_g$, take any $i \geqslant f, g$, and define $[a_f] + [b_g] = [[\phi_{if} a_f + \phi_{ig} b_g]]$. It suffices to show that if also $j \geqslant f, g$, that then $\phi_{if} a_f + \phi_{ig} b_g \sim \phi_{jf} a_f + \phi_{jg} a_g$. To do this, take $k \geqslant i, j$. Then

$$\phi_{ki}(\phi_{if} a_f + \phi_{ig} b_g) = \phi_{kk}(\phi_{kf} a_f + \phi_{kg} b_g)$$
$$\Rightarrow \phi_{if} a_f + \phi_{ig} b_g \sim \phi_{kf} a_f + \phi_{kg} b_g.$$

Similarly also $\phi_{kf} a_f + \phi_{kg} b_g \sim \phi_{jf} a_f + \phi_{jg} a_g$. Hence we are done by transitivity. Thus E is an R-module.

If in addition the maps ϕ_{ij} are homomorphisms of rings then $[a_f][b_g] = [(\phi_{if} a_f)(\phi_{ig} a_g)]$ for any $i \geqslant f, g$ defines a ring structure on E. That this is independent of i follows by a parallel argument, where the multiplicative ring homomorphism property of the ϕ_{ik} is used:

$$\phi_{ki}[(\phi_{if} a_f)(\phi_{ig} b_g)] = (\phi_{ki} \phi_{if} a_f)(\phi_{ki} \phi_{ig} b_g)$$
$$= (\phi_{kf} a_f)(\phi_{kg} b_g) = \phi_{kj}[(\phi_{jf} a_f)(\phi_{jg} b_g)].$$

16-7   Exercises

1. In Mod_R, a functor $T: \text{Mod}_R \to \text{Mod}_R$ is a *subfunctor of the identity* functor, if and only if for any module M, and any morphism $f: M \to N$ of modules, $TM \leqslant M$, $f(TM) \subseteq TN$, and Tf is the restriction and corestriction of f to $TM \xrightarrow{f} TN$. For two subfunctors T and F of the identity define $T \leqslant F$ if

$TM \subseteq FM$ for all modules. Show that "$\leqslant$" is a partial order. Show that $T^2 \leqslant T$. If $T = T^2$, T is called *idempotent*.

2. Let $\rho: \mathrm{Mod}_R \to \mathrm{Mod}_R$ be a subfunctor of the identity; ρ is said to be a *torsion preradical* if for any modules $N < M$, $\rho N = N \cap \rho M$. A torsion preradical ρ is called a *torsion radical* if in addition $\rho[M/\rho M] = 0$ for all modules M. (i) Show that any torsion preradical is idempotent. (ii) Prove that a subfunctor ρ of the identity is a torsion preradical if and only if ρ is left exact.

3. In Mod_R, for a module M, let $Z(M)$ and $\mathrm{soc}(M)$ be the singular submodule of M, and the socle of M. Show that Z and soc are torsion preradicals but not torsion radicals.

4. For a right R-module M, define $Z_2 M$ by $Z[M/(ZM)] = (Z_2 M)/(ZM)$. Then $Z_2 M$ is called the *second singular submodule* of M. Show that Z_2 is a functor. Show that Z is a subfunctor of Z_2. Generalize this to any subfunctor of the identity functor on Mod_R.

Definition. A set $\mathscr{F}$ of right ideals of a ring R is a *prefilter* if

(i) $\forall A, B \in \mathscr{F} \Rightarrow A \cap B \in \mathscr{F}$;

(ii) $A \in \mathscr{F}$, $A \subseteq B \leqslant R \Rightarrow B \in \mathscr{F}$; and

(iii) $\forall r \in R$, $\forall A \in \mathscr{F} \Rightarrow r^{-1} A \in \mathscr{F}$.

A prefilter is called a *hereditary filter* if in addition

(iv) $\forall A \leqslant R$, if $\exists B \in \mathscr{F}$ such that $\{b^{-1}A \mid b \in B\} \subseteq \mathscr{F}$, then $A \in \mathscr{F}$.

5. Prove that there is a one-to-one correspondence between torsion preradicals ρ and prefilters $\mathscr{F}$ given by the following: $\rho \to \mathscr{F}_\rho = \{A \leqslant R \mid \rho(R/A) = R/A\}$; $\mathscr{F} \to \rho_{\mathscr{F}}$ where $\rho_{\mathscr{F}} M = \{m \in M \mid m^{\perp} \in \mathscr{F}\}$.

6. Prove that there is a one-to-one correspondence between torsion preradicals ρ and classes of right R-modules $\mathscr{C}$, which are closed under submodules, quotient modules, and direct sums given by: $\rho \to \mathscr{C}_\rho = \{M = M_R \mid \rho M = M\}$; $\mathscr{C} \to \rho_{\mathscr{F}}$ where $\mathscr{F} = \{A \leqslant R \mid R/A \in \mathscr{C}\}$.

7*. A class $\mathscr{C}$ of modules is closed under extensions if for any short exact sequence $0 \to A \to B \to C \to 0$ with $A, C \in \mathscr{C}$ it follows that also $B \in \mathscr{C}$. Prove that exercises 5 and 6 give a bijective correspondence between the following three objects: torsion radicals, hereditary filters, and classes of modules closed under submodules, quotients, direct sums, and extensions.

8. For any module M, $Z_2(M/(Z_2 M)) = 0$, and $Z_2(EM) = E(Z_2 M)$; $Z_2 M < M$ is a complement submodule. (See exercise 4).

9. Prove that the functor Z_2 is a torsion radical, and that it is the

unique smallest (see exercise 1) torsion radical whose associated hereditary filter contains all the large right ideals of R.

10*. For any fixed injective right R-module W, for any module M, define σM to be the intersection of all the kernels of all homomorphisms of M into W. Prove that σ is a torsion radical and that every torsion radical is of this type.

11. For any R and M_R, let $R = N_0 \supseteq N_1 \supseteq \cdots \supseteq N_k = 0$ be a chain of left ideals. Define $M_i = \{m \in M \mid mN_i = 0\}$ and form $(0) = M_0 \subseteq M_1 \subseteq \cdots \subseteq M_k = M$. Prove that $M_{i+1}/M_i \cong \operatorname{Hom}_R(N_i/N_{i+1}, M)$ as right R-modules if M is injective. (Rosenberg and Zelinsky 1959).

12. For R-modules A_R, $_RB$, C_R let A_S, $_SB$, C_S be the S-modules induced by an identity preserving ring homomorphism $\varphi : S \to R$. Prove that there are abelian group homomorphisms induced by φ

$$\varphi_* : A_S \underset{S}{\otimes} {}_SB \to A_R \underset{R}{\otimes} {}_RB, \qquad a \underset{R}{\otimes} b \to a \underset{S}{\otimes} b;$$

$$\varphi^* : \operatorname{Hom}_R(A_R, C_R) \to \operatorname{Hom}_S(A_S, C_S).$$

If φ is surjective, then φ_* is an isomorphism and φ^* is the identity.

13. For right R-module P and I prove that

(i) P is projective $\Leftrightarrow \operatorname{Hom}_R(P, \cdot)$ is an exact functor.
(ii) I is injective $\Leftrightarrow \operatorname{Hom}_R(\cdot, I)$ is an exact functor.

14. For rings R and S suppose that A_R, C_S, D_S are modules $f : C_S \to D_S$ is a right S-homomorphism, and $_RB_S$ is a bimodule. Show that there is a commutative diagram of abelian groups induced by f, where the vertical maps are isomorphisms:

$$
\begin{array}{ccc}
\operatorname{Hom}_R(A_R, \operatorname{Hom}_S({}_RB_S, C_S)) & \longrightarrow & \operatorname{Hom}_R(A_R, \operatorname{Hom}_S({}_RB_S, D_S)) \\
\downarrow & & \downarrow \\
\operatorname{Hom}_S\left(A \underset{R}{\otimes} B_S, C_S\right) & \longrightarrow & \operatorname{Hom}_S\left(A \underset{R}{\otimes} B_S, D_S\right)
\end{array}
$$

15. In exercise 14 assume in addition that A_R is a projective R-module, while B_S is projective as an S-module. Prove that $A \otimes_R B_S$ is a projective S-module. (Hint: In exercise 14, let f be onto. Then use exercise 13 to show that $\operatorname{Hom}_S(A \otimes B, \cdot)$ is an exact functor.)

16. For a right R-module M, let M^* be the left R-module $M^* = \operatorname{Hom}_R(M, R)$. (i) Prove that the abelian group

$\Lambda = M \otimes_R M^*$ becomes an associative ring where the multiplication rule $(x \otimes \alpha)(y \otimes \beta) = x\alpha(y) \otimes \beta$, $x, y \in M$, $\alpha, \beta \in M^*$, is extended to all of Λ by linearity and distributivity. (ii) Let $\varphi: \Lambda \to \operatorname{Hom}_R(M, M)$ be defined by $\varphi(y \otimes \beta)(z) = y\beta(z)$, $z \in M$. (iii) Prove that φ is a ring homomorphism. (iv) If $R = F$ is a division ring, and M is a right vector space over F, show that φ is monic, and that $\varphi(\Lambda)$ consists exactly of all finite rank F-linear operators on M. (v) Generalize (iv) to the case when M is a free R-module.

17. For M and M^* as in the previous exercise, let M^* be the right R-module $M^{**} = \operatorname{Hom}_R(M^*, R)$. For $m \in M$, define $\hat{m}: M^* \to R$ by $\hat{m}(\alpha) = \alpha(m)$, $\alpha \in M^*$, $m \in M$; and let $\hat{M} = \{\hat{m} \mid m \in M\}$. (i) Show that $\hat{M} < M^{**}$ as a right R-submodule of M^{**} and that $m \to \hat{m}$, $M \to \hat{M}$ is an R-module homomorphism. (ii) If M is free, prove that $M \cong \hat{M}$.

16-8   Exercises on direct and inverse limits

Let $\langle (I, \leqslant), \{A_i \mid i \in I\}, \{\phi_{ji} \mid i \leqslant j \in I\} \rangle$ and $\langle I, \{b_i\}, \{\psi_{ji}\} \rangle$ be direct systems of modules. Let $A = \varinjlim A_i$ and $B = \varinjlim B_i$ with associated maps $\alpha_i: A_i \to A$ and $\beta_i: B_i \to B$. A *map* of *direct systems* is a set $\{f_i \mid i \in I\}$ of module homomorphisms $f_i: A_i \to B_i$ such that $\psi_{ji} f_i = f_j \phi_{ji}$ for all $i \leqslant j$.

1. Prove that a map of direct systems induces a unique module homomorphism $f: A \to B$ such that $f\alpha_i = \beta_i f_i$ for all $i \in I$.

2. Formulate the definition of a *map* of *inverse systems* and prove a similar result.

3. For a map of direct systems as above assume that I is upper directed and that $fa = 0$ for $a \in A$. Show that there exists an $i \in I$ such that for any $q \geqslant i$, $a = \alpha_i a_i = \alpha_q \phi_{qi} a_i$ and $0 = f_i a_i = \beta_q f_q \psi_{qi} a_i$ for some $a_i \in A_i$.

4. For a direct system $\langle I, \{A_i\}, \{\phi_{ji}\} \rangle$ of modules with I upper directed, let $A = \varinjlim A_i$ and $\alpha_i: A_i \to A$ as in 16-5.5. Prove that if all ϕ_{ji} are monic, then all α_i are also monic. Formulate and prove an analogous result for an inverse limit of modules.

5. For a map f of direct systems as above, find hypotheses which will guarantee that f is monic.

6. Let $\{A_i \mid i \in I\}$ be an indexed set of R-modules where $I = \{0 < 1 < 2 < \cdots\}$, and $f_i: A_i \to A_{i+1}$ be R-maps for $0 \leqslant i$. Let $0 \leqslant i \leqslant j \leqslant k$ be arbitrary. Define $\phi_{ii} = 1$ and $\phi_{ij} = f_j f_{j-1} \cdots f_{i+1} f_i: A_i \to A_j$. Let $e_i: A_i \to \oplus A_i$ be the inclusion map. Define $H = \langle \{-e_i a_i + e_j \phi_{ji} a_i \mid 0 \leqslant i \leqslant j, a_i \in A\} \rangle$. Let $A_* = \oplus A_i / H$ and

$\alpha_i \colon A_i \to A_*$, $\alpha_i a_i = e_i a_i + H$.

Let $B = B_R$ and $g_i \colon A_i \to B$ be R-maps such that $g_i = g_{i+1} f_i$ for all $0 \leqslant i$. Prove the following:

(a) $H = \langle \{ -e_i a_i + e_{i+1} f_i a_i \,|\, 0 \leqslant i, \, a_i \in A_i \} \rangle$.
(b) $g_i = g_j \phi_{ji}$.
(c) $\{ A_i, \phi_{ij}; \, i \leqslant j \in I \}$ is a direct system.
(d) Give a direct proof that $A_* = \varinjlim A_i$ by showing that there exists a unique R-map $\beta \colon A_* \to B$ such that $g_i = \beta \alpha_i$, $0 \leqslant i$.

7. Suppose that M is a topological R-module ($M \times M \to M$ and $M \times R \to M$ are continuous) and that $\{ N_i \,|\, i \in I \}$ is a base of neighborhoods at $0 \in M$, where each N_i is a submodule of M. Define $i \leqslant j$ if $N_j \subseteq N_i$, and $\phi_{ij} \colon M/N_j \to M/N_i$ as the natural map. (i) Prove that $\langle (I, \leqslant), \{ M/N_i \}, \{ \phi_{ij} \} \rangle$ is an inverse system. (ii) Show that $\varprojlim M_i = \{ (m_i + N_i)_{i \in I} \in \prod_{i \in I} M/N_i \,|\, \forall i \leqslant j,\ m_j - m_i \in N_i \}$. (iii) Assume now that M is Hausdorff. Show that $M \subseteq \varprojlim M_i$, where equality holds if and only if M is topologically complete.

8. Let $\langle (I, \leqslant), \{ A_i \,|\, i \in I \}, \{ \pi_{ji} \,|\, i \leqslant j \in I \} \rangle$ be an inverse system of modules with inverse limit A. For an arbitrary $i_0 \in I$, set $I_0 = \{ i \in I \,|\, i_0 \leqslant i \}$. Check that $\langle (I_0, \leqslant), \{ A_i \,|\, i \in I_0 \}, \{ \pi_{ji} \,|\, i \leqslant j \in I_0 \} \rangle$ is still an inverse system with the same inverse limit A.

9. Let $F[x]$ be the polynomial ring over a field F and $\mathbf{N} = 1, 2, \ldots$ with the natural order. Let $(x^j) = F[x] x^j$. For $i \leqslant j$, map $\pi_{ij} \colon F[x]/(x^j) \to F[x]/(x^i)$ by $r + (x^j) \to r + (x^i)$, $r \in F[x]$. Verify that $\langle \mathbf{N}, \{ F[x]/(x^i) \}, \{ \pi_{ij} \} \rangle$ is an inverse system whose direct limit is the formal power series ring $F[[x]]$.

10. For $i \leqslant j \in \mathbf{N} = 1, 2, \ldots$, let $p \in \mathbf{Z} = 0, \pm 1, \pm 2, \ldots$ be a prime. Map $\pi_{ij} \colon \mathbf{Z}/(p^j) \to \mathbf{Z}/(p^i)$ by $r + (p^j) \to r + (p^i)$, $r \in \mathbf{Z}$. Verify that $\langle \mathbf{N}, \{ \mathbf{Z}/(p^i) \}, \{ \pi_{ij} \} \rangle$ is an inverse system whose direct limit is the ring of p-adic integers J_p.

11. Let M_R be a module over a discrete ring R. Then M is said to have a *linear topology* if M is a topological space which has a neighborhood basis $\mathcal{N}(0) = \{ \mathcal{N} \}$ at 0 consisting of submodules $N < M$, and if the neighborhood basis of an arbitrary point $m \in M$ consists of the cosets $\mathcal{N}(m) = \{ m + N \,|\, N \in \mathcal{N}(0) \}$. (i) Show that M is a topological module, i.e. the functions "$-$": $M \to M$, "$+$": $M \times M \to M$, and $M \times R \to M$ are continuous. (ii) M is Hausdorff if and only if $\cap \mathcal{N}(0) = \{ (0) \}$. (iii) If $K < M$ is a closed submodule, then K is open. (Closed submodules in general need not be open.) (iv) A coset $a + K$ is closed if and only if K is closed.

12. (Continuation of 11). Verify that for the given rings R and modules M, the following define linear topologies. (i) Given modules M_i, $i \in I$ with linear topologies, then the product topology on $\Pi\{M_i | i \in I\}$ is a linear topology. (ii) $R = \mathbf{Z}$, M is an abelian group $\mathcal{N}(0) = \{nM | n \in \mathbf{Z}\}$. (iii) Either $R = \mathbf{Z}$ or J_p the p-adic integers, $M = J_p$; $\mathcal{N}(0) = \{p^k J_p | k = 0, 1, 2, \dots\}$. Prove that J_p is compact (iv) R is commutative, M any R-module; $\mathcal{N}(0) = \{Mr | 0 \neq r \in R\}$. Find hypotheses so that M is not discrete. (v) In examples (i), (ii), and (iv), describe the elements of the completion of M by the direct limit method of exercise 7.

13. A module M with a linear topology is said to be *linearly compact* if whenever $\{x_i + K_i | i \in I\}$ is a set of closed cosets of M such that any finite number of them have a nonempty intersection, then $\cap\{x_i + K_i | i \in I\} \neq \emptyset$. For $R = \mathbf{Z}$ and $M = \mathbf{Z}(p^\infty)$ with the discrete topology, show that $\mathbf{Z}(p^\infty)$ is linearly compact, but not compact. (For more about topological rings, see [Eckstein 74]).

The next sequence of exercises outlines how direct and inverse limits can be made into functors.

14. For any categories $\mathscr{A}$, $\mathscr{B}$ and any object Y of $\mathscr{B}$ show that the following defines a functor $|\ |: \mathscr{A} \to \mathscr{B}$, called a *constant functor*: $|A| = Y$ for all $A \in \mathit{Ob}\mathscr{A}$, and all morphisms of $\mathscr{A}$ are sent by $|\ |$ to the identity morphism of Y.

15. Any quasi-ordered set I is a category with objects the elements of I, and where there exists a unique morphism $i \to j$ if and only if $i \leqslant j \in I$, with $i \to i$ being the identity morphism. For any direct system of modules $\langle (I, \leqslant), \{A_i | i \in I\}, \{\phi_{ji} | i \leqslant j \in I\}\rangle$, define $D(i) = A_i$ and $D(i \to j) = \varphi_{ji}$. Prove that $D: I \to \mathrm{Mod}_R$ is a functor.

For any functor $F: \mathscr{A} \to \mathscr{B}$, a *direct root* of F is a constant functor $|\ |: \mathscr{A} \to \mathscr{B}$ and a natural transformation $\alpha: F \to |\ |$ having the following universal property. For any constant functor $C: \mathscr{A} \to \mathscr{B}$ and natural transformation $\gamma: F \to C$, there exists a unique natural transformation $\beta: |\ | \to C$ such that $\beta\alpha = \gamma$, i.e. such that $F \xrightarrow{\alpha} |\ | \xrightarrow{\beta} C$ is $F \xrightarrow{\gamma} C$.

16. For the direct system in the last exercise 13, let $|\ |: I \to \mathrm{Mod}_R$ be the constant functor defined by the object $A = \varinjlim A_i$. Let α_i be the maps $\alpha_i: A_i \to A$ given by the direct limit. Define $\alpha: D \to |\ |$ by $\alpha[D(i)] = \alpha(A_i) = \alpha_i$. (i) Verify that α is a natural transformation of functors. (ii) Suppose that $\gamma: D \to C$ is a natural transformation where C is a constant functor. Show that any such γ is uniquely

determined by an object B in Mod_R and maps $\beta_i: A_i \to B$ such that if $i \leqslant j$, then $\beta_i = \beta_j \phi_{ji}$. (iii) Prove that every natural transformation $\beta: |\ | \to C$ is determined by a morphism $\beta: A \to B$ in Mod_R such that $\beta \alpha_i = \beta_i$ for all $i \in I$. (iv) Prove that $\alpha: D \to |\ |$ is a direct root of D.

17. For a second direct system of modules $\langle I, \{B_i\}, \{\psi_{ji}\} \rangle$ let $\Delta: I \to \text{Mod}_R$ be the associated functor as in exercise 13. For any natural transformation $f: D \to \Delta$, define $f[D(i)] = f[A_i] = f_i: A_i \to B_i$. Then $\{f_i | i \in I\}$ is a map of direct systems, i.e. $\psi_{ji} f_i = f_j \phi_{ji}$ for all $i \leqslant j \in I$. Conversely, any map of direct systems defines in this way a functor $D \to \nabla$.

18. Formulate the definitions of direct and inverse limits in any category. If a direct limit exists (in general categories it need not exist), verify that in the previous exercises Mod_R may be replaced by an arbitrary category.

19. Formulate precise analogous definitions for inverse systems such as an *inverse root*, and prove the appropriate analogous results.

CHAPTER 17

Flat modules

Introduction

A left R-module $_RD$ is flat if the functor $\_\otimes_R D : \mathcal{M}od_R \to \mathcal{A}b$ is exact. (It is always right exact.) For $R = \mathbf{Z} = 0, \pm 1, \pm 2, \ldots$ the flat abelian groups are exactly the torsion free ones. Somewhat more generally, if R is a domain in which every finitely generated right ideal is principal, then the flat left R-modules are the torsion free modules $(rd = 0, 0 \neq r \in R, d \in D \Rightarrow d = 0)$.

The character module D_R^* of $_RD$ is the right R-module $\mathrm{Hom}_\mathbf{Z}(D, \mathbf{Q}/\mathbf{Z})$, where $\mathbf{Q}$ are the rationals. Character modules are not only used to study flat modules but will also be used in the next chapter to study pure submodules. In fact, they will provide a bridge between flat modules and pure submodules. Moreover, by use of character modules we obtain a conceptually simple way to embed any right R-module in an injective module (see 17-1.10).

17-1   Character modules

Whenever definitions or results are for right modules, it is up to the reader to formulate obvious left side analogues. Both right and left R-modules are used later.

17-1   Definition and notation. For a ring R with identity, for $\mathbf{Z} = 0, \pm 1, \pm 2, \ldots$ and $\mathbf{Q}$ the rationals, and for any right R-module M, define the *character module* of M to be $M^* = \mathrm{Hom}_\mathbf{Z}(M, \mathbf{Q}/\mathbf{Z})$. Then $M^* = {}_R M^*$ is a left R-module; for $\varphi \in M^*$, $r \in R$ the element $r\varphi \in M^*$ is defined by $(r\varphi)(m) = \varphi(mr)$, $m \in M$.

If $L = {}_R L$ is a left module, $\xi \in L^*$, $v \in L$, and $a \in R$, then $\xi \cdot a \in L^*$ is defined by $(\xi \cdot a)(v) = \xi(av)$.

An R-map $f : N \to M$ induces an R-homomorphism $f_* : M^* \to N^*$ of left R-modules by $f_*(\varphi) = f_*\varphi = \varphi$ of, where $(f_*\varphi)(n) = \varphi(f(n))$, $n \in N$. Thus the assignment $M \to M^*$ and $f \to f_*$ defines a contravariant functor $\mathscr{M}od_R \to {}_R\mathscr{M}od$.

17-1.2   Lemma. *For any right R-module M and $0 \neq m \in M$, there exists a $x \in M^*$ such that $\chi(m) \neq 0$.*

Proof. Define $\chi : \mathbf{Z}m \to \mathbf{Q}/\mathbf{Z}$ by $\chi m = 1/j + \mathbf{Z}$ if order $m = j$; and $\chi m = 1/15 + \mathbf{Z}$ if order of m is infinite. Then x extends to $x : M \to \mathbf{Q}/\mathbf{Z}$ by the divisibility of the latter.

17-1.3   Corollary. *If $B_R < A_R$ and $a \in A \backslash B$, then there exists a $\varphi \in A^*$ such that $\varphi B = 0$ and $\varphi a \neq 0$.*

Proof. For $M = A/B$ and $m = a + B \neq 0$, there is a $\chi \in M^*$ with $\chi(a + B) \neq 0$. If $\pi : A \to A/B$, then $\varphi = \chi \circ \pi$.

17-1.4   Corollary. *Let $A_1 < B$ and $A_2 < B$. Then $A_1 = A_2 \Leftrightarrow \forall b^* \in B^*$, $b^*A_1 = 0$ if and only if $b^*A_2 = 0$.*

Proof. $\Leftarrow$: If $m \in A_i \backslash A_j$, $i \neq j$, then there exists a $b^* \in B^*$ with $b^*m \neq 0$ and $b^*A_j = 0$, a contradiction.

17-1.5   Corollary. *For any $N = N_R$, there is a canonical inclusion $N \leqslant (N^*)^* = N^{**}$ of N as a right submodule of N^{**}. For any R-module map $f : N \to M$, the restriction and corestriction of $f_{**} : N^{**} \to M^{**}$ to N is $f_{**} | N = f$.*

Proof. For $n \in N$, let $\hat{n} : N^* \to \mathbf{Q}/\mathbf{Z}$ by $\hat{n}(\chi) = \chi(n)$ for $\chi \in N^*$. By 17-1.2, $\hat{n} = 0$ if and only if $n = 0$. Clearly, $\hat{N} = \{\hat{n} \mid n \in N\} \subset N^{**}$ is an abelian subgroup. Also for $r \in R$, $(\hat{n} \cdot r)(\chi) = \hat{n}(r\chi) = r\chi(n) = \chi(nr) = (nr)^\wedge(\chi)$ by repeated use of 17-1.1. Thus $N \cong \hat{N} \leqslant N^{**}$. Lastly

$$(f_{**}(\hat{n}))(\chi) = (\hat{n} \circ f_*)(\chi) = \hat{n}(f_*(\chi)) = \hat{n}(\chi \circ f)$$

$$= (\chi \circ f)(n) = \chi(f(n)) = (f(n))^\wedge(\chi).$$

Thus $f_{**}(\hat{n}) = (f(n))^\wedge$.

17-1.6 Remark. $\mathbf{Q}/\mathbf{Z}$ is the smallest injective cogenerator in the category of $\mathbf{Z}$-modules. Everything here would work if $\mathbf{Q}/\mathbf{Z}$ were replaced by any other injective cogenerator for abelian groups.

17-1.7 Lemma. *Let $f : N \to M$ be a homomorphism of right R-modules, and let $f_* : M^* \to N^*$. Then*

    (i)   *f is epic $\Leftrightarrow f_*$ is monic;*
    (ii) *f is monic $\Leftrightarrow f_*$ is epic.*

Proof. (i) $\Rightarrow$: If $f_*(\varphi) = 0$ for $\varphi \in M^*$, then $(f_*\varphi)N = \varphi(f(N)) = \varphi(M) = 0$. Hence $\varphi = 0$. (i) $\Leftarrow$: If not, take $0 \neq m + fN \in M/fN$ and $\chi \in (M/fN)^*$ with $\chi(m + fN) \neq 0$. Let $\pi : M \to M/fN$. Then $0 \neq \pi_*\chi = \chi \circ \pi \in M^*$, and $f_*(\chi \circ \pi) = \chi \circ \pi \circ f = 0$ contradicts $\chi \circ \pi(m) \neq 0$.
(ii) $\Rightarrow$: Let $\chi \in N^*$. Then $fN \leqslant M$, and define $\chi_1 : fN \to \mathbf{Q}/\mathbf{Z}$ by $\chi_1(fn) = \chi n$, $n \in N$. By the divisibility of $\mathbf{Q}/\mathbf{Z}$, χ_1 extends to $\chi_2 : M \to \mathbf{Q}/\mathbf{Z}$. But then $f_*(\chi_2)n = \chi_2(f(n)) = \chi n$, and $f_*\chi_2 = \chi$. (ii) $\Leftarrow$: Suppose $fn = 0$, $0 \neq n \in N$. Let $\chi \in N^*$, $\chi n \neq 0$. There exists $\varphi \in M^*$, $f_*\varphi = \chi$. Then $0 \neq \chi n = (f_*\varphi)n = \varphi(fn) = 0$ is a contradiction.

17-1.8 Lemma. *Consider any sequence whatever of right R-modules and R maps (1), and then form the sequence (2) of left R-module homomorphisms. Then (1) is exact $\Leftrightarrow$ (2) is exact.*

    (1) $0 \longrightarrow A \xrightarrow{\ \alpha\ } B \xrightarrow{\ \beta\ } C \longrightarrow 0;$
    (2) $0 \longrightarrow {}_RC^* \xrightarrow{\ \beta_*\ } {}_RB^* \xrightarrow{\ \alpha_*\ } {}_RA^* \to 0.$

Proof. It suffices to verify exactness in the category of $\mathbf{Z}$-modules. For $c^* \in C^*$, $\alpha_*\beta_*(c^*) = c^* \circ (\beta\alpha)$. Thus $\alpha_*\beta_* = 0$ if and only if $\beta\alpha = 0$ by 17-1.2.
$\Rightarrow$: By 17-1.7, β_* is monic and α_* epic. Since $\alpha_*\beta_* = 0$, it suffices to prove that kernel $\alpha_* \subseteq$ image β_*. So take $b^* \in B^*$ with $0 = \alpha_*(b^*) = b^* \circ \alpha$. Thus $b^* : B \to \mathbf{Q}/\mathbf{Z}$ with $\alpha A \subseteq$ kernel b^*. For b, $b_2 \in B$, if $\beta b = \beta b_2$, then $b - b_2 \in \alpha A$, and $b^*(b) = b^*(b_2)$. Thus define $c^* : C \to \mathbf{Q}/\mathbf{Z}$ by $c^*(\beta b) = b^*(b)$ for any $\beta b \in C$. Hence $\beta_*(c^*) = c^* \circ \beta = b^*$, and kernel $\alpha_* = \beta^* C^*$.
$\Leftarrow$: By 17-1.7, α is monic and β epic. Since $\beta\alpha = 0$, it only remains to show that kernel $\beta \subseteq$ image α. If not, then $b \in B \backslash \alpha A$ with $\beta b = 0$. Let $\pi : B \to B/\alpha A$. Since $0 \neq b + \alpha A \in B/\alpha A = C$, there exists a $c^* \in C^*$ with $c^*(\pi b) \neq 0$. Hence $c^* \circ \pi : B \to \mathbf{Q}/\mathbf{Z}$, $c^* \circ \pi \in B^*$, with $(c^* \circ \pi)(\alpha A) = 0$.

Therefore $[\alpha^*(c^* \circ \pi)](A) = (c^* \circ \pi \circ \alpha)(A) = 0$, and $c^* \circ \pi \in$ kernel $\alpha_* =$ image β_*. Thus $c^* \circ \pi = \beta_*(b^*) = b^* \circ \beta$ for some $b^* \in B^*$. Consequently $0 \neq (c^* \circ \pi)(b) = (b^* \circ \beta)(b) = b^*(\beta b)$ contradicts that $b \in$ kernel β.

17-1.9   Proposition. *If $_R F$ is free, then F_R^* is injective.*

Proof. Let F be free on $\{e_i \,|\, i \in I\} \subset F = \oplus Re_i$. Given are $B < R$ and $\alpha : B \to F_R^*$ as in Baer's criterion. Let $\langle \cdot, \cdot \rangle : F^* \times F \to \mathbf{Q}/\mathbf{Z}$ by $\langle \chi, f \rangle = \chi(f)$, $\chi \in F^*$, $f \in F$. Thus $\langle \alpha(br), f \rangle = \langle (\alpha b) \cdot r, f \rangle = \langle \alpha b, rf \rangle$ for $b \in B$, $r \in R$.

Now BF is an additive subgroup of F. For any $\xi \in BF$, there is a unique expression $\xi = \Sigma \, b_i e_i$, $0 \neq b_i \in B$. Define $\beta : BF \to \mathbf{Q}/\mathbf{Z}$ by $\beta(\xi) = \Sigma \langle \alpha b_i, e_i \rangle$.

For $f \in F$ and $b \in B$, it is asserted that $\beta(bf) = \langle \alpha b, f \rangle$. For if $f = \Sigma \, r_i e_i$, $r_i \in R$, then the unique expression for bf is obtained from $bf = \Sigma \, br_i e_i$. Hence

$$\beta(bf) = \Sigma \langle \alpha(br_i), e_i \rangle = \Sigma \langle \alpha b, r_i e_i \rangle = \langle \alpha b, \Sigma \, r_i e_i \rangle = \langle \alpha b, f \rangle.$$

Extend β to an abelian group homomorphism $\gamma : RF = F \to \mathbf{Q}/\mathbf{Z}$. Define $\hat{\alpha} : R \to F^*$, where for $y \in R$, $\hat{\alpha}(y) : F \to \mathbf{Q}/\mathbf{Z}$ is defined by $\langle \hat{\alpha}(y), f \rangle = \gamma(yf)$ for $f \in F$. To verify that $\hat{\alpha}$ is a right R-module homomorphism it suffices to show that $\hat{\alpha}(yr) = (\hat{\alpha}y) \cdot r$ for $r \in R$. This follows from $\langle (\hat{\alpha}y) \cdot r, f \rangle = \langle \hat{\alpha}y, rf \rangle = \gamma(y(rf)) = \langle \hat{\alpha}(yr), f \rangle$. For any $b \in B$, $\langle \hat{\alpha}b, f \rangle = \gamma(bf) = \beta(bf) = \langle \alpha b, f \rangle$. Thus $\hat{\alpha} \,|\, B = \alpha$, and F_R^* is injective.

17-1.10   Corollary. *For any right R-module M, there is a free left R-module F such that M is R-isomorphic to a submodule of F_R^*; F_R^* is injective.*

Proof. As in 17-1.5, $M \cong \hat{M} \leqslant M^{**}$ canonically. There is an epimorphism $\pi : _R F \to _R M^*$ of a free left R-module F onto M^*. Consequently $\pi_* : M_R^{**} \to F_R^*$ is monic, and $M \cong \pi_* \hat{M} < F_R^*$.

17-2   Flat module basics

Here R is a ring with $1 \in R$, and $\_\otimes\_ = \_\otimes_R\_$, or $\_\otimes B = \_\otimes_R B$. Although flat left modules are discussed, parallel to statements hold for flat right R-modules. As before $\mathbf{Z} \subset \mathbf{Q}$ are the integers and rationals.

17-2.1   Definition. A left R-module B is *flat* if $\_\otimes B$ is left exact. Alternatively, $_R B$ is flat if for any monic homomorphism of right R-modules $f : N \to M$, $f \otimes 1 : N \otimes B \to M \otimes B$ is monic. A third equivalent definition is that B is flat if for any $N < M$, also $N \otimes B \subset M \otimes B$.

17-2.2 Observation. For $A = A_R$ and $M = {}_R M$ by adjoint associativity 17-4.3 there exist natural abelian group isomorphisms φ and ψ with $\psi = \varphi^{-1}$:

$$(A \otimes M)^* = \text{Hom}_Z(A \otimes M, \mathbf{Q}/\mathbf{Z}) \underset{\psi}{\overset{\varphi}{\longleftarrow\!\!\!\longrightarrow}} \text{Hom}_R(A, \text{Hom}_Z(M, \mathbf{Q}/\mathbf{Z}))$$

$$= \text{Hom}_R(A, M_R^*)$$

17-2.3 Proposition. *For a left R-module M, ${}_R M$ is flat $\Leftrightarrow M_R^*$ is injective.*

Proof. Let $0 \to A_R \xrightarrow{\alpha} B_R$ be any exact sequence. Now form the sequences

(1) $\quad 0 \to A \otimes M \xrightarrow{\alpha \otimes 1} B \otimes M,$
(2) $\quad (B \otimes M)^* \to (A \otimes M)^* \to 0,$
(3) $\quad \text{Hom}_R(B, M_R^*) \to \text{Hom}_R(A, M_R^*) \to 0.$

By 17-1.8 (with $C = 0$), (1) is exact $\Leftrightarrow$ (2) is exact. In view of the last observation, (2) is exact if and only if (3) is. Thus M^* is injective $\Leftrightarrow$ (3) is exact $\Rightarrow$ (1) is exact.

Although we already know from 16-5.2 that a free module is flat, nevertheless this follows now very neatly from the last proposition.

17-2.4 Corollary. *For a left module M, M is free $\Rightarrow M_R^*$ is injective $\Rightarrow {}_R M$ is flat. In particular, ${}_R R$ is flat.*

17-2.5 Lemma. *For modules $A_R < B_R$ and ${}_R M$, there is a natural isomorphism of abelian groups*

$$B \otimes M/[A \otimes M] \cong (B/A) \otimes M,$$

where

$$[A \otimes M] \subset B \otimes M$$

is the subgroup generated in $B \otimes M$ by all the elements $a \otimes m$, $a \in A$, $m \in M$.

Proof. Let $\pi: B/A \times M \to (B/A) \otimes M$ be the usual R-bilinear function $\pi(b + A, m) = (b + A) \otimes m$, $b \in B$, $m \in M$. Let φ be the well defined R-bilinear map $\varphi: B/A \times M \to B \otimes M/[A \otimes M]$ defined by

$$\phi(b + A, m) = b \otimes m + A \otimes M.$$

By the universal property of the tensor product there exists a unique abelian group homomorphism $\bar{\varphi}: B/A \otimes M \to B \otimes M/[A \otimes M]$ such that $\varphi = \bar{\varphi}\pi$. Then $\psi[\Sigma b \otimes m + A \otimes M] = \Sigma(b + A) \otimes m$ is a well-defined abelian group homomorphism $\psi: B \otimes M/[A \otimes M] \to B/A \otimes M$. Next, we verify that $\psi\bar{\phi} = 1$ and $\bar{\phi}\psi = 1$ by using that the inner triangle below commutes.

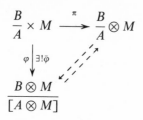

In general for any module $_RM$ and any $A \leqslant R$, there always is a natural abelian group homomorphism $A \otimes M \to AM$, $a \otimes m \to am$.

17-2.6   Proposition. *For any left R-module M.*

$$_RM \text{ is flat} \Leftrightarrow \forall A < R, A \otimes M \cong AM \text{ naturally.}$$

Furthermore, the above remains valid if A ranges over the finitely generated right ideals of R only.

Proof. There is always a left R-module isomorphism $\mu: R \otimes M \to M$, $\mu(r \otimes m) = rm$. $\Rightarrow$: Since $A < R$ and M is flat, $A \otimes M \subset R \otimes M$, and the image of $A \otimes M$ under the monic map μ is AM, so $A \otimes M \cong AM$.

$\Leftarrow$: Assume $\forall A < R$, $A \otimes M \cong AM$. Since $AM \leqslant M$ the composite map f of $A \otimes M \xrightarrow{\cong} AM \twoheadrightarrow M$ is a monomorphism $f: A \otimes M \to M$. Hence $f_*: M^* \to (A \otimes M)^*$ is epic. For simplicity first identity $M^* = (R \otimes M)^*$ and then use observation 17-2.2 to obtain a commutative diagram

$$(A \otimes M)^* \cong \operatorname{Hom}_R(A, M_R^*)$$
$$f_* \uparrow$$
$$(R \otimes M)^* \cong \operatorname{Hom}_R(R, M_R^*).$$

The latter means that every R-homomorphism $A \to M^*$ lifts to $R \to M^*$, for any $A < R$. By Baer's criterion, M_R^* is injective, and by a previous proposition, $_RM$ is flat.

Lastly assume that $A \otimes M \cong AM$ for all finitely generated $A < R$ only.

Let $B < R$ be arbitrary. We assert that still $B \otimes M \to M$ is monic. Let $\xi = \Sigma_{i=1}^n b_i \otimes m_i \in B \otimes M$, $b_i \in B$, $m_i \in M$ and suppose that $\mu(\xi) = \Sigma_{i=1}^n b_i m_i = 0$. Form the finitely generated right ideal $A = \Sigma_{i=1}^n b_i R$. Then $\xi \in A \otimes M$, and $\mu : A \otimes M \to AM$ is monic by hypothesis. Since $\mu\xi = 0$, $0 = \xi \in A \otimes M$ already, and hence automatically in $B \otimes M$.

17-2.7 Lemma. *For left R-modules $\oplus\{M_i \mid i \in I\}$ is flat $\Leftrightarrow \forall\ i \in I$, M_i is flat.*

Proof. Always $\mathrm{Hom}_Z(\oplus M_i, \mathbf{Q}/\mathbf{Z}) = \Pi\,\mathrm{Hom}_Z(M_i, \mathbf{Q}/\mathbf{Z})$. By 17-2.3, $\oplus M_i$ is flat $\Leftrightarrow \Pi M_i^*$ is injective. $\Leftrightarrow \forall\ i \in I$, M_i^* is injective $\Leftrightarrow \forall\ i \in I$, M_i is flat.

17-2.8 Corollary. *Every free module and every projective module is flat.*

Proof. By 17-1.9 and 17-2.3, every free module F is flat. Alternatively, since trivially $_R R$ is flat, and F is a direct sum of R's, by the last lemma, F is flat. Any projective module P is a direct summand of a free module $F = P \oplus Q$. Again, by the last lemma, P is flat.

17-3 Exercises

1. A left R-module M is flat if and only if for any large right ideal $A < R$, $A \otimes M \cong AM$.
2. Let R be a domain in which every finitely generated right ideal is principal. Then a left R-module M is flat if and only if $m^\perp = 0$ for every $0 \neq m \in M$. (*Hint:* Use 17-2.6. Note that for any $\xi = \in A \otimes M$, and for $\mu : A \otimes M \to M$, $\mu(\xi) = am$ for some $a \in A$, and $m \in M$.)
3*. For a left R-module $M = F/K$ where F is flat the following are all equivalent. (a) M is flat. (b) $\forall A < R$, $AF \cap K = AK$. (c) $\forall$ finitely generated $A < R$, $AF \cap K = AK$.
4. Assuming the results of Exercise 3, suppose that R is a ring such that every finitely generated right ideal is principal, and that $_R M = F/K$ where $_R F$ is free. Then M is flat if and only if $rF \cap K = rK$ for any $r \in R$.
5. Use Exercise 4 to show that every left R-module is flat $\Leftrightarrow R$ is a regular ring.
6. A module $_R M$ is flat if and only if for any free module F_R and any $G < F$, the induced map $G \otimes M \to F \otimes M$ is monic.
7*. A module $_R M$ is flat if and only if for any integer n and any $x = (x_1, \ldots, x_n) \in M^n$ and any $c = (c_1, \ldots, c_n) \in R^n$ with $c_1 x_1 +$

$\cdots + c_n x_n = 0$ it follows that there exists an integer m, an $n \times m$ matrix $\|r_{ij}\|$, $r_{ij} \in R$, and a $y = (y_1, \ldots, y_m) \in M^m$ such that $x = \|r_{ij}\| y^T$ and $c\|r_{ij}\| = 0$. Here y^T is the transpose of y.

8*. Suppose that $0 \to K \to F \to M \to 0$ is an exact sequence of left R-modules where $_R F$ is free and $K < F$. Prove that the following are all equivalent. (a) $_R M$ is flat. (b) $\forall k \in K$, $\exists$ a homomorphism $\gamma : F \to K$ such that $\gamma k = k$. (c) $\forall n$, $\forall k_1, \ldots, k_n \in K$, $\exists$ a homomorphism $\gamma : F \to K$ such that $\gamma k_i = k_i$ for all i.

9. Let $_R F = \bigoplus_1^\infty R e_i$ be free on $e_1, e_2, \ldots$. Let $b_1, b_2, \ldots \in R$ be arbitrary. Define K to be the left submodule of F generated by $e_1 - b_1 e_2, e_2 - b_2 e_3, \ldots$. Then
   (i) K is free with basis $\{e_i - b_i e_{i+1} \mid i = 1, 2, \ldots\}$.
   (ii) F/K is a flat left R-module.
   (*Hint*: for (ii), assume the results of Exercise 8.)

CHAPTER 18

Purity

Introduction

For an abelian group B, a subgroup $A \subset B$ is pure in B if whenever an equation $nx = c$ with $c \in A$ and $n \in \mathbf{Z}$ has a solution $x = b \in B$ in the big group B, then it already has a solution $x = a \in A$ in the small subgroup A. Equivalently, $A \subset B$ is pure if $A \cap nB = nA$ for all $n \in \mathbf{Z}$. Direct summands of B are pure in B, and purity is a generalization of a direct summand. This chapter generalizes purity to a module context over an arbitrary ring R, where $1 \in R$ throughout this chapter.

In Section 1, an introduction to systems of equations over a module is given, which is a topic of independent interest and usefulness.

An exact sequence of modules $0 \to A \to B \to C \to 0$ is pure exact if the image of A is pure in B, and a module is pure projective, by definition, if it has projective property relative to all pure exact sequences. This chapter studies pure projective modules and pure exact sequences.

Suppose that $0 \to A \to B \to C \to 0$ is a short exact sequence of modules and we tensor it with a left R-module U to obtain $0 \to A \otimes U \to B \otimes U \to C \otimes U \to 0$. In the last chapter we saw that those modules U for which the last sequence is always exact were the flat modules. Here we ask the following converse question. Which short exact sequences have the property that the tensored sequence remains exact for every left R-module U? The answer is that they are precisely the pure exact sequences.

Pure exact sequences are useful as tools for proving other results which do not explicitly involve purity. Also there are connections between flat modules and pure exactness. Some facts about flat modules are most easily provable using pure exact sequences. For these and other reasons, Section 3 shows that the pure exact sequences are exactly the direct limits of splitting exact sequences.

Sections 4 and 5 develop the theory of pure injective modules. Several

367

equivalent characterizations of pure injective modules are given. Section 5 shows that modules have pure injective hulls with properties somewhat analogous to injective hulls. In the model theory of modules and rings, pure injective modules are absolutely vital and perhaps more natural than injective modules. Roughly speaking, topologically compact or complete modules sometimes tend to be pure injective (see Theorem 18-4.6 and Examples 18-5.11).

Throughout this chapter, $\_\otimes_R\_$ is abbreviated as $\_\otimes\_$, and the notation of the previous chapter is used for duals of modules.

18-1   Systems of equations in modules

Fundamental properties of systems of equations, such as compatibility and equivalence, are defined and reinterpreted as module theoretic properties of free modules.

18-1.1   Definitions and notation. For a right R-module M, for

$$r_1,\ldots,r_n \in R \qquad \text{and} \qquad c \in M,$$

an equation in M is an expression $x_1 r_1 + \cdots + x_n r_n = c$, where the x_i are to be thought of as unknowns. A solution in M is an indexed sequence $b_1,\ldots,b_n \in M$ such that $b_1 r_1 + \cdots + b_n r_n = c$; we write $x_1 = b_1,\ldots,$ $x_n = b_n$.

Suppose that I and J are finite or infinite index sets, and that $\|r_{ij}\|$ is an $I \times J$ matrix with entries $r_{ij} \in R$ such that each row has only a finite number $|\{r_{ij} \mid j \in J\}| < \infty$ of nonzero entries. Such a matrix $\|r_{ij}\|$ is called *row finite*; *column finite* is defined analogously. Let $c_j \in M$, ${}^j\in I$. Then the set of expressions

$$\sum_{i \in I} x_i r_{ij} = c_j, \qquad j \in J \tag{I}$$

is called a system of equations in M with unknowns $\{x_i \mid i \in I\}$. Similarly, an indexed set $\{b_i \mid i \in I\}$, $b_i \in M$, is a solution in M if $\Sigma b_i r_{ij} = c_j$ for all $j \in J$. In this case we say that $x_i = b_i \in M$, $i \in I$, is a solution of (I). In view of the next definition the latter is also called a global solution of (I).

The system (I) is *finitely solvable* if for every finite subset $F \subset J$, there exist $b_i \in M$ such that for $j \in F$ only $\dot\Sigma b_i r_{ij} = c_j$. For this, only a finite set $\{b_i \mid r_{ij} \neq 0, j \in F\}$ of nonzero b_i's is required.

Now form the free right R-module $F = \oplus\{x_i R \mid i \in I\}$ with free basis $\{x_i \mid i \in I\}$. Now $\Sigma_i x_i r_{ij} \in F$. Suppose that $\{s_j \mid j \in J\} \subset R$ where only a finite number of s_j are $\neq 0$. Then

$$\sum_j \left(\sum_i x_i r_{ij} \right) s_j = 0 \Leftrightarrow \forall i, \quad \sum_j r_{ij} s_j = 0.$$

The system of equations (I) is called *compatible* if for any choice of the s_j as above, necessarily also $\Sigma_j c_j s_j = 0$. If a system of equations is not compatible it is called *incompatible*. An incompatible system of equations cannot possibly have a solution.

For the arbitrary index sets I, J, K and the same indeterminates $\{x_i \mid i \in I\}$, consider a second system of equations in M

$$\sum_i x_i s_{ik} = d_k \in M; \; s_{ik} \in R, \; \|s_{ik}\| \text{ is row finite}; \; (k \in K). \tag{II}$$

Then systems of equations (I) and (II) are defined to be *equivalent* if there exist a $J \times K$ column finite matrices $\|p_{jk}\|$ and $\|q_{kj}\|$ over R such that

$$\sum_{j \in J} r_{ij} p_{jk} = s_{ik}, \; \sum_{j \in J} c_j p_{jk} = d_k;$$

$$\sum_{k \in K} s_{ik} q_{kj} = r_{ij}, \; \sum_{k \in K} d_k q_{kj} = c_j.$$

Next, define $G = \langle \{\Sigma_i x_i r_{ij} \mid j \in J\} \rangle < F = \oplus\{x_i R \mid i \in I\}$ to be the right R-submodule generated by the left sides of I. Then define a correspondence $f : \{\Sigma_i x_i r_{ij} \mid j \in J\} \to M$ by $f(\Sigma_i x_i r_{ij}) = c_j \in M$. Let $G_{\text{II}} < F$ and a function f_{II} of sets be defined analogously from (II).

18-1.2 The above two systems of Equations (I) and (II) in 18-1.1 are equivalent $\Leftrightarrow G = G_{\text{II}}$ and $f = f_{\text{II}}$.

18-1.3 The system of Equations (I) in 18-1.1 is compatible $\Leftrightarrow f$ extends to a module homomorphism $g : F \to M$.

18-1.4 Terminology. A system of Equations (I) satisfying the above 18-3.3 will frequently be denoted by (F, G, g), where $g : G \to M$ is a module homomorphism, and will be referred to as a *system*. (Note that if for some $i(0) \in I$, all $r_{i(0),j} = 0$, that then $x_{i(0)}$ does not appear in G.)

18-1.5 A compatible system (F, G, g) is solvable $\Leftrightarrow \exists$ an extension $h : F \to M$ of g with $h \mid G = g$.

Thus the system of equations in Equations (I), 18-1.1 is solvable if and only if the set function f extends to a module homomorphism $h: F \to M$ with $h(\Sigma_i x_i r_{ij}) = c_j, j \in J$.

Reformulation of familiar concepts in terms of equations in modules allows us to use these known concepts together with facts about systems, and thus increasing their applicability. Secondly, it sometimes suggests alternative proofs.

18-1.6   A right R-module M is injective $\Leftrightarrow$ every compatible system of equations over M is solvable in M.

Proof. $\Rightarrow$: trivial. $\Leftarrow$: Let $G < R$ and $g: G \to M$ be given as in Baer's criterion. Take $|I| = 1$, $F = xR$ with $x = 1 \in R$, and any set of generators of $G = \langle \{r_{ij} | j \in J\} \rangle$ as a right R-module. The system of equations $x r_{1j} = g(r_{1j}) \in M$, $j \in J$ in M with one indeterminate x is compatible because $F = xR = R$ is free on $\{x\}$ and g is an R-homomorphism. By hypotheses and 18-1.5, g extends to an R-homomorphism $h: R \to M$.

18-2   Pure projectives and pure exact sequences

Now we specialize our previous general framework to finite compatible systems of equations in a finite number of indeterminates.

18-2.1   Definition. Let $M < N$ be right R-modules. Then M is *pure in N* (i.e. $M < N$ is a *pure extension*) if for any finite system $\Sigma x_i r_{ij} = c_j \in M$, $1 \leqslant i \leqslant n$, $1 \leqslant j \leqslant m$; of equations in M which is solvable in N, this system is also solvable in M.

18-2.2   Consequences. (0) Trivially, $0 < N$ and $N \leqq N$ are pure.

(1) Direct summands are pure.

(2) (a) Suppose that $\mathscr{C} = \{M_\alpha\}$ is a chain of pure submodules $M_\alpha < N$ of a fixed module N. Then $\cup \mathscr{C} = \cup_\alpha M_\alpha \leqq N$ is pure.

(b) If $\{H_\alpha\}$ is a chain of modules and $D < H_\alpha$ is pure for all α, then $D < \cup \mathscr{C} = \cup_\alpha H_\alpha$ is pure.

(3) Assume that $P < M < N$ are any modules. Then the purity relation satisfies the following.

(a) Transitive: $P < M$ pure, $M < N$ pure $\Rightarrow P < N$ pure.

(b) Hereditary: $M < N$ pure $\Rightarrow M/P < N/P$ is pure.

(c) Partial converse of hereditary: if $M/P < N/P$ is pure, and in addition $P < N$ is pure, then $\Rightarrow M < N$ is pure.

Proof. (a) Any system of equations in P which is solvable in N, can first be viewed as a system in M. The rest is clear.

(b) Suppose that $\Sigma\, x_i r_{ij} = c_j + P \in M/P$ is a finite system of equations in M/P which has a solution $x_i = b_i + P \in N/P$, where $1 \leqslant i \leqslant n$, $1 \leqslant j \leqslant m$. Then $\Sigma\, b_i r_{ij} = c_j + p_j$ for some $p_j \in P$. Thus $\Sigma_i x_i r_{ij} = c_j + p_j \in M$ is a system of equations in M which has a solution $x_i = b_i \in N$. Since $M < N$ is pure, there is a solution $x_i = a_i \in M$, $1 \leq i \leq n$. But then $\Sigma_i (a_i + P) r_{ij} = c_j + P$, and $x_i = a_i + P \in M/P$ is a solution of the original equations in M/P.

(c) We are given a finite system of equations

$$\sum_i x_i r_{ij} = d_j \in M \;(1 \leq i \leq n,\, 1 \leq j \leq m)$$

which has a solution $x_i = c_i \in N$. The above system of equations modulo P has a solution $x_i = c_i + P \in N/P$, and hence a solution $x_i = b_i + P$ where $b_i \in M$. Thus $\Sigma_i b_i r_{ij} = d_j + p_j$ for some $p_j \in P$. Then $\Sigma_i (b_i - c_i) r_{ij} = p_j$ shows that the second system $\Sigma_i x_i r_{ij} = p_j$ of equations in P has a solution $x_i = b_i - c_i \in N$. Hence the second system of equations has a solution $x_i = a_i \in P$ with $\Sigma_i a_i r_{ij} = p_j$. Thus $\Sigma_i (b_i - a_i) r_{ij} = d_j$ and $x_i = b_i - a_i \in M$ solves our original equations in M.

18-2.3   **Definition.** A sequence of R-modules $0 \to A \xrightarrow{\alpha} B \xrightarrow{\beta} C \to 0$ is *pure exact* if first, it is a short exact sequence, and secondly, if the image $\alpha A < B$ of α is pure in B.

A module P has the *projective property* relative to this short exact sequence if for any homomorphism $f : P \to C$ there exists a homomorphism $g : P \to B$ such that $f = \beta g$. An R-module P is *pure projective* if it has the projective property relative to all pure exact sequences. Hence every projective module is automatically pure projective.

A module I has the *injective property* relative to the above sequence if for any R-map $f : I \to A$ there exists a homomorphism $g : I \to B$ such that $g = \alpha f$. The module I is *pure injective* if it has the injective property relative to all pure exact sequences. Every injective module is necessarily also pure injective.

18-2.4   **Definition.** A module M is *finitely presented* if there exists some finitely generated free module F and a finitely generated submodule G of F such that $M \cong F/G$.

18-2.5   **Definition and notation.** Any R-module M is of the form $M \cong F/G$, where $F = \oplus\{x_i R \,|\, i \in I\}$ is free on the x_i, and $G < F$. Consequently G

has a set of generators such that $G = \langle\{\Sigma_i x_i r_{ij} | j \in J\}\rangle$, where $r_{ij} \in R$, and the $I \times J$ matrix $\|r_{ij}\|$ is column finite. Define y_i to be the image of x_i under $F/G \cong M$. Then also $\Sigma_i y_i r_{ij} = 0$. Let $s_i \in R$, $s_i = 0$ for *almost all i*. In general, the latter means that either all $s_i = 0$, or that only a finite number of s_j are unequal to zero. Then

$$\sum_{i \in I} y_i s_i = 0 \Rightarrow \begin{cases} \exists t_j \in R, \; t_j = 0 & \text{for almost all } j, \\ s_i = \sum_{j \in J} r_{ij} t_j & \text{for all } i \in I. \end{cases}$$

Write $M = \langle y_i, \; i \in I \,|\, \Sigma_i y_i r_{ij} = 0; \; j \in J \rangle$. The latter is called a *presentation* of M, that is the submodule generated by the y_i, $i \in I$, subject only to the indicated relations.

For any module C whatever, and any $c_i \in C$, $i \in I$, the correspondence $fy_i = c_i$ extends to an R-module homomorphism, provided that $\Sigma_i c_i r_{ij} = 0$ for all $j \in J$. For suppose that $\Sigma_i y_i s_i = 0$ for s_i as above. Then as above

$$f\left(\sum y_i s_i\right) = \sum c_i s_i = \sum_{j \in J}\left(\sum_{i \in I} c_i r_{ij}\right) t_j = 0.$$

If M is finitely presented, for any finite $I = \{1, 2, \dots, n\}$ there exists a finite set $J = \{1, 2, \dots, m\}$ such that

$$M = \left\langle y_1, \dots, y_n \,\middle|\, \sum_1^n y_i r_{ij} = 0; \; j = 1, \dots, m \right\rangle.$$

18-2.6 Construction. Any module M is isomorphic to a direct limit of finitely presented modules.

Proof. Take any presentation $M = \langle y_i, \; i \in I \,|\, \Sigma_i y_i r_{ij} = 0; \; j \in L \rangle$ of M as in the last definition. For any finite subset $F_2 \subset J$, define $F_1 \subset I$ by $F_1 = \{i \in I \,|\, \exists_i y_i r_{ij} = 0 \text{ with } j \in F_2 \text{ and } r_{ij} \neq 0\}$. Write $F_2 = \{j(1), \dots, j(m)\}$ and $F_1 = \{i(1), i(2), \dots, i(n)\}$. Note that F_1 is completely determined by F_2.

Define $F = F_1 \times F_2 \subset I \times J$ and similarly $G = G_1 \times G_2 \subset I \times J$. Let $\Gamma = \{F, G, \dots\}$ be the set of all such finite subsets of $I \times J$. Note that $F_1 \times F_2, F_1 \times F_2' \in \Gamma$ with $F_2 \neq F_2'$ is possible. Then partially order Γ by defining $F \leqslant G$ if $F_1 \subseteq G_1$ and $F_2 \subseteq G_2$. Note that $F \leqslant G$ if and only if $F_2 \subseteq G_2$. Define a finitely presented module M_F by

$$M_F = \left\langle x_{i(1)}^F, \dots, x_{i(n)}^F \,\middle|\, \sum_{k=1}^n x_{i(k)}^F r_{i(k),j} = 0; \; j \in F_2 \right\rangle.$$

Define an R-map $\eta_F x_i^F = y_i$. Note that η_F need not be monic. The reader may prefer to think of M_F as simply the module that would be generated by $y_{i(1)}, \ldots, y_{i(n)} \in M$ if the latter where subjected only to the finite subset of relations $\Sigma_{k=1}^{n} y_{i(k)} r_{i(k),j} = 0$ singled out by $j \in F_2$, where now η_F subjects these n generators of M to all the possible relations. If $F = \varnothing$, set $M_{\varnothing} = (0)$ and $\eta_{\varnothing} = 0$. From now on, let $F \leqslant G \in \Gamma$. Define an R-map $\phi_{GF} : M_F \to M_G$ by $\phi_{GF} x_i^F = x_i^G$. In the notation of 16-6.2,

$$\langle (\Gamma, \leqslant), \{M_F | F \in \Gamma\}, \{\varphi_{GF} | F \leqslant G \in \Gamma\} \rangle$$

is a direct system of modules. Let $M_* = \oplus \{M_F | F \in \Gamma\}/H$ be its direct limit and $M_F \cong e_F M_F \subset \oplus M_F$. By 16-6.5, there are R-maps $\delta_F : M_F \to M_*$ defined by $\delta_F x_i^F = e_F x_i^F + H \in M_*$. Then $\delta_F = \delta_G \varphi_{GF}$. We next show that M_* is isomorphic to M in a way which will allow us to replace the δ_F's by η_F's.

First define a correspondence ψ of generators of M_* with those of M by $\psi(e_F x_i^F + H) = y_i = \eta_F x_i^F$. Then for $-e_F x_i^F + e_G \phi_{GF} x_i^F = -e_F x_i^F + e_G x_i^G \in H$, the equation $\psi(e_F x_i^F + H) = \psi(e_G x_i^G + H) = y_i$ shows that ψ extends to a well defined R-module epimorphism $\psi : M_* \to M$. Every element of M_* is of the form $e_F w + H$ for some sufficiently large $F \in \Gamma$, and $w = \Sigma_i x_i^F s_i \in M_F$, $s_i \in R$. Furthermore $\psi(e_F w + H) = \eta_F w = \Sigma_i y_i s_i$. If $\psi(e_F w + H) = 0$, by replacing F by some larger element of Γ we may assume by 18-2.5 that $s_i = \Sigma \{r_{ij} t_j | j \in F_2\}$. Hence $w = \Sigma_i x_i^F s_i = \Sigma_j(\Sigma_i x_i^F r_{ij}) t_j = 0$ because each inner sum is already zero by the definition of M_F. Thus ψ is an isomorphism. Moreover,

$$\psi \delta_F x_i^F = \psi(e_F x_i^F + H) = y_i = \eta_F(x_i^F)$$

shows that $\psi \delta_F = \eta_F$ for all $F \in \Gamma$. Hence from $\delta_F = \delta_G \phi_{GF}$ we get that $\eta_F = \psi \delta_F = \psi \delta_G \phi_{GF} = \eta_G \phi_{GF}$. Thus M together with the maps η_F is the direct limit $M \cong \varinjlim M_F = \varinjlim \langle (\Gamma, \leqslant), \{M_F | F \in \Gamma\}, \{\varphi_{GF} | F \leqslant G \in \Gamma\} \rangle$ of finitely presented modules M_F. There is a commutative diagram.

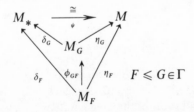

18-2.7 Theorem. *A short exact sequence of modules* $0 \to A \to B \to C \to 0$ *is pure exact* $\Leftrightarrow$ *every finitely presented R-module M has the projective property relative to this sequence, i.e. for any* $f : M \to C$ *there exists a*

$g: M \to B$ and a commutative diagram

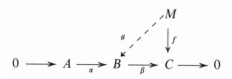

Proof. Without loss of generality, let $A \subset B$. $\Rightarrow$: Let

$$M = \left\langle y_1, \ldots, y_n \middle| \sum_1^n y_i r_{ij} = 0; j = 1, \ldots, m \right\rangle$$

be finitely presented. Let $\beta b_i = f y_i \in C$ for some $b_i \in B$. From

$$0 = f \sum_1^n y_i r_{ij} = \beta \sum_1^n b_i r_{ij}$$

it follows from the exactness of the given sequence that

$$\sum_1^n b_i r_{ij} = a_j \in A.$$

Now $\sum_1^n x_i r_{ij} = a_j$ is a finite system of equations in A with a solution $x_i = b_i \in B$ in B. Since $A \subset B$ is pure, the system has a solution $x_i = z_i \in A$, $\sum_i^n z_i r_{ij} = a_j$. It follows from 18-1.3 and $\sum_1^n (b_i - z_i) r_{ij} = 0$ that the correspondence $g y_i = b_i - z_i \in B$ gives a well defined module homomorphism $g: M \to B$. From $\beta g y_i = \beta(b_i - z_i) = \beta b_i = f y_i$ it follows that $f = \beta g$.

$\Leftarrow$: We are given a finite system $\sum_1^n x_i r_{ij} = a_j \in A$ of equations in A with a solution $x_i = b_i \in B$ in B. Define a finitely presented module

$$M = \left\langle y_1, \ldots, y_n \middle| \sum_1^n y_i r_{ij} = 0; j = 1, \ldots, m \right\rangle.$$

Then by 18-2.5 the assignment $f y_i = \beta b_i$ extends to an R-map $f: M \to C$, because

$$0 = -\beta a_j = \; = \sum_{i=1}^n (\beta b_i) r_{ij} = f \left(\sum_{i=1}^n y_i r_{ij} \right).$$

By hypotheses, there exists an R-map $g: M \to B$ with $f = \beta g$. Set $g y_i = z_i \in B$. Thus $\beta z_i = \beta g y_i = f y_i = \beta b_i$, and $\beta(b_i - z_i) = 0$. Hence $b_i - z_i \in A$. The following shows tht $b_i - z_i$ is a solution of the given system of equations:

$$\sum_{i=1}^n (b_i - z_i) r_{ij} = \sum_{i=1}^n b_i r_{ij} - \sum_{i=1}^n g y_i r_{ij}$$

$$= \sum b_i r_{ij} - 0 = a_j.$$

18-2.8   Corollary. *The following modules are pure projective:*

projective modules;	(0)
finitely presented modules;	(i)
direct sums of finitely presented modules;	(ii)
summands of direct sums of finitely presented modules.	(iii)

Proof. (ii) and (iii). More generally, the class of all modules having the projective property relative to any given short exact sequence is closed under the operations of (ii) and (iii).

18-2.9   Proposition. *For any R-module M, there exists a pure exact sequence*

$$0 \to K \to P \to M \to 0$$

with P a pure projective module (and $K < P$ a pure submodule).

Proof. Any fixed presentation $M = \langle y_i, i \in I \mid \Sigma\, y_i r_{ij} = 0, j \in J \rangle$ of M determines a set $\Gamma = \{F, G, \ldots\}$ and homomorphisms $\eta_F : M_F \to M$ as in the proof of 18-2.6. The set $\mathscr{S}$ of all possible presentations (18-2.5) of M can thus be identified with the set $\mathscr{S}$ of all such Γ's, i.e. $\mathscr{S} = \{\Gamma\}$. Define $P = \bigoplus_{\Gamma \in \mathscr{S}} \{\bigoplus_{F \in \Gamma} M_F\}$, where the outer sum ranges over all possible presentations of M. The module P is pure projective (18-2.8(ii)). The maps $\eta_F : M_F \to M$ as in 18-2.6 define by the universal property of the direct sum an epimorphism $f : P \to M$.

Set $K = \ker f$. Suppose that $\Sigma_1^n x_i r_{ij} = k_j \in K, j = 1, \ldots, m$; is a finite system of equations with a solution $x_i = a_i \in P$, $\Sigma_1^n a_i r_{ij} = k_j$. Hence $0 = fk_j = \Sigma_1^n (fa_i) r_{ij} = 0$. There exists a presentation of M in $\mathscr{S}$ such that $fa_1, fa_2, \ldots, fa_n$ are among the generators, and $\Sigma_1^n (fa_i) r_{ij} = 0$, $j = 1, \ldots, m$; are among the relations.

That is to say,

$$M = \left\langle y_i, i \in I \,\middle|\, \sum_i y_i s_{ij} = 0;\ j \in J,\ s_{ij} \in R \right\rangle,$$

where

$$F_1 = \{1, \ldots, n\} \subset I,\ F_2 = \{1, \ldots, m\} \subset J,$$

and

$$s_{ij} = r_{ij} \text{ for } 1 \leqslant i \leqslant n,\ 1 \leqslant j \leqslant m.$$

(Note that if, say $fa_2 = 0$, then there is a trivial relation $(y_2) \cdot 1 = 0$.) Corresponding to the above relations, let Γ be as in the proof of 2.6.

Without loss of generality assume that no row of the matrix $\|r_{ij}\|$ is zero. This guarantees that $F = F_1 \times F_2 \in \Gamma$.

Now as in the proof of 2.6, we have

$$M_F = \left\langle x_1^F, \ldots, x_n^F \middle| \sum_i x_i^F r_{ij} = 0; j = 1, \ldots, m \right\rangle,$$

and $\eta_F : M_F \to M$ with $\eta_F x_i^F = fa_i$. Hence M_F is a direct summand of P with $fx_i^F = \eta x_i^F = fa_i \in M$. Thus $f(a_i - x_i^F) = 0$, or $a_i - x_i^F \in K$. Moreover,

$$\sum_1^n (a_i - x_i^F) r_{ij} = \sum_1^n a_i r_{ij} - \sum_1^n x_i^F r_{ij} = k_j - 0.$$

Thus $a_i - x_i^F$ is the required solution in K of the given finite system, which proves that $K < P$ is pure.

The next theorem tells us exactly what are all pure projective modules and allows us to construct many examples of such.

18-2.10 Theorem. *An R-module M is pure projective $\Leftrightarrow M$ is a direct summand of a direct sum of finitely presented R-modules.*

Proof. $\Leftarrow$: See 18-2.8(ii) and (iii). $\Rightarrow$: There is a commutative diagram

$$0 \longrightarrow K \longrightarrow P \xrightarrow{\ f\ } M \longrightarrow 0.$$

where the bottom row is the pure exact sequence given by the last lemma, and g exists by the hypothesis that M is pure projective. Thus $M \cong gM$ is a direct summand of P, which is a direct sum of finitely presented modules.

Why are pure exact sequences so important that we are studying them? The next theorem tells us that they are precisely those short exact sequences which remain short exact after tensoring by any arbitrary module. The next theorem also obtains several necessary and sufficient conditions in order for an exact sequence to be pure exact. Recall that throughout this section we abbreviate $\_\otimes_R\_$ as just $\_\otimes\_$ and omit the R.

18-2.11 Theorem. *Consider an arbitrary left R-module $U = {}_R U$, a fixed sequence (a) of right R-modules, and then form (b):*

$$0 \to A \xrightarrow{\alpha} B \xrightarrow{\beta} C \to 0; \tag{a}$$
$$0 \to A \otimes U \to B \otimes U \to C \otimes U \to 0. \tag{b}$$

Then

(i) (a) *is pure exact* $\Rightarrow$ (b) *is exact. Conversely if*

(ii) (b) *is exact for any finitely presented* U, *then* $\Rightarrow$ (a) *is pure exact. Consequently*

(iii) (a) *is pure exact* $\Leftrightarrow \forall U$, (b) *is exact.*

(iv) (a) *is pure exact* $\Leftrightarrow \forall$ *finitely presented* U, (b) *is exact.*

Proof. (i) We only have to show that $\alpha \otimes 1$ is monic. Suppose that

$$\sum_1^n a_j \otimes u_j \in A \otimes U \quad \text{with} \quad 0 = \sum_1^n \alpha a_j \otimes u_j \in B \otimes U.$$

By 16-5.3, there exists an $s \times (n + p)$ matrix $\|r_{ij}\|$ with $r_{ij} \in R$, and $u_{n+1}, \ldots, u_{n+p} \in U$, and $b_1, \ldots, b_s \in B$ such that for all $i = 1, \ldots, s$

(1) $\displaystyle\sum_{j=1}^{n+p} r_{ij} u_j = 0,$

(2) $\displaystyle\sum_{i=1}^{s} b_i r_{ij} = \begin{cases} \alpha a_j & j = 1, \ldots, n; \\ 0 & j = n+1, \ldots, n+p. \end{cases}$

In view of (2), since $\alpha A < B$ is pure and since α is monic, the equations below have a solution $x_i = \alpha c_i \in \alpha A$, $i = 1, \ldots, s$; such that the following hold:

$$\sum_{i=1}^{s} x_i r_{ij} = \begin{cases} \alpha a_j \\ 0 \end{cases} \quad \sum_{i=1}^{s} c_i r_{ij} = \begin{cases} a_j & j = 1, \ldots, n; \\ 0 & j = n+1, \ldots, n+p. \end{cases}$$

Thus

$$\sum_{j=1}^{n} a_j \otimes u_j = \sum_{j=1}^{n+p} \sum_{i=1}^{s} c_i r_{ij} \otimes u_j = \sum_{i=1}^{s} c_i \otimes \sum_{j=1}^{n+p} r_{ij} u_j = 0.$$

Thus (b) is exact.

(ii) The sequence (a) is exact because (b) with $U = {}_R R$ is (a). Let

$$\sum_{i=1}^{n} x_i r_{ij} = \alpha a_j \in \alpha A \qquad (j = 1, \ldots, m)$$

be a finite system with a solution $x_i = b_i \in B$. Define

$$_R U = \left\langle u_1, \ldots, u_m \,\middle|\, \sum_{j=1}^{m} r_{ij} u_j = 0 \quad i = 1, \ldots, n \right\rangle.$$

In

$$B \otimes U, \quad \sum_{j=1}^{m} \alpha a_j \otimes u_j = \sum_{j=1}^{m} \sum_{i=1}^{n} b_i r_{ij} \otimes u_j = 0.$$

By hypothesis already $0 = \Sigma_{j=1}^{m} a_j \otimes u_j \in A \otimes U$. Use of 16-5.4 gives an $s \times (m+p)$ matrix with $s_{ij} \in R$, elements $u_{m+1}, \ldots, u_{m+p} \in U$, and $y_1, \ldots, y_s \in A$ such that for all $k = 1, \ldots, s$

(1) $\quad \sum_{j=1}^{m+p} s_{kj} u_j = 0,$

(2) $\quad \sum_{k=1}^{s} y_k s_{kj} = \begin{cases} \alpha a_j & j = 1, \ldots, m; \\ 0 & j = m+1, \ldots, m+p. \end{cases}$

By definition of U, in 18-2.5, $s_{kj} = \Sigma_{i=1}^{s} t_{ki} r_{ij}$ for some $t_{ki} \in R$. Consequently our original system has the following solution in A

$$\alpha a_j = \sum_{k=1}^{s} y_k s_{kj} = \sum_{i=1}^{s} \underbrace{\left(\sum_{k=1}^{s} y_k t_{ki} \right)}_{\varepsilon A} r_{ij}.$$

18-2.12 Observation. An exact sequence of right R-modules

$$0 \longrightarrow C \xrightarrow{\beta} B \xrightarrow{\alpha} A \longrightarrow 0$$

splits $\Leftrightarrow \forall M = M_R$, the induced map $\mathrm{Hom}_R(M, B) \to \mathrm{Hom}_R(M, A)$ is onto. The analogous result also holds for left R-modules.

Proof. $\Rightarrow$: Clear. $\Leftarrow$: With $M = A$, the identity map $1_A : A \to A$ factors through B as $1_A = \alpha g$, $g : A \to B$.

18-2.13 Consider an exact sequence (1) of right R-modules, and its dual (2):

(1) $\quad 0 \to A \xrightarrow{\alpha} B \xrightarrow{\beta} C \to 0;$

(2) $\quad 0 \to {}_RC^* \to {}_RB^* \to {}_RA^* \to 0.$

Then (1) is pure exact $\Leftrightarrow$ (2) is splitting exact.

Proof. From 17-1.8, (2) is exact if (1) is. For any finitely presented left R-module U, form the following sequence

(3) $\quad 0 \to \mathrm{Hom}_R(U, C^*) \to \mathrm{Hom}_R(U, B^*) \to \mathrm{Hom}_R(U, A^*) \to 0.$ By adjoint associativity (16-4.4), there are abelian group isomorphisms

$$\mathrm{Hom}_Z(A \otimes U, \mathbf{Q}/\mathbf{Z}) \cong \mathrm{Hom}_R(U, \mathrm{Hom}_Z(A, \mathbf{Q}/\mathbf{Z})).$$

Consequently the above sequence (3) is isomorphic to (4),

(4) $0 \to (C \otimes U)^* \to (B \otimes U)^* \to (A \otimes U)^* \to 0.$

Lastly form

(5) $0 \to A \otimes U \to B \otimes U \to C \otimes U \to 0.$

By 18-2.11(iii), (1) is pure exact $\Leftrightarrow$ for any U, the sequence (5) is exact. By 18-1.8, (5) is exact $\Leftrightarrow$ (4) is exact $\Leftrightarrow$ (3) is exact. By 18-2.12, for any U the sequence (3) is exact $\Leftrightarrow$ (2) is splitting exact. Thus (1) is pure exact $\Leftrightarrow$ (2) is splitting exact.

The next corollary generalizes to arbitrary rings the fact that for abelian groups $A < B$, if B/A is torsion free, then A is pure in B.

18-2.14 Corollary. *If* $0 \to A \to B \to C \to 0$ *is an exact sequence of modules and C is flat, then this sequence is pure exact.*

Proof. The module C is flat if and only if C^* is injective (17-2.3). But if C^* is injective, then $0 \to C^* \to B^* \to A^* \to 0$ is splitting exact. Now by 18-2.13, the sequence $0 \to A \to B \to C \to 0$ is pure exact.

For sake of emphasis, we summarize and reformulate equivalent characterizations of a pure exact sequence.

18-2.15 Pure exactness. An exact sequence of right R-modules

$$0 \longrightarrow A \overset{\alpha}{\longrightarrow} B \overset{\beta}{\longrightarrow} C \longrightarrow 0$$

is pure exact if and only if one of the following conditions (a)–(d) hold.

(a)   If the system $\Sigma_i \, x_i r_{ij} = a_j \in A$, $1 \leqslant i \leqslant n$ ($j = 1, \ldots, m$) is solvable in B, then it is also solvable in A.

(b) $\forall$ finitely presented module $M = M_R$, the induced map

$$\mathrm{Hom}_R(M, B) \to \mathrm{Hom}_R(M, C) \text{ is epic.}$$

(c) $\forall$ finitely presented module $_R U$, $1 \otimes \alpha : A \otimes U \to B \otimes U$ is monic.

($\bar{c}$) $\forall$ module $_R U$, $\alpha \otimes 1 : A \otimes U \to B \otimes U$ is monic.

(d) $0 \to {_R C^*} \overset{\beta_*}{\longrightarrow} {_R B^*} \overset{\alpha_*}{\longrightarrow} {_R A^*} \longrightarrow 0$ is splitting exact.

18-3   Direct limits

First it will be shown that direct limits of short exact sequences are exact. Direct limits of short exact sequences are not only used here in

subsequent theorems, but are occasionally useful in other module theoretic proofs. It is shown that a direct limit of flat modules is flat. Then the
pure exact sequences are characterized as precisely the direct limits of
short exact sequences.

18-3.1   Let I be a partially ordered upper directed set, and

$$0 \to A_i \xrightarrow{\;f_i\;} B_i \xrightarrow{\;g_i\;} C_i \longrightarrow 0 \qquad i \in I$$

an indexed family of short exact sequences. Suppose that for $i \leqslant j$ there
are module maps α_{ji}, β_{ji}, and γ_{ji} so that $f_j \alpha_{ji} = \beta_{ji} f_i$ and $g_j \beta_{ji} = \gamma_{ji} g_i$.
Furthermore, suppose that $\langle (I, \leqslant), \{A_i \,|\, i \in I\}, \{\alpha_{ji} \,|\, i \leqslant j \in I\} \rangle$, $\langle B_i, \beta_{ji} \rangle$,
and $\langle C_i, \gamma_{ji} \rangle$ are individually three direct systems (16-6.2(i), (ii)). Then the
above short exact sequences and connecting maps are called a *direct
system of short exact sequences*.

Now let $A_* = \varinjlim A_i$, $B_* = \varinjlim B_i$, and $C_* = \varinjlim C_i$ and let $\alpha_i : A_i \to A_*$,
$\beta_i : B_I \to B_*$, and $\gamma_i : C_i \to C_*$ be the canonical homomorphisms 16-6.5(i).
Then for the families of maps $\beta_i f_i : A_i \to B$, and $\gamma_i g_i : B_i \to C$ for $i \in I$, by
the universal property of the direct limit, there exist unique maps
$0 \to A_* \xrightarrow{f} B_* \xrightarrow{g} C_* \to 0$. It is asserted that the latter sequence is exact.

By 16-6.5(i) and (ii), $C_* = \cup\{\gamma_i C_i \,|\, i \in I\}$, and similarly for A_* and B_*.
Thus for any $\gamma_i c_i \in \gamma_i C_i \subseteq C_*$, let $g_i b_i = c_i$ for some $b_i \in B_i$. But $\gamma_i g_i = g \beta_i$
by 16-6.2. Hence $\gamma_i c_i = \gamma_i g_i b_i = g(\beta_i b_i) \in g(B_*)$. Thus g is onto.

Next, let $b = \beta_i b_i \in B_*$ with $gb = 0$. Since $\gamma_i g_i = g \beta_i$ by 16-6.5(iii), we
get that $\gamma_i(g_i b_i) = 0$. Set $z = g_i b_i$. By 16-6.5(iii), $\gamma_{qi} z = 0$ for some $q \geqslant i$.
But $\gamma_{qi} g_i = g_q \beta_{qi}$. Thus $0 = \gamma_{qi} z = \gamma_{qi} g_i b_i = g_q(\beta_{qi} b_i)$. But then $\beta_{qi} b_i = f_q a_q$ for some $a_q \in A_q$ because the qth sequence is exact by assumption.
Finally, $\beta_q f_q = f \alpha_q$ and $\beta_i = \beta_q \beta_{qi}$ for $q \geqslant i$, and hence

$$b = \beta_i b_i = \beta_q \beta_{qi} b_i = \beta_q f_q a_q = f(\alpha_q a_q) \in f(A_*).$$

Thus kernel $g = $ image f.

Suppose that $a = \alpha_i a_i \in A_*$ with $fa = 0$. From $f \alpha_i = \beta_i f_i$ we get that
$f \alpha_i a_i = \beta_i(f_i a_i) = 0$. Hence by 16-6.5(iii), there exists a $q \geqslant i$ such that
$\beta_{qi}(f_i a_i) = 0$. But $\beta_{qi} f_i = f_q \alpha_{qi}$, and thus $f_q(\alpha_{qi} a_i) = 0$. Since f_q is monic,
$\alpha_{qi} a_i = 0$. But then since $\alpha_i = \alpha_q \alpha_{qi}$ for $q \geqslant i$, $a = \alpha_i a_i = \alpha_q \alpha_{qi} a_i = 0$. Thus
f is monic. Lastly, for any $\alpha_i a_i \in A_*$, $gf(\alpha_i a_i) = g \beta_i(f_i a_i) = \gamma_i g_i f_i a_i = 0$
because $g_i f_i = 0$. Hence $gf = 0$, and the direct limit sequence is exact.

18-3.2   Proposition. *If* $\langle (I, \leqslant), \{M_i \,|\, i \in I\}, \{\phi_{ji} \,|\, i \leqslant j \in I\} \rangle$ *is a direct
system of flat modules* M_i *where* I *is upper directed, then their direct limit*
$M = \varinjlim M_i$ *is flat.*

Proof. Let $0 \to A \to B \to C \to 0$ be any exact sequence of left R-modules. For $i \leqslant j \in I$, the maps $\phi_{ji} : M_i \to M_j$ from the direct system, after tensoring with the identity maps on A, B, and C, give a direct system of short exact sequences $0 \to A \otimes M_i \to B \otimes M_i \to C \otimes M_i \to 0$ $(i \in I)$. The direct limit of the latter by 18-3.1 is exact, and by 16-3.4, direct limits commute with tensor products, thus resulting in the short exact sequence

$$0 \to A \otimes M \to B \otimes M \to C \otimes M \to 0.$$

Thus M is flat.

18-3.3   Theorem. *An exact sequence* $0 \to A \to B \to C \to 0$ *is pure exact* $\Leftrightarrow$ *it is a direct limit of splitting exact sequences.*

18-3.4   Corollary. *A direct limit of pure exact sequences is pure exact.*

Proof of 3.3 $\Leftarrow$ *and* 18-3.4. Let $0 \to A_i \to B_i \to C_i \to 0$ $(i \in I)$, be any direct system (as in 18-3.1) of pure exact sequences and let $0 \to A \to B \to C \to 0$ be their direct limit as in 18-3.1. We have replaced $A = A_*$, $B = B_*$, and $C = C_*$. By 18-2.11(iii) for any left R-module $U = {}_R U$, the sequences $0 \to A_i \otimes U \to B_i \otimes U \to C_i \otimes U \to 0$ $(i \in I)$, are all exact. Since by 16-3.4(ii), the tensor product commutes with taking direct limits, the vertical arrows in the commutative diagram below are natural isomorphisms

$$0 \to \varinjlim (U \otimes A_i) \to \varinjlim (U \otimes B_i) \to \varinjlim (U \otimes C_i) \to 0$$
$$\downarrow \qquad\qquad \downarrow \qquad\qquad \downarrow$$
$$0 \to U \otimes \varinjlim A_i \to U \otimes \varinjlim B_i \to U \otimes \varinjlim C_i \to 0$$

The first sequence is exact by 18-3.1, and hence so is also the second. But then it follows from 18-2.11(iii) that $0 \to A \to B \to C \to 0$ is pure exact.

$\Rightarrow$. Conversely, let $0 \to A \xrightarrow{f} B \xrightarrow{g} C \to 0$ be pure exact. By 18-2.6, write $C = \varinjlim \langle (I, \leqslant), \{C_i | i \in I\}, \{\gamma_{ji} | i \leqslant j \in I\} \rangle = \varinjlim C_i$ of finitely presented modules C_i. Let $\gamma_i : C_i \to C$ be the maps given by the direct limit. By Theorem 18-2.7, there is a map $h_i : C_i \to B$ with $gh_i = \gamma_i$ for all $i \in I$. Next, let $g_i : B_i \to C_i$ and $\beta_i : B_i \to B$ be the pullback of $\gamma_i : C_i \to C$ and $g : B \to C$. By 10-2.2, $B_i \cong \ker g \oplus C_i$, and we may take $B_i = A \oplus C_i$. From the proof of 10-2.2 it follows that g_i is the natural projection $g_i : A \oplus C_i \to C_i$, and that the kernel of g_i is A. Thus if $f_i : A \to B_i$ is the inclusion map, then $0 \to A \xrightarrow{f_i} B_i \xrightarrow{g_i} C_i \to 0$ is an exact sequence with $\beta_i f_i = f$ for all i.

Let $i \leqslant j$. Since $\gamma_j(\gamma_{ji}g_i) = g\beta_i$, by the pullback property of B_j, there exists a unique map β_{ji} so that the first pullback diagram below commutes. In view of the first diagram, in order to show that the third diagram commutes, all that remains is to show that $f_j = \beta_{ji}f_i$. Since $\gamma_j g_j f_j = g\beta_j f_j = gf = 0$, by the pullback property of B_j, there is a unique map δ which makes the second diagram commute. Since $\beta_j f_j = f$, this diagram commutes with $\delta = f_j$. If however $\delta = \beta_{ji}f_i$, then $g_j\delta = g_j\beta_{ji}f_i = \gamma_{ji}g_i f_i = 0 = gf_j$. Thus $f_j = \beta_{ji}f_i$, and the third diagram below commutes.

It is asserted that $\langle(I, \leqslant), \{B_i \,|\, i \in I\}, \{\beta_{ji} \,|\, i \leqslant j \in I\}\rangle$ is a direct system. By definition of β_{ii}, $\beta_{ii} = 1$. Suppose that $i \leqslant r \leqslant j$ for some $r \in I$. In order to show that $\beta_{ji} = \beta_{jr}\beta_{ri}$, it suffices to show that the first diagram also commutes when β_{ji} is replaced by $\beta_{jr}\beta_{ri}$. But

$$\beta_j\beta_{jr}\beta_{ri} = \beta_r\beta_{ri} = \beta_i,$$

and

$$(g_j\beta_{jr})\beta_{ri} = \gamma_{jr}(g_r\beta_{ri}) = (\gamma_{jr}\gamma_{ri})g_i = \gamma_{ji}g_i = g_j\beta_{ji} = \gamma_{ji}g_i,$$

where all the equalities follow from the previous definitions of the various maps involved. Thus $0 \longrightarrow A \xrightarrow{\,f_i\,} B_i \xrightarrow{\,g_i\,} C_i \longrightarrow 0$ together with the maps 1_A, β_{ji}, and γ_{ji} for $i \leqslant j$ is a direct system of short exact sequences. Let $0 \longrightarrow A \xrightarrow{\,f_*\,} B_* \xrightarrow{\,g_*\,} C \longrightarrow 0$ be their direct limit as in 18-3.1, where $\phi_i : B_i \to B_*$ are the natural maps associated with this direct limit.

Our objective is to show that there exists an isomorphism $h : B_* \to B$ such that Diag. 4 commutes. This would say that we may simply take $B_* = B$ and $\phi_i = \gamma_i$. Since by definition of β_{ji}, $\beta_i = \beta_j\beta_{ji}$, by 16-6.2, there exists a unique map $h : B_* \to B$ such that $\beta_i = h\phi_i$.

Since f_i and $f_j = \beta_{ji}f_i$ are inclusion maps, it follows that the restrictions of β_i, β_{ji}, f and f_* to A are all equal to the identity. Thus $f = hf_*$.

Next, we show why $gh = g_*$. In 18-3.1, g_* was defined as the unique map such that $g_*\phi_i = \gamma_i g_i$ for all i. But also $\gamma_i g_i = g\beta_i$, $\beta_i = h\phi_i$, and consequently $\gamma_i g_i = (gh)\phi_i$. By uniqueness, $g_* = gh$.

Thus in Diag. 4 we have a commutative diagram with exact rows. In an abelian category this implies that h is an isomorphism. For simplicity, we give an elementwise proof. Since $gh = g_*$, kernel $h \subseteq$ kernel $g_* = A$. But $h \,|\, A =$ identity, so kernel $h = 0$. In order to show that h is onto, take any $b \in B$. Since $g_* B_* = C$, there is a $b_* \in B_*$ with $gb = g_* b_*$, and $gb = ghb_*$. Thus $g(b - hb_*) = 0$, and hence $b - hb_* = a \in A$. Therefore $b = hb_* - a = hb_* - ha = h(b^* - a)$, and h is an isomorphism.

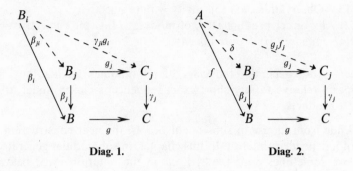

Diag. 1.   Diag. 2.

$$0 \longrightarrow A \xrightarrow{f_i} B_i \xrightarrow{g_i} C_i \longrightarrow 0$$

$$\|\qquad \beta_{ji}\downarrow \qquad \downarrow\gamma_{ji}$$

$$0 \longrightarrow A \xrightarrow{f_j} B_j \xrightarrow{g_j} C_j \longrightarrow 0$$

$$\|\qquad \beta_j\downarrow \qquad \downarrow\gamma_j$$

$$0 \longrightarrow A \xrightarrow{f} B \xrightarrow{g} C \longrightarrow 0$$

Diag. 3.

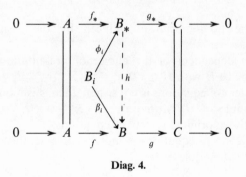

Diag. 4.

18-4 Pure injectives

Recall that a module was defined to be pure injective if it had the injective property relative to all pure exact sequences (see 18-2.3). Equivalent characterizations of pure injectivity are given. It is shown that every module can be embedded as a pure submodule of a pure injective module.

18-4.1   Observations. (1) Injectivity ⇒ pure injectivity.

(2) Any direct product ΠM_i of modules M_i is pure injective ⇔ each direct summand M_i is pure injective.

Proof. More generally the class of all modules having the injective property relative to any short exact sequence is closed under arbitrary direct products.

Aside from its use in subsequent proofs, the next construction might be of independent interest. In this chapter, in submodules generated by a set, we sometimes omit the "{ }" as in the notation in (i) below. The reader may wish to read 18-4.3 first, before going into the details of proofs in 18-4.2.

18-4.2   Construction. For a module D, let $\Sigma x_i r_{ij} = d_j \in D$ $(i \in I, j \in J)$ be a completely arbitrary system of equations over D. Take any presentation $D = \langle g_\alpha, \alpha \in I_D \mid \Sigma_\alpha g_\alpha s_{\alpha\beta} = 0, \ \beta \in J_D \rangle$ of D. Then write $d_j = \Sigma_\alpha g_\alpha k_{\alpha j}$, $k_{\alpha j} \in R$. Form a module H as a quotient of a free module with free generators $\bar{g}_\alpha, \bar{x}_i$ as follows:

$$H = \frac{\oplus\{\bar{g}_\alpha R \mid \alpha \in I_D\} \oplus [\oplus\{\bar{x}_i R \mid i \in I\}]}{G}$$

$$G = \left\langle \sum_\alpha \bar{g}_\alpha s_{\alpha\beta}, \sum_i \bar{x}_i r_{ij} - \sum_\alpha \bar{g}_\alpha k_{\alpha j} \mid \beta \in J_D, j \in J \right\rangle.$$

Let $\varphi : D \to H$ be the homomorphism induced by the assignment $\varphi g_\alpha = \bar{g}_\alpha + G$. Then

(i) H and φ are independent of the choice of presentation of D. Therefore write or replace H with $H \cong \langle D, x_i, i \in I \mid \Sigma_i x_i r_{ij} = d_j, j \in J \rangle$.

(ii) If the above system of equations is consistent, then φ is monic, and hence D is a submodule of H. Moreover, $x_i = \bar{x}_i + G \in H$, $i \in I$, is a solution in H of the above original system of equations.

(iii) If the above system of equations is finitely solvable, then, moreover, $D < H$ is a pure submodule of H.

Proof. (i) Since later (i) will never be used, its proof is omitted. (ii) Suppose that $\varphi(\Sigma_\alpha g_\alpha s_\alpha) = 0$, $\Sigma_\alpha \bar{g}_\alpha s_\alpha \in G$, or there exist $t_\beta, t'_j \in R$ almost all zero such that

$$\sum_\alpha \bar{g}_\alpha s_\alpha = \sum_\beta \sum_\alpha \bar{g}_\alpha s_{\alpha\beta} t_\beta + \sum_{j \in J} \left(\sum_i \bar{x}_i r_{ij} - \sum_\alpha \bar{g}_\alpha k_{\alpha j} \right) t'_j. \tag{1}$$

In the very last sum there is a finite subset $F \subset J$ such that $t'_j = 0$ for

$j \in J \backslash F$. This is a relation among the free generators $\bar{g}_\alpha$ and $\bar{x}_i$ of a free module, hence

$$s_\alpha = \sum_\beta s_{\alpha\beta} t_\beta - \sum_j k_{\alpha j} t'_j \quad \forall \alpha, \tag{2}$$

$$\sum_{j \in J} r_{ij} t'_j = \sum_{j \in F} r_{ij} t'_j = 0. \tag{3}$$

Thus (2) implies that

$$\sum_\alpha g_\alpha s_\alpha = \sum_\beta \left(\sum_\alpha g_\alpha s_{\alpha\beta} \right) t_\beta - \sum_j \left(\sum_\alpha g_\alpha k_{\alpha j} \right) t'_j$$

$$= 0 - \sum_j d_j t'_j.$$

The fact that the system of equations $\sum_i x_i r_{ij} = d_j$ ($j \in J$), is consistent, together with the fact that $\sum_j \sum_i x_i r_{ij} t'_j = 0$, by definition of 'consistent', implies that also $\sum_{j \in J} d_j t'_j = 0$. By the equation immediately above $\sum_\alpha g_\alpha s_\alpha = 0$, and φ is monic. Lastly if $x_i = \bar{x}_i + G$, then

$$\sum_i x_i r_{ij} = \sum_i \bar{x}_i r_{ij} + G = \sum_\alpha \bar{g}_\alpha k_{\alpha j} = \varphi d_j$$

shows that the original system of equations has a solution in H.

(iii) Identify $D \equiv \langle \{\bar{g}_\alpha + G \mid \alpha \in I_D\} \rangle$. We work in the free group. Let

$$\sum_{k=1}^n (y_k + G) a_{kp} = z_p + G; \quad p = 1, 2, \ldots, m;$$

be a finite system of equations in unknowns $y_k + G$ and $z_p = \sum_\alpha \bar{g}_\alpha b_{\alpha p}$; a_{kp}, $b_{\alpha p} \in R$. Suppose that these equations have a solution in H also denoted by $y_k + G$, with $y_k = \sum_\alpha \bar{g}_\alpha c_{\alpha k} + \sum_{i \in I} \bar{x}_i c'_{ik}$; $c_{\alpha k}, c'_{ik} \in R$ where $\|c_{\alpha k}\|$ and $\|c'_{ik}\|$ are column finite matrices. Thus we get the following equation modulo G

$$\sum_{k=1}^n \sum_\alpha \bar{g}_\alpha c_{\alpha k} a_{kp} + \sum_{k=1}^n \sum_{i \in I} \bar{x}_i c'_{ik} a_{kp} - \sum_\alpha \bar{g}_\alpha b_{\alpha p} \equiv 0 \pmod{G}. \tag{4}$$

Thus the left side of this equation equals a typical element of G as in the right side of (1).

$$\sum_\alpha \bar{g}_\alpha \left(\sum_{k=1}^n c_{\alpha k} a_{kp} - b_{\alpha p} \right) + \sum_{i \in I} \bar{x}_i \sum_{k=1}^n c'_{ik} a_{kp}$$

$$= \sum_\alpha \sum_\beta \bar{g}_\alpha s_{\alpha\beta} t_\beta + \sum_{j \in F} \left(\sum_{i \in I} \bar{x}_i r_{ij} - \sum_\alpha \bar{g}_\alpha k_{\alpha j} \right) t'_j. \tag{5}$$

Since this is a relation among free generators of a free module, upon equating the coefficients of $\bar{g}_\alpha$ and $\bar{x}_i$, we get

$$\sum_{k=1}^{n} c_{\alpha k} a_{kp} = b_{\alpha p} + \sum_{\beta} s_{\alpha \beta} t_\beta - \sum_{j \in F} k_{\alpha j} t'_j; \tag{6}$$

$$\sum_{k=1}^{n} c'_{ik} a_{kp} = \sum_{j \in F} r_{ij} t'_j. \tag{7}$$

As before let $\Sigma_{i \in I} a_i r_{ij} = d_j$ for all $j \in F$ with $a_i \in D$. We now verify that $z_k + G$ is a solution of our given finite system of equations where $z_k = \Sigma_\alpha \bar{g}_\alpha c_{\alpha k} + \Sigma_{i \in I} a_i c'_{ik}$. Modulo G, the left side becomes

$$\sum_{k=1}^{n} z_k a_{kp} = \sum_{a} \bar{g}_\alpha \sum_{k=1}^{n} c_{\alpha k} a_{kp} + \sum_{i \in I} a_i \sum_{k=1}^{n} c'_{ik} a_{kp}.$$

Use of (6) and (7) in the first and second terms on the right respectively gives the following,

$$\sum_{k=1}^{n} z_k a_{kp} = \sum_{\alpha} \bar{g}_\alpha b_{\alpha p} + \sum_{\beta} \left(\sum_{\alpha} \bar{g}_\alpha s_{\alpha \beta} \right) t_\beta - \sum_{j \in F} \left(\sum_{\alpha} \bar{g}_\alpha k_{\alpha j} \right) t'_j + \sum_{j \in F} \left(\sum_{i \in I} a_i r_{ij} \right) t'_j$$

$$= z_p + \sum_{\beta} \left(\sum_{\alpha} \bar{g}_\alpha s_{\alpha \beta} \right) - \sum_{j \in F} d_j t'_j + \sum_{j \in F} d_j t'_j \in z_p + G,$$

where the second term is in G. Thus $D < H$ is pure.

In the previous construction $D < H$ is pure, where H is a quotient of a free module. For some purposes the latter additional property is not needed and a simpler construction can be used as in the next corollary.

18-4.3   Corollary. *For a module D, let $\Sigma_i x_i r_{ij} = d_j \in D$ ($i \in I, j \in J$) be any finitely solvable system. Let $C = \langle \Sigma_i \bar{x}_i r_{ij} - d_i \,|\, i \in I, j \in J \rangle < \oplus_{i \in I} \bar{x}_i R$, where the latter module is free with free generators $\bar{x}_i$. Then*

$$D \to \frac{D \oplus \left[\displaystyle\bigoplus_{i \in I} \bar{x}_i R \right]}{C} = B$$

$$d \to (d, 0) + C \quad (d \in D);$$

embeds D as a pure submodule of B. The above system of equations has a global solution $x_i = \bar{x}_i + C$ ($i \in I$), in B.

Proof. In the notation of 18-4.2, set

$$F = \left[\bigoplus_{\alpha \in I_D} \bar{g}_\alpha R \right] \oplus \left[\bigoplus_{i \in I} \bar{x}_i R \right],$$

$$E = \left\langle \left\{ \sum_\alpha \bar{g}_\alpha s_{\alpha\beta} \mid \alpha \in I_D,\ \beta \in J_D \right\} \right\rangle < G \cap \left[\bigoplus_\alpha \bar{g}_\alpha R \right].$$

Then

$$G/E \cong C, \qquad \left[\bigoplus_a \bar{g}_a R \right] \Big/ E \cong D, \qquad F/E \cong D \oplus \left[\bigoplus_i \bar{x}_i R \right],$$

and

$$H = \frac{F}{G} \cong \frac{F/E}{G/E} \cong B.$$

Since $\Sigma_i(\bar{x}_i + C)r_{ij} = d_j + C$, $x_i = \bar{x}_i + C$ gives a solution of the whole system above.

18-4.4 Theorem. *For a module D, the following are all equivalent.*

(1) *D is pure injective.*

(2) *D is a direct summand in every module which contains it as a pure submodule.*

(3) *Any system of equations in D which is finitely solvable in D, has a global solution in D.*

Proof. (1) $\Rightarrow$ (2). If $D < B$ is pure, the identity map $1_D : D \to D$ (by the pure injectivity of the last D) extends to a retraction $g : B \to D$ with $g \mid D = 1_D$. Hence $B = D \oplus (1_B - g)B$.

(2) $\Rightarrow$ (3). Suppose that $\Sigma_{i \in I} x_i r_{ij} = d_j$ $(j \in J)$ is any (in general infinite) system of equations which is finitely solvable in D. Define

$$H = \langle D, x_i \ (i \in I) \mid \Sigma x_i r_{ij} = d_j \ (j \in J) \rangle$$

as in the last construction, where $D < H$ is pure. By hypothesis, $H = D \oplus G$ for some G. Write $x_i = y_i + z_i$, $y_i \in D$, $z_i \in G$. Then $\Sigma_{i \in I} x_i r_{ij} = d_j$ for all $j \in J$ as a relation inside H, becomes

$$\sum_i y_i r_{ij} - d_j = -\sum_i z_i r_{ij} \in D \cap G = 0.$$

Then $y_i \in D$ and $\Sigma_i y_i r_{ij} = d_j$ for all $j \in J$ gives a global solution.

(3) $\Rightarrow$ (1). Given are a pure exact sequence $0 \to A \subset B \to B/A \to 0$ and a homomorphism $\varphi : A \to D$. Select $b_i \in B$ $(i \in I)$ such that the set A together with the b_i's generate B, i.e. $B = \langle A \cup \{b_i \mid i \in I\} \rangle$. List all existing relationships of the form $\Sigma_i b_i r_{ij} = a_j \in A$ $(j \in J)$. Note that these include all the relationships among the b_i, in which case $a_j = 0$.

Consider the system of equations $\Sigma_{i \in I} x_i r_{ij} = \varphi a_j$ $(j \in J)$, over D. Let

$F \subset J$ be a completely arbitrary finite subset. Since $A \subset B$ is pure (and since $\Sigma_i b_i r_{ij} = a_j$ for all $j \in J$) there is a solution $x_i = c_i \in A$ of this finite system, i.e. $\Sigma_i c_i r_{ij} = a_j$ for $j \in F$. Upon applying φ to the latter we get a solution $\Sigma_{i \in I} (\varphi c_i) r_{ij} = \varphi a_j$ $(j \in F)$, in D. By hypothesis (3), there exists a global solution $x_i = d_i \in D$, i.e. $\Sigma_{i \in I} d_i r_{ij} = \varphi a_j$ for all $j \in J$. Set $\bar{\varphi} b_i = d_i$, and let $\bar{\varphi} | A = \varphi$. Wheneverd $\Sigma_i b_i r_{ij} = a_j$, we get $\Sigma_i (\bar{\varphi} b_i) r_{ij} = \varphi a_j$ for all $j \in J$. Thus $\bar{\varphi}$ is well defined, $\bar{\varphi} | A = \varphi$, and then $\bar{\varphi}: B \to D$ is the required extension of φ.

In order to show that a module is pure injective by use of criterion (3) of the previous theorem, systems of equations of arbitrarily large cardinality have to be considered. This leads to the next definition.

18-4.5 Definition. For an infinite cardinal $\aleph$, a module M is $\aleph^<$-*equationally compact* if for any finitely solvable system of linear equations $\Sigma_{i \in I} x_i r_{ij} = d_j \in M$ $(j \in J)$ over M with $|J| < \aleph$, there exists a global solution in M. The module M is *equationally compact* if it is $\aleph^<$-equationally compact for all infinite cardinals $\aleph$. Thus by the last theorem, equational compactness is equivalent to pure injectivity.

Not only does the next theorem give a new characterization of pure injectivity, but it also will be required later in the construction of pure injective hulls of modules.

18-4.6 Theorem. *For a right R-module M, set $\aleph = [max(|R|, \aleph_0)]^+$, i.e. the successor cardinal of $|R| + \aleph_0$. Then the following are equivalent.*

(1)   *M is equationally compact;*
(2)   *M is $\aleph^<$-equationally compact;*
(3)   *M is pure injective.*

Proof. $(3) \Leftrightarrow (1) \Rightarrow (2)$ follows from 18-4.4. $(2) \Rightarrow (3)$. Let $A < B$ be pure, and let $\varphi: A \to M$ be an R-module homomorphism. Let $\mathscr{S} = \{(V, f)\}$ be the set of all pairs (V, f) where $A \leqslant V \leqslant B$, and where $f: V \to M$ is an R-map with $f | A = \varphi$, and where the following additional property (of finite character) holds:

For any $v_1, \ldots, v_m \in V$; $b_1, \ldots, b_n \in B$; and $r_{ij} \in R$ such that $\Sigma_{i=1}^n b_i r_{ij} = v_j, j = 1, \ldots, m$; it follows that there exist $m_1, \ldots, m_n \in M$ such that $\Sigma_{i=1}^n m_i r_{ij} = f(v_j)$, $j = 1, \ldots, m$. This can be equivalently rephrased in terms of equations. For any finite system $\Sigma_{i=1}^n x_i r_{ij} = v_j \in V$, $j = 1, \ldots, m$; over V which has a solution $x_i = b_i \in B$ $(i = 1, \ldots, n)$, it follows that the induced system $\Sigma_{i=1}^n X_i r_{ij} = f(v_j) \in M$ over M has a solution $X_i = m_i \in M$ $(i = 1, \ldots, n)$.

Suppose that $A \leqslant V \leqslant B$, $f : V \to M$ with $f \mid A = \varphi$, and where $V \leqslant B$ is pure. It will be shown that in this case $(V, f) \in \mathcal{S}$. For if $\Sigma_i b_i r_{ij} = v_j \in V$, $b_i \in B$ are as above, then there exist $w_i \in V$ with $\Sigma_i w_i r_{ij} = v_j$. Hence $\Sigma_i (\varphi w_i) r_{ij} = \varphi v_j$ as required. In particular, $(A, \varphi) \in \mathcal{S} \neq \phi$. Define $(V, f) \leqslant (V', f') \in \mathcal{S}$ if $V \subseteq V'$ and $f' \mid V = f$. By Zorn's lemma, let $(V, f) \in \mathcal{S}$ be a maximal element. Suppose that $V \neq B$ and let $w \in B \backslash V$. Set $W = V + wR$.

Consider the set $\mathcal{F} = \{(\Sigma_{i=1}^n x_i r_{ij} - x r_j = v_j, j = 1, \dots, m)\}$ of all possible finite systems of equations for any n, m, where r_{ij}, $r_j \in R$, and $v_j \in V$ such that there exists at least one solution of the form $x_1 = b_1$, $x_2 = b_2, \dots, x_n = b_n$; $b_i \in B$; $x = w$. Note that $|\mathcal{F}| \leqslant (|R| + \aleph_0 + |V|)$. For each ordered set of coefficients $(\|r_{ij}\|, r_j)_{i=1,\dots,n; j=1,\dots,m}$ that appears in $\mathcal{F}$ (possibly many times) select only one m-tuple $(v_1, v_2, \dots, v_m)$, $v_j \in V$ such that $\Sigma_i b_i r_{ij} - w r_j = v_j$, $j = 1, \dots, m$; for some $b_i \in B$. Let $\mathcal{G} \subset \mathcal{F}$ denote the resulting set of equations (by replacing the b_i with x_i and w with x). Unlike $\mathcal{F}$, each element of $\mathcal{G}$ is uniquely determined by $(\|r_{ij}\|, r_j)$. Hence $|\mathcal{G}| \leqslant |R| + \aleph_0$.

Consider any finite subset of $\mathcal{G}$ consisting of a finite number of systems of equations $\{(\Sigma_i x_i r_{ij} - x r_j = v_j), (\Sigma_i x_i' r_{ij}' - x r_j' = v_j'), (\Sigma_i x_i'' r_{ij}'' - x r_j'' = v_j''), \dots\} \subset \mathcal{G}$. First of all, solve each of the finite number of systems appearing in this finite subset of $\mathcal{F}$ separately with the same value $x = w$:

(0) $\Sigma_i b_i r_{ij} - w r_j = v_j$; $x_i = b_i \in B$;
(1) $\Sigma_i b_i' r_{ij} - w r_j' = v_j'$; $x_i' = b_i' \in B$;
(2) $\Sigma_i b_i'' r_{ij} - w r_j'' = v_j''$; $x_i'' = b_i'' \in B$;
⋮   ⋮

Combine these finite number of separate systems into one single system of equations by treating all the x_i, x_i', $x_i'', \dots$ as distinct with only the variable x in common among the finite number of systems. The above shows that this new single amalgamated system also belongs to $\mathcal{G}$.

Next, consider the reduced set $\bar{\mathcal{G}}$ of equations over M obtained from $\mathcal{G}$. That is any element $(\Sigma_{i=1}^n x_i r_{ij} - x r_j = v_j; \ 1 \leqslant i \leqslant n, \ 1 \leqslant j \leqslant m) \in \mathcal{G}$, determines an element $(\Sigma_{i=1}^n X_i r_{ij} - X r_j = f v_j \in M; \ 1 \leqslant i \leqslant n, \ 1 \leqslant j \leqslant m) \in \bar{\mathcal{G}}$. Since our maximal element $(V, f) \in \mathcal{S}$, by definition of $\mathcal{S}$, the associated system $\Sigma_i X_i r_{ij} - X r_j = f v_j$ has a solution $X_i = m_i$, $X = m_0 \in M$; $i = 1, \dots, n, j = 1, \dots, m$. Exactly as was the case for $\mathcal{G}$, any finite number of separate systems of $\bar{\mathcal{G}}$ (having no variables in common except X) can be coalesced into one single system of equations over M. Hence $\bar{\mathcal{G}}$ is finitely solvable in M. At this point we invoke the $\aleph^<$-equational compactness of M to conclude that there exists a global solution of all the equations of all the systems of $\bar{\mathcal{G}}$ in M. In particular for this global solution, $X = m^* \in M$.

Define g: $W \to M$ by $g[v + wr] = fv + m^*r$ for $v + wr \in V + wR$. To see that g is well defined suppose that $v + wr = 0$. Then the singleton system of equations $xr = -v$ is in $\mathscr{F}$, because $x = w \in B$ is a solution. Hence there exists a singleton system of equations $xr = v_1$ in $\mathscr{G}$ (with the same r) which satisfies $wr = v_1$. Thus $v_1 = -v$. Hence the one equation system $Xr = -fv$ belongs to $\bar{\mathscr{G}}$, and thus $m^*r = -fv$. Thus $fv + m^*r = 0$, and g is well defined.

So far $V \subset W$, $g: W \to M$, and $g|V = f$. It remains to verify that $(W, g) \in \mathscr{S}$. Given are a finite set of equations $\Sigma_{i=1}^n x_i r_{ij} = w_j \in W$ $(j = 1, \ldots, m)$, which have a solution $x_i = b_i \in B$. Write $w_j = v_j + wr_j$, for some $v_j \in V$ and $r_j \in R$. We have to prove that $\Sigma_{i=1}^n X_i r_{ij} = gw_j = fv_j + m^*r_j$ $(j = 1, \ldots, m)$ has a solution in M. Since the system $\Sigma_i x_i r_{ij} - xr_j = v_j$ belongs to $\mathscr{F}$, there is a corresponding system $\Sigma_i x_i r_{ij} = \tilde{v}_j + xr_j$ $(j = 1, \ldots, m)$ in $\mathscr{G}$, with $\Sigma_i \tilde{b}_i r_{ij} = \tilde{v}_j + wr_j$ for a particular choice of $\tilde{v}_j \in V$, and some $\tilde{b}_i \in B$. Since the system $\Sigma_i X_i r_{ij} - Xr_j = f\tilde{v}_j$ belongs to $\bar{\mathscr{G}}$, our global solution $X_i = m_i^*$, $X = m^*$ satisfies it, i.e. $\Sigma_i m_i^* r_{ij} - m^*r_j = f\tilde{v}_j$. Next, $\Sigma_{i=1}^n x_i r_{ij} = v_j - \tilde{v}_j$ $(j = 1, \ldots, m)$ has the solution $x_i = b_i - \tilde{b}_i \in B$. Since $(V, f) \in \mathscr{S}$, there exist $\tilde{m}_i \in M$ such that $\Sigma_i \tilde{m}_i r_{ij} = fv_j - f\tilde{v}_j$. Finally, $\Sigma_i X_i r_{ij} = fv_j + m^*r_j$ $(j = 1, \ldots, m)$ has the solution $X_i = m_i^* + \tilde{m}_i$, because $\Sigma_i(m_i^* + \tilde{m}_i)r_{ij} = m^*r_j + f\tilde{v}_j + fv_j - f\tilde{v}_j$. Thus $(V, f) < (W, g) \in \mathscr{S}$ is a contraction, and M is pure injective.

18-4.7   Theorem. *Every module D can be embedded as a pure submodule $D < G$ of a pure injective module G.*

Proof. Set $\aleph = [\max(|R|, \aleph_0)]^+$. Index the set of all finitely solvable systems of equations over D containing strictly less than $\aleph$-equations by ordinals $\alpha < \nu$:

$$\sum_i x_i^\alpha r_{ij}^\alpha = d_j^\alpha \in D, \ i \in I^\alpha, j \in J^\alpha, |I^\alpha| \leqslant |J^\alpha| < \aleph; \ \alpha \in [0, \nu).$$

Note that

$$|\nu| \leqslant \left(\sup_{\alpha < \nu} |I^\alpha| |J^\alpha| \right) |R| |D| \aleph_0 \leqslant \aleph\aleph |R| |D| \aleph_0 = \aleph|D|.$$

By 18-4.2, $D < H_0 = \langle D, x_i^0, i \in I^0 | \Sigma_i x_i^0 r_{ij}^0 = d_j^0, j \in J^0 \rangle$ is pure, and the 0-th system is globally solvable in H^0. By ordinal induction assume that there is some $\beta < \nu$ such that for all $\alpha < \beta$, H_α has already been defined having the two properties that $D < H_\alpha$ is pure, and that for any $\gamma \leqslant \alpha$, the γth system of equations is globally solvable in H_α. By 18-2.2 (2)(b),

$D < \cup\{H_\alpha \mid \alpha < \beta\}$ is pure. By 18-2.2(3)(a) and 18-4.2,

$$D < H_\beta = \left\langle \bigcup_{\alpha < \beta} H_\alpha, x_i^\beta, i \in I^\beta \,\middle|\, \sum x_i^\beta r_{ij}^\beta = d_j^\beta, j \in J^\beta \right\rangle$$

is also pure, and for all $\gamma \leqslant \beta$, the γth equation can be solved in H_β. The induction is complete, and as before $D < \cup\{H_\alpha \mid \alpha < \nu\} = D_1$ is pure, and all finitely solvable systems containing less than $\aleph$-equations over D have global solutions in D_1.

Let θ be the cardinal ($=$ ordinal) number $\theta = \aleph^+$. Now repeat the above procedure but with D replaced by D_1 to obtain $D_1 < D_2$ pure, and such that any finitely solvable system of less than $\aleph$-equations over D_1 has a solution in D_2. By ordinal induction repeat this procedure, where for limit ordinals $\alpha < \theta$, D_α is the union of all of its predecessors. The result is a chain $D < D_1 \cdots < D_\alpha < \cdots$, $\alpha < \theta$, where $D < D_\alpha$ is pure, and for any $\alpha < \beta < \theta$, $D_\alpha < D_\beta$ is pure. Set $G = D_\theta = \cup\{D_\alpha \mid \alpha < \theta\}$. Suppose that we are given any finitely solvable system of equations over G indexed by $[0, \delta)$, where δ is a cardinal $\delta < \aleph$.

It is shown next that this whole system of equations is contained in some D_β where $\beta < \theta$. For each $\alpha < \delta$, the αth equation is contained in some $D_{\nu(\alpha)}$ for some $\nu(\alpha) < \theta$. Define $\beta = \sup\{\nu(\alpha) \mid \alpha < \delta\}$. The whole system is contained in $\cup\{D_{\nu(\alpha)} \mid \alpha < \delta\} \leqslant D_\beta$. At this point we have to quote the known fact from set theory that $\beta < \theta$. (See [Hrbacek and Jech 84; pp. 192–194, Theorem 2.3].) But then our system has a solution in $D_{\beta + 1}$, and hence in G. By the last Theorem 18-4.6, G is pure injective, and by 18-2.2(2)(b), $D < G$ is pure.

18-5   Pure injective hull

It will be shown that every module has a pure injective hull which will be unique up to isomorphism. An essential step in this is the result (18-4.7) that any module D can be embedded as a pure submodule in a pure injective module.

18-5.1   Definition. For any modules $K < M$, define $E(K, M) = \{L \mid L < M$; (i) $K \cap L = 0$; and (ii) $(K + L)/L < M/L$ is pure$\}$. Note that $(0) \in E(K, M)$ if and only if $K < M$ is pure. The extension $K < M$ is a *pure essential extension* if $E(K, M) = \{(0)\}$.

The next lemma merely reformulates the previous definition.

18-5.2   Lemma. *Let $K < M$. Then $K < M$ is pure essential* $\Leftrightarrow \forall R$-map

$\varphi: M \to N$ such that $\varphi K \leqslant \varphi M$ is pure, and also such that $\varphi \mid K$ is monic, necessarily φ is monic.

Proof. Without loss of generality we may assume that the above φ is surjective and that hence $\varphi: M \to N = M/\ker \varphi$ is the natural projection. Let $\mathscr{S} = \{\varphi \mid \varphi: M \twoheadrightarrow M/\ker \varphi, \ \varphi K \leqslant \varphi M$ is pure, $\varphi \mid K$ is monic$\}$. There is a bijection $E(K, M) \leftrightarrow \mathscr{S}$ of sets. For $L \in E(K, M)$, map $L \to \varphi$, where $\varphi: M \to M/L$ is the natural projection. Conversely, for $\varphi \in \mathscr{S}$, $\varphi \to \ker \varphi = L$ is the inverse of this map. Thus $E(K, M) = \{(0)\}$ translates into the fact $\mathscr{S} = \{\varphi\}$ is a singleton, where $\varphi = 1_M$ is the identity.

The second observation follows from the first one.

18-5.3 Observations. Let $K \oplus L \leqslant M$. Then

(1) $(K+L)/L \leqslant M/L$ is pure $\Leftrightarrow \forall \ \Sigma_{i=1}^n m_i r_{ij} = a_j + b_j$, $m_i \in M$, $r_{ij} \in R$, $a_j \in K$, $b_j \in L$ it follows that $\exists \ k_i \in K$ such that $\Sigma_{i=1}^n k_i r_{ij} = a_j$ for $j = 1, \ldots, m$.

Proof. (1) $\Rightarrow$: Set $\bar{m}_i = m_i + L$, $\bar{a}_j = a_j + L$, etc. From $\Sigma_{i=1}^n \bar{m}_i r_{ij} = \bar{a}_j$ and the purity of $(K + L)/L \leqslant M/L$ we conclude that there exist $\bar{k}_i = k_i + L \in (K + L)/L$, with $k_i \in K$, such that $\Sigma_{i=1}^n \bar{k}_i r_{ij} = \bar{a}_j$. Due to our specialized choice of coset representatives k_i, $a_j \in K$, we have that $\Sigma_{i=1}^n k_i r_{ij} - a_j \in K \cap L = 0$ for all j.

(1) $\Leftarrow$: Given $\Sigma_{i=1}^n x_i r_{ij} = \bar{c}_j \in (K + L)/L$, $j = 1, \ldots, m$; and a solution $x_i = \bar{m}_i = m_i + L \in M/L$, write $\bar{c}_j = a_j + L$, where a_j is the unique coset representative of $\bar{c}_j$ with $a_j \in K$. Then define b_j's by $b_j = \Sigma_{i=1}^n m_i r_{ij} - a_j \in L$. Now by hypothesis, $\Sigma_{i=1}^n k_i r_{ij} = a_j$ for some $k_i \in K$. Thus $\Sigma_{i=1}^n \bar{k}_i r_{ij} = \bar{c}_j$ for all j, and hence $(K + L)/L \leqslant M/L$ is pure.

(2) $E(K, M)$ is closed under submodules and unions of chains of submodules.

(3) Let $L \in E(K, M)$ be arbitrary. Then

$$\forall L \leqslant N \leqslant M, \ N/L \in E((K + L)/L, M/L) \Rightarrow N \in E(K, M).$$

Proof. First, $L = N \cap (K \oplus L) = (N \cap K) \oplus L$, or $N \cap K = 0$. Secondly, $[(K + L)/L] \oplus (N/L) \leqslant M/L$, and the middle extension below is pure:

$$\frac{K \oplus N}{N} \cong \frac{\dfrac{(K + L)}{L} \oplus \dfrac{N}{L}}{\dfrac{N}{L}} \leqslant \frac{\dfrac{M}{L}}{\dfrac{N}{L}} \cong \frac{M}{N}.$$

Hence $N \in E(K, M)$.

18-5.4   Lemma. *Let $K < M$ be pure and let $L \in E(K, M)$ be any maximal submodule. (By 18-5.3(2), there exist such.) Then $K \cong (K + L)/L < M/L$ is a pure essential extension.*

Proof. For any $N/L \in E((K + L)/L, M)$, the last observation 18-5.3(3) shows that $N \in E(K, M)$. Hence by the maximality of L, $N = L$. Therefore $E((K + L)/L, M/L) = \{(\bar{0})\}$, and $(K + L)/L < M/L$ is pure essential.

18-5.5   Lemma. *For modules $K < M$, suppose that $\{F_\alpha\}_\alpha$ is a chain of submodules $K \leqslant F_\alpha < M$ and set $G = \cup F_\alpha$. If each $K \leqslant F_\alpha$ is pure essential, then $K \leqslant G$ is also a pure essential extension.*

Proof. First, $K \leqslant G$ is pure by 18-2.2(2), and so $(0) \in E(K, G)$. If $0 \neq L \in E(K, G)$, then $L \cap F_\alpha \neq 0$ for some α. By 18-5.3(2), $L \cap F_\alpha \in E(K, G)$, i.e. $(L \cap F_\alpha) \cap K \subseteq L \cap K = 0$, and $(K + L \cap F_\alpha)/L \cap F_\alpha \leqslant G/L \cap F_\alpha$ is pure. By 18-2.1, also $(K + L \cap F_\alpha)/L \cap F_\alpha \leqslant F_\alpha/L \cap F_\alpha$ is pure. Consequently $L \cap F_\alpha \in E(K, F_\alpha) = \{(0)\}$, a contradiction. Thus $K \leqslant G$ is pure essential.

Zorn's lemma immediately gives the next corollary.

18-5.6   Corollary. *For any submodule $K < M$ of any module M whatever, there exists a maximal pure essential extension $K \leqslant G$ of K with $G \leqslant M$.*

18-5.7   Lemma. *Let $K < M$ be pure essential, and D a pure injective module such that $K < D$ is pure. Then the identity map $1_K: K \to K$ extends to a monic map $\varphi: M \to D$, with $\varphi | K = 1_K$, i.e. the diagram below commutes.*

$$0 \longrightarrow K \xrightarrow{\text{inc}} M$$

with vertical inclusion map inc. from K down to D and dashed diagonal map φ.

Proof. By the pure injectivity of D, there exists $\varphi: M \to D$ with $\varphi | K = \text{identity}$. Thus $K = \varphi K < \varphi M \leqslant D$. Since $K < D$ is pure, also $\varphi K < \varphi M$ is pure. By Lemma 18-3.2, φ is monic.

18-5.8   Definition. For a module K, an *absolutely maximal pure essential extension* F of K is a (i) pure essential extension $K \leqslant F$ such that (ii) for any module G, if $F \leqslant G$, and if $K \leqslant G$ is a pure essential extension of K, that then necessarily $F = G$.

18-5.9 Proposition. *For every pure essential extension $K \leqslant F$ of modules, there exists an absolutely maximal pure essential extension $K \leqslant G$ of K with $K \leqslant F \leqslant G$.*

Proof. Suppose not. Set $F_0 = F$. Then there exists a pure essential extension $K < F_1$ with $F_0 < F_1$. Assume that for some ordinal β, F_α has already been defined for all $\alpha < \beta$. If β has a predecessor $\beta - 1$, then by our hypothesis there exists a proper pure essential extension $K < F_\beta$ with $F_{\beta-1} < F_\beta$. If β is a limit ordinal, then $F_\beta = \cup\{F_\alpha \mid \alpha < \beta\}$ is a pure essential extension of K by Lemma 18-5.5.

Let D be a pure injective module containing $F < D$ as a pure submodule. It exists by Theorem 18-4.7. By Lemma 18-5.7, for ech α, the identity map $1_K: K \to K$ of K extends to a monic map $\varphi_\alpha: F_\alpha \to D$. Let the cardinality $|\beta|$ of β satisfy $|\beta| > |D|$. Then $|F_\beta| \geqslant |\beta|$, and $F_\beta \cong \varphi_\beta(F_\beta) \leqslant D$ is a contradiction. Thus there exists an ordinal β such that $F \leqslant F_\beta = G$ is an absolutely maximal pure essential extension. By 18-2.2(3)(a), $K \leqslant G$ is pure.

18-5.9 Theorem. *For an extension $K < G$ of modules the following are equivalent*:

   (i)   *G is an absolutely maximal pure essential extension of K;*
  (ii)   *G is a minimal pure injective module containing K as a pure submodule.*

Proof. First it will be shown that any absolutely maximal pure essential extension G of K is itself pure injective. It suffices to show that for any pure extension $G < H$, G is a direct summand of H. By 18-5.4 applied to $E(K, H)$ there is a surjective map $\pi: H \to \pi H$ such that $K \cong \pi K < \pi H$ is a pure essential extension. Thus $\pi K < \pi G$ is pure, and by Lemma 18-5.2, $\pi \mid G$ is monic. But then $\pi K < \pi G$ is an absolutely maximal pure essential extension. Since $\pi K < \pi G \leqslant \pi H$, $\pi G = \pi H$. Let $\rho: \pi G \to G$ be the inverse of the restriction $\pi \mid G$ of π to G. Thus ρ is an isomorphism, $\rho \pi H = G$, $\rho \pi H = G$, and $H = G \oplus (1_H - \rho \pi)H$. Hence G is pure injective.

(i) $\Rightarrow$ (ii). Suppose that $K < H \leqslant G$, where $K < H$ is pure, and H is pure injective. By applying 18-5.7 (with $K \leftrightarrow K$, $M \leftrightarrow G$, and $D \leftrightarrow H$), we obtain a monic map $\varphi: G \to H$. Thus $K < \varphi H \leqslant \varphi G \leqslant H \leqslant G$. But $K < \varphi G$ is also an absolutely maximal pure essential extension of K. Hence $\varphi G = G$, and necessarily $H = G$.

(ii) $\Rightarrow$ (i). By 18-5.8, there exists an absolutely maximal pure essential extension M of K with $K < G \leqslant M$. An application of 18-5.7 (with $D \leftrightarrow G$) gives a monic map $\varphi: M \to G$ with $\varphi \,|\, K = 1_K$. Thus $K < \varphi M \leqslant G \leqslant M$. Both $K < \varphi M$ and $K < M$ are absolutely maximal pure essential extensions of K. Hence $\varphi M = M$ and $G = M$. (Alternatively, by the first part (i) $\Rightarrow$ (ii) of the proof, both $K < \varphi M$ and $K < M$ are minimal pure injective modules containing K as a pure submodule. Hence again $\varphi M = M$, and $G = M$.) Thus G is an absolutely maximal pure essential extension of K.

18-5.10 Definition. For any right R-module K, the *pure injective hull of* K is any pure injective module G containing K as a pure submodule such that $K \leqslant G$ is a minimal pure injective extension of K.

The pure injective hull of a module and its injective hull can be of different cardinalities. Either cardinality can be the bigger one. A module need not be essential in its pure injective hull.

18-5.11 Theorem. *For any right R-module K all pure injective hulls of K are isomorphic over K.*

Proof. Let $K \leqslant G$ and $K \leqslant H$ be two pure injective hulls of K. Both of these are pure essential extensions. Then 18-5.7 gives a monic map $\varphi: G \to H$ with $\varphi \,|\, K = 1_K$. Either 18-5.9(i) or (ii) applied to $K \leqslant \varphi G \leqslant H$ implies that $H = \varphi G$, and that φ is an isomorphism.

18-5.12 Examples. Let $R = \mathbf{Z} = 0, \pm 1, \pm 2, \ldots, p$ a prime, $\mathbf{Z}(p^\infty)$ the injective hull of $\mathbf{Z}/p^n\mathbf{Z}$, for $n = 1, 2, \ldots,$ and J_p the p-adic integers. Then $\mathbf{Z}/p^n\mathbf{Z}$ and J_p are pure-injective abelian groups which are not injective. A necessary and sufficient condition that an abelian group is pure injective is that it is a direct summand of a product of $\mathbf{Z}(p^\infty)$'s. (See [Fuchs 1970; Vol. I, pp. 127, 160, and 164].)

18-5.13 Remarks. Much of the material here on pure projectives and pure injectives has been greatly influenced by Laszlo Fuchs, both by his book ([L. Fuchs 1970; Vol. I, pp. 120–128, and pp. 170–174]) as well as his lectures and courses. Particularly, the construction of the pure injective hull of a module presented here is essentially as in [L. Fuchs, 1970, Vol. I, p. 170–173].

The method of proof in Theorem 18-4.6 that (2) $\Rightarrow$ (3) is adapted from [Eklof, P. C. and Mekler, A. H, 1990; pp. 199–122].

18-6   Exercises

In the first seven exercises, $R = \mathbf{Z}$ and modules are abelian groups.

1. For abelian groups $H \subset G$, traditionally H has been defined to be pure as an abelian group in G if $H \cap nG = nH$ for every integer n. Prove that $H \subset G$ satisfies the latter condition if and only if H is pure in G as a $\mathbf{Z}$-module in the sense of Definition 18-2.1. For this exercise assume as known Kulikov's Theorem: If $H \subset A$ is pure as an abelian group, and A/H is a direct sum of cyclics, then H is a direct summand of A.

   (Hint: Given are $\Sigma_{i=1}^{n} x_i r_{ij} = d_j \in H$, $r_{ij} \in \mathbf{Z}$, $j = 1, \ldots, m$; with a solution $x_1 = g_1, \ldots, x_n = g_n$ in G. Define $A = \langle H, \{g_1, \ldots, g_n\}\rangle$. Show $H \subset A$ is a pure abelian subgroup. Write $A = H \oplus B$, and $g_i = h_i + b_i$, $h_i \in H$, $b_i \in B$. Show $x_i = h_i$ is a solution.)

2. If $H \subset G$ is pure in G then G/H is torsion free. If G is torsion free, then the converse holds.

3. A subgroup H of G is pure in G if and only if every coset of G module H can be represented by a coset representative in G of the same order as the coset.

4. In torsion free groups, intersections of pure subgroups are pure. Give an example to show that the latter fails in torsion groups.

   (Hint: Let $G = \langle a \rangle \oplus \langle b \rangle$ where a is of order p and b is of order p^2. Set $H_1 = \langle b \rangle$ and $H_2 = \langle a + b \rangle$.)

5. The groups $\mathbf{Q}$ and $\mathbf{Z}(p^\infty)$ have only trivial pure subgroups.

6. The torsion subgroup of any group G is pure in G.

7. An infinite cyclic group is not $\aleph_1^<$-equationally compact.

   (Hint: $x_0 = 2x_1$, $x_1 = 2x_2$, $x_2 = 2x_3, \ldots$ is finitely solvable but not globally.)

8. Let $\mathbf{C} = \{M_\alpha\}$ be a chain of R-modules and let $N \leqslant \cap M_\alpha$. If for each α, $N \leqslant M_\alpha$ is pure, then $N < \cup\mathbf{C} = \cup_\alpha M_\alpha$ is also pure.

9. Give a direct proof of Corollary 18-4.3 along the lines of 18-4.2.

10. Give a direct proof that for any modules $D < M$, there exists a pure submodule $C \leqslant M$ with $D \subset C$.

11*. For a right R-module M, the following are all equivalent.

   (i) Any short exact sequence of modules $0 \to M \to B \to C \to 0$, with C finitely presented splits.

   (ii) Every finite consistent system of equations over M has a solution in M.

   (iii) M is a pure submodule of an injective module.

12. For a prime p, let $R = \mathbf{Z}_{(p)} = \{a/b \mid gcd(b, p) = 1, 0 \neq b, a \in \mathbf{Z}\}$. Then the only indecomposable pure injective R-modules (up to isomorphism) are: (i) $\mathbf{Z}/p^n\mathbf{Z}$, $n \geqslant 1$, (ii) $\mathbf{Z}(p^\infty)$, (iii) $\mathbf{Q}$, and (iv) J_p, the p-adic integers.

13. A submodule $D < M$ is a direct summand of M if and only if every system of equations over D which is solvable in M is also solvable in D.

14. Prove that the following are equivalent conditions on a submodule $D < M$ of a module M.

    (i) D is a direct summand of M.
    (ii) For a free module F and any submodule $G < F$, if $g: G \to D$ is a homomorphism which is extendible to a homomorphism $f: F \to M$, then there exists an extension h of G, $h: F \to D$.
        (Hint: Choose $\{g_i \mid i \in I\} \subset M$ such that $\langle D \cup \{g_i\}\rangle = M$. Let $F = \oplus\{x_i R \mid i \in I\}$ be free on $\{x_i\}$.

15. Exercise for logicians. Let M be a module, $D \subseteq M$ a subset, and $n \geqslant 1$ a fixed integer. Consider equations $\Sigma_{i=1}^n x_i r_{ij} = d_j \in D$ ($j = 1, \ldots, m$) where $r_{ij} \in R$, $d_j \in D$ may be arbitrary. If $x_1, \ldots, x_n$ satisfy these equations, write $p(x_1, \ldots, x_n)$ and abbreviate $p = p(x_1, \ldots, x_n)$; if not, $]p =]p(x_1, \ldots, x_n)$. Similar notation is used for other similar sets of equations with the same $x_1, \ldots, x_n$ (but different r_{ij}'s and d_j's): $q = q(x_1, \ldots, x_n)$, $]q$. Now let ψ be a well formed usual propositional calculus type logical formula obtained by applying a finite number of times the following logical symbols to a finite number of p's and q's: $p \wedge q$ (p and q), $p \vee q$ (either p or q), and $]$ (negation). Write $\psi = \psi(x_1, \ldots, x_n)$. Let $\mathscr{B}$ be the set of all subsets $Y \subseteq M^n = M \times M \times \cdots \times M$ which can be defined by some such formula ψ, i.e. $Y = \{(y_1, \ldots, y_n) \mid \psi(y_1, \ldots, y_n) \text{ holds}\}$.

    (i) Show that $\mathscr{B} \subseteq \mathscr{P}(M^n)$ is a Boolean subring of the Boolean ring $\mathscr{P}(M^n)$ of all subsets of M^n.
    (ii) Let $a = \{a_1, \ldots, a_n\} \in M^n$ be fixed. Define $\mathscr{F} = \{Y \in \mathscr{B} \mid a \in Y\}$. Prove that $\mathscr{F}$ is an ultrafilter in $\mathscr{B}$.

APPENDIX A

Basics

Introduction

The reader might wish to skip this appendix, referring back to it when necessary. The most elementary properties of the axiom of choice, Zorn's lemma, ordinals, and cardinals are outlined as briefly as possible on an intuitive level. Quasi-ordered and partially ordered sets, equivalence relations, and functions are discussed.

One aim of this book is to make the reader aware that the material presented here might hold in a broader framework. For this reason some definitions and the isomorphism theorems are reviewed in the context of nonassociative rings, although the latter are not really used in this book. Another reason is that hopefully this chapter may say something new also to those readers who do not need a review.

The goal here is to supply self-contained proofs of the facts about cardinal numbers needed later in the book, and the one such result which makes this possible is that for any infinite set X, the size or cardinality of $X \times X$ equals that of X, i.e. $|X \times X| = |X|$ (A-1.11). This implies that if "$\sim$" is an equivalence relation on an infinite set X whose equivalence classes are finite, that then the cardinality of the set of equivalence classes $X/\sim$ is the same as that of X, that is $|X/\sim| = |X|$ (A-1.12). We will already use the latter fact in Chapter 2 to show that for infinite rank free modules the rank is well defined.

A-1 Sets, symbols, and functions

A-1.1 The usual set theoretic symbols are used: $\subset, \subseteq, \supset, \supseteq, \cup, \cap, \in$ (belongs to, or is a member of), $\exists$ (there exists), $\exists !$ (there exists a unique);
398

! – unique; ∀ (for any; for every; for all), $\emptyset$ is the empty set; $Y \times Z$ is the direct product of sets; if $Y \subseteq X$, $X \backslash Y = \{x \mid x \in X, x \notin Y\}$.

$A \Leftarrow B$ or $A \rightarrow B$ (A implies B). $A \Leftrightarrow B$ or $A \leftrightarrow B$ (A holds if and only if B holds). "If and only if will sometimes be abbreviated as "iff". In longer proofs where there are main implications (like theorems) and minor implications (lemmas), it is sometimes convenient to use "$\Rightarrow$" for major, and "$\rightarrow$" for minor implications, all perhaps in the same proof.

Some texts such as the Bourbaki use only $\subset$ (where $A \subset B$ means $A = B$ or A is contained in B). Usually the case when $A = B$ is either trivial, uninteresting, or it just does not make any difference whether $A = B$ or $A \subseteq B$ for the purposes of the argument. In this book the symbols "$\subseteq$" and "$\subset$" will be used more or less the same way as the usual order. Relations "$\leqslant$" (less than or equal to), and "$<$" (strictly less than). Here, the following modus operandi will be adopted. "Do not become a slave to the symbols but rather let the symbols be your servants." Therefore, on those rare occasions where this difference is actually important, the reader will be warned by something like this: $A \subset B$ but $A \neq B$. On the other hand, in a small number of places where a certain proof is trivial in case $A = B$, "$A \subset B$" might be used, even though strict logic would dictate "$A \subseteq B$".

A slash through a symbol indicates negation: $\notin$, $\not\subset$, $\not\subseteq$, $\not>$, $\neq$, $\nexists$ (there does not exist), $\nRightarrow$ (does not imply).

A-1.2   Functions. Let $A, B \subseteq X$ and $C, D \subseteq Y$ be sets, and $f: X \rightarrow Y$ a function. As usual

$$Y \backslash D = \{y \in Y \mid y \notin D\}; \ f^{-1}(D) = f^{-1}D = \{x \in X \mid f(x) = fx \in D\}.$$

Here $D \backslash f(X) \neq \emptyset$ is allowed. The restriction of f to A is written $f \mid A$, thus $f \mid A : A \rightarrow Y$. The following hold.

(i) $f^{-1}(C \cap D) = (f^{-1}C) \cap (f^{-1}D)$.
(ii) $f(A \cap B) \subseteq (fA) \cap (fB)$; $(fA) \cap (fB) \backslash f(A \cap B)$
      $= \{y \in Y \mid \exists (a, b) \in (A \backslash B) \times (B \backslash A), \ y = fa = fb\}$.

A-1.3   Order relations. A *relation* on a set X is a subset of $X \times X$. However if (a, b) belongs to this relation we will instead write "$a \leqslant b$".

A relation "$\leqslant$" on a set X is upper *directed* if for every $a, b \in X$, there exists a $c \in X$ with $a \leqslant c$ and $b \leqslant c$.

A *quasi-ordered* set is a pair $(X, \leqslant)$ consisting of a set X and a relation "$\leqslant$" satisfying the following for all $a, b, c \in X$:

(i) $a \leqslant a$ (reflexive);

(ii) $a \leqslant b, b \leqslant c \to a \leqslant c$ (transitive).

A partially ordered set (abbreviated *poset*) is a quasi-ordered set $(X, \leqslant)$ where in addition to (i) and (ii), also (iii) holds for all $a, b \in X$,

(iii) $a \leqslant b, b \leqslant a \to a = b$ (antisymmetric).

A poset is linearly or totally ordered if any two elements are comparable; and a totally ordered set X is *well ordered* if every possible subset of X actually contains within itself a smallest element.

The axiom of choice, Zorn's lemma, and the well orderability of any set are all equivalent. All three will be used later.

A-1.4   Axiom of choice. If $\varnothing \neq X_i, i \in I$, are non-empty sets indexed by a set I, then there exists a function $f : I \to \cup \{X_i \mid i \in I\}$ such that $f(i) \in X_i$ for each i.

A-1.5   Zorn's lemma. *Suppose that X is a partially ordered set satisfying the following:*

(i) $X \neq \varnothing$;
(ii)   *every totally ordered subset of X has an upper bound,*
(iii)   *and that upper bound is an element of X.*

Then X contains at least one maximal element.

A-1.6   Every set can be well ordered.

A-1.7   Ordinals. The ordinal numbers may be viewed as sets: $0 = \varnothing$, $1 = \{\varnothing\}$, $2 = \{\varnothing, \{\varnothing\}\}$, $3 = \{\varnothing, \{\varnothing\}, \{\varnothing, \{\varnothing\}\}\}$, etc. For any two ordinals α and β, either $\beta \subseteq \alpha$, or $\alpha \subsetneqq \beta$. In the later case by definition, $\alpha \ngeqslant \beta$. Any ordinals α, β satisfy the following:

(i) $\alpha \subsetneqq \beta \Leftrightarrow \alpha \ngeqslant \beta \Leftrightarrow \alpha \in \beta$;
(ii) $\beta = \cup \{\alpha \mid \alpha \ngeqslant \beta\}$;
(iii) $\beta = \{\alpha \mid \alpha < \beta\} \equiv [0, \beta)$.

Every ordinal α has a successor $\alpha + 1$, and α is called the predecessor (or ancestor) of $\alpha + 1$. A limit ordinal is one not of the form $\alpha + 1$.

A-1.8   Cardinals. If X is a finite set, then the cardinality of X is simply the number of elements in X. For any set X the cardinality of X can be defined as the class of all sets Z such that there exists some bijective function $h: X \to Z$. For any two sets define "$|X| \leqslant |Y|$" to mean that there exists a monic function $f: X \to Y$. Define $|X| < |Y|$ if $|X| \leqslant |Y|$, but *not* $|Y| \leqslant |X|$. Lastly $|X| = |Y|$ is defined to mean that there is a bijection between X and Y. If $|X| \leqslant |Y|$ and $|Y| \leqslant |X|$, then it is a consequence of the next theorem that $|X| = |Y|$.

A-1.9 Schröeder-Bernstein theorem. *If* $f: X \to Y$ *and* $g: Y \to X$ *are monic functions of sets X and Y, then there exists a bijection* $h: X \to Y$.

A-1.10 An *initial ordinal* γ is one such that for any ordinal $\delta < \gamma$, $|\delta| < |\gamma|$. An initial ordinal is always a limit ordinal. The initial ordinals are customarily written as $\omega_0 = \omega < \omega_1 < \cdots < \omega_\alpha < \cdots$ where the subscript α ranges over all the ordinals.

For any infinite set X, there exists a unique initial ordinal ω_α and a bijective function $X \to \{\gamma \,|\, \gamma < \omega_\alpha\} = [0, \omega_\alpha) = \omega_\alpha$. If Y is another infinite set with a corresponding bijection $Y \to \omega_\beta$ then $|X| \leqslant |Y|$ if and only if $\omega_\alpha \leqslant \omega_\beta$ if and only if $[0, \omega_\alpha) \subseteq [0, \omega_\beta)$ (if and only if $\alpha \subseteq \beta$). In view of the above one-to-one correspondence $|X| \leftrightarrow \omega_\alpha$ it is convenient to view the cardinal numbers as that subset of the ordinal numbers consisting of all finite ordinals and the initial ordinals ω_α. Yet we wish to keep the two separate. Thus if $X \leftrightarrow \omega_\alpha$, then we will write $|X| = |\omega_\alpha|$. In other words, ω_α is an ordinal while $|\omega_\alpha|$ is the corresponding cardinal. Note that $|0| = |\varnothing| = 0$.

The following general situation occurs frequently and is sometimes hidden because of various additional specialized hypotheses.

The statement (but not the proof) of the next proposition will be needed for the proof of A-1.13 and A-1.14.

A-1.11 Proposition. *If X is any infinite set, then*

(1) $\quad |X \times X| = |X|$.

Next, let $(X, \leqslant)$ be well ordered. Define a relation $\prec$ on $(X \times X) \times (X \times X)$ as follows. For $a = (a_1, a_2)$, $b = (b_1, b_2) \in X \times X$ with $a \neq b$, define $a \prec b$ if and only if one of the following hold

(i) $\max\{a_1, a_2\} < \max\{b_1, b_2\}$; or
(ii) $\max\{a_1, a_2\} = \max\{b_1, b_2\}$ and $a_1 < b_1$; or lastly
(iii) $\max\{a_1, a_2\} = \max\{b_1, b_2\}$, $a_1 = a_2$, but $a_2 < b_2$.

Then

(2) $\prec$ *defines a well ordering of* $X \times X$.

Proof. (2) For any arbitrary $a \neq b \in X \times X$, exactly one of the above three mutually exclusive alternatives (i), (ii), or (iii) must hold. Thus $a \prec b$, or $b \prec a$.

To see that $\prec$ is transitive, take $a \prec b$, and $b \prec c = (c_1, c_2) \in X \times X$. Without loss of generality, let $a \neq b$ and $b \neq c$. By (1)(i), (ii), and (iii) above, $\max\{a_1, a_2\} \leqslant \max\{b_1, b_2\} \leqslant \max\{c_1, c_2\}$. If (i) $\max\{a_1, a_2\} < \max\{c_1, c_2\}$, then $a \prec c$. So assume $\max\{a_1, a_2\} = \max\{b_1, b_2\} = \max\{c_1, c_2\}$. Then in view of (ii) and (iii), $a_1 \leqslant b_1 \leqslant c_1$, and thus $a_1 \leqslant c_1$. When $a_1 < c_1$, then $a \prec c$. Otherwise $a_1 = b_1 = c_1$, as well as $\max\{a_1, a_2\} = \max\{b_1, b_2\} = \max\{c_1, c_2\}$. Since $a \neq b$ and $b \neq c$, necessarily we have $a_2 < b_2 < c_2$. Hence $a \prec c$ once more.

To prove that $\prec$ well orders $X \times X$, we have to show that any subset $\varnothing \neq Y \subset X \times X$ contains an $\prec$-smallest element. Since X is well ordered, the subset $\{\max(x_1, x_2) | (x_1, x_2) \in Y\}$ has a unique smallest element d. Define $Z = \{(z_1, z_2) \in Y | \max\{z_1, z_2\} = d\}$. Thus $Z \neq \varnothing$. For $y = (y_1, y_2) \in Y \backslash Z$, $d < \max\{y_1, y_2\}$ and hence $z < y$ for every $z \in Z$. This means that if Y contains a smallest element, then it is in Z. Thus it suffices to show that Z contains a smallest element.

Let e be the smallest element of $\{z_1 | \exists (z_1, z_2) \in Z\}$. Define $\varnothing \neq W \subset Z$ by $W = \{(x_1, x_2) \in Z | x_1 = e\}$. For any $(e, x_2) \in W$, and any $(z_1, z_2) \in Z \backslash W$, since $\max\{e, x_2\} = \max\{z_1, z_2\} = d \in X$, and $e \nleqq z_1$, we have $(e, x_2) \prec (z_1, z_2)$. Thus so far $W \subset Z \subset Y \subset X \times X$, and Y contains a smallest element if and only if W does. Finally let f be the smallest element contained in the set $\{x_2 | (e, x_2) \in W\}$. Then $(e, f) \in W$, and (e, f) is the smallest element of W. Thus $(e, f) \in Y$ and (e, f) is the smallest element of Y. Hence $X \times X$ is well ordered by $\prec$.

(1) By well ordering X, we may assume that $X = \omega_\alpha$ for some initial ordinal ω_α. If X is countable, then a well-known diagonal argument shows that $|\omega \times \omega| = |\omega|$. By ordinal induction assume that $|\omega_\beta \times \omega_\beta| = |\omega_\beta|$ for all $\beta < \alpha$. It has to be shown that $|\omega_\alpha \times \omega_\alpha| = |\omega_\alpha|$. By way of contradiction, suppose that $|\omega_\alpha \times \omega_\alpha| > |\omega_\alpha|$. Since $(\omega_\alpha, <)$ is order isomorphic to an initial segment of $(\omega_\alpha \times \omega_\alpha, \prec)$, there exists an $(a_1, a_2) \in \omega_\alpha \times \omega_\alpha$ such that the set

$$Y = \{(y_1, y_2) \in \omega_\alpha \times \omega_\alpha | (y_1, y_2) \prec (a_1, a_2)\}$$

has cardinality $|Y| \geqslant |\omega_\alpha|$. Define a_3 to be the successor ordinal of $\max(a_1, a_2)$, i.e. a_3 is the smallest ordinal such that $a_1 < a_3$ and $a_2 < a_3$. Since ω_α is a limit ordinal, $a_3 \in \omega_\alpha$. For any $(y_1, y_2) \in Y$, $\max\{y_1, y_2\} \leqslant$

$\max\{a_1, a_2\} < a_3$. Consequently $y_1 < a_3$ and $y_2 < a_3$. By A-1.7, $y_1 \in a_3$, $y_2 \in a_3$ and thus $(y_1, y_2) \in a_3 \times a_3$. Hence $Y \subseteq a_3 \times a_3$. Since ω_α is an initial ordinal (see A-1.10), $|a_3| < |\omega_\alpha|$. Let $|a_3| = |\omega_\beta|$ where $\beta < \alpha$. The cardinality of $a_3 \times a_3$ is also the same as the cardinality of $\omega_\beta \times \omega_\beta$. Then $|Y| \leqslant |a_3 \times a_3| = |\omega_\beta \times \omega_\beta| \leqslant |\omega_\beta|$ by induction. Thus $|\omega_\alpha| \leqslant |Y| \leqslant |\omega_\beta| < |\omega_\alpha|$, where $\beta < \alpha$, is a contradiction. Thus $|X \times X| = |X|$.

The diagram below illustrates the well ordering $\prec$ on $\omega \times \omega$.

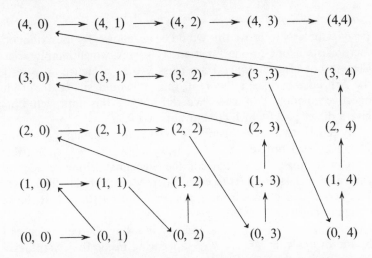

A-1.12 Suppose that $f: X \to Y$ is any function whatever. For $x, z \in X$, define an equivalence relation "$\sim$" on X by $x \sim z$ if $f(x) = f(z)$. Let $\bar{x} = \{z \mid x \sim z\}$ denote the equivalence class containing x. If $X/\sim = \{\bar{x} \mid x \in X\}$ is the set of equivalence classes, let $\pi: X \to X/\sim$ be the so-called projection map $\pi x = \bar{x}$. The function $g: X/\sim \to Y$ defined by $g(\bar{x}) = g(x)$ for any $x \in \bar{x}$ is injective, that is one to one. Thus for any function f whatever, there is an associated diagram

which is commutative, i.e. $f = g\pi$.

In view of the bijection $g: X/\sim \to f(X)$, it is useful to know that every equivalence class $\bar{x}$ is uniquely of the form $\bar{x} = f^{-1}(\{y\}) \equiv f^{-1}(y)$ for some unique $y \in f(X)$. If $Y \neq f(X)$, and $w \in Y \backslash f(X)$ is arbitrary, then $\varnothing = f^{-1}(w) \notin X/\sim$.

Although stated independently, the following lemma is sometimes used in the preceding context involving a function $f: X \to Y$.

A-1.13 Lemma. *Suppose that X is an infinite set, that "$\sim$" is an equivalence relation in X such that every equivalence class is finite. Then*

$$|X/\sim| = |X|.$$

Proof. One way to prove this would be as follows. By the axiom of choice there exists a one-to-one function $X/\sim \to X$ which simply selects one element out of each equivalence class $\bar{x} \to z \in x$. Hence $|X/\sim| \leqslant |X|$. By the Schröder-Bernstein theorem, then it would suffice to produce an injective function $X \to X/\sim$. We will not use this approach, but rather give a direct proof which circumvents A-1.9.

By A-1.10, there is a bijection $X \to [0, \tau)$ for some limit ordinal τ. Thus X can be well ordered $X = \{x_0, x_1, x_2, \ldots, x_\alpha, \ldots \mid \alpha < \tau\}$. Furthermore, we can arrange X so that all the elements belonging to the same equivalence class are next to each other. More precisely the subscripts of every equivalence class form a convex subset of $[0, \tau)$. List the smallest element of every equivalence class $x_0 = x_{n(0)}, x_{n(1)}, x_{n(2)}, \ldots, x_{n(\alpha)}, \ldots$, with $n(0) < n(1) < \cdots < n(\alpha) < \ldots$, thereby indexing the equivalences classes by the ordinals $0, 1, 2, \ldots, \alpha, \ldots$ appearing inside the parenthesis. Thus $X/\sim = \{\tilde{x}_{n(0)}, \bar{x}_{n(1)}, \bar{x}_{n(2)}, \ldots, \bar{x}_{n(\alpha)}, \ldots \mid$ all $\alpha, n(\alpha) < \tau\}$. Let $x \in X$ be arbitrary. Then for some α, $x \in \bar{x}_{n(\alpha)} = \{x_{n(\alpha)}, x_{n(\alpha)+1}, x_{n(\alpha)+2}, \ldots, x_{n(\alpha)+m}\}$ with $x = x_{n(\alpha)+i}$ for $0 \leqslant i \leqslant m$, where the equivalence class of x contains $m + 1$ elements. Define a one-to-one function $f: X \to (X/\sim) \times \omega = (X/\sim) \times \{0, 1, 2, \ldots\}$ by $f(x) = (\bar{x}_{n(\alpha)}, i)$. Thus $|X| \leqslant |(X/\sim) \times \omega| \leqslant |(X/\sim) \times (X/\sim)| = |X/\sim|$ where the last equality follows by the last proposition. Thus $|X/\sim| = |X|$.

Next A-1.12 and A-1.13 are combined to obtain a fact which is used later.

A-1.14 Suppose that $f: X \to Y$ is any function of an infinite set X into a set Y and assume that the inverse image $f^{-1}(\{y\}) \subset X$ of every $y \in Y$ is a finite set $|f^{-1}(\{y\})| \ngtr \infty$. Then

(i) $|Y| = \infty$;
(ii) $|X| = |f(X)| \leqslant |Y|$; and in particular
(iii) if $f(X) = Y$, then $|X| = |Y|$.

Proof. In the notation of A-1.12, there is a bijection $g: X/\sim \to f(X) \subseteq Y$

and hence $|Y| \geqslant |f(X)| = |X/\sim| = |X|$, where the last equality follows from A-1.13.

A-2   Background review

Those readers familiar with basic notations like ring, subring, ideal, quotient ring, and ring homomorphism are advised to go directly to Chapter 1, which at the expense of some very slight repetition is self contained and independent of this section. Basic notations from commutative or abelian group theory are assumed, like subgroup, quotient group, group homomorphism, and the isomorphism theorems.

A-2.1   An associative binary composition law on a set S is a function satisfying the following

$$S \times S \to S, (a, b) \to ab \in S; a, b, c \in S, (ab)c = a(bc).$$

Here the composition law has been written multiplicatively. A set S equipped with such an associative composition law is called a *semigroup*. A *monoid* is a semigroup S which contains an identity element $e \in S$ such that for all $a \in S$, $ae = ea = a$; a multiplicative identity is also called a unity element.

If G is either a commutative or noncommutative group, then $G \times G \to G, (a, b) \to ab^{-1}$ is an example of a nonassociative composition law, if G is nontrivial.

A-2.2   Definition. An additive abelian group R, $+$ together with a binary composition law $R \times R \to R, (a, b) \to ab \in R; a, b \in R$ is a *ring* if it satisfies the following associative and distributive axioms for all a, b and $c \in R$

(i) $(ab)c = a(bc)$;
(ii) $c(a + b) = ca + cb$;
(iii) $(a + b)c = ac + bc$.

An element u of a ring R is called a *left identity* element if $ur = r$ for all $r \in R$. An element $e \in R$ is the *identity* element or the *unity* element if e is both a left and a right identity, $ea = ae$ for all $a \in R$, usually denoted simply as $e = 1$.

For sake of a wider perspective, the reader may find it illuminating that some parts of this book dealing with associative rings, hold in a more general context. Nonassociative rings not only form an important branch of ring theory, but sometimes associative ring can be constructed out of nonassociative ones.

A-2.3   Definition. A *nonassociative* ring is an additive group R and binary composition law $R \times R \to R$ satisfying (ii) and (iii) of the previous definition.

A clearer name but one that is not used because of its length is "a not necessarily associative ring". Every associative ring is automatically also a nonassociative ring. Also the term "associative ring" is almost never used. A nonassociative ring may or may not satisfy (i) of A-2.2, whereas a ring always does.

A-2.4   Definition. A subset $S \subseteq R$ of a nonassociative ring is a *subring* of R if S is closed under the addition and multiplication inherited from R. That is, for every choice of a, b, c, and $d \in S$, also $a + b \in S$ and $cd \in S$.

An additive subgroup B of a nonassociative ring R is a *right ideal* if $br \in B$ for every $b \in B$ and $r \in R$; a left ideal is defined similarly. An additive subgroup I of R that is simultaneously both a right and a left ideal is called an *ideal*. Ideals in any ring will sometimes be denoted by "$\lhd$", e.g. $I \lhd R$, where equality is allowed.

A left or a right ideal is sometimes called a one-sided ideal, whereas the term two-sided ideal is rarely used. Among very many ring theorists the word "ideal" automatically means two sided. This historically traditional terminology is unfortunate in the sense that one-sided ideals seem to be used more frequently than ideals, violating the rule that the most frequently used words should be the shortest.

A-2.5   A *homomorphism* $\phi: R \to S$ of nonassociative rings R and S is a group homomorphism ϕ of the additive groups of R and S such that $\phi(ab) = (\phi a)(\phi b)$ for all $a, b \in R$. The *kernel* of ϕ is the ideal kernel $\phi = \ker \phi = \{r \in R \,|\, \phi r = 0\}$. The homomorphism ϕ is *one-to-one* or *monic* if $\phi a = \phi b$ for $a, b \in R$ implies that $a = b$; ϕ is monic if and only if $\ker \phi = \{0\}$. The *image* of ϕ is the subring image $\phi = \operatorname{im} \phi = \phi R = \{s \in S \,|\, s = \phi a \text{ for some } a \in R\} \subseteq S$. The map ϕ is *onto* if $\phi R = S$. An *isomorphism* is a monic onto map, in which case R and S are said to be isomorphic, sometimes denoted by $R \cong S$.

The homomorphism ϕ is said to preserve identities if (i) both R and S possess identity elements 1_R and 1_S and (ii) $\phi(1_R) = 1_S$.

The analogues of the isomorphism theorems will not be repeated for right R-modules over an associative ring, which are defined in the next

chapter. The next four isomorphism theorems will hold verbatim if
systematically throughout every occurrence of the word "ideal" is replaced by "right ideal" and "isomorphism of nonassociative rings" by
"isomorphism of right R-modules".

A-2.6   Definition. If I is an ideal in a nonassociative ring R, the *quotient
ring* of R modulo I is the ordinary abelian quotient group R/I whose
elements are cosets $a + I = \{a + i \mid i \in I\}$, and which is equipped with the
multiplication defined by coset representatives, $a, b \in R$ as follows:

$$\frac{R}{J} \times \frac{R}{J} \to \frac{R}{I}, \qquad (a + I)(b + I) = ab + I.$$

The *natural or quotient map* $\pi: R \to R/I$, $\pi r = r + I$, $r \in R$, is a
homomorphism of nonassociative rings R and R/I.

Remark. If N is a normal subgroup of any commutative or noncom-
mutative group G and $\bar{x}, \bar{y} \in G/N$, then the product $\bar{x}\bar{y}$ in G/N can be
defined naturally as a product of sets, $\bar{x}\bar{y} \equiv \{ab \mid a \in \bar{x}, b \in \bar{y}\}$ without
reference to coset representative of $\bar{x}$ and $\bar{y}$. For quotient rings, this is no
longer possible; if $\bar{x}, \bar{y} \in R/I$ then the inclusion $\{ab \mid a \in \bar{x}, b \in \bar{y}\} \subseteq \bar{x}\bar{y}$ may
very well be proper.

A-2.7   1st isomorphism theorem. If $\phi: R \to S$ is any homomorphism of
nonassociative rings R and S, let $\pi: R \to R/\ker \phi$ be the natural map
$\pi r = r + \ker \phi$, $r \in R$. Then $\bar{\phi}: R/\ker \phi \to \phi R$ defined by $\bar{\phi}(r + \ker \phi) = \phi r$
is an isomorphism such that $\phi = \bar{\phi}\pi$, and there is a commutative diagram

$$
\begin{array}{ccc}
R & \xrightarrow{\phi} & S \\
{\scriptstyle\pi}\downarrow & & \uparrow{\scriptstyle\text{inc.}} \\
\dfrac{R}{\ker \phi} & \xrightarrow{\bar{\phi}} & R
\end{array}
$$

A-2.8   2nd isomorphism theorem. If $I, J \subseteq R$ are ideals of a nonassociative ring R, with $J \subseteq I$, then I/J is an ideal of R/J and there is a ring
isomorphism:

$$\frac{R/J}{I/J} \cong \frac{R}{I}, \qquad (r + J) + \frac{I}{J} \to r + I; \quad r \in R.$$

A-2.9   3rd isomorphism theorem. *If I and J are ideals of a nonassociative ring R, then $I + J = \{i + j \mid i \in I, j \in J\}$ is an ideal of R, J is an ideal of the nonassociative ring $I + J$ and there is an isomorphism*

$$\frac{I + J}{J} \to \frac{I}{I \cap J}, \quad i + j + J \to i + I \cap J; \quad i \in I, j \in J.$$

Although infrequently used, the last isomorphism theorem is stated only for sake of completeness. Below, the second isomorphism is obtained by systematically interchanging A and B in the first.

A-2.10   4th isomorphism theorem. *If A^*, A, B^*, and B are four subrings of a nonassociative ring R such that $A^* \lhd A$ and $B^* \lhd B$ are ideals, then there are isomorphisms of nonassociative rings*

$$\frac{A^* + A \cap B}{A^* + A \cap B^*} \cong \frac{A \cap B}{A \cap B^* + A^* \cap B} \cong \frac{B^* + A \cap B}{B^* + A^* \cap B}$$

$$a^* + c + [A^* + A \cap B^*] \to c + [A \cap B^* + A^* \cap B];$$

$$b^* + c + [B^* + A^* \cap B] \to c + [A \cap B^* + A^* \cap B];$$

$$a^* \in A, \ c \in A \cap B, \ b^* \in B^*;$$

where the numerators are subrings of R, and the denominators are ideals in the numerators.

A-3   Exercises

Notation. Let $f: X \to Y$ be a function of sets and $\{X_\alpha\}$ and $\{Y_\beta\}$ arbitrary indexed families of subsets of X and Y where α and β range over some index sets which have been suppressed.

1. For $A, B \subseteq Y$, show that $f^{-1}[A \setminus B] = (f^{-1}A) \setminus (f^{-1}B)$.
2. Prove that

$$f[\cup X_\alpha] = \cup (f X_\alpha) \qquad f^{-1}(\cup Y_\beta) = \cup (f^{-1} Y_\beta)$$
$$f[\cap X_\alpha] \subseteq \cap (f X_\alpha) \qquad f^{-1}(\cap Y_\beta) = \cap (f^{-1} Y_\beta)$$

3. Establish generalized De Morgan's laws:

$$X \setminus \cup X_\alpha = \cap (X \setminus X_\alpha)$$
$$Y \setminus \cap Y_\beta = \cup (Y \setminus Y_\beta)$$

4. Consider the natural numbers $0 = \emptyset$, $1 = \{0\} = \{\emptyset\}$, $2 = \{0, 1\} = \{\emptyset, \{\emptyset\}\}$, $3 = \{0, 1, 2\} = \{\emptyset, \{\emptyset\}, \{\emptyset, \{\emptyset\}\}\}, \ldots \ n =$

$\{0, 1, \ldots, n - 1\}$. Show that (i) for any $a, b \in n$, either $a \in b$, $a = b$, or $b \in a$; (ii) and $n + 1 = n \cup \{n\}$.

5. A set X is *transitive* if for all $y \in X$, $y \subset X$. Show the following
   (a) Every ordinal is transitive.
   (b) Every transitive set containing two elements is of the form $\{y, \{y\}\}$.
   (c) There exists a transitive set built up out of "$\varnothing$" and "$\{\ \}$" with three elements, but unequal to 3.
   (d) The set in (c) does not have the property (i) of problem 4.
   (e) Is the set $\{\varnothing, \{\varnothing\}, \{\{\varnothing\}\}, \{\{\{\varnothing\}\}\}, \{\varnothing, \{\varnothing\}\}, \{\varnothing, \{\{\varnothing\}\}\}\}$, transitive? Is it a natural number?

6. Let $\mathbb{N} = 0, 1, 2, \ldots$ Give several different proofs that $|\mathbb{N} \times \mathbb{N}| = |\mathbb{N}|$. (Hint: If $f(i, j) = 2^i(2j + 1) - 1$, then $f : \mathbb{N} \times \mathbb{N} \to \mathbb{N}$ is a one-to-one and onto mapping.)

7. Let X be any set, $\mathscr{P}(X)$ the set of all subsets of X (including $\varnothing$, $X \in \mathscr{P}(X)$), and $f : \mathscr{P}(X) \to \mathscr{P}(X)$ any function such that whenever, $A \subset B \subseteq X$, then $f(A) \subseteq f(B)$. Let $\mathscr{F} = \{C \in \mathscr{P}(X) \mid C \subseteq f(C)\}$. Set $D = \cup \mathscr{F}$. Prove that $f(D) = D$.

8. Use the last problem to prove the Schröder-Bernstein Theorem (A.1.9) as follows. Let $f : X \to Y$ and $g : Y \to X$ both be one-to-one functions (i) For any $A \subseteq X$, define $A_* = X \setminus g[Y \setminus f(A)]$. Show that if $A \subset B$, then $A_* \subseteq B_*$. (ii) Show that there exists a subset $D \subseteq X$ such that

$$D = X \setminus g[Y \setminus f(D)] \qquad \text{and} \qquad X \setminus D = g[Y \setminus f(D)]$$

Let $h : X \to Y$ by $h \mid D = f$ and $h \mid (X \setminus D) = g^{-1}$. Show that $h : X \to Y$ is a well-defined one-to-one function. Hint:

X Y

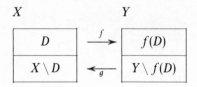

Definition. Let X be any infinite set and $\mathscr{P}(X)$ the set of all subsets of X with $\varnothing$, $X \in \mathscr{P}(X)$. A set $\mathscr{F}$ of subsets of X (i.e. $\mathscr{F} \subseteq \mathscr{P}(X)$) is a *filter* on X if the following hold:
(a) $\varnothing \notin \mathscr{F}$;
(b) If $E_1, E_2 \in \mathscr{F}$, then $E_1 \cap E_2 \in \mathscr{F}$.
(c) If $E \in \mathscr{F}$ and $E \subset \bar{E} \subseteq X$, then $\bar{E} \in F$.

An *ultrafilter* is a maximal filter, i.e. one that is not properly contained in a bigger filter on X.

A filter $\mathcal{F}$ on X is *prime* if for any E_1, $E_2 \subseteq X$ such that $E_1 \cup E_2 \in \mathcal{F}$, it follows that at least one of E_1 or E_2 belongs to $\mathcal{F}$.

9. Prove that $\{X\}$ is a filter.

10. Prove that a filter $\mathcal{F}$ is an ultrafilter $\Leftrightarrow \forall Y \subset X$, either $Y \in \mathcal{F}$, or $X \setminus Y \in F$.

11. Let $G \subseteq X$. Show that $\{E \in \mathcal{P}(X) \mid G \subseteq E\}$ is a filter. It is called a *principal filter*, or the principal filter generated by G. Prove that it is an ultrafilter if and only if G is a singleton.

12. Prove that every set of subsets of X which is closed under finite intersections is contained in a filter on X.

13. Prove that every filter $\mathcal{F}$ on X is contained in an ultrafilter on X. Show that there exist nonprincipal ultrafilters.

14. Suppose that E_1, $E_2 \subseteq X$ and that $E_1 \cup E_2 \in \mathcal{F}$ which is a filter. Show that (i) either $E_1 \cap F \neq \varnothing$ for all $F \in \mathcal{F}$, or (ii) that $E_2 \cap F \neq \varnothing$ for every $F \in \mathcal{F}$. Now use the above to show that every ultrafilter is prime.

15. Prove that $\mathcal{P}(X)$ is a ring under the following operations:

$$A \cdot B = A \cap B, \; A + B = A \cup B \setminus A \cap B$$
$$1 = X, \, 0 = \varnothing; \, A, \, B \in \mathcal{P}(X).$$

(The above set $A + B$ is sometimes called the symmetric difference.) Show that the following hold.

(i) $A^2 = A$ and $A + A = 0$; $A \cup B = A + B + A \cdot B$.
(ii) $\forall Y \subset X$, $\mathcal{P}(Y) \lhd \mathcal{P}(X)$ is an ideal.
(iii) $\mathcal{P}(X)/\mathcal{P}(Y) \cong \mathcal{P}(X \setminus Y)$.

Definition. Let X be an infinite set. A *set ideal* on X is a subset $\mathcal{I} \subseteq \mathcal{P}(X)$ satisfying the following

(a*) $X \notin \mathcal{I}$;
(b*) If G_1, $G_2 \in \mathcal{I}$, then $G_1 \cup G_2 \in \mathcal{I}$.
(c*) If $G \in \mathcal{I}$, and $X \supset \tilde{G} \subset G$, then $\tilde{G} \in \mathcal{I}$.

16. Let $\mathcal{F}$ be any filter and $\mathcal{I}$ any set ideal on X. (i) Show that $\mathcal{F}^* \equiv \{X \setminus E \mid E \in \mathcal{F}\}$ is a set ideal on X (called the dual ideal of $\mathcal{F}$). (ii) Show that $\mathcal{I}^* \equiv \{X \setminus G \mid G \in \mathcal{I}\}$ is a filter on X.

17. Prove that any set ideal $\mathcal{I}$ on X is an ideal $\mathcal{I} \lhd \mathcal{P}(X)$ in the ring $\mathcal{P}(X)$ as in exercise 15. Conversely, given any ideal $\mathcal{J} \lhd \mathcal{P}(X)$, show that it is a set ideal on X.

18. Show that $\mathcal{F} = \{E \subset X \mid |X \setminus E| < \infty\}$ is a filter. Show that $\mathcal{P}(X)/\mathcal{F}^*$ does not have any nonzero minimal ideals.

19. Let $\emptyset \neq \mathcal{S} \subset \mathcal{P}(X)$ be a subset such that $X \notin \mathcal{S}$, and for any A, $B \in \mathcal{S}$ there exists a $C \in \mathcal{S}$ such that $A \cup B \subset C$. Prove that the set $\mathcal{I} = \{D \subset X \mid \exists A \in \mathcal{S}, D \subseteq A\}$ is an ideal in the ring $\mathcal{P}(X)$, and moreover the smallest ideal containing $\mathcal{S}$. Formulate and prove a dual problem for $\mathcal{S}^*$ and $\mathcal{I}^*$.

20. Suppose that X is not countable and define $\mathcal{I}_\omega = \{A \subset X \mid A \text{ is countable}\}$. (i) Prove that $\mathcal{I}_\omega \lhd \mathcal{P}(X)$. (ii) Now suppose that κ is some infinite cardinal number such that $\kappa < |X|$. Show that $\mathcal{I}_\kappa \equiv \{B \subset X \mid |B| \leqslant \kappa\} \lhd \mathcal{P}(X)$.

21. Prove: $\mathcal{F} \subset \mathcal{P}(X)$ is an ultrafilter $\Leftrightarrow \mathcal{F}^* \lhd \mathcal{P}(X)$ is a proper maximal ideal.

22. Let $\mathcal{F}$ be the principal filter generated by a set of the form $X \setminus Y$ for some $Y \subset X$, $Y \neq X$. Describe $\mathcal{F}^*$. When is $\mathcal{F}^*$ a proper maximal ideal of $\mathcal{P}(X)$?

APPENDIX B

Certain important algebras

Introduction

The bare essentials of free, tensor, and exterior algebras were given in Chapter 5. This appendix gives two additional applications of finite dimensional exterior algebras. The first one is a necessary and sufficient condition on a finite set of vectors in the underlying vector space V to be independent. The other one is the Laplace expansion of a determinant. From there on we drop the restrictive hypothesis on V that it be finite dimensional, and give a different and far more detailed account of the tensor algebra on V. Then an alternate unified construction of all the basic algebras as quotients of the tensor algebra is given. Thus the tensor algebra serves as a central focus for building various algebras, and in this way this appendix gives a new perspective.

B-2 Exterior algebras

Since this appendix is an addendum to and also a continuation of Chapter 5, we simply will continue the numbering of paragraphs from Chapter 5, and assume that the reader is familiar with 5-1.11 through 5-2.20.

The symbol S_r will be used to denote the group of all permutations of $\{1, 2, \ldots, r\}$; S_r is also frequently called the symmetric group on r elements.

B-2.1   For $1 \leqslant r \leqslant N$ and any r vectors $h_1, \ldots, h_r \in V$

$$h_1, \ldots, h_r \text{ are dependent} \Leftrightarrow h_1 h_2 \cdots h_r = 0$$

Proof. $\Rightarrow$. The notation can be so chosen that $h_1 = c_2 h_2 + \cdots + c_r h_r$, $c_i \in F$. Then

$$h_1 h_2 \cdots h_r = \sum_{i=2}^{r} c_i h_i h_2 \cdots h_i \cdots h_r = 0$$

$\Leftarrow$. If not, $h_1, \ldots, h_r$ are linearly independent. Express $h_i = \Sigma\, c_{ik(i)} u_{k(i)}$ where the summation index $k(i)$ ranges over $1 \leqslant k(i) \leqslant N$, and $c_{ik(i)} \in F$. If we were dealing with only one h_i we would not have to index the dummy summation variable with i. However

$$0 = h_1 h_2 \cdots h_r$$

$$= \sum_{k(1)=1}^{N} \sum_{k(2)=1}^{N} c \sum_{k(r)=1}^{N} c_{1k(1)} c_{2k(2)} \cdots c_{rk(r)} u_{k(1)} u_{k(2)} \cdots u_{k(r)}$$

Fix $T = \{j(1) < \cdots < j(r)\}$ and collect together in the inner sum below all those terms above which are a permutation of T, $u_{k(1)} u_{k(2)} \cdots u_{k(r)} = u_{j(1\sigma)} u_{j(2\sigma)} \cdots u_{j(r\sigma)}$, where $\sigma \in S_r$. Then

$$0 = h_1 h_2 \cdots h_r$$

$$= \sum_{\substack{T \subset X, |T|=r \\ T=\{j(1)<\cdots<j(r)\}}} \sum_{\sigma \in S_r} c_{1j(1\sigma)} c_{2j(2\sigma)} \cdots c_{rj(r\sigma)} u_{j(1\sigma)} u_{j(2\sigma)} \cdots u_{j(r\sigma)}$$

$$= \sum_{T \subseteq X, |T|=r} u_T \det C_{\{1\cdots r\} \times \{j(1)\cdots j(r)\}}\,,$$

where

$$C_{\{1,\ldots,r\} \times \{j(1),\ldots,j(r)\}} = \begin{vmatrix} C_{1j(1)} C_{1j(2)} \cdots C_{1j(r)} \\ C_{2j(1)} C_{2j(2)} \cdots C_{2j(r)} \\ C_{rj(1)} \, C_{rj(2)} \, \cdots C_{rj(r)} \end{vmatrix}$$

Recall the very elementary fact that for any $r \times N$ matrix such as $C = \|c_{ij}\|$, the row rank of C equals the column rank of C which equals the size of the largest square submatrix of C of nonzero determinant.

If $h_1, \ldots, h_r$ are linarly independent, then the row rank of C equals r. Hence there exists an $r \times r$ submatrix of C of the above form $C_{\{1,\ldots,r\} \times \{j(1),\ldots,j(r)\}}$ with nonzero determinant. Since the set

$$\{u_T \mid T \subseteq X, |T| = r\}$$

is linearly independent, the last equation above is a contradiction. Thus $h_1, \ldots, h_r$ are dependent.

B-2.2   Laplace expansion of a determinant. Fix r, $1 \leqslant r \leqslant N - 1$, and fix a subset $S = \{i(1) < i(2) < \cdots i(r)\} \subset \{1, \ldots, N\}$; S^c will denote the complement of S, $S^c = \{1, \ldots, N\} \backslash S = \{i(r+1) < \cdots < i(N)\}$. Thus $S \cap S^c = \varnothing$ and $S \cup S^c = \{1, \ldots, N\}$. In any product of the u_i an interchange is the operation of interchanging any two (not necessarily adjacent) u_i, e.g. $\cdots u_p \cdots u_i \cdots u_q \to \cdots \to u_q \cdots u_i \cdots u_p$. A transposition is the interchange of two adjacent u's, e.g. $\cdots u_i u_j \cdots \to \cdots u_j u_i \cdots$. In order to keep track of minus signs, it will suffice to determine only the parity of either the total number of transpositions or interchanges. Write $Lu_{i(1)} = \Sigma a_{i(1)k(1)}u_{k(1)}$, where $k(1)$ ranges over $1 \leqslant k(1) \leqslant N$, and similarly for the other $u_{i(p)}$ and $k(p)$. Then

$$\hat{L}u_S = \hat{L}(u_{i(1)}\cdots u_{i(r)}) = (Lu_{i(1)})(Lu_{i(2)}) \cdots (Lu_{i(r)})$$

$$= \sum_{k(1)=1}^{N} \cdots \sum_{k(r)=1}^{N} a_{i(1)k(1)}a_{i(2)k(2)}\cdots a_{i(r)k(r)}u_{k(1)}u_{k(2)}\cdots u_{k(r)}$$

Fix $T = \{j(1) < \cdots < j(n)\} \subset X$. There are many terms above such that $\{k(1), k(2), \ldots, k(r)\} = T$. Thus

$$\hat{L}u_S = \sum_{\substack{T \subset X, |T|=r \\ T = \{j(1) < \cdots < j(r)\}}} \sum_{\sigma \in S_r} a_{i(1)j(1\sigma)}\cdots a_{i(r)j(r\sigma)}u_{j(1\sigma)}\cdots u_{j(r\sigma)}$$

$$= \sum_{j(1) < \cdots < j(r)} \sum_{\sigma \in S_r} a_{i(1)j(1\sigma)}a_{i(2)j(2\sigma)}\cdots u_{i(r)j(r\sigma)}(\text{sgn }\sigma)u_T$$

$$= \sum_{\substack{T \\ |T|=r}} u_T \det L_{S,T}.$$

where $L_{S,T}$ is the $r \times r$ submatrix of $\|a_{ij}\|$ determined by S and T. Similarly

$$\hat{L}u_{S^c} = \sum_{\substack{P \\ |P|=N-r}} u_P \det L_{S^c,P}.$$

Now the Laplace expansion is derived by comparing the results of computing the same expression two ways.

(1) $\hat{L}(u_S u_{S^c}) = \hat{L}[(-1)^{\varepsilon_{s,s^c}}u_1 u_2 \cdots u_N]$
$= (-1)^{\varepsilon_{s,s^c}}(Lu_1)(Lu_2)\cdots(Lu_N)$
$= (-1)^{\varepsilon_{s,s^c}}(\det L)u_1 u_2 \cdots u_N.$

(2) $\hat{L}(u_S u_{S^c}) = (\hat{L}u_S)(\hat{L}u_{S^c})$
$= \left[\sum_{T,|T|=r} (\det L_{S,T})u_T\right]\left[\sum_{P,|P|=N-r} (\det L_{S^c,P})u_P\right].$

After cross multiplication of the last two sums most of the terms are zero,

$$u_T u_P = \begin{cases} 0 & P \neq T^c \\ u_T u_{T^c} = (-1)^{\varepsilon_{T,T^c}} u_1 u_2 \cdots u_N. \end{cases}$$

Thus (2) becomes

(2) $\hat{L}(u_S u_{S^c}) = \displaystyle\sum_{T=\{j(1)<\cdots<j(r)\}} (-1)^{\varepsilon_{T,T^c}} (\det L_{S,T})(\det L_{S^c,T^c})$

Comparison of the exponents ε_{S,S^c} and ε_{T,T^c} of minus one in (1) and (2) gives

$$\varepsilon_{S,S^c} = (i(1) - 1) + (i(2) - 2)\cdots(i(r) - r)$$

$$= i(1) + i(2) + i(r) - \frac{r(r+1)}{2}$$

$$\varepsilon_{T,T^c} = (j(1) - 1) + (j(2) - 2)\cdots(j(r) - r)$$

$$= j(1) + j(2) + j(r) - \frac{r(r+1)}{2}.$$

Since the exponent of -1 can be changed modulo two, comparison of (1) and (2) shows that

$$(-1)^{i(1)+\cdots+i(r)-\frac{r(r+1)}{2}} (\det L) u_1 u_2 \cdots u_N$$

$$= \sum_{T=\{j(1)<\cdots<j(r)\}} (-1)^{j(1)+\cdots+j(r)-\frac{r(r+1)}{2}} (\det L_{S,T})(\det L_{S^c,T^c}),$$

$$\det L = \sum_{1 \leqslant j(1)<\cdots<j(r)\leqslant N, T=\{j(1)<\cdots<j(r)\}} (\det L_{S,T})(\det L_{S^c,T^c})(-1)^{i(1)+\cdots+i(r)+j(1)+\cdots+j(r)}$$

Since the determinant of the transpose of a matrix is equal to the determinant of its transpose, an analogous expansion holds if we fix r columns $T = \{j(1) < \cdots < j(r)\}$, and sum above on

$$S = \{i(1) < \cdots < i(r)\}.$$

If $r = 1$, the usual expansion of a determinant according to row $i(1)$ is recovered.

B-3   A unified approach

Here an alternate basis free construction of the tensor algebra will be given, and then the symmetric and exterior algebras will be realized as quotients of the tensor algebra constructed in this particular intrinsic way. In this section the underlying vector space V could be either finite or infinite dimensional. An attempt will be made to keep this section independent of the previous sections on free, tensor, and exterior algebras. Thus those readers who wish a short, direct, and concrete construction of the tensor and finite dimensional exterior algebras will find it in 5-1 and 5-2. However, those readers who prefer a more abstract unified construction, need only read this appendix to Chapter 5, and may omit the rest of Chapter 5.

Thus the tensor algebras on V is the key ingredient from which other useful algebras can be constructed, not only the symmetric and exterior algebras, but some others which are not treated here in depth but are merely mentioned; such as the Clifford algebra, and the Lie enveloping algebra if V is a Lie algebra.

B-3.1   Terminology. Here F will be a fixed field and all algebras will be algebras over F with identity. Throughout this section, all algebra homomorphisms will be F-linear ring homomorphisms and will map identity elements to identity elements, i.e. will preserve the identity. They will be written on the left.

First a generally useful elementary lemma is noted.

B-3.2   Lemma. *For any direct sum of R-modules of the form*

$$H = \oplus\{H_i \,|\, i \in I\} \subset G = \bigoplus\{G_i \,|\, i \in I\},$$

where $H_i \subseteq G_i$ is a submodule for all i, there exists a natural isomorphism $G/H \cong \oplus\{G_i/H_i \,|\, i \in I\}$.

Proof. For any $g = \{g_i\}_{i \in I} \in G$, map $g \to \{g_i + H_i\}_{i \in I}$.

B-3.3   Definition. Let Γ be any (finite or infinite) additive abelian group with zero element $\theta \in \Gamma$. A Γ-*graded algebra* is an algebra A which is a vector space direct sum $A = \Sigma\,\{A_\gamma \,|\, \gamma \in \Gamma\}$ of subspaces A_γ such that $A_\alpha A_\beta \subseteq A_{\alpha+\beta}$ for all $\alpha,\ \beta \in \Gamma$. That is $x \in A_\alpha$, $y \in A_\beta$ implies that $xy \in A_{\alpha+\beta}$.

Suppose that $A = \Sigma\,A_\gamma$ and $B = \Sigma\,B_\gamma$ are two algebras graded by the same group Γ. A linear map $f : A \to B$ is *homogeneous of degree* γ, if $fA_\alpha \subseteq B_{\alpha+\gamma}$ for all $\alpha \in \Gamma$. A homomorphism of graded algebras $\phi : A \to B$ is one, which in addition maps all A_γ into $\phi A_\gamma \subseteq B_\gamma$. In other words an algebra homomorphism ϕ is a homomorphism of graded algebras, provided in addition ϕ is homogeneous of degree θ.

A vector subspace $C \subseteq A$ of a graded algebra A is *homogeneous* if $C = \Sigma C \cap A_\gamma$. An element $x \in A$ is homogeneous if $x \in A_\gamma$ for some $\gamma \in \Gamma$, in which case x is said to be homogeneous of degree γ.

Whenever Γ is the additive group of integers (or any totally ordered group), then the *degree* of $0 \neq a \in A$ is $\deg a = n$ if

$$a \in \oplus\{A_i \mid i \leqslant n\} \backslash \oplus \{A_i \mid i < n\}.$$

In the next paragraphs, $A = \Sigma\{A_\gamma \mid \gamma \in \Gamma\}$ is an algebra graded by the group Γ.

B-3.4   Lemma. *The identity element of a graded algebra A is homogeneous of degree θ; moreover so are also the scalars $F = F \cdot 1 \subseteq A_\theta$.*

Proof. Write $1 = \Sigma e_\gamma$, $e_\gamma \in A_\gamma$. For any $\alpha \in \Gamma$ and an arbitrary $x \in A_\alpha$, $x = \Sigma x e_\gamma = \Sigma e_\gamma x \in A_\alpha$ implies that $x = e_\theta x = x e_\theta$. Thus $1 = e_\theta$. Since A_θ is an F-subspace, $F = F \cdot 1 \subseteq A_\theta$.

B-3.5   Lemma. *If $H \lhd A$ is a homogeneous algebra ideal, then A/H also becomes a Γ-graded algebra in a natural way.*

Proof. In order to show that $\Sigma[(A_\gamma + H)/H]$ is a vector space direct sum, take any finite number $n + 1$ of distinct elements $\gamma(0) = 0$, $\gamma(1) = 1, \ldots, \gamma(n) = n \in \Gamma$.

Any $z \in (A_0 + H) \cap (A_1 + \cdots + A_n + H)$ is of the form

$$z = -a_0 + h' = a_1 + \cdots + a_n + h'', \; a_i \in A_i; \; h', h'' \in H.$$

Then $h = h'' - h' = h_0 + h_1 + \cdots + h_n$, $h_i \in H \cap A_i$. Thus $(a_0 + h_0) + (a_1 + h_1) + \cdots + (a_n + h_n) = 0$, and the directness of $\Sigma A_\gamma = \oplus A_\gamma$ shows that $a_i + h_i = 0$ for all $i = 0, 1, \ldots, n$. Hence $z = -a_0 + h' = h_0 + h' \in H$ as required.

It is instructive to give an alternate proof of the latter based on a double application of the modular law. First, since

$$H \subset A_1 + \cdots + A_n + H, (A_1 + \cdots + A_n + H) \cap (H + A_0)$$
$$= H + (A_1 + \cdots + A_n + H) \cap A_0.$$

It suffices to show that the latter intersection is contained in H. Set

$$H^* = H \cap \sum \{A_\gamma \mid \gamma \neq 0, 1, \ldots, n\}.$$

Then by the homogeneity of H, $H = (A_1 + \cdots + A_n) \cap H + H \cap A_0 + H^*$, and consequently

$$(A_1 + \cdots + A_n + H) \cap A_0 = (A_1 + \cdots + A_n + H \cap A_0 + H^*) \cap A_0.$$

Since $H \cap A_0 \subset A_0$, by the modular law the latter intersection equals $H \cap A_0 + (A_1 + \cdots + A_n + H^*) \cap A_0 = H \cap A_0$ as required..

The above two proofs that $A/H = \Sigma[(A_y + H)/H] = \oplus[(A_y + H)/H]$ is a vector space direct sum did not use the fact that H is an ideal, but merely that $H \subset A$ is homogeneous vector subspace. The fact that $H \lhd A$ is an ideal is only now required to conclude that the above direct sum makes A/H into a Γ-graded algebra.

B-3.6   Corollary. *For $H \lhd A$ as above, the natural abelian group isomorphism $(A_y + H)/H \cong A_y/(H \cap A_y)$ endows the external direct vector space sum $\oplus[A_y/(H \cap A_y)]$ with a Γ-graded algebra structure from its identification with A/H.*

In particular, if

$$\hat{x} = \sum(x_y + H \cap A_y) \quad and \quad \hat{y} = \sum(y_y + H \cap A_y) \in \oplus[A_y/(H \cap A_y)]$$

are arbitrary elements, then their product $\hat{x}\hat{y}$ is

$$\hat{x}\hat{y} = \sum \left[\sum \{x_\alpha y_\beta \,|\, \alpha + \beta = \gamma\} + H \cap A_y \right]$$

independent of the particular choice of coset representatives $x_y, y_y \in A_y$ of $\hat{x}$ and $\hat{y}$.

B-3.7   Notation. Now V will be a vector space of arbitrary finite or infinite dimension over a field F. All tensor products (denoted by "$\_\otimes\_$") are over F unless otherwise stated. An "F-map" means an F-linear transformation, and the conventions of B-3.1 will be continued.

At this point the following should be recalled: Definition 5-1.3 of a tensor algebra (T, ρ) on V, $\rho : V \to T$; and its uniqueness 5-1.4 over V.

5-3.8   Tensor algebra. Define an F-vector space TV and subspaces $T^nV \subset TV$ as follows: $T^0V = F$, $T^1V = V$, $T^2V = V \otimes V, \ldots,$ $T^nV = V \otimes T^{n-1}V$, etc., and $TV = \oplus_0^\infty T^nV$ as their vector space direct sum.

For any $1 \leqslant i, j$ if

$$t^i = a_1 \otimes \cdots \otimes a_i \in T^iV; \quad t^j = b_1 \otimes \cdots \otimes b_j \in T^jV; \quad a_p, b_q \in V;$$

and

$$k \in F \quad \text{then } kt^i = (ka_1) \otimes \cdots \otimes a_i = a_1 \otimes (ka_2) \otimes \cdots \otimes a_i = \cdots$$

is already defined. Define the product $t^i t^j \in TV$ to be

$$t^i t^j = a_1 \otimes \cdots \otimes a_i \otimes b_1 \otimes \cdots \otimes b_j \otimes T^{i+j}V.$$

In order to be able to use the above definition of $t^i t^j$ even if one or both of i, j are zero, for any vector space A, identify $F \otimes A = A = A \otimes F$ as usual by $k \otimes a = ka = a \otimes k$. If it is assumed that $t^i t^j$ is well defined, then by linearity and distributivity this extends to a binary composition law on TV. The next argument shows that this is a well defined multiplication which turns TV into an associative algebra.

For any vector spaces A, B and C, identify $(A \otimes B) \otimes C = A \otimes (B \otimes C)$. Take $\alpha \in T^i V$, $\beta \in T^j V$, $\gamma \in T^k V$ arbitrarily. Then $T^i V \otimes T^j V = T^{i+j}V$, and $\alpha \otimes \beta$ is a well defined element of $T^{i+j}V$. Since

$$(T^i V \otimes T^j V) \otimes T^k V = T^i V \otimes (T^j V \otimes T^k V),$$

also $(\alpha \otimes \beta) \otimes \gamma = \alpha \otimes (\beta \otimes \gamma)$. Thus TV is an associative algebra.

Furthermore TV is graded by the additive group of integers $\mathbb{Z} = 0, \pm 1, \ldots$, provided that for $i < 0$, we define $T^i V = 0$; clearly $T^i V T^j V = T^{i+j}V$ for all $i, j \in \mathbb{Z}$.

Note that a typical element $\alpha \in T^i V$ is a sum of elements of the above form t^i, where the coefficient of each t^i may be taken to be 1. If dim ≥ 2, then there are elements α of $T^2 V$ which are not equal to any element of the form $a_1 \otimes a_2$ for any choice of $a_i \in V$. Note that in general $a_1 \otimes a_2 \neq a_2 \otimes a_1$. In fact, $a_1 \otimes a_2 = a_3 \otimes a_4$ for $0 \neq a_i \in V$ if and only if there are scalars $c, d \in F$ such that $a_1 = ca_3$, $a_2 = da_4$, and $cd = 1$.

For any right module M over any ring R and any subsets $X \subseteq M$, $Y \subseteq R$, define $X * Y \subseteq M$ to be the subset $X * Y = \{xy \mid x \in X, y \in Y\}$. Thus $X * Y \subseteq XY$, where the latter of course consists of all finite sums of elements of $X * Y$. Define $V^{*2} = V * V$, and $V * V * \cdots * V = V^{*i} \subset T^i V$ for i factors.

B-3.9 In particular, if $M = R = A$ is an algebra and $X = Y$ above, similarly define

$$Y^{*0} = \{1\}, \ Y^{*1} = Y, \ Y^{*2} = Y * Y, \ldots,$$

and

$$Y * Y * \cdots * Y = Y^{*i}$$

for i factors.

Now assume that the set $Y \cup \{1\}$ "generates A as an algebra" which means that the smallest subalgebra of A containing $Y \cup \{1\}$ is all of A, or alternatively $A = \Sigma \{FY^{*i} \mid i = 0, 1, 2, \ldots\}$, which happens if and only if

$\{1\} \cup \{Y^{*i} \,|\, i = 1, 2, \ldots\}$ contains a vector space basis of A.

(i) Then if $H \lhd A$ is any algebra ideal, the set

$$\{1 + H\} \cup \{y + H \,|\, y \in Y\}$$

generates A/H as an algebra.

(ii) Any algebra homomorphism $f : A \to B$ of A into any algebra B is completely determined by the values fY of f on Y. In particular, if $f, g : A \to B$ are two algebra maps which agree $f \,|\, Y = g \,|\, Y$ on Y, then necessarily $f = g$.

(iii) For $A = TV$ the set $Y = V$ has the above two desirable properties (i) and (ii). Furthermore, if $Z \subset TV$ is any set of the form $Z = \cup_1^\infty Z_i$ where each $Z_i \subset T^i V$, then the smallest algebra ideal $H \lhd TV$ containing Z is homogeneous, and in this case TV/H is also a graded algebra.

B-3.10 Lemma. *For any vector space V over F (of arbitrary dimension), the above algebra TV (in 3.8) together with the natural inclusion $\rho : V \to TV$, $\rho V = V$ is the tensor algebra on V.*

Proof. In order to verify that $V \subset TV$ satisfies the universal property 5-1.3, suppose that $\phi : V \to A$ is any linear map of V into an algebra A. If $\{u_\alpha\} \subset V$ is any F-vector space basis of V, then

$$\{t^i = u_1 \otimes \cdots \otimes u_i \,|\, u_j \in \{u_\alpha\}\}$$

is a basis of $T^i V$ for $1 \leqslant i$. First, set

$$\bar{\phi} 1 = 1_A \in A \quad \text{and} \quad \bar{\phi} t^i = (\phi u_1)(\phi u_2) \cdots (\phi u_i);$$

secondly, extend $\bar{\phi}$ to a linear map $\bar{\phi} : T^i V \to A$ for all $0 \leqslant i$; and lastly $\phi : \oplus_0^\infty T^i V \to A$ to a linear map which by its construction is an algebra homomorphism.

By definition of $\bar{\phi}$, its restriction to $V \subset TV$ is $\bar{\phi} \,|\, V = \phi$. If $f : TV \to A$ is another algebra homomorphism such that $f \,|\, V = \phi$, then $f = \bar{\phi}$, because the smallest subalgebra of TV containing V is all of TV.

B-3.11 Definition. A *symmetric* algebra on a vector space V is a pair (S, μ), where S is an algebra and $\mu : V \to S$ is a monic linear map satisfying the following universal property.

For any algebra A and any linear map $\phi : V \to A$ such that $(\phi v)(\phi w) = (\phi w)(\phi v)$ for all $v, w \in V$, it follows that there exists a unique algebra homomorphism $\tilde{\phi} : S \to A$ such that $\tilde{\phi} \mu = \phi$, i.e. the diagram below commutes.

An equivalent definition is that an algebra S is the symmetric algebra on V, if $V \subset S$, and if for any linear map $\phi: V \to C$ into any commutative algebra C, there exists a unique algebra homomorphism $\tilde{\phi}: S \to C$ which extends ϕ, $\tilde{\phi} \mid V = \phi$.

Another frequently used name for the symmetric algebra is the polynomial algebra on V. Clearly, any two symmetric algebras on V are isomorphic over V.

B-3.12 Symmetric algebra. Let H be the smallest algebra ideal of TV containing the set $Y = \{v \otimes w - w \otimes v \mid v, w \in V\}$, i.e. H is the intersection of all the algebra ideals of TV containing Y. Set $SV = TV/H$, and let $\pi: TV \to SV$ be the natural projection. Let $\mu = \pi\rho: V \to SH$, where as before $\rho: V \to TV$ is the natural inclusion map. Since $H \subseteq \oplus_2^\infty T^i V$ and $V \cap H = 0$, $\mu V = (V + H)/H \cong V$, and μ is monic. Note that the algebra SH is commutative, because

$$(v + H)(w + H) = v \otimes w + H$$

$$= v \otimes w - v \otimes w + w \otimes v + H$$

$$= (w + H)(v + H) \; \forall \, v, w \in V.$$

By 5-3.9(iii), SV is a graded algebra.

B-3.13 Lemma. *The previous algebra (SV, μ) is the symmetric algebra on V.*

Proof. If $\phi: V \to A$ is given as in B-3.11, then ϕ extends to $\bar{\phi}: TV \to A$. Since

$$\bar{\phi}(v \otimes w - w \otimes v) = (\bar{\phi}v)(\bar{\phi}w) - (\bar{\phi}w)(\bar{\phi}v)$$

$$= (\phi v)(\phi w) - (\phi w)(\phi v) = 0 \text{ for all } v, w \in V,$$

$\bar{\phi}$ vanishes on Y and hence on H. Thus $\bar{\phi}$ factors uniquely through π, i.e. there is an algebra homomorphism $\tilde{\phi}: SV \to A$ such that $\bar{\phi} = \tilde{\phi}\pi$. Then $\tilde{\phi}\mu = \tilde{\phi}\pi\rho = \bar{\phi}\rho = \phi$. By B-3.9, the extension $\tilde{\phi}$ is unique.

B-3.14 Corollary. *If dimension* $V = \dim_F V = N$ *is finite, then* $SV \cong F[x_1, \ldots, x_N]$ *where the latter is the commutative polynomial ring over F in N algebraically independent variables* $x_1, \ldots, x_N$, *and where* $\pi V = Fx_1 + \cdots + Fx_N$.

Proof. If $\{u_1, \ldots, u_N\}$ is any basis of V, then clearly from its construction, the tensor algebra TV can be taken as the free algebra $F\{u_1, \ldots, u_N\}$ in N noncommuting indeterminates u_i, and where $\rho : V \to F\{u_1, \ldots, u_N\}$ is the inclusion map induced by $\rho u_i = u_i$. But then H is the ideal containing the N^2 elements $\{u_i u_j - u_j u_i \,|\, 1 \leqslant i, j \leqslant N\}$. Thus as before

$$SV = F\{u_1, \ldots, u_N\}/H = F[x_1, \ldots, x_N],$$

where $x_i = u_i + H$, and $\mu u_i = \mu \rho u_i = x_i$.

B-3.15 Examples. If $\dim_F V = 0$ or $\dim_F V = 1$, then in B-3.12 the set $Y = \{0\} = H$ and $TV \cong SV$. More explicitly, if

   (1) $V = (0)$, $TV = SV = F$; and if
   (2) $\dim_F V = 1$, $V = Fx$, $x \in V$, then $TV = SV = F[x]$

is the ordinary polynomial ring over F in one transcendental indeterminate x.

B-3.16 Exterior algebra. We define five subsets of the tensor algebra TV; first Y^i, Y, $\bar{Y}$, and then H, and $\bar{H}$.
   (1) For any $i \geqslant 2$, Y^i is the set of all elements τ of the form

$$\tau = v_1 \otimes \cdots \otimes v_k \otimes \cdots \otimes v_s \cdots \otimes v_i,$$

all $v_j \in V$, $k \neq s$, but $v_k = v_s = v$.
   Alternatively,

$$\tau = a \otimes v \otimes b \otimes v \otimes c$$

where

$$a \in V^{*p} \subseteq T^p V, \, b \in V^{*q} \subseteq T^q V, \, c \in V^{*r} \subseteq T^r V,$$

$$p + q + r = i - 2 \geqslant 0, \, 0 \leqslant p, q, r.$$

   (2) Set $Y = \cup_2^\infty Y^i$; and
   (3) $\bar{Y} = \{(pv)^2 = v^2 \,|\, v \in V\} \subset V^{*2} \subset T^2 V$. These sets generate homogeneous algebra ideals:
   (4) $H \lhd TV$, H is the ideal generated by Y; and

(5) $\bar{H} \lhd TV$ the one generated by $\bar{Y}$. It will be shown that $TV/H = ,TV/\bar{H}$ is the exterior algebra.

Note that above in (4), H is simply the vector space spanned by Y.

B-3.17 Lemma. *In the above notation, $H = \bar{H}$.*

Proof. Since $Y^2 = \bar{Y} = \{v \otimes v \,|\, v \in V\} \subset Y$, $\bar{H} \subseteq H$. For the converse, it suffices to show that $Y \subset \bar{H}$. For any v, $w \in V$, $v \otimes w = -w \otimes v + [(v + w) \otimes (v + w) - v \otimes v - w \otimes w]$, where the last term in the square brackets belongs to $\bar{H}$. The word τ in B-3.16(1) — by successive interchanges of two adjacent letters — now becomes

$$\tau = (-1)^{\deg b} a \otimes b \otimes v \otimes v \otimes c + \bar{h}$$

for some $\bar{h} \in \bar{H}$. Thus $\tau \in \bar{H}$, also $Y \subset \bar{H}$, and $H = \bar{H}$.

B-3.18 Lemma. *Let A be any F-algebra, and $\phi : V \to A$ an F-vector space map. Assume that $(\phi v)^2 = 0$ for any $v \in V$. Then $(\phi v)(\phi w) = (-\phi w)(\phi v)$ for all $v, w \in V$.*

Proof. Set $\bar{v} = \phi v$. Thus $0 = [\phi(v + w)]^2 = (\bar{v} + \bar{w})^2 = \bar{v}\bar{w} + \bar{w}\bar{v}$.

B-3.19 Theorem. *If $\rho : V \to TV$ is the tensor algebra, and if $\bar{H} \lhd TV$ is the smallest algebra ideal containing $\{(\rho v)^2 \,|\, v \in V\}$, set $EV = TV/\bar{H}$, let $\pi : TV \to EV$ be the natural quotient map, and define $\eta : V \to EV$ as $\eta = \pi\rho$. Then*

(1) *$\bar{H}$ consists of all F-linear combinations of elements τ of the form*

$$\tau = a \otimes v \otimes b \otimes v \otimes c, \; v \in V; \; a, b, c \in TV;$$

$\bar{H}$ is homogeneous, and hence EV is graded.

(2) *η is monic; $(\eta v)\eta w = -(\eta w)\eta v$ for any $v, w \in V$.*

(3) *$\eta : V \to EV$ is the exterior algebra on V.*

Proof. (1) Conclusion (1) follows from $\bar{H} = H$ as established in 3.16 and 3.17.

(2) Since $\bar{H} \subseteq \oplus_2^\infty T^i V$, $V \cong \pi\rho V = (V + H)/H$, and η is monic. Since $(\eta v)^2 = (\pi\rho v)\pi\rho v = \pi[(\rho v)^2] = 0$, by Lemma B-3.18, $(\eta v)\eta w = -(\eta w)\eta v$.

(3) Given a linear map $\phi : V \to A$ into an algebra A with $(\phi v)^2 = 0$ for all $v \in V$, it has to be shown that there exists an algebra homomorphism $\hat{\phi} : EV \to A$ unique with the property that $\phi = \hat{\phi}\eta$. The plan of a two step proof will be to first show in step one that there exist algebra homomor-

phisms $\bar\phi$ and $\hat\phi$, and three commutative diagrams

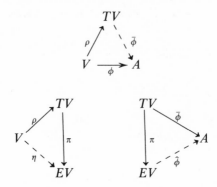

The second diagram commutes simply by the definition of η as $\eta = \pi\rho$. The first one commutes by the universal property of TV, which gives $\bar\phi$ with $\phi = \bar\phi\rho$. Since $\bar Y = \{(\rho v)^2 \mid v \in V\} \subset$ kernel $\bar\phi$, also $\bar H \subseteq \ker \bar\phi$. Thus $\bar\phi$ factors as $\bar\phi = \hat\phi\pi$ for an algebra map $\hat\phi : EV \to A$, giving the third commutative diagram. The second step of the proof now is to show that the above three diagrams coalesce into a single commutative diagram

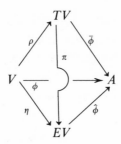

or very simply that $\phi = \hat\phi\eta$. Replacement of $\bar\phi$ in $\phi = \bar\phi\rho$ with $\bar\phi = \hat\phi\pi$ gives $\phi = \hat\phi\pi\rho$. However, $\pi\rho = \eta$, and thus $\phi = \hat\phi\eta$ as required. Since any algebra homomorphism $f : EV \to A$ with $\phi = f\eta$ agrees with $\hat\phi$ on ηV — i.e. $f\eta = \phi = \hat\phi\eta$, or $f \mid (\eta V) = \hat\phi \mid \eta V$ — it follows easily from B-3.9(i) and (ii) that $f = \hat\phi$.

B-3.20 Alternate description of EV. For $H = \bar H$ as before, set $H^i = H \cap T^iV$. Thus $H = \oplus_0^\infty H^i \lhd TV$, with $H^0 = H^1 = 0$, where H^i consists exactly of all F-linear combinations of elements of the form $\tau = a \otimes v \otimes b \otimes v \otimes c$ as in B-3.16 with $\deg \tau = i$.

By way of the natural abelian group isomorphism $EV \cong \oplus_0^\infty T^iV/H^i$, the latter inherits a unique ring multiplication from the former. Let $\Lambda^iV = T^iV/H^i$ for $2 \leqslant i$. Since $H^0 = H^1 = 0$, we may take $\Lambda^0V = F$ and $\Lambda^1V = V$. Whereas in this section EV will stand for cosets $EV = TV/H$ modulo H, define ΛV to be $\Lambda V = \oplus_0^\infty \Lambda^iV$, where the associative ring multiplication in ΛV will be denoted by "$\ldots \wedge \ldots$"; in TV — by "$\ldots \otimes \ldots$", and in EV — by juxtaposition.

Take $k \in F$; $v, w \in V$; $t^i = a_1 \otimes \cdots \otimes a_i \in T^iV$ and $t^j = b_1 \otimes \cdots \otimes b_j \in T^jV$ for $1 \leqslant i, j$ with $a_p, b_q \in V$. Let $\phi : EV \to \Lambda V$ be the unique natural algebra isomorphism given by B-3.2. The elements of degrees 0 and 1 are mapped and identified by π and ϕ as follows

i	TV	EV	$\oplus_0^\infty T^iV/H^i$	$\Lambda V = F \oplus V \oplus (\oplus_2^\infty \Lambda^2 V)$
0	F	$\dfrac{F+H}{H} \equiv F$	$\dfrac{F+H^0}{H^0} \equiv F$	F
	k	$k + H \equiv k$	$k + (0) \equiv k$	k
1	V	$\dfrac{V+H}{H}$	$\dfrac{V+H^1}{H^1} = \dfrac{V}{(0)}$	V
	v	$v + H \equiv v$	$v + (0)$	v

Note that $k \otimes v = v \otimes k = kv$ necessarily implies that $k \wedge v = v \wedge k = kv$. Furthermore

$$\phi(v \otimes w + H) = v \otimes w + H^2$$
$$\phi[(v+H)(w+H)] = [\phi(v+H)] \wedge [\phi(w+H)] = v \wedge w,$$

and hence

$$v \otimes w + H^2 = v \wedge w.$$

Similarly

$$\phi(t^i + H) = \phi[(a_1 + H)(a_2 + H) \cdots (a_i + H)]$$
$$= [\phi(a_1 + H)] \wedge [\phi[(a_2 + H)] \wedge \cdots \wedge [\phi(a_i + H)]$$
$$= a_1 \wedge a_2 \wedge \cdots \wedge a_i,$$
$$a_1 \otimes \cdots \otimes a_i + H^i = a_1 \wedge a_2 \wedge \cdots \wedge a_i.$$

Note, however, that the ring multiplication "$\ldots \wedge \ldots$" is not limited to V:

$$\phi[(t^i + H)(t^j + H)] = [\phi(t^i + H)] \wedge [\phi(t^j + H)]$$
$$= [t^i + H^i] \wedge [t^j + H^j],$$

and

$$[a_1 \otimes \cdots \otimes a_i + H^i] \wedge [b_1 \otimes \cdots \otimes b_j + H^j]$$
$$= a_1 \wedge \cdots \wedge a_i \wedge b_1 \wedge \cdots \wedge b_j.$$

Symbols and notation

Chapter 1

R	ring
$M = M_R,\ M_R$	right R-module
$V = {}_RV,\ {}_RV$	left R-module
$<,\ \leqslant$	right R-submodule
$N < M,\ N \leqslant M$	N is a submodule of M
$\rightarrowtail$	one to one map
$\twoheadrightarrow$	onto map
$\mathrm{Hom}_R(M, W)$	homomorphisms $M \to W$
$\mathrm{End}_R M$	endomorphism ring of M
$\mathbf{Z}$	integers
$\mathbf{Q}$	rationals
$\mathbf{R}$	reals
$\mathbf{C}$	complex numbers
$\mathbf{Z}(p^\infty)$	the group of p-th roots of unity
$\mathrm{im}\, f$	image of f
$\ker f$	kernel of f
$\mathrm{coker}\, f$	cokernel of f
$Y^\perp$	right annihilator of set Y
$m^\perp$	right annihilator of element m
$F[x]$	polynomial ring over F
$\Pi\{M_i \mid i \in I\},\ \Pi M_i$	direct product
$(x_i)_{i \in I},\ (x_i)$	element of a direct product
$(\ldots, x_i, \ldots)$	element of a direct product
$\oplus\{M_i \mid i \in I\},\ \oplus M_i$	direct sum
$\oplus f_i$	map of direct sums
$s \oplus \{N \mid I\} \to N$	sum map

$s \circ (\oplus f_i)$	composite of above
$\langle T \rangle$	right R-submodule generated by set T
$\langle x \rangle$	right R-submodule generated by element x
$\Sigma\{M_i \mid i \in I\}, \Sigma M_i$	sum of submodules
R^1	$R^1 = R$ if $1 \in R$; $R^1 = \mathbf{Z} \times R$ if $1 \notin R$
$\langle (I, \leqslant), \{A_i \mid i \in I\}, \{\phi_{ji} \mid i \leqslant j \in I\} \rangle$	direct system
$\langle I, A_i, \phi_{ji} \rangle$	direct system
$\varinjlim A_i$	direct limit
$\langle (I, \leqslant), \{A_i \mid i \in I\}, \{\pi_{ij} \mid i \leqslant j \in I\} \rangle$	inverse system
$\varprojlim A_i$	inverse limit

Chapter 2

$\oplus\{R_t \mid t \in T\}$	free R-module on T
e_t	t-th coordinate vector, or t-th basis vector

Chapter 3

$\oplus\{Q \mid X\}$	direct sum of X copies of Q
$\Pi\{Q \mid X\}$	direct product of X copies of Q
$f * a$	$a \in R$ multiplies $f: R \to D$
1_M	identity map of M
$E(M), EM, \hat{M}$	injective hull

Chapter 4

$A \oplus_F B$	tensor product
$A \otimes_R B$	tensor product
(p)	ideal in $\mathbf{Z}$ generated by p
$\mathbf{Z}/(p)$	integers modulo p
$\mathbf{Z}_n$	integers modulo n

Chapter 5

$F\{X\}$	free algebra on X
$E(V), EV$	exterior algebra on V
$T(V), TV$	tensor algebra on V

Chapter 6

$R:L$	annihilator of R/L
$N(L)$	idealizer of right ideal
$[E, F; G, H]$	matrix like module
$\wedge$	missing term
$Y^{\#}$	annihilator in a module of a subset Y of a ring
$\overset{\circ}{+}$	vector space direct sum
$\dim_D W$	dimension of W over D
E_{ij}, e_{ij}	matrix units
$\mathrm{diag}(N, N, \ldots)$	diagonal matrix with entries in N

Chapter 7

$x \circ y$	quasi-product
$U(R^1)$	multiplicative group of units of R^1
x_R^{-1}	right quasi-inverse of x
x_L^{-1}	left quasi-inverse of x
x^0	quasi-inverse of x
$\Sigma(R), \Sigma$	classes of simple R-modules
$\mathscr{M}(R), \mathscr{M}$	regular maximal right ideals of R
$\mathscr{P}(R), \mathscr{P}$	primitive ideals of R
$J(R)$	Jacobson radical of R
J	Jacobson radical of R
$\|w_{ij}\|$	matrix
$N(R)$	sum of nilpotent right ideals of R
C_P	a prime ideal localization of C at P

Chapter 8, 9, 10

$\mathrm{rad}\, R$	prime radical of R
$\oplus f_i$	direct sum of maps
K_2	2×2 matrices over K
$\begin{vmatrix} K & K \\ 0 & 0 \end{vmatrix}$	right ideal of K_2
$M_n(D)$	$n \times n$ matrix ring over D
$J(M)$	radical of module M
J	radical of module M

Chapter 11

R_x	right multiplication by x
L_y	left multiplication by y
$C_R(Y)$, $C(Y:R)$, $C(Y)$	centralizer of Y in R
R^{op}	opposite ring
$[A:F]$	dimension of A over F

Chapter 15

$\mathscr{C}$	category
$Ob\mathscr{C}$	object of $\mathscr{C}$
Morph $\mathscr{C}$	morphisms of $\mathscr{C}$
$\mathscr{C}(A, B)$	set of morphisms $A \to B$ in $\mathscr{C}$
$\langle Ob\mathscr{C}, \text{Morph}\,\mathscr{C}, \mathscr{C}(\,\cdot\,,\,\cdot\,), \circ \rangle$	category
1_B	identity morphism of B
$F:\mathscr{C} \to \mathscr{D}$	functor between categories $\mathscr{C}$ and $\mathscr{D}$
$1_{\mathscr{D}}$	identity functor
$\mathscr{M}od_R$	category of right R-modules
$_R\mathscr{M}od$	category of left R-modules
$\mathscr{A}b$	category of abelian groups
$\mathscr{S}ets$	category of sets
$\mathscr{R}ing$	category of rings
$\mathscr{R}ing^{(1)}$	category of rings with identity and identity preserving homomorphisms
$\mathscr{C}(G,\,\cdot\,)$	functor of $\mathscr{C}$ to Sets
$\mathscr{C}(\,\cdot\,, C)$	contravariant functor of $\mathscr{C}$ to Sets
$\mathscr{C}^{op}$	opposite category of $\mathscr{C}$
$\mathscr{B}(E\,\cdot\,,\,\cdot\,)$	functor of $\mathscr{A}^op \times \mathscr{B}$ to Sets
$\mathscr{A}(\,\cdot\,, F\,\cdot\,)$	functor of $\mathscr{A}^op \times \mathscr{B}$ to Sets
$\ker(f)$, $\ker f$	kernel object
$\text{Ker}(f)$, $\text{Ker}\, f$	kernel morphism
$\text{coker}(f)$, $\text{coker}\, f$	cokernel object
$\text{Coker}(f)$, $\text{Coker}\, f$	cokernel morphism
$\text{im}\, f$	image object
$\text{Im}\, f$	image morphism

Chapter 16

$\_\otimes\_$, $\_\otimes_R\_$	tesnor product functor
$\_\otimes M$, $\_\otimes_R M$	functor
$\alpha \otimes M$	above functor applied to α

$-\otimes 1_M$	functor
$\text{Hom}_S(\cdot,\cdot)$	hom functor
$\text{Hom}_S(B,\cdot)$	functor

Chapter 17, 18

M^*	character module
$\text{Hom}_Z(M,\mathbf{Q}/\mathbf{Z})$	character module
$\langle y_i, i\in I \mid \Sigma_i y_i r_{ij}=0; j\in J\rangle$	presentation of a module
$E(K,M)$	certain set of submodules of M, where $K < M$

Appendix A and Appendix B

$\in$	is a member of
$\subset, \subseteq, \supset, \supseteq$	is contained in
$\exists$	there exists
$\exists!$	there exists a unique
$\forall$	for any (for every, or for all)
$\varnothing$	empty set
$Y\times Z$	direct product
$X\backslash Y$	difference set
$\Rightarrow$	left implies right
$\Leftarrow$	right implies left
$\Leftrightarrow$	if and only if
$\notin, \not\subset, \nexists$	negation
$f\mid A$	restriction of a function to A
$(X,\leqslant)$	quasi-ordered set
$\cup\{x_i\mid i\in I\}$	union of sets
$\lvert X\rvert$	cardinality of set X
ω_α	α-th initial ordinal
$X/\sim$	equivalence classes
$\triangleleft$	ideal, two sided

Bibliography

Andrunakievic, V., Modular ideals, radicals and semi-simnplicity of rings, *Usp. Mat. Nauk. SSSR*, 12, No. 3, **75** (1957), 133–139, (Russian).

Andrunakievic, V., Radicals in associative rings I, *Mat. Sb.* **44** (1958), 179–212, (Russian).

Artin, E., Zur Theorie der hyperkomplexen Zahlen, *Abh. Math. Sem. Univ.* Hamburg **5** (1927), 251–260.

Artin, E., *Geometric Algebra*, Interscience Publications; Wiley, 1957.

Artin, E. and Nesbitt, C. J. and Thrall, R. M., *Rings with Minimum Condition*, University of Michigan Press, *Ann. Arobor*, 1968.

Barshay, J., *Topics in Ring Theory*, W. A. Benjamin, Inc., New York, 1969.

Beachy, J. and Blair, W., *Abstract Algebra with a Concrete Introduction*, Prentice Hall, Englewood Cliffs, New Jersey, 1990.

Behrens, E. A., *Ring Theory*, Academic Press, New York, 1972.

Bergman, G. M., A ring primitive on the right but not on the left, *Proc. Amer. Math. Soc.* **15** (1964), 473–475.

Bhattacharya, P. B. and Jain, S. K., and Nagpaul, S. R., *Basic Abstract Algebra*, Cambridge University Press, Cambridge, England, 1986.

Birkhoff, G. and Pierce, R. S., Lattice-ordered rings, *An. da Acad.* Brasileira de Ciéncias, **28** (1956), 41–69.

Bitzer, C. V., Inverses in rings with unity, *Amer. Math.* **70** (1963), 315.

Brauer, R. and Weiss, E., *Noncommutative Ring Theory*, Harvard Lecture Notes, 1962.

Bumby, R. T., Modules which are isomorphic to submodules of each other, *Arch. der Math.* **16** (1965), 184–185.

Burgess, W. D. and Stewart, P. N., The characteristic ring and the "best" way to adjoin a one, *Jour. Australian Math. Soc. Ser.* A **47** (1989), 483–496.

Cartan, H. and Eilenberg, S., *Homological Algebra*, Princeton University Press, Princeton, 1956.

Chase, S. U., A generalization of the ring of triangular matrices, *Nagoya Math. Jour.* **18** (1961), 13–25.

Chatters, A. W. and Hajarnavis, C., *Rings with Chain Conditions*, Pitman, London, 1980.

Chevalley, C., *The Construction and Study of Certain Important Algebras*, Math. Soc. Japan, Herald Printing Co., Tokyo, 1955.

Clifford, A. H. and G. B. Preston, *The Algebraic Theory of Semigroups*, Amer. Math. Soc., Math. Surveys, No. 7, Vol. I, 1961; and Vol. II, 1967.

Cohn, P. M., *Algebra*, Vol. 2, Wiley, New York and London, 1977.

432

Cozzens, J. and Faith, C., *Simple Noetherian rings*, Cambridge Univ. Press, Cambr. Tracts in Math., Cambridge, England 1975.

Curtis, C. W. and Reiner, I., *Representation Theory of Finite Groups and Associative Algebras*, Wiley-Interscience Publication, Wiley, New York (1962).

Dauns, J., Chains of modules with completely reducible quotients, *Pacific J. Math.* **17** (1966), 235–242.

Dauns, J., Multiplier rings and primitive ideals, *Trans. Amer. Math. Soc.* **145** (1969), 125–158.

Dauns, J., Simple modules and centralizers, *Trans. Amer. Math. Soc.* **166** (1972), 457–477.

Dauns, J., Prime modules, *Jour. Reine Angwdte Math.* **298** (1978), 156–181.

Dauns, J., *A Concrete Approach to Division Rings*, Heldermann Verlag, Berlin 1982.

Dauns, J., Generalized skew polynomial rings, *Trans. Amer. Math. Soc.* **271** (1982), 575–586.

Dauns, J., Metrics are Clifford algebra involutions, *International Jour. Theoretical Physics*, **27** (1988), 183–192.

Dauns, J. and Hofmann, K. H., The representation of biregular rings by sheaves, *Math. Zeitschr.* **91** (1966), 103–123.

Dauns, J., Representations of Rings by sections, *Amer. Math. Soc. Memoir No.* **83** (1968).

Deavours, C. A., The quaternion calculus, *Amer. Math. Monthly* **80** (1973), 995–1008.

Dirac, P. A. M., Applications of quaternions to Lorentz transformations, *Proc. Royal Irish Academy*, Sect. A **50** (1945), 261–270.

Divinsky, N. J., *Rings and Radicals*, Mathematical Expositions No. 14, University of Toronto Press, 1965.

Eckstein, F., Topological rings of quoteints and rings without open left ideals, *Comm. Algebra* **1** (1974), 365–376.

Eklof, P. C. and Mekler, A. H., *Almost Free Modules Set-theoretic Methods*, North-Holland, Vol. 46, New York, 1990.

Faith, P. C., Modules finite over endomorphism rings, *Tulane Ring Year Lecture Notes*, Lecture Notes Math. 246, Springer Verlag, New York, 1972.

Faith, C., *Algebras: Rings, Modules and Categories I*, Springer Verlag, New York, 1976.

Faith, C. and Page, S., *FPF Ring Theory Faithful Modules and Generators of mod-R*, Cambridge Univ. Press, 1984.

Freyd, P., *Abelian Categories*, Harper Row, Pub., New York, 1964.

Fuchs, L., *Infinite Abelian Groups*, Vol. I and II, Academic Press, New York, 1970.

Golan, J., *Localization of Noncommutative Rings*, Marcel Dekker, New York, 1975.

Golan, J., *Torsion Theories*, Pitman monographs and surveys in pure and applied math. 29, Longman Scientific & Technical, England, 1986; copublished in the U.S. with John Wiley and Sons, Inc., New York, 1986.

Goldman, O., A characterization of semi-simple rings with the descending chain condition *P.A.M.S.* **52** (1946), 1021–1027.

Goldman, O., Semi-simple extensions of rings, *P.A.M.S.* **52** (1946), 1028–1032.

Goodearl, K. R., *Ring Theory: Nonsingular Rings and Modules*, Marcel Dekker, New York, 1976.

Goodearl, K. R., *Von Neumann Regular Rings*, Pitman, London, 1979.

Goodearl, K. R. and Warfield, R. B., *An Introduction to Noncommutative Noetherian Rings*, London Math. Society Student Texts 16, Cambridge Univ. Press, 1989, Cambridge, New York.

Gordon, R. and Robson, J. C., Krull Dimension, *Amer. Math.* Society Memoir No. **133** (1973).

Head, T., *Modules a Primer of Structure Theorems*, Brooks/Cole Pub. Co., a division of

Wadsworth Pub. Co., Inc., Monterey, California, 1974.

Helmer, O., Divisibility properties of integral functions, *Duke Math. J.* **6** (1940), 345–356.

Holland, C., A totally ordered integral domain with a convex left ideal which is not an ideal, *Proc. Amer. Math. Soc.* **11** (1960), 703.

Hopkins, C., Rings with minimal conditions for left ideals, *Ann. of Math.* **40** (1939), 712–730.

Hrbacek, K. and Jech, T., *Introduction to Set Theory*, Marcel Dekker, Inc., New York (1984).

Jacobson, N., *The Theory of Rings*, Math. Surveys No. II, Amer. Math. Society, New York, 1943.

Jacobson, N., Structure theory of simple rings without finiteness assumptions, *Trans. Amer. Math. Soc.* **57** (1945), 228–245.

Jacobson, N., *Structure of Rings*, Amer. Math. Society Colloquium Publications, Vol. XXXVII, 1956, 1964.

Jacobson, N., *Basic Algebra II*, W. H. Freeman and Co., San Francisco, 1980.

Jans, J. P., *Rings and Homology*, Holt, Rinehart, and Winston, 1964.

Jategaonkar, A. V., *Left Principal Ideal Rings*, Lecture Notes in Math. **123**, Springer Verlag, New York, 1970.

Jones, B., *Modern Algebra*, Macmillan Publ. Co., Inc., New York, 1975.

Kasch, F., *Moduln und Ringe*, Mathematische Leitfäden, B. G. Teubner Stuttgart, 1977.

Kasch, F., *Modules and Rings*, London Math. Society Monographs 17, Academic Press, New York, 1982.

Kaplansky, I., Projective modules, *Ann. Math.* **68** (1958), 372–377.

Kaplansky, L., On the dimension of modules and algebras, *Nagoya Math. Jour.* **13** (1958), 85–88.

Koh, K., Quasi-simple modules and other topics in ring theory, *Tulane Ring Year Lecture Notes*, Lecture Notes Math. **246**, Springer Verlag, New York, 1972.

Koh, K., On the complete ring of quotients, *Can. Math. Bull.* **17** (1974), 285–288.

Lam, T. Y., *A First Course in Noncommutative Rings*, Springer Verlag, New York, 1991.

Lambek, J. and Michler, G., The torsion theory at a prime ideal of a right Noetherian ring, *J. Alg.* **25** (1973), 364–382.

MacLane, S., *Categories for the Working Mathematician*, Springer Verlag, New York, 1972.

McCoy, N. H., *Rings and Ideals*, Mathematical Association of America, Providence, Rhode Island, 1948.

McCoy, N. H., *The Theory of Rings*, Macmillan Co., New York, 1964.

Năstăsescu, C. and Oystaeyen, F. van, *Dimensions of Ring Theory*, D. Reidel Pub. Co., Dordrecht, 1987.

Noether, E., Hyperkomplexe Grössen und Darstellungstheorie, *Math. Zeitsckrift* **30** (1929), 641–692.

Osofsky, B. L., On ring properties of injective hulls, *Canadian Math. Bull.* **7** (1964), 405–413.

Passman, D. S., *The Algebraic Structure of Group Rings*, Wiley-Interscience Publication, Wiley, New York, 1977.

Pierce, R. S., *Introduction to the Theory of Abstract Algebras*, Holt, Rinehart, and Winston, New York, 1968.

Prest, M., *Model Theory and Modules*, London Math. Soc. Lecture Note Series 130, Cambridge University Press, 1988.

Queblemann, H. G., Schiefkörper als Endomorphismenringe einfacher Moduln über einer Weyl Algebra, *Jour. Algebra* **59** (1979), 311–312.

Ribenboim, P., *Rings and Modules*, Wiley-Interscience Publication, Wiley, New York (1969).

Rinehart, G., Note on the global dimension of a certain ring, *Proc. Amer. Math. Soc.* **13** (1962), 341–346.

Robson, J. C., Do simple rings have a unity element? *J. Algebra* **12** (1967), 140–143.

Rosenberg, A. and Zelinsky, D., Finiteness of the injective hull, *Math. Zeitschr.* **70** (1959), 372–380.

Rotman, J., *An Introduction to Homological Algebra*, Academic Press, New York, 1979.

Rowen, L., *Ring Theory*, Vol. I and II, Academic Press, New York, 1988.

Sandomierski, F. L., Some examples of right self injective rings which are not left self-injective, *Proc. Amer. Math. Sco.* **26** (1970), 244–245.

Sands, A. D., Prime ideals in matrix rings, *Proc. Glasgow Math. Assoc.* **2** (1956), 191–195.

Sasiada, E. and Cohn, P. M., An example of a simple radical ring, *J. Algebra* **5** (1967), 373–377.

Schaefer, R. D., *An Introduction to Nonassociative Algebras*, Pure and Applied Math. **22**, Academic Press, New York, 1966.

Schafarewich, J., *Algebra-1, Series Some Problems of Mathematics*, No. 11, Akademia Nauk U.S.S.R., Moscow, 1986, (Russian).

Sharpe, W. D. and Vamos, P., Injective Modules, Cambridge University Press, Cambridge, England, 1972.

Small, L. W., An example in Noetherian rings, *Proc. Nat. Acad. Sci. U.S.A.* **54** (1965), 1035–1036.

Small, L. W., Hereditary rings, *Proc. Nat. Acad. Sci. U.S.A.* **55** (1966), 25–27.

Small, L. W., On some questions in Noetherian rings, *Bull. Amer. Math. Soc.*, **72** (1966), 853–857.

Spivak, M., *Calculus on Manifolds*, W. A. Benjamin, Inc., New York, 1965.

Stenström, B., *Rings of Quotients*, Springer Verlag, New York, 1975.

Storrer, H., On Goldman's primary decomposition, *Tulane Ring Year Lecture Notes*, Lecture Notes Math. 246, Springer Verlag, New York, 1972.

Szasz, F. A., *Radicals of Rings*, Wiley Interscience, John Wiley & Sons, New York, 1981.

Szele, T., On ordered skew fields, *Proc. Amer. Math. Soc.* **3** (1952), 410–413.

Wedderburn, J. H. M., On hypercomplex numbers, *Proc. London Math. Soc.* **6** (1908), 77–117.

Subject index

Author index

442